EARTHQUAKE-RESISTANT DESIGN OF STRUCTURES

Second Edition

S.K. DUGGAL

Department of Civil Engineering
Motilal Nehru National Institute of Technology
Allahabad

Oxford University Press is a department of the University of Oxford.
It furthers the University's objective of excellence in research, scholarship,
and education by publishing worldwide. Oxford is a registered trade mark of
Oxford University Press in the UK and in certain other countries.

Published in India by
Oxford University Press
YMCA Library Building, 1 Jai Singh Road, New Delhi 110001, India

First Edition published in 2007
Second Edition published in 2013
Fourth impression 2015

ISBN-13: 978-0-19-808352-8
ISBN-10: 0-19-808352-1
Typeset in Times New Roman
by Recto Graphics, Delhi 110096
Printed in India by Replika Press Pvt. Ltd

Preface to the Second Edition

The main objective of a course on earthquake-resistant design of structures is to introduce students to the phenomenon of earthquakes and the process, measurements, and factors that affect the design of structures in seismic areas. This objective is achieved by understanding the fundamentals of the theory of vibrations necessary to comprehend and analyse the mutual dynamics of earthquakes and structures. The student is also familiarized with the codal provisions as well as aseismic design methodology.

According to *Earthquakes Today*, around 500,000 earthquakes occur each year; 100,000 of these can actually be felt. The magnitude and destructive consequences of earthquakes have serious implications. The destruction and damage of constructed and natural environment and the loss and impairment of human life are of prime concern. A notable feature of the disaster caused by earthquakes is that harm to life is associated almost entirely with man-made structures, e.g., collapse of buildings, bridges, dams.

During an earthquake, when the ground shakes at a building site, the building's foundations vibrate in a manner that is similar to the surrounding ground. To overcome this effect, research has spawned numerous innovations now common in earthquake engineering, including ductile detailing of concrete structures, improved connections for moment frames, base-isolation technology, energy-dissipation technology, and computing tools. Current research activities are focused on three areas: performance-based design, development of damage-resistant systems, and improvement in the ability to predict the occurrence and intensity of earthquakes.

About the Book

Our knowledge about earthquakes has advanced enormously during the past couple of decades. Corresponding advances have also been made in our understanding of the behaviour of structures, and consequently new and stringent specifications have been introduced and enforced. In this second edition, the dynamics of structures is dealt with in detail to give the book completeness. Further, with the revision of the code IS 800 in 2007, the design philosophy recommended is limit state design. Consequently, the treatment of steel structures in Chapter 9 has been overhauled completely, focusing on the new code IS 800: 2007.

While working on the second edition, the goal has been to retain the organization of the book and to develop and update the topics more carefully and logically. The text has been reviewed closely and objectively modified. An attempt has been made to provide a simple, balanced, and exhaustive coverage of the information needed for the design of earthquake-resistant structures.

The first edition of the book was prepared with an understanding that students undertaking a course on earthquake-resistant design of structures must already have the basic knowledge of dynamics of structures. However, feedback and suggestions from students, faculty, and reviewers suggest that the book must cover dynamics of structures in greater detail. Although textbooks are available on structural dynamics as well as earthquake-resistant design of structures, this book is an effort to cover a range of structures in a balanced way, which was lacking in other books.

Numerous solved examples have been added to assist the reader in developing a full understanding of the application of the theory involved. Moreover, each of these examples has been carefully chosen to supplement and extend the ideas and concepts given in the text.

New to the Second Edition

- Comprehensive and thoroughly revised text
- Completely revised chapters on dynamics of structures and seismic response and steel buildings
- New illustrations, problems, and review questions
- New sections including the Indian plate and Himalayan earthquakes, response spectrum, hybrid seismic control systems, confined masonry construction, and lateral load transfer in timber buildings
- Two new appendices—Determination of natural frequencies and mode shapes, and Bauschinger effect

Key Features

- Elucidates the theoretical as well as practical aspects of earthquake-resistant designs
- Includes a case study on the 2001 Bhuj earthquake as an insight into the effects of seismic forces on structures
- Provides standard tables both in the text and in the appendices for ready reference
- Incorporates pedagogical features such as solved problems, review questions, chapter-end exercises, and highlighted information on design specifications and norms

Extended Chapter Material

Chapter 1 has been expanded. It now introduces the movements of the Indian plate and the Himalayan earthquakes. The graphical method to locate an earthquake and the method used to determine the annual frequency of earthquakes are described. The chapter classifies earthquakes, expands the two measures of earthquakes, and includes the MSK-64 intensity scale. The section on magnitude has been expanded and rewritten. The influence of local site conditions on the dynamic characteristics of ground motion—local site effects—is presented in detail.

Chapter 2 has been completely rewritten. The methods of solving the differential equation of motion have been introduced and their suitability for linear and non-linear systems discussed. SI units have been introduced and followed as far as possible. A simple method to estimate damping has been introduced. Response to forced vibrations including harmonic vibration and general loading is described.

Chapter 3 includes soil models, methods of analysis of soil-structure interaction, and testing of soil characteristics.

Chapter 5 has been expanded, including detailed discussions on seismic response control and systems. Passive, active, hybrid, and semi-active control systems are described along with their strategic application in structures. The relative merits, demerits, and limitations of each of these systems are highlighted.

Chapter 6 now discusses confined masonry construction.

Chapter 7 elaborates upon lateral load transfer in timber buildings, floors and roofs, and ductile behaviour of joints.

Chapter 9 is completely rewritten, as the code of practice, IS 800: 2007, for steel structures has been revised and recommends the use of limit state design. It also contains provisions and specifications for seismic design of steel structures. Although this is an elementary treatment, the reader will find the text quite rigorous and comprehensive.

Two new appendices—Determination of natural frequencies and mode shapes and Bauschinger effect—have been added to support the text.

Content and Structure

Chapter 1 begins with an introduction to the earthquake phenomenon, including the causes, occurrence, and properties of earthquakes. It then goes on to explain the characteristics of seismic waves, the effect they have on structures, and how seismic design theory attempts to combat the effects of seismic forces on buildings and structures.

Chapter 2 deals with the dynamics of structures and their seismic response. The concepts of mechanics involved in the design of structures and in modelling a structure for the study of seismic forces are the highlights of this chapter.

Chapter 3 elucidates the behaviour of soils and soil elements and the analysis of soil–structure systems. Soil modes and testing of soil characteristics are also integrated in this chapter.

Chapter 4 elaborates upon the scientific and economical arrangement of structural members to support the anticipated seismic forces, the lateral load transfer mechanism, the effects of asymmetry, and irregularities in plan and elevation and their effects.

Chapter 5 discusses the analysis and design of common structures in general. The two methods of analysis—the equivalent lateral force method and the response spectrum method—are described in detail.

Chapter 6 discusses the behaviour of unreinforced and reinforced masonry walls. The chapter also provides an insight into the behaviour of infill walls, load combinations, and permissible stresses. The methods for seismic design of walls and bands and improvement of seismic behaviour of masonry buildings are also elucidated.

Chapter 7 discusses the seismic behaviour and design of timber buildings. It includes detailed discussions on structural form, site response, fire resistance, and decay of timber buildings. The construction methods described in this chapter include brick-nogged timber frame construction, timber shear panel construction, and restoration and strengthening of timber buildings.

Chapter 8 describes the behaviour of reinforced cement concrete buildings under seismic forces and details the seismic design requirements for RC buildings. The topics covered in this chapter include behaviour of structural elements, joints, and shear walls; seismic design of structural elements; pre-stressed and pre-cast concrete; and retrofitting and strengthening of RC buildings.

Chapter 9 deals with steel buildings and covers important details of steel construction in seismically active areas. The chapter opens with an introduction to the behaviour of steel and steel frames and goes on to discuss frame members and flexural members, connection design and joint behaviour, and steel panel zones and bracing members.

Chapter 10 provides information on non-structural elements, including topics such as failure mechanisms of non-structures, dynamic and static analyses of non-structures, and methods for prevention of damage to non-structures.

Chapter 11 is a case study of the Bhuj 2001 earthquake. It includes a list of earthquake parameters and geological effects and analyses the behaviour of different types of buildings in the most-affected zones. The case study is supplemented with images from the affected area.

The 11 appendices at the end of the book provide additional information to support the text.

Acknowledgements

I am deeply grateful to my colleague, Dr Rama Shankar, who has provided all tangible and intangible assistance in the preparation of this edition of the book. I am thankful to Prof. K.K. Shukla, Prof. A.K. Sachan, Prof. R.K. Srivastava, and Dr Kumar Pallav, Scientific Officer, IIT Guwahati, for their invaluable support and technical suggestions as and when needed. Thanks are due to my Ph D scholar Mr S.T. Rushad, postgraduate student Mr Jemy, and undergraduate students who worked hard to help in logically checking the text and solved examples and exercises.

The scope of this book is restricted to the field of earthquake-resistant design and books on earthquakes and structural dynamics must be referred in order to gain a well-balanced view of the entire field.

All suggestions and feedback for further improvement of the text can be sent to shashikantduggal@rediffmail.com.

S.K. Duggal

Preface to the First Edition

Earthquakes are perhaps the most unpredictable and devastating of all natural disasters. They not only cause great destruction in terms of human casualties but also have a tremendous economic impact on the affected area. The concern about seismic hazards has led to an increasing awareness and demand for structures designed to withstand seismic forces. In such a scenario, the onus of making the buildings and structures safe in earthquake-prone areas, lies on the designers, architects, and engineers who conceptualize these structures. Codes and recommendations, postulated by the relevant authorities, study of the behaviour of structures in past earthquakes, and understanding the physics of earthquakes are some of the factors that help in the designing of an earthquake-resistant structure.

About the Book

This book introduces and explains all aspects of earthquake-resistant design of structures. Designed as a textbook for undergraduate and graduate students of civil engineering, practising engineers and architects will also find the book equally useful. It has been assumed that the reader is well acquainted with structural analysis, structural dynamics, and structural design.

The design of earthquake-resistant structures is an art as well as science. It is necessary to have an understanding of the manner in which a structure absorbs the energy transmitted to it during an earthquake. The book provides a comprehensive coverage of the basic principles of earthquake-resistant design with special emphasis on the design of masonry, reinforced concrete, and steel buildings. The text is focussed on the design of structural and non-structural elements in accordance with the BIS codes (456, 800, 875, 1893, 1905, 4326, 13828, 13920, and 13935).

This book contains 11 chapters, which comprehensively discusses the design of earthquake-resistant structures. Starting with the elements of earthquake theory and seismic design, dynamics of structures and soils and their seismic response, the book goes on to elucidate the conceptualization and actualization of the design of earthquake-resistant structures. Detailed seismic analyses of different types of buildings, such as masonry, timber, reinforced concrete, and steel buildings, follow. Finally, a comprehensive discussion of the behaviour of non-structural elements under seismic forces and an analysis of the 2001 Bhuj earthquake are presented as concluding chapters.

Suitable figures and diagrams have been provided to ensure an easy understanding of the concepts involved. The chapters are divided into small sections that are independent in themselves. A large number of solved problems, which encompass the topics introduced in the chapter, have been integrated at the end of relevant chapters. A summary of the salient features of each chapter has also been consolidated into the body of the chapter. Another unique feature of the book is the grey screens interspersed throughout the text, which highlight important design norms and considerations. Bibliographic references, both empirical and theoretical, are listed at the end of the book for those interested in further reading. A list of websites that have been referred to is also provided therein. A case study of the 2001 Bhuj earthquake has been included as the final chapter of the book. A special attempt has been made to cover all relevant topics of the discipline and make the book a self-contained course in earthquake-resistant design of structures.

Acknowledgements

I extend my sincere thanks to Prof. C.V.R. Murthy of IIT Kanpur for his valuable suggestions for the preparation and organization of the text. I am grateful to Prof. Sudhir K. Jain of IIT Kanpur for his support. I am also thankful to the Earthquake Engineering Research Institute at Oakland, California, for their permission to use the photographs of damage caused by the Bhuj 2001 earthquake. I acknowledge my colleagues Prof. K.K. Shukla, Dr. P.K. Mehta, K. Venkatesh, and G. Ghosh for their support and suggestions. I am also thankful to Shubhendu Singh, P. Mohan Rao, Meenakshi Dewangan, and S. Chakravarthi G. for their assistance in the typing of the manuscript and the preparation of the figures.

All suggestions and feedback for further improvement of the text are welcome.

S.K. Duggal

Brief Contents

Detailed Contents

CHAPTER 1

Earthquakes and Ground Motion

From time immemorial, nature's forces have influenced human existence. Even in the face of catastrophic natural phenomena, human beings have tried to control nature and coexist with it. Of all the natural disasters, for example, earthquakes, floods, tornadoes, hurricanes, droughts, and volcanic eruptions, the least understood and the most destructive are earthquakes. Although the average annual losses due to floods, tornadoes, hurricanes, etc., exceed those due to earthquakes, the total, unexpected, and nearly instantaneous devastation caused by a major earthquake has a unique psychological impact on the affected. Thus, this significant life hazard demands serious attention.

An *earthquake* may be defined as a wave-like motion generated by forces in constant turmoil under the surface layer of the earth (the lithosphere), travelling through the earth's crust. It may also be defined as the vibration, sometimes violent, of the earth's surface as a result of a release of energy in the earth's crust. This release of energy can be caused by sudden dislocations of segments of the crust, volcanic eruptions, or even explosions created by humans. Dislocations of crust segments, however, lead to the most destructive earthquakes. In the process of dislocation, vibrations called *seismic waves* are generated.

These waves travel outwards from the source of the earthquake at varying speeds, causing the earth to quiver or ring like a bell or tuning fork.

During an earthquake, enormous amounts of energy are released. The size and severity of an earthquake is estimated by two important parameters—intensity and magnitude. The magnitude is a measure of the amount of energy released, while the intensity is the apparent effect experienced at a specific location. The concept of strong ground motion, its characteristics, and the response of structures to strong earthquake motion are discussed. The need of seismic zoning and general principles to be observed in the earthquake-resistant design of structures are also discussed. For a better understanding of the causes of earthquakes and the proposed theories, some important aspects are explained in Section 1.1.

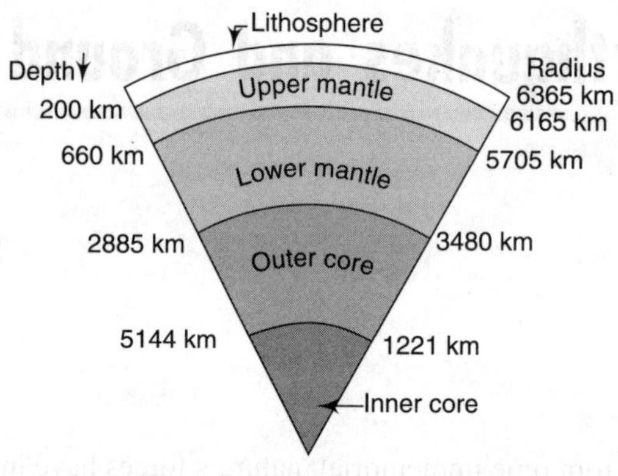

Fig. 1.1 Interior of the earth

1.1 The Interior of the Earth

The earth is conceived to be composed of a sequence of shells or layers called *geospheres*, the heaviest of which forms the core as shown in Fig. 1.1. Various geospheres that constitute the earth are discussed further.

Barysphere Also known as the *core*, it is the densest central part of the earth. It is composed of the inner and outer cores. The inner core, 1221 km in radius, is composed mainly of nickel and iron. Its density is 16,000 kg/m^3 and it behaves like a solid mass. The outer core surrounding the inner core is 2259 km thick and is composed of an alloy of nickel, iron, and silica. The outer core exists as a liquid of density 12,000 kg/m^3. The temperature at the core is about 2500°C and the pressure is about 4×10^6 atm.

Asthenosphere Also known as the *mantle*, it is 2685 km thick, surrounding the core. It is composed of hot, dense ultrabasic igneous rock in a plastic state with a density of 5000–6000 kg/m^3.

Lithosphere Also known as the *crust*, it is the thinnest outer solid shell. It is 200 km thick with a density of 1500 kg/m^3. The temperature of the crust is about 25°C and the pressure within it is 1 atm.

Convection Currents

The high pressure and temperature gradients between the crust and the core cause convection currents to develop in the viscous mantle as shown in Fig. 1.2(a). The energy for these circulations is derived from the heat produced from the incessant

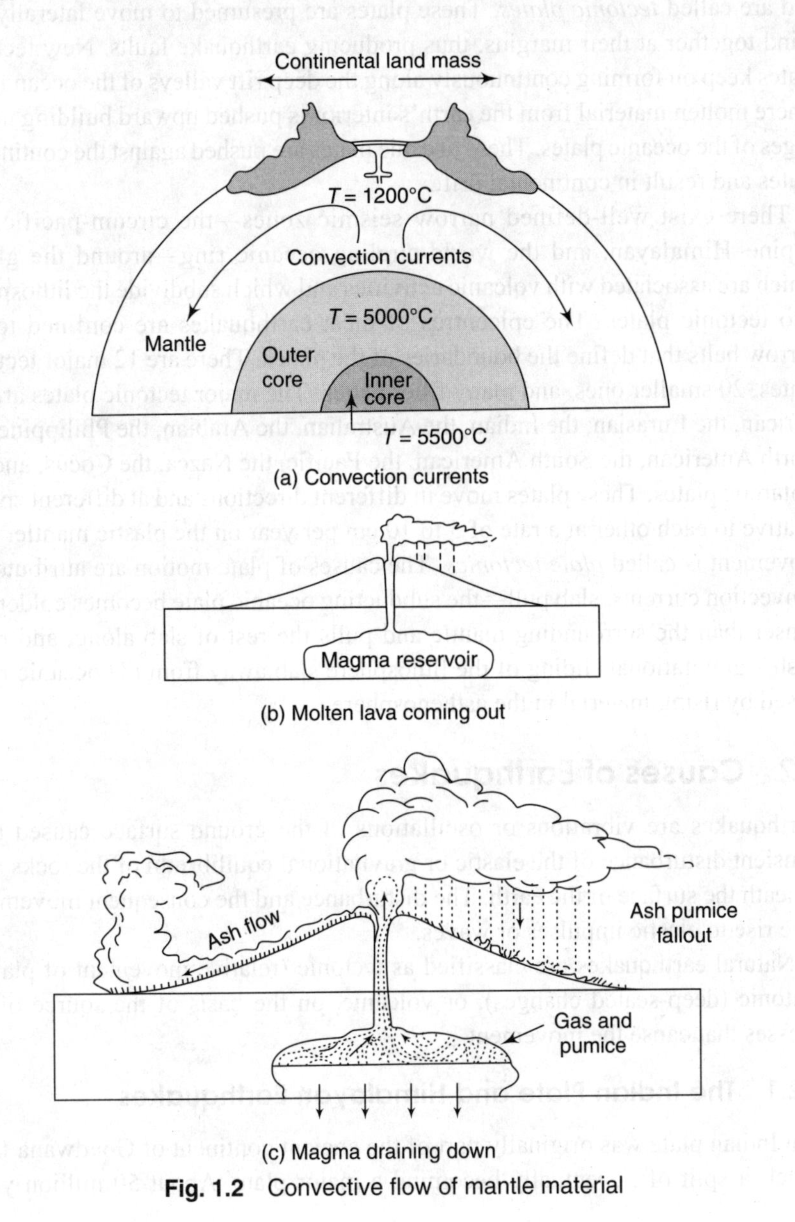

(a) Convection currents

(b) Molten lava coming out

(c) Magma draining down

Fig. 1.2 Convective flow of mantle material

decay of radioactive elements present in rocks throughout the earth's interior. These convection currents cause the earth's mass to circulate; as a result, hot molten lava comes out [Fig. 1.2(b)] and the cool rock mass goes down into the earth [Fig. 1.2(c)], where it melts and becomes a part of the mantle. The convective flow of the mantle material causes the crust and some portion of the mantle to slide on the hot molten outer core. The earth's crust, therefore, is not static but subjected to motion. It consists of several gigantic rigid rock plates of about 80 km in thickness that float in slow motion on the viscous (partially plastic) mantle and are called *tectonic plates*. These plates are presumed to move laterally and grind together at their margins, thus producing earthquake faults. New tectonic plates keep on forming continuously along the deep rift valleys of the ocean floor, where molten material from the earth's interior is pushed upward building up the edges of the oceanic plates. These oceanic plates are pushed against the continental plates and result in continental drift.

There exist well-defined narrow seismic zones—the circum-pacific, the Alpine–Himalayan, and the world-circling oceanic ring—around the globe, which are associated with volcanic activities and which subdivide the lithosphere into tectonic plates. The epicentres of most earthquakes are confined to the narrow belts that define the boundaries of the plates. There are 12 major tectonic plates, 20 smaller ones, and many filler plates. The major tectonic plates are the African, the Eurasian, the Indian, the Australian, the Arabian, the Philippine, the North American, the South American, the Pacific, the Nazca, the Cocus, and the Antarctic plates. These plates move in different directions and at different speeds relative to each other at a rate of 5 to 10 cm per year on the plastic mantle. This movement is called *plate tectonics*. The causes of plate motion are attributed to convection currents, slab pull—the subducting oceanic plate becomes colder and denser than the surrounding mantle and pulls the rest of slab along, and ridge push—gravitational sliding of the lithosphere slab away from the oceanic ridge raised by rising material in the asthenosphere.

1.2 Causes of Earthquakes

Earthquakes are vibrations or oscillations of the ground surface caused by a transient disturbance of the elastic or gravitational equilibrium of the rocks at or beneath the surface of the earth. The disturbance and the consequent movements give rise to elastic impulses or waves.

Natural earthquakes are classified as tectonic (relative movement of plates), plutonic (deep-seated changes), or volcanic, on the basis of the source of the stresses that cause the movement.

1.2.1 The Indian Plate and Himalayan Earthquakes

The Indian plate was originally part of the ancient continent of Gondwana from which it split off, eventually becoming a major plate. About 50 million years ago, it fused with the adjacent Australian plate. It includes most of south Asia and portion of basin under the Indian ocean including parts of south China and eastern Indonesia. The 2004 Indian ocean earthquake of moment magnitude of about 9.2 was caused because of release of stresses built up along the subduction zone where the Indian plate is sliding under the Burma plate in the eastern Indian ocean at a rate of 5 cm/year.

The collision with the Eurasian plate along the boundary between India and Nepal created the Tibetan plateau and the Himalayas. The Indian plate is currently moving north-east at 5 cm/year, while the Eurasian plate is moving north at only 2 cm/year. This is causing the Eurasian plate to deform, and the Indian plate to compress at a rate of 4 mm/year. The mechanism of recent earthquakes in peninsular India is consistent with the stresses induced in the Indian plate fixed by its collision with Tibet. A region of abnormally high seismicity in western India appears to be caused by local convergence across the Rann of Kutch and possibly due to other rift zones of India. The strain rate in the Indian plate is less than 3 nano-strain/year. It has been found that faults are active both on the subcontinent and the Himalayan boundary. India is currently penetrating into Asia at a rate of 45 mm/year and rotating slowly anticlockwise. This rotation and translation is resulting in left-lateral transform slip in Baluchistan at approximately 42 mm/year and right-lateral slip relative to Asia in the Indo-Burma approximately at 55 mm/year. Deformations within Asia reduce India's convergence with Tibet to approximately 18 mm/year. Further, since Tibet is extending east–west, convergence along the Himalayas is approximately normal to the arc. Arc-normal convergence across the Himalayas results in the development of potential slip available to drive large thrust earthquakes beneath the Himalayas.

The Indian plate is bent downwards by 4–6 km beneath the southern edge of the Himalayas, attaining depths of 18 km beneath the southern edge of Tibet. Stresses within the plate vary from tensile above the neutral axis to compressional below it. The presence of both flexural stresses and plate boundary slip allows possible mechanism of earthquakes to occur below the lesser Himalayas. At depths of 4–18 km, large thrust earthquakes with shallow northerly dip occur infrequently, which permit the northward descent of the Indian plate beneath the sub-continent. Earthquakes in the Indian plate beneath this thrust event range from tensile just below the plate interface to compressional and strike-slip at depths of 30–50 km. A belt of micro and moderate earthquakes beneath the greater Himalayas on the southern edge of Tibet indicate a transition from stick-slip faulting to probably a seismic creep at around 18 km.

The origin or causes of tectonic earthquakes can be explained by the following theories.

1.2.2 Elastic Rebound Theory

The elastic rebound theory, first proposed by M.F. Reid in 1906, attributes the occurrence of tectonic earthquakes to the gradual accumulation of strain in a given zone and the subsequent gradual increase in the amount of elastic forces stored. It has been discussed earlier that new-formed oceanic plates push against the continental plates resulting in continental drift. Where the plates collide, they may be locked in place, that is, these may be prevented from moving because of the frictional resistance along the plate boundaries. This causes building up of stresses along the plate edges until sudden slippage due to elastic rebound or fracture of the rock occurs, resulting in sudden release of strain energy that may cause the upper crust of the earth to fracture along a certain direction and form a fault. This is the origin of an earthquake. The gradual accumulation and subsequent release of stress and strain is described as *elastic rebound.* The elastic rebound theory postulates that the source of an earthquake is the sudden displacement of the ground on both sides of the fault, which is a result of the rupturing of the crustal rock.

The upper parts of the earth's crust and lithosphere are very strong and brittle. When this rock is subjected to deformation, it actually bends slightly (Fig. 1.3). However, it is able to withstand very light stress with only slight bending or strain. The elastic rebound theory requires the strain to build up rapidly up to the elastic limit of the rock. Beyond this point, the earth's crust ruptures due to the formation of a fault and the bent rock snaps back to regain its original shape, releasing the stored energy in the form of rebounding and violent vibrations (elastic waves). These vibrations shake the ground; the maximum shaking effect is felt along the fault. After the earthquake, the process of strain build-up at this modified interface between the rocks starts all over again (Kutch earthquake, 1819). Most earthquakes occur along the boundaries of the tectonic plates and are called *interplate earthquakes* (Great Assam earthquake, 1950). The others occurring within the plates themselves, away from the plate boundaries (Latur earthquake, 1993), are called *intraplate earthquakes*. In both types, slips are generated during

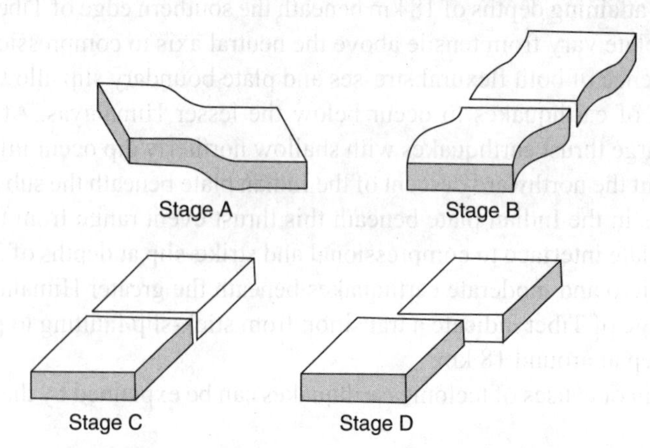

Fig. 1.3 Elastic strain build-up and rupture—Stage A: Slow deformation of rock in the vicinity of a plate boundary, Stage B: Rupture of the rock due to strain built up beyond elastic limit, Stage C: Bent rock regains its original shape after the release of strain energy, and Stage D: The displaced rock after the earthquake

the earthquake at the fault along both horizontal and vertical directions, known as *dip slip* [Fig. 1.4(a)], and the lateral direction, known as *strike slip* [Fig. 1.4(b)]. Some instances of major earthquakes in India are listed in Appendix I.

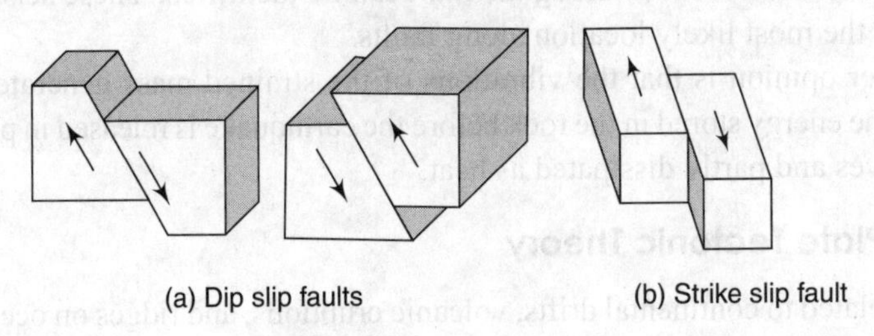

Fig. 1.4 Types of faults

The phenomenon of earthquakes caused by a sudden displacement along the sides of a fault can be summarized as follows:

- Strain that has accumulated in the fault for a long time reaches its maximum limit [Fig. 1.5(a)].
- A slip occurs at the fault and causes a rebound [Fig. 1.5(b)].
- A push and pull force initiates at the fault [Fig. 1.5(c)].
- The situation is equivalent to two pairs of coupled forces acting suddenly [Fig. 1.5(d)].
- This action causes radial wave propagation.

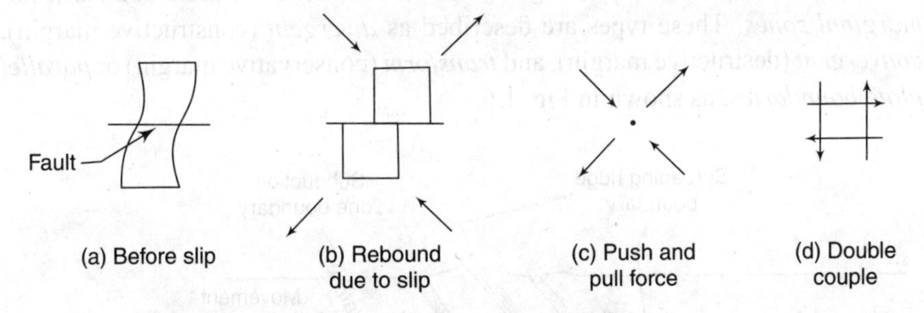

Fig. 1.5 Mechanism of earthquakes caused by displacement along sides of faults

The moment of each couple is known as the *earthquake moment* or *seismic moment.* Recently, the seismic moment has been used as a measure of earthquake size.

The elastic rebound theory implies that an earthquake relieves the accumulated stresses along the portion of the fault on which rupture occurs. Further, this segment will not rupture again until the stresses build up again which, of course, will take its own time. Therefore, earthquakes can reoccur only after some period of time and that, perhaps, depends on the amount of energy released in the earthquake.

The probability of occurrence of an earthquake is more likely along the fault where no seismic activity has been observed for some time. By plotting fault movement and historical earthquake activity along a fault, the gaps in seismic activity at certain locations along the fault can be identified. These seismic gaps represent the most likely location along faults.

Another opinion is that the vibrations of the strained mass generate seismic waves. The energy stored in the rock before the earthquake is released in producing these waves and partly dissipated as heat.

1.2.3 Plate Tectonic Theory

Studies related to continental drifts, volcanic eruptions, and ridges on ocean floors have led to development of the theory of plate tectonics. According to that, the earth's crust consists of a number of large rigid blocks called *crustal plates*. These plates bear the loads of land masses, water bodies, or both and are in constant motion on the viscous mantle, overriding, plunging beneath one another, colliding with each other, or brushing past one another. Some segments of adjacent plates, however, remain immovable and locked together for years, only to break free in great lurches (faulting) and produce seismic vibrations along boundaries, causing destruction (Uttarkashi earthquake, 1991). Plate tectonics is responsible for features such as *continental drift*—in which the two plates move away from each other, *mountain formation*—in which the front plate is slower so that the rear plate collides with it, *volcanic eruptions*, and *earthquakes*. The plates may also move side by side along the same direction or in opposite directions. The relative motion of crustal plates gives rise to three kinds of plate boundaries or *marginal zones*. These types are described as *divergent* (constructive margin), *convergent* (destructive margin), and *transform* (conservative margin) or *parallel plate boundaries*, as shown in Fig. 1.6.

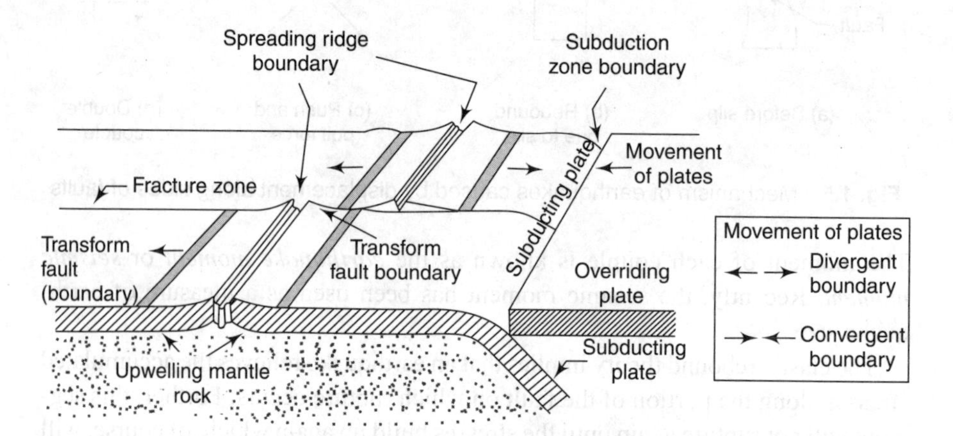

Fig. 1.6 Types of plate boundaries

Zones of Divergence (Constructive Margin)

Zones of divergence are rift or spreading zones, that is, divergent boundaries in continental regions. These are zones of tension in which the lithosphere splits, separates, and moves apart as hot magma wells up through cracks, solidifies, and deposits new material onto the edges of oceanic plates, forming oceanic ridges; hence the term *constructive margin*. This process is also known as *sea-floor spreading*. This seismicity (occurrence of earthquakes) is associated with volcanic activity along the axes of ridges. A well-known divergent boundary is the Mid-Atlantic Ridge.

The stretching caused by this process is not uniform all along the oceanic ridges. The differential stretching is a result of the plates moving along a pole of rotation, with minimum velocity at the poles and increasing towards the equator. Thus, the oceanic ridges are offset by many transform faults. Movement along these transform faults generates earthquakes that have shallow foci (2–8 km). Because of this, the strain build-up at these boundaries is not enough to cause earthquakes with magnitude greater than 6.

Zones of Convergence (Destructive Margin)

Zones of convergence are boundaries along which the edge of one plate overrides the other. Plates are said to converge when two plates from opposite directions come together and collide. Upon collision, the leading edge of the higher density plate may bend downwards, causing it to descend beneath the other plate. The plunging plate enters the hot asthenosphere, gets heated, melts, and assimilates completely within the material of the upper mantle forming new magma. This process is known as *subduction*. The new magma rises to the surface and erupts, forming a chain of volcanoes around the edges of the plate boundary areas, known as *subduction zones*. These narrow plate boundary areas are associated with the creation of deep ocean trenches and major earthquakes. When, upon collision, the two plates are pushed upwards against each other, they form major mountain systems such as the Himalayas. Since one of the plates is destroyed here, such a boundary is known as a *destructive margin*.

Subduction zones are the sites of the most widespread and intense earthquakes. Besides volcanism and shallow-to-deep focus earthquakes, these boundaries also produce deep trenches, basins, and folded mountain chains. When an oceanic plate collides with a continental plate, it slides beneath the continental plate forming a *deep oceanic trench*.

Although the surface characteristics of earthquakes associated with oceanic trenches and island arcs are varied, a majority of such earthquakes appear to be confined to a narrow dipping zone. *Tensional earthquakes* occur on the oceanic side of the trench, where normal faulting occurs due to tensional stresses generated by the initial bending of the plate. *Shallow earthquakes* are produced by dip-slip motion resulting in thrust faulting, as descending plates slide beneath the overlying plates. This type of activity persists up to a depth of 100 km.

At intermediate depths, earthquakes are caused by extension or compression, depending on the specific characteristics of the subduction zone. Extension and normal faulting result when a descending slab that is denser than the surrounding mantle sinks due to its own weight. Compression results when the mantle resists the downward motion of the descending plate. The zone of deep earthquakes shows compression within the descending zone of the lithosphere, indicating that the mantle material at that depth resists the movement of the descending plate.

Transform Zones (Conservative Margin)

Transform zones are also known as *transformed faults* or *fracture zones*. In these zones, the lithosphere plates slide past each other horizontally without any creation or destruction. The edges of the two plates scrape each other closely, creating tension along the boundaries associated with shallow focus seismic events, unaccompanied with volcanic activity. This boundary is, thus, also called a *parallel* or *transform fault boundary*. The transform faults move roughly parallel to the direction of plate movement. Most transform faults are found on the ocean floor.

Note: The theory of tectonics explains well the earthquakes along existing plate boundaries. However, it does not explain mid-point earthquakes far distant from plate margins.

1.2.4 Causes of Volcanic Earthquakes

Volcanic earthquakes are a special feature of explosive eruption, small in energy and seldom damaging. There is an emerging realization that volcanoes and earthquakes may have a common origin in the deep movement of mantle materials. The coincidence of belts of major earthquake activity with belts that include active volcanoes supports this idea. The most obvious common cause of seismic and volcanic activity relates to plate interactions, in the process of which fracture zones allow volcanic material to well up from the lower crust of the mantle. These boundaries are also areas in which earthquakes would naturally occur due to plate interactions in zones of convergence or divergence, or areas where two plates slide past one another along the parallel boundaries.

1.3 Nature and Occurrence of Earthquakes

When there is a sudden localized disturbance in rocks, waves similar to those caused by a stone thrown into a pool spread out through the earth. An earthquake generates a similar disturbance. The maximum effect of an earthquake is felt near its source, diminishing with distance from the source (earthquakes shake the ground even hundreds of kilometres away). The vibrations felt in the bedrock are called *shocks*. Some earthquakes are preceded by smaller *foreshocks* and larger earthquakes are always followed by *aftershocks*. Foreshocks are usually interpreted as being caused by plastic deformation or small ruptures. Aftershocks are usually due to fresh ruptures or readjustment of fractured rocks.

The point of generation of an earthquake is known as the *focus*, *centre*, or *hypocentre*. The point on the earth's surface directly above the focus is known as *epicentre*. The depth of the focus from the epicentre is known as the *focal depth*. The distance from the epicentre to any point of interest is known as the *focal distance* or *epicentral distance* (Fig. 1.7). Seismic destruction propagates from the focus through a limited region of the surrounding earth's body, which is called the *focal region*. The line joining locations experiencing equal earthquake intensity is known as the *isoseismal line* and the line joining locations at which the shock arrives simultaneously is known as the *homoseismal line*.

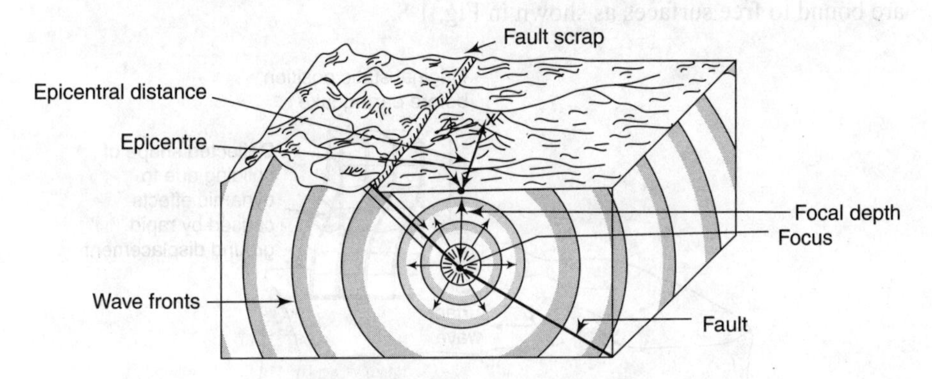

Fig. 1.7 Occurrence of earthquake

The location of an earthquake's focus is important because it indicates the depth at which rupture and movement occur. Although movement of material within the earth occurs throughout the mantle and core, earthquakes are concentrated in the upper 700 km only. *Shallow-focus earthquakes* are most frequent and originate from up to a depth of 70 km from the surface of the earth. *Intermediate-focus earthquakes* occur between 70 and 300 km. Earthquakes having a focal depth of more than 300 km are classified as *deep-focus earthquakes*. The maximum energy released by an earthquake progressively tends to become smaller as the focal depth increases. Also, seismic energy from a source deeper than 70 km gets largely dissipated by the time it reaches the surface. Therefore, the main consideration in the design of earthquake-resistant structures is shallow-focus earthquakes. The focus of an earthquake is calculated from the time that elapses between the arrival of three major types of seismic waves.

The movement caused by an earthquake at a given point of the ground surface may be resolved into three translations, parallel to the three mutually perpendicular axes. There are also three rotations about these axes, which, being small, may be neglected. The translations (or displacements) are measured by seismographs, discussed in Section 1.7.5.

1.4 Seismic Waves

The large strain energy released during an earthquake travels in the form of seismic waves in all directions (Fig. 1.8), with accompanying reflections from earth's surface as well as reflections and refractions as they traverse the earth's interior (Fig. 1.9). These waves can be classified as *body waves* travelling through the interior of the earth—consisting of *P*-waves (primary, longitudinal, or compressional waves) and *S*-waves (secondary, transverse, or shear waves), and *surface waves* resulting from interaction between body waves and surface layers of earth—consisting of *L*-waves (love waves) and Rayleigh waves (Fig. 1.10). Body waves travel through the interior of elastic media and surface waves are bound to free surfaces as shown in Fig. 1.8.

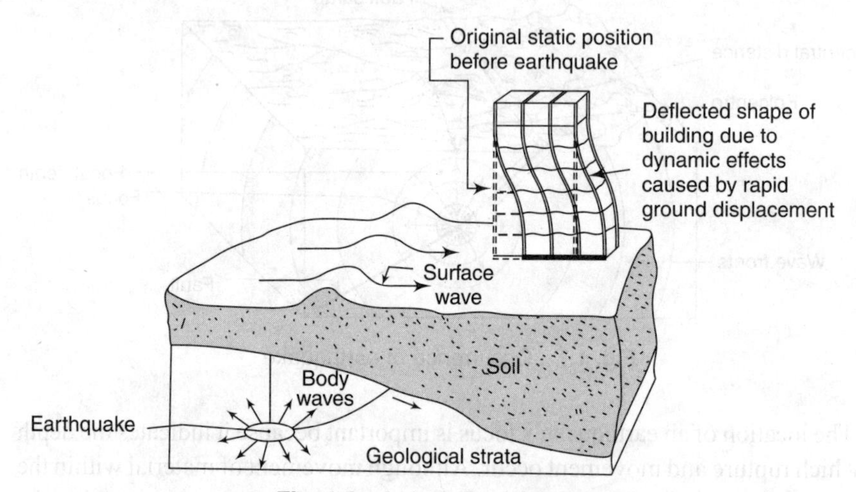

Fig. 1.8 Arrival of seismic wave

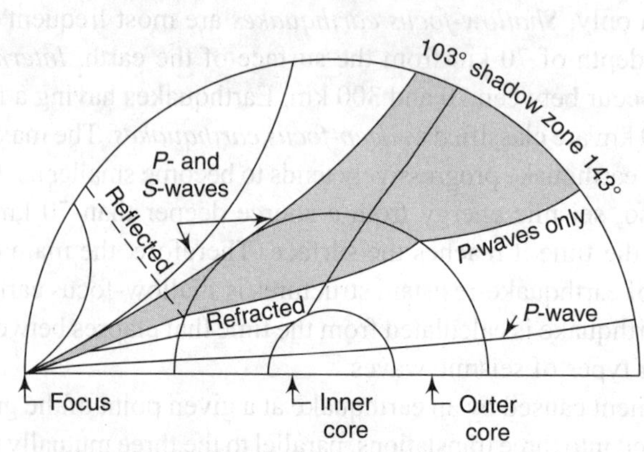

Fig. 1.9 Seismic wave paths (Richter 1958)

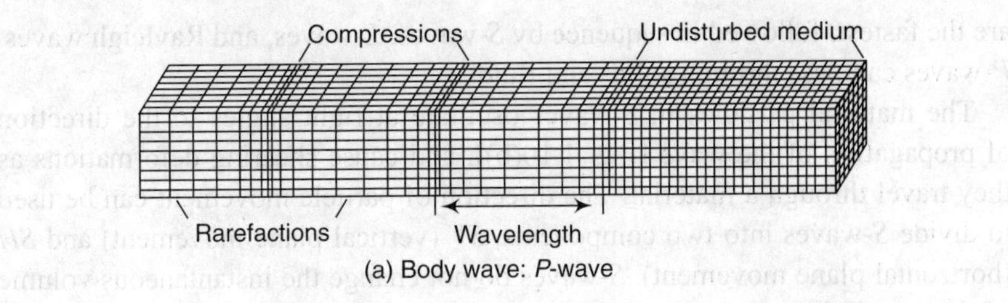

(a) Body wave: *P*-wave

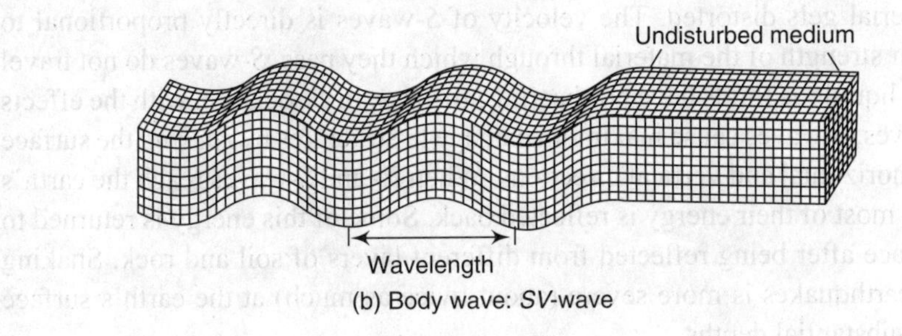

(b) Body wave: *SV*-wave

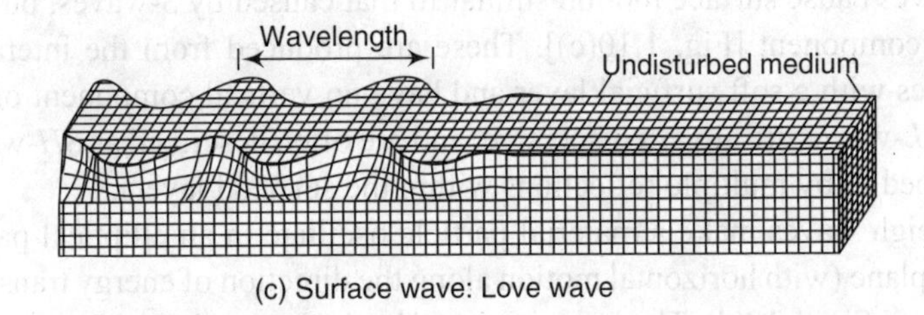

(c) Surface wave: Love wave

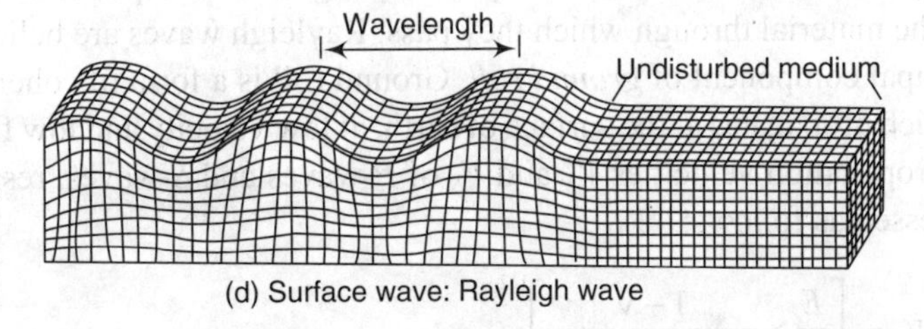

(d) Surface wave: Rayleigh wave

Fig. 1.10 Deformations produced by seismic waves (Bolt 2004)

In *P*-waves, the material particles oscillate back and forth in the direction of propagation of the wave and cause alternate compression (push) and tension (rarefaction of material; pull) of the medium as shown in Fig. 1.10(a). These waves cause a momentary volume change in the material through which they pass without any concomitant momentary shape change in the material. *P*-waves are similar to sound waves and obey all the physical laws of science and acoustics. Since geological materials are stiffer in volumetric compression, the *P*-waves are the fastest, followed in sequence by *S*-waves, *L*-waves, and Rayleigh waves. *P*-waves can pass through solids and fluids.

The material particles in *S*-waves oscillate at right angles to the direction of propagation of the wave [Fig. 1.10(b)], and cause shearing deformations as they travel through a material. The direction of particle movement can be used to divide *S*-waves into two components, *SV* (vertical plane movement) and *SH* (horizontal plane movement). *S*-waves do not change the instantaneous volume of the material through which they pass. However, the instantaneous shape of the material gets distorted. The velocity of *S*-waves is directly proportional to the shear strength of the material through which they pass. *S*-waves do not travel through liquids as fluids have no shearing stiffness. In association with the effects of *L*-waves, *S*-waves cause maximum damage to structures by *rocking* the surface in both horizontal and vertical directions. When *P*- and *S*-waves reach the earth's surface, most of their energy is reflected back. Some of this energy is returned to the surface after being reflected from different layers of soil and rock. Shaking due to earthquakes is more severe (about twice as much) at the earth's surface than at substantial depths.

L-waves cause surface motion similar to that caused by *S*-waves, but with no vertical component [Fig. 1.10(c)]. These are produced from the interaction of *SH*-waves with a soft surficial layer and have no vertical component of particle motion. *L*-waves are always dispersive, and are often described as *SH*-waves that are trapped in by multiple reflections within the surficial layers.

Rayleigh waves make a material particle oscillate in an elliptical path in the vertical plane (with horizontal motion along the direction of energy transmission) as shown in Fig. 1.10(d). These are produced by the interaction of *P*- and *SV*-waves with the surface of earth. The velocity of Rayleigh waves depends on Poisson's ratio of the material through which they pass. Rayleigh waves are believed to be the principal component of *ground roll*. Ground roll is a form of coherent linear noise which propagates at the surface of earth, at low velocity and low frequency.

The propagation velocities V_P and V_S of *P*-waves and *S*-waves, respectively, are expressed as follows:

$$V_P = \left[\frac{E}{\rho} \times \frac{1-\nu}{(1+\nu)(1-2\nu)}\right]^{1/2} \tag{1.1}$$

$$V_S = \left[\frac{G}{\rho}\right]^{1/2} = \left[\frac{E}{\rho} \times \frac{1}{2(1+\nu)}\right]^{1/2} \tag{1.2}$$

where E is the Young's modulus, G is the shear modulus, ρ is the mass density, and ν is the Poisson's ratio (0.25 for the earth). From Eqns (1.1) and (1.2)

$$V_P = \sqrt{3}\, V_S$$

Near the surface of the earth, $V_P = 5\text{–}7$ km/s and $V_S = 3\text{–}4$ km/s.

The time interval between the arrival of a *P*-wave and an *S*-wave at the observation station is known as *duration of primary tremors*, T_{SP} and is given by

$$T_{SP} = \left(\frac{1}{V_S} - \frac{1}{V_P}\right)\Delta_{P\text{–}S} \tag{1.3}$$

where $\Delta_{P\text{–}S}$ is the distance from the focus to the observation point. The epicentre can thus be located and the depth of the focus obtained graphically if earthquake records are made at least at three different observation points.

1.5 Graphical Method of Locating Earthquakes

Location of an earthquake implies location of its epicentre. The preliminary location is based on the relative arrival times of *P*- and *S*-waves with the help of three seismographs, which is then refined using sophisticated techniques. The following step-by-step procedure is used.

1. The speed of the *P*-waves (3–8 km/s) is more than that of *S*-waves (2–5 km/s). The epicentral distance is given by

   $$d = \frac{\Delta t_{P\text{-}S}}{\dfrac{1}{V_S} - \dfrac{1}{V_P}}$$

   where $\Delta t_{P\text{–}S}$ is the difference between first *P*- and *S*-waves arrival to the seismograph
   V_P, V_S are the *P*- and *S*-wave velocities.
   Thus the epicentral distances d_1, d_2, and d_3, at the three seismographs, can be determined.
2. With the epicentral distance d_1 as radius, a circle can be drawn at the point of seismograph location, say *A* (Fig. 1.11). A second circle with radius d_2 when

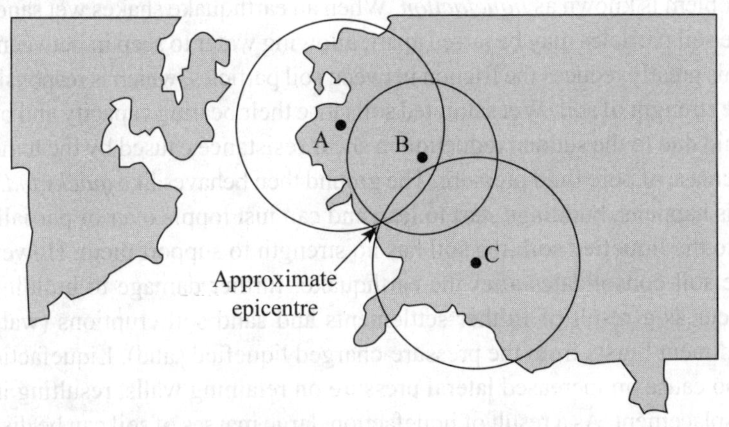

Fig. 1.11 Preliminary location of epicentre from differential time measurement at seismographs at A, B, and C

drawn with the point of seismograph location B, the two circles will intersect at two points P and Q. When a third circle is drawn with the point of seismograph location C, it must pass through P or Q if the three epicentral distances were correct. Thus the common point of intersection of the three circles will be the likely location of the epicentre.

3. In case there is some error and the three circles do not intersect at one point, a refined estimate of the epicentre is made by using multiple seismographs and numerical optimization techniques.

1.6 Effects of Earthquakes

Earthquakes are major hazards and can cause catastrophic damage. They have two types of effects—direct and indirect. Direct effects cause damages directly and include ground motion and faulting. Indirect effects cause damages indirectly, as a result of the processes set in motion by an earthquake.

Direct effects The direct effects of earthquakes are as follows:

(a) Seismic waves, especially surface waves, through surface rock layers and regolith result in ground motion. Such motion can damage and, sometimes, completely destroy buildings. If a structure, such as a building or a road, straddles a fault, then the ground displacement that occurs during an earthquake will seriously damage or rip apart that structure.

(b) In regions consisting of hills and steep slopes, earthquake vibration may cause landslides and mudslides and cliffs to collapse, which can damage buildings and lead to loss of life.

(c) Soil vibration can either shake a building off its foundation, modify its supports, or cause its foundation to disintegrate.

(d) Ground shaking may compound the problem in areas with very wet ground—infilled land, near the coast, or in locations that have a high water table. This problem is known as *liquefaction*. When an earthquake shakes wet sandy soil, the soil particles may be jarred apart, allowing water to seep in between them. This greatly reduces the friction between soil particles, which is responsible for the strength of soil. Wet saturated soils lose their bearing capacity and become fluid due to the sudden reduction in shear resistance caused by the temporary increase of pore fluid pressure. The ground then behaves like *quicksand*. When this happens, buildings start to lean and can just topple over or partially sink into the liquefied soil; the soil has no strength to support them. However, as the soil consolidates after the earthquake, further damage to buildings can occur as a result of further settlements and sand soil eruptions (water and sediment bursts from the pressure-charged liquefied sand). Liquefaction can also cause an increased lateral pressure on retaining walls, resulting in their displacement. As a result of liquefaction, large masses of soil can be displaced laterally, termed *lateral spreading*, with serious consequences. The displaced ground suffers cracks, rifting, and buckling. Lateral spreading disrupts the foundations of buildings built across the fault, and causes bridges to buckle and service pipelines to break.

(e) Strong surface seismic waves make the ground heave and lurch and damage the structure.

Indirect or consequential effects The following are indirect effects of an earthquake:

(a) If the epicentre of an earthquake is under the sea, one side of the ocean floor drops suddenly, sliding under the other plate and, in doing so, creates a vertical fault. The violent movement of the sea floor results in series of sea waves with extremely long time periods. These waves are called *tsunamis* (Fig. 1.12). These usually take place along the subduction zone and are very common in the Pacific ocean. In open sea, a tsunami is only an unusually broad swell on the water surface. Like all waves, tsunamis only develop into breakers as they approach the shore and the undulating water touches the bottom. Near shores, the energy of a tsunami gets concentrated in the vertical direction (due to reduction in water depth) as well as in the horizontal direction (because of shortening of wavelength due to reduction in velocity). The breakers associated with tsunamis can easily be over 15 m high in case of larger earthquakes, and their effects correspondingly dramatic. Several such breakers may crash over the coast in succession; between waves, the water may be pulled swiftly seaward, emptying a harbour or bay and, perhaps, pulling unwary onlookers along. Tsunamis can travel very quickly—speeds of 1,000 km/h are not uncommon. The velocity of tsunami waves with large wavelengths may be estimated using the following expression.

$$V_t = \sqrt{gh}$$

where g is the acceleration due to gravity and h is the depth of the water.

Notes: 1. The first indication of a tsunami is generally a severe recession of the water, which is shortly followed by a returning rush of water that floods inland a distance depending on the height of the wave.

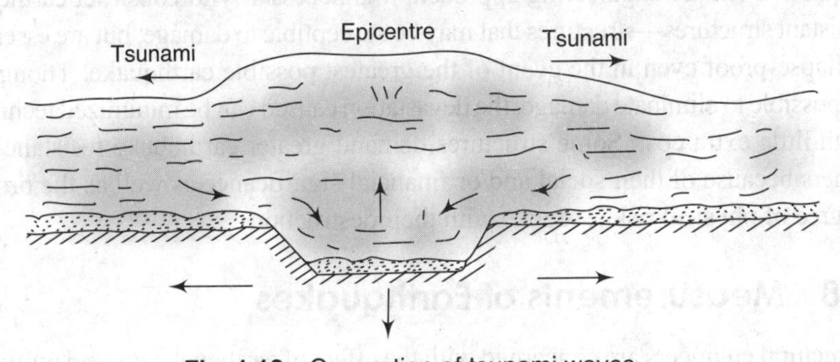

Fig. 1.12 Generation of tsunami waves

This recession and return of the water continues at intervals as each wave of the usual series arrives at the coast.

2. The tsunamis are long period waves that may travel long distances from their point of generation.

(b) Since a tsunami occurs because of sudden displacement of a large body of water, this displacement may be caused by
   - *undersea landslides* whereby large amount of sediment is dislodged from the sea floor, displacing a water column and potentially generating a localized tsunami;
   - *surface land sliding into the ocean* due to earthquake, resulting in local tsunami; and
   - *volcanic eruptions* in or near the ocean which may cause tsunami, but are not usual.

(c) *Sieches*, similar to small tsunamis, occur as a result of the sloshing of enclosed water in reservoirs, lakes, and harbours shaken by earthquakes.

(d) Earthquakes can cause fire by damaging gas lines and snapping electric wires.

(e) Earthquakes can rupture dams and levees (raised river embankments), causing floods, resulting in damage to structures and considerable loss of life.

1.7 Consequences of Earthquake Damage

The magnitude and destructive consequences of earthquakes have serious implications. The consequences are unique in many respects. The destruction and damage of constructed and natural environments, and the loss and impairment of human life are of prime concern. Moreover, loss of life, failure of infrastructure, and psychological fear of the region being earthquake prone results in decline in economic growth of the area.

An important feature of the disaster caused by earthquakes is that harm to life is associated almost entirely with man-made structures, for example, collapse of buildings, bridges, dams, and so on. It is well understood that earthquakes cannot be predicted accurately. Therefore, an attempt to mitigate the effects of an earthquake requires a unique engineering approach, it is necessarily to construct earthquake-resistant structures—structures that may be susceptible to damage, but are essentially collapse-proof even in the event of the greatest possible earthquake. Though it is impossible to eliminate damage, the devastation caused can be minimized technically with little extra cost. Some structures demand greater earthquake resistance than others because of their social and/or financial significance as well as the probable degree of physical risk associate with their destruction.

1.8 Measurements of Earthquakes

Structural engineers are concerned with the effect of earthquake ground motions on structures, that is, the amount of damage inflicted on the structures. This damage (stress and deformation) potential depends to a large extent on the size (severity) of the earthquake.

The severity of an earthquake can be assessed in the following ways—(i) quantifying its magnitude in terms of the energy released—measuring the amplitude, frequency, and location of the seismic waves; (ii) evaluating the intensity—considering the destructive effect of shaking ground on people, structures, and natural features. It is easier to measure the magnitude because, unlike the intensity, which can vary with location and has no mathematical backing, the magnitude of a particular earthquake remains constant.

1.8.1 Intensity

The size of an earthquake can be described by its intensity, the oldest measure of earthquake size. The intensity, or destructive power, of an earthquake is an evaluation of the severity of the ground motion at a given location and is represented by a numerical index. It is measured in relation to the effect of the earthquake on human life. Generally, destruction is described in terms of the damage caused to buildings, dams, bridges, and so on, as reported by witnesses. It is not a unique, precisely defined characteristic of an earthquake. Intensity is a somewhat subjective (qualitative) measure in that it is based on direct observation by individuals, rather than on instrumental measurements. Different observers of the same earthquake may assign different intensity values to it. Intensity is represented by roman capital numerals (see Table 1.1).

Table 1.1 Qualitative assignment of numerical value to earthquake intensity

Intensity	Observed effects
I	Not felt, except by very few people under special conditions. Detected mostly by instruments.
II	Felt by a few people, especially those on upper floors of buildings. Suspended objects may swing.
III	Felt noticeably indoors. Parked automobiles may rock slightly.
IV	Felt by many people indoors, by a few outdoors. At night, some are awakened. Dishes, windows, and doors rattle.
V	Felt by nearly everyone. Many are awakened. Some dishes and windows are broken. Unstable objects are overturned.
VI	Felt by everyone. Many people are frightened and run outdoors. Some heavy furniture is moved. Some plaster falls.
VII	Most people are alarmed and run outside. Damage is negligible in buildings of good construction.
VIII	Damage is slight in specially designed structures, considerable in ordinary buildings, great in poorly built structures. Heavy furniture is overturned.
IX	Damage is considerable in specially designed structures. Buildings shift from their foundations and partly collapse. Underground pipes are broken.

(*Contd*)

Table 1.1 (*Contd*)

X	Some well-built wooden structures are destroyed. Most masonry structures are destroyed. T he ground is badly cracked. Considerable landslides occur on steep slopes.
XI	Few, if any, masonry structures remain standing. Rails are bent. Broad fissures appear in the ground.
XII	Virtually total destruction. Waves are seen on the ground surface. Objects are thrown in the air.

The intensity of an earthquake at a specific location depends on a number of factors. Foremost among these are the total amount of energy released, the distance from the epicentre, and the type of rock and degree of consolidation. In general, the wave amplitude and extent of destruction are greater in soft, unconsolidated material than in dense, crystalline rock. The intensity is greatest close to the epicentre. Several dozen intensity scales are in use worldwide based on three features of shaking—perception by people and animals, performance of buildings, and changes to natural surroundings.

Two intensity scales, the Modified Mercalli (MM) intensity scale, 1931, and the Medvedev-Spoonheuer-Karnik (MSK-64) intensity scale, 1964, are generally used.

Modified Mercalli Earthquake Intensity Scale

The MM intensity of an earthquake is usually assessed by distributing questionnaires to or interviewing persons in the affected areas, in addition to the observations of the earthquake's effects by experienced personnel. It is apparent that the determination of intensity at a particular site involves considerable subjective judgements and is greatly influenced by the type and quality of structures as well as the geology of the area. Thus, comparisons of intensity ratings made by different people in different countries or under different conditions can be misleading. From the point of view of the structural engineer, the reported MM intensities (Table 1.2) can be considered only as a very crude quantitative measure of the destructiveness of an earthquake, since they do not provide specific information on damage corresponding to structures of interest in terms of the relevant structural parameters.

Medvedev-Spoonheuer-Karnik Intensity Scale

The MSK-64 intensity scale is prevalent in India, and central and eastern Europe. The qualitative nature of MSK-64 scale is apparent from the description of each intensity level which takes into account the type of structure, grade of damage to the structure, and the description of characteristic effects (Table 1.3). The scale is more comprehensive than MM intensity scale and describes the intensity of earthquakes more precisely.

Table 1.2 Abridged MM earthquake intensity scale and magnitude of earthquakes

Intensity by scale	Mercalli intensity	Description of characteristic effects	Richter magnitude corresponding to highest intensity reached
I	Instrumental	Detected only by seismographs	—
II	Feeble	Noticed only by sensitive people	3.5–4.2
III	Slight	Like the vibrations due to a passing lorry; felt by people at rest, especially on upper floors.	3.5–4.2
IV	Moderate	Felt by people while walking, rocking of loose objects, including parked vehicles.	4.3–4.8
V	Rather strong	Felt generally; most people sleeping are awakened and bells ring	
VI	Strong	Trees sway and all suspended objects swing; damage by overturning and falling loose objects	4.9–5.4
VII	Very strong	General alarm; walls crack, plaster falls	5.5–6.1
VIII	Destructive	Car drivers seriously disturbed; masonry fissures; chimneys fall; poorly constructed buildings damaged	6.2–6.9
IX	Ruinous	Some houses collapse where ground begins to crack, and pipes break open	
X	Disastrous	Ground cracks badly; many buildings get destroyed and railway lines get bent; landslides on steep slopes	7.0–7.3
XI	Very disastrous	Few buildings remain standing; bridges get destroyed; all services (railway, pipes, and cables) are put out of action; great landslides and floods	7.4–8.1
XII	Catastrophic	Total destruction; objects thrown into air; ground rises and falls in waves	> 8.1 (maximum known 8.9)

Table 1.3 MSK-64 Intensity Scale

I Noticeable	The intensity of the vibration is below the limits of sensibility: the tremor is detected and recorded by seismograph only.
II Scarcely noticeable (very slight)	Vibration is felt only by individuals at rest in houses, especially on upper floors of buildings.
III Weak, partially observed only	The earthquake is felt indoors by a few people, outdoors only in favourable circumstances. The vibration is similar to the passing of a light truck. Attentive observers notice a slight swinging of hanging objects, somewhat more heavily on upper floors.

(*Contd*)

Table 1.3 (*Contd*)

IV	Largely observed	The earthquake is felt indoors by many people, outdoors by few, but no one is frightened. The vibration is similar to the passing of a heavily loaded truck. Windows, doors, and dishes rattle. Floors and walls crack. Furniture begins to shake. Hanging objects swing slightly. Liquid in open vessels are slightly disturbed. In parked motor cars, the shock is noticeable.
V	Awakening	(i) The earthquake is felt indoors by all, outdoors by many. Many people awake. A few run outdoors. Animals become uneasy. Buildings tremble throughout. Hanging objects swing considerably. Pictures knock against walls or swing out of place. Occasionally, pendulum clocks stop. Unstable objects overturn or shift. Open doors and windows are thrust open and slam back again. Liquids spill in small amounts from well-filled open containers. The sensation of vibration is similar to heavy objects falling inside the buildings. (ii) Slight damages to buildings of Type A are possible. (iii) Sometimes change in flow of springs is noticeable.
VI	Frightening	(i) Felt by most people indoors and outdoors. Many people in buildings are frightened and run outdoors. A few persons lose their balance. Domestic animals run out of their stalls. In few instances, dishes and glassware may break, and books fall down. Heavy furniture may possibly move and small steeple bells may ring. (ii) Damage of Grade 1 is sustained in a single building of Type B and in many of Type A. Damage in few buildings of Type A is of Grade 2. (iii) In few cases, cracks up to widths of 1 cm are possible on wet grounds. In mountains occasional landslips, and change in flow of springs and in level of well water is observed.
VII	Damage of buildings	(i) Most people are frightened and run outdoors. Many find it difficult to stand. The vibration is noticed by persons driving motor cars. Large bells ring. (ii) In many buildings of Type C damage of Grade 1 is caused; in many buildings of Type B damage is of Grade 2. Most buildings of Type A suffer damage of Grade 3, few of Grade 4. In single instances, landslides of roadway on steep slopes, crack in roads, seams of pipelines are damaged, and cracks in stone walls are observed. (iii) Waves are formed on water, and is made turbid by mud stirred up, Water levels in wells change and the flow of springs changes. Sometimes dry springs have their flow resorted and existing springs stop flowing. In isolated instances parts of sand and gravel banks slip off.
VIII	Destruction of buildings	(i) Fright and panic; also persons driving motor cars are disturbed; branches of trees break off. Even heavy furniture moves and partly overturns. Hanging lamps are damaged in part.

(*Contd*)

Table 1.3 (*Contd*)

	(ii)	Most buildings of Type C suffer damage of Grade 2, and few of Grade 3, Most buildings of Type B suffer damage of Grade 3. Most buildings of Type A suffer damage of Grade 4. Occasional breaking of pipe seams occurs. Memorials and monuments move and twist. Tombstones overturn. Stone walls collapse.
	(iii)	Small landslips in hollows and on banked roads on steep slopes; cracks in ground upto widths of several centimetres are observed. Water in lakes becomes turbid. New reservoirs come into existence. Dry wells refill and existing wells become dry. In many cases, change in flow and level of water is observed.
IX General damage of buildings	(i)	General panic; considerable damage to furniture. Animals run to and fro in confusion, and cry.
	(ii)	Many buildings of Type C suffer damage of Grade 3, and a few of Grade 4. Many buildings of Type B show a damage of Grade 4 and a few of Grade 5. Many buildings of Type A suffer damage of Grade 5. Monuments and columns fall. Considerable damage to reservoirs is caused and underground pipes are partly broken. In individual cases, railway lines are bent and roadways damaged.
	(iii)	On flat land overflow of water, sand, and mud is often observed. Ground cracks to widths of up to 10 cm, and on slopes and river banks more than 10 cm are noticed. Furthermore, a large number of slight cracks in ground; fall of rocks, many landslides, and earth flows; large waves in water are observed. Dry wells renew their flow and existing wells dry up.
X General destruction of building	(i)	Many buildings of Type C suffer damage of Grade 4 and a few of Grade 5. Many buildings of Type B show damage of Grade 5. Most of Type A has destruction of Grade 5. Critical damage to dykes and dams. Severe damage to bridges. Railway lines are bent slightly. Underground pipes are bent or broken. Road paving and asphalt show waves.
	(ii)	In ground, cracks up to widths of several centimetres, sometimes up to 1 m, Broad fissures occur parallel to water courses. Loose ground slides from steep slopes. From river banks and steep coasts, considerable landslides are possible. In coastal areas, displacement of sand and mud; change of water level in wells; water from canals, lakes, rivers, etc., thrown on land. New lakes form.
XI Destruction	(i)	Severe damage even to well-built buildings, bridges, water dams, and railway lines. Highways become useless. Underground pipes destroyed.
	(ii)	Ground considerably distorted by broad cracks and fissures, as well as movement in horizontal and vertical directions. Numerous landslips and fall of rocks occur. The intensity of the earthquake requires to be investigated specifically.

(*Contd*)

Table 1.3 (*Contd*)

XII Landscape changes	(i)	Practically, all structures above and below the ground are greatly damaged or destroyed.
	(ii)	The surface of the ground is radically changed. Considerable ground cracks with extensive vertical and horizontal movements are observed. Falling of rock and slumping of river banks over wide areas, lakes are dammed, waterfalls appear, and rivers are deflected. The intensity of the earthquake requires special investigation.

Notes:

1. *Type of structures (buildings)*

Type A	Building in field-stone, rural structures, unburnt-brick houses, clay houses
Type B	Ordinary brick buildings, buildings of large blocks and prefabricated type, half-timbered structures, buildings in natural hewn stone
Type C	Reinforced buildings, well-built wooden structures

2. *Definition of quantity:*

Single, few	About 5%
Many	About 50%
Most	About 75%

3. *Classification of damage to buildings*

Grade 1 Slight damage	Fine cracks in plaster, fall of small pieces of plaster
Grade 2 Moderate damage	Small cracks in plaster, fall of fairly large pieces of plaster, pantiles slip off, cracks in chimneys, parts of chimney fall down
Grade 3 Heavy damage	Large and deep cracks in plaster, fall of chimneys
Grade 4 Destruction	Gaps in walls; parts of buildings may collapse; separate parts of the buildings lose their cohesion; inner walls collapse
Grade 5 Total damage	Total collapse of the buildings

1.8.2 Magnitude

The magnitude of an earthquake is a measure of the amount of energy released. It is a *quantitative measure* of the actual size or strength of the earthquake and is a much more precise measure than intensity. Earthquake magnitudes are based on direct measurements of the size (amplitude) of seismic waves, made with recording instruments, rather than on subjective observations of the destruction caused. The total energy released by an earthquake can be calculated from the amplitude of the waves and the distance from the epicentre.

The amount of ground shaking is related to the magnitude of the earthquake. Earthquake magnitude is most often reported using the Richter magnitude scale. The Richter magnitude is based on the energy release of the earthquake, which is closely related to the length of the fault on which the slippage occurs. A magnitude number is assigned to an earthquake on the basis of the amount of ground displacement or vibration it produces, as measured by a *seismograph* (Section 1.7.5). (The reading at a given station is adjusted for the distance of the instrument from the earthquake's epicentre, because ground vibration naturally decreases with increasing distance from the site of the earthquake, as the energy is dissipated.) The *Richter scale* is a logarithmic scale, meaning that an earthquake of magnitude 4 (M_4) causes 10 times as much ground movement as one of magnitude 3,100 times as much as one of magnitude 2, and so on. An earthquake of M_2 is the smallest normally felt by human beings. $M_{8.9}$ is the largest earthquake magnitude ever recorded on the earth—Lisbon (Portugal) earthquake, 1755. Hundreds of thousands of smaller earthquakes occur each year. Technically, there is no upper limit to the Richter scale.

Note: It must be noted that the Richter number is not a good measure of the damage potential of an earthquake; the longer duration of shaking would probably produce somewhat greater damage to structures.

The magnitude M of an earthquake is given by Eqn (1.4) when a standard seismometre shows a maximum amplitude of A μm at a point 100 km from the epicentre.

$$M = \log_{10} A \tag{1.4}$$

However, a standard seismometre is not always set at a point 100 km from the epicentre, in which case one may use the logarithmic form of Richter magnitude scale M (also sometimes referred as M_L) given as

$$M = \log_{10} A - \log_{10} A_0 \tag{1.5}$$

where A is the maximum recorded trace amplitude for a given earthquake at a distance and A_0 is that for a particular earthquake selected as a standard. Since magnitude is a measure of the seismic energy released, which is proportional to $(A/T)^2$, the general form of the Richter magnitude scale [Eqn (1.4)] may be modified as follows.

$$M = \log_{10}\left(\frac{A}{T}\right)_{\max} + \sigma(\Delta, h) + C_r + C_s \tag{1.6}$$

where A and T are the ground displacement amplitude and the period of the considered wave, respectively, $\sigma(\Delta, h)$ is the distance correction factor at epicentral distance Δ and focal depth h, C_r is the regional source correction factor, and C_s is the station correction factor.

The length of the fault L in kilometres and the slip U in the fault are related to the magnitude M as

$$M = 0.98 \log_{10} L + 5.65 \tag{1.7}$$

and

$$M = 1.32 \log_{10} U + 4.27 \tag{1.8}$$

The idea behind the Richter magnitude scale was a modest one at its inception. The type of seismic wave to be used was not specified; the only condition was that the wave chosen—whether *P*-, *S*-, or surface wave—be the one with the largest amplitude. The Richter local magnitude is the best known magnitude scale, but is not always the most appropriate scale for description of earthquake size. Currently, several magnitude scales other than the original Richter magnitude scale are in use.

An uncomplicated earthquake record clearly shows a *P*-wave, an *S*-wave, and a train of Rayleigh waves. Now, if Richter's procedure for determining the local magnitude were followed, we would measure the amplitude of the largest of the three waves and then make some adjustment for epicentral distance and the magnification of the seismograph. It has become a routine in seismology to measure the amplitude of the *P*-wave, which is not affected by the focal depth of the source, and thereby determine a *P*-wave magnitude (called body wave magnitude, m_b). For shallow earthquakes, a surface wave train is also present. It is common practice to measure the amplitude of the largest swing in the surface wave train that has a period of nearly 20 s. This value yields the surface-wave magnitude (M_s). Neither of these two magnitudes, m_b and M_s, is the Richter magnitude, but each has an important part in describing the size of an earthquake. However, M_s correlates much more closely with the size of an earthquake than does m_b.

$$M_s = \log_{10}\left(\frac{A_s}{T}\right)_{\max} + 1.66\log_{10}\Delta + 3.3 \qquad (1.9)$$

where A_s is the amplitude of the horizontal ground motion in μm, T is the period (20 ± 2 s), and Δ is the epicentral distance in degrees; (360° corresponding to the circumference of the earth). It may be noted that the surface wave magnitude is based on the maximum ground displacement amplitude rather than the trace amplitude of a particular seismograph.

$$m_b = \log_{10}\left(\frac{A_s}{T}\right)_{\max} + \sigma(\Delta, h) \qquad (1.10)$$

where σ(Δ, *h*) is the correction factor.

Gutenberg and Richter gave the following relationship between the energy of the seismic waves, *E*, and the magnitude of surface waves, M_s

$$\log_{10}E = 4.4 + 1.5M_s \qquad (1.11)$$

The amount of energy released rises even faster with increasing magnitude, by about a factor of 32 ($10^{1.5}$) for each unit of magnitude, and 1,000 times for an increase of 2 in *M*.

Note: All magnitude scales M_L, M_S, and M_W have been designed to give numerically similar results. However, m_b scale gives somewhat different earthquake sizes.

Additional magnitude scales, such as moment magnitude (M_W), have been introduced to further improve the standards used for expressing earthquake magnitude. This is described in the subsection that follows.

1.8.3 Moment Magnitude

In the course of development of the science of seismology, highly precise terms have emerged to completely describe the magnitude of an earthquake. As mentioned earlier, the first such term was *seismic intensity*, introduced as early as 1857 by Mallet. Although variations of the intensity scale are still in use, they are not a true mechanical measure of source size, like force or energy. Rather they indicate the strength of the vibrations relative to a standard.

The first index of the size of an earthquake, based on measured wave motion, is earthquake magnitude which has been described earlier. This magnitude refers to the maximum ground-wave amplitude on a seismogram, projected back to the focus of the earthquake by allowing for attenuation. Such peak values, however, do not directly measure the overall mechanical power of the source.

Seismologists favour the *seismic moment* as a measure for estimating the size of seismic sources. This has been found to yield a consistent scale for the size of the earthquake. The concept has been adopted from the theory of mechanics. The elastic rebound along a rupturing fault can be thought of as being produced by force couples along and across it. The seismic moment is quite independent of any frictional dissipation of energy along the fault surface or as the waves propagate away from it. When these seismic moment values are correlated with the magnitude, they define another variety of magnitude called the *moment magnitude*, M_W. This is the moment released during an earthquake rupture. For scientific purposes, moment magnitude has proved to be the best measure, since the seismic moment (on which moment magnitude is based) is a measure of the whole dimension of the fault, which, for great earthquakes, may extend to hundreds of kilometres.

Moment magnitude may be calculated using the relationship given by Kanamori.

$$M_W = \frac{2}{3}[\log M_o - 16] \qquad (M_o = \mu A d) \qquad (1.12)$$

where M_W is the moment magnitude, μ is the rigidity, A is the area of rupture, d is the displacement, and M_o is the seismic moment (in dyn-cm).

Notes:

1. Seismic moment is proportional to the area of fault rupture, the average slip on the fault plane and the rigidity of the crust. Thus it measures the physical size of the event. The moment magnitude is derived from it empirically as a quantity without unit, and is just a number to conform to M_S scale. While the other magnitudes are derived from a simple measurement of the amplitude of a specifically defined wave, a spectral analysis is required to obtain seismic moment.
2. It is important to note that the ground shaking characteristics do not necessarily increase in proportion to the amount of energy released during an earthquake. In fact, for strong earthquakes the ground shaking characteristics become less sensitive to the earthquake than for small earthquakes. This is referred to as *saturation*. The

values for saturation for M_L and m_b range between 6 to 7 and that for M_S is about 8. All scales, except M_W, saturate at a certain size and are unable to distinguish the sizes of large earthquakes. This means that they are based on the amplitudes of the waves which have wavelength shorter than the rupture length of the earthquakes. Since the moment magnitude M_W is independent of saturation, it is the most suitable magnitude scale for describing the size of very large earthquakes.

3. Bolt (1989) suggests that m_b may be used for shallow earthquakes of magnitude 3 to 7, M_S for magnitudes 5 to 7.5, and M_W for magnitudes greater than 7.5.

The total seismic energy released during an earthquake is often estimated using the relationship (Gutenberg and Richter 1956), as follows.

$$\log E = 11.8 + 1.5\, M_S$$

The seismic wave energy (in ergs) may also be computed using the Kanamori relationship as follows.

$$E = \frac{\text{moment}}{20{,}000}$$

The unit change in magnitude corresponds to a $10^{1.5}$ or 32-fold increase in seismic energy.

1.8.4 Magnitude and Intensity in Seismic Regions

Often the question is posed whether a particular building can withstand an earthquake of a certain magnitude, say, 6.5. Now, an $M_{6.5}$ earthquake causes different shaking intensities at different locations and the damage induced in buildings at these locations is also different. Thus, it is a particular level of intensity of shaking that a structure is designed to resist, and not so much the magnitude of an earthquake. The peak ground acceleration experienced by the ground during shaking is one way of quantifying the severity of the ground vibration. Approximate empirical correlations between the MM intensities and the percentage ground acceleration (PGA) are given in Table 1.4.

Table 1.4 Peak ground acceleration

MMI	V	VI	VII	VIII	IX	X
PGA (g)	0.03–0.04	0.06–0.07	0.10–0.15	0.25–0.30	0.50–0.55	>0.6

In 1956, Gutenberg and Richter gave an approximate correlation between the local magnitude M_L of an earthquake and the intensity I_0 sustained in the epicentre area as follows.

$$M_L = \left(\frac{2}{3}\right) I_0 + 1 \tag{1.13}$$

For using Eqn (1.13) the Roman numbers of intensity are replaced with numerals, for example, VII with 7.

Esteva and Rasenblueth gave the relationship between seismic intensity, magnitude, and a short epicentral distance, *r,* as

$$I = 8.16 + 1.45M - 2.46 \ln r \quad (1.14)$$

where r is in kilometres. The limitation of Eqn (1.14) is that it cannot be used for cases in which r reaches the same order as the focal region.

1.8.5 Seismographs

A *seismograph* is an instrument used to measure the vibration of the earth. *Seismographs* are used to measure relatively weak ground motions. The records produced are called *seismograms*. Strong ground motions are measured by *accelerographs*. The principle of the seismograph is that ground motion is measured by the vibration record of a simple pendulum hanging from a steady point. A schematic diagram of a typical seismograph is shown in Fig. 1.13. It has three components: *the sensor*—consisting of the pendulum mass, string, magnet, and support; the *recorder*—consisting of the drum, pen, and chart paper; and the *timer*—the motor that rotates the drum at constant speed. When the supporting frame is shaken by earthquake waves, the intertia of the mass causes it to lag behind the motion of the frame. This relative motion can be recorded as a wiggly line by pen and ink on paper wrapped around a rotating drum. The earthquake records so obtained are called seismograms.

One such instrument is required in each of the two orthogonal horizontal directions. Of course, for measuring vertical oscillations, the string pendulum (Fig. 1.13) is replaced with a spring pendulum oscillating about a fulcrum. Some

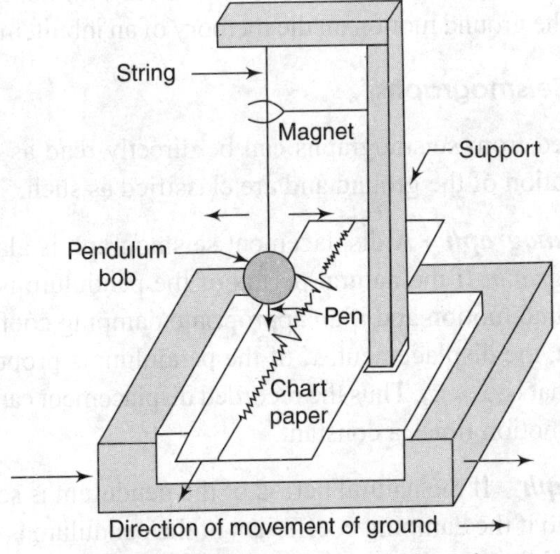

Fig. 1.13 Schematic diagram of a seismograph

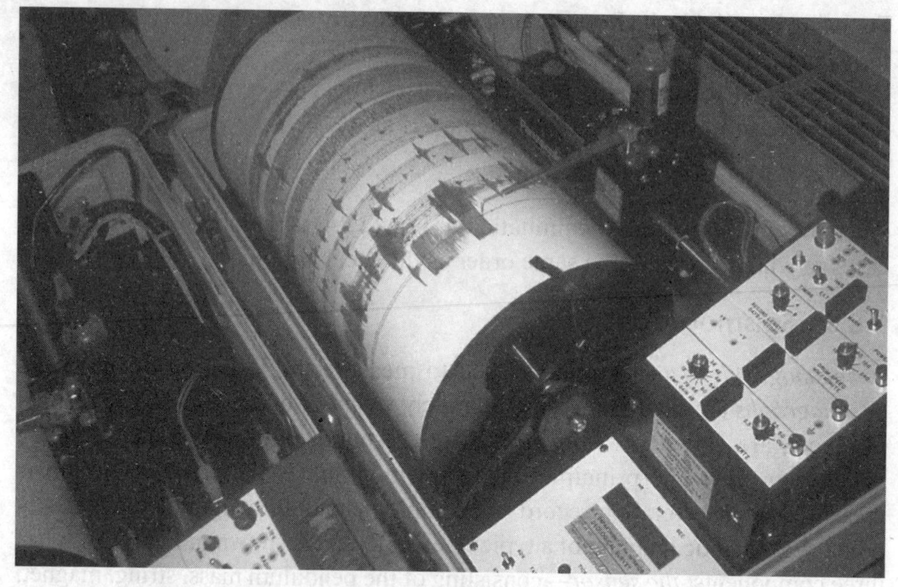

Typical seismograph
(*Source:* http://www.ema.gov.au/www/ema/schools.nsf/Page/Get_The_FactsEarthquakes)

instruments do not have a timer device (i.e., the drum holding the chart paper does not rotate). Such instruments provide only the maximum extent (or scope) of motion during an earthquake; for this reason they are called *seismoscopes*. In modern seismographs, the relative motion between the pendulum and the frame produces an electrical signal which is electronically magnified (1000X) before it is used to drive an electric stylus to produce the seismogram depicting even very weak seismic waves. Analog instruments have evolved over time; but today digital instruments using modern computer technology are more commonly used. Digital instruments record the ground motion on the memory of an inbuilt microprocessor.

Classification of Seismographs

The records obtained from seismographs can be directly read as displacement, velocity, or acceleration of the ground and are classified as such.

Displacement seismograph A displacement seismograph is also known as a *long-period seismograph*. If the natural period of the pendulum is long relative to the period of ground motion and if an appropriate damping coefficient for the pendulum is chosen, the displacement, x, of the pendulum is proportional to the ground motion, x_g, that is, $x \propto x_g$. Thus the recorded displacement can be expressed in terms of ground motion times a constant.

Velocity seismograph If the natural period of the pendulum is set close to that of ground motion and if the damping coefficient of the pendulum is large enough, then x is proportional to $\dot{x}_g$, and the ground velocity, $\dot{x}_g$, can be determined.

Acceleration seismograph An acceleration seismograph is also known as a *short-period seismograph*. These are also called *accelerographs* or *accelerometres*; the latter use electronic transducers. If the period of the pendulum is set short enough relative to that of ground motion, by means of an appropriate value of the pendulum's damping coefficient, $x = \ddot{x}_g$ is obtained. Thus the ground acceleration $\ddot{x}_g$ can also be recorded.

1.9 Strong Ground Motion

The earth vibrates continuously at periods ranging from milliseconds to days and the amplitudes may vary from nanometres to metres. It is pertinent to note that most vibrations are quite weak to even be felt. Such microscopic activity is important for seismologists only. The motion that affects living beings and their environment is of interest for engineers and is termed as *strong ground motion*. As already mentioned in Section 1.3, the ground motion at a particular instant of time can be completely defined by the three orthogonal components of translation; the three components of rotation being quite small, are neglected.

Vibration of the earth's surface is a net consequence of motions, vertical as well as horizontal, caused by seismic waves that are generated by energy release at each material point within the three-dimensional volume that ruptures at the fault. These waves arrive at various instants of time, have different amplitudes, and carry different levels of energy. Thus, the motion at any site on the ground is random in nature, its amplitude and direction varying randomly with time.

Large earthquakes at great distances can produce weak motions that may not damage structures or even be felt by humans. However, from an engineering viewpoint, strong motions that can possibly damage structures are of interest. This may occur with earthquakes in the vicinity or even with high-intensity earthquakes at medium to large distances.

Characteristics of Ground Motion

The motion of the ground can be described in terms of displacement, velocity, or acceleration. The variation of ground acceleration with time, recorded at a point on the ground during an earthquake, is called an *accelerogram* (Fig. 1.14). The ground velocity and displacement can be obtained by direct integration of an accelerogram. Typical ground motion records are called *time histories*—the acceleration, velocity and displacement time histories. From an engineering viewpoint, the amplitude, the frequency, and the duration of motion are the three important characteristics of the ground motion parameters. For structural engineering purposes, acceleration gives the best measure of an earthquake's intensity. From Fig. 1.14 it can be seen that during a short initial period, the intensity of ground acceleration increases to strong shaking, followed by a strong acceleration phase, which is followed by a gradual decreasing motion. The ground velocity is directly related to the energy

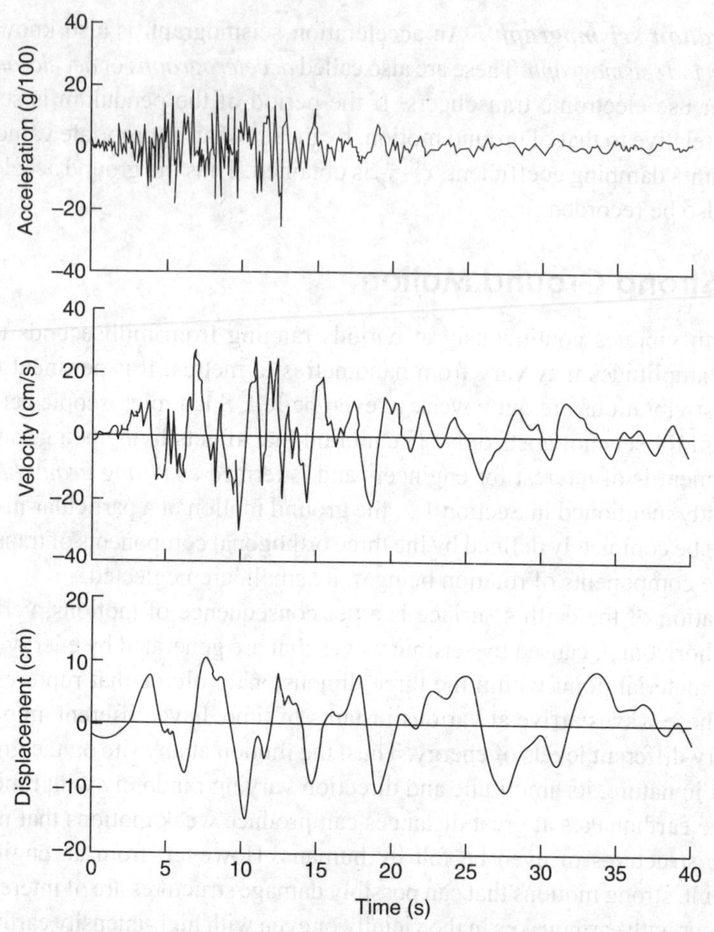

Fig. 1.14 Time-based earthquake strong motion records

transmitted to the structures and the intensity of damage caused. The ground displacement may be of interest for the design of underground structures.

Although measured ground accelerations do form a basis for dynamic analyses of structures, they do not have any direct correlation with the seismic coefficients that are used in engineering design. The most important factor in determining the values of these coefficients is observation by competent engineers of the actual damage of the structures. Further, these values are only a fraction of the peak ground accelerations that strong motion instruments would measure during an earthquake.

The nature of accelerograms may vary depending on the energy released at the source, the type of slip at the fault rupture, the geology along the travel path from the fault rupture to the earth's surface, and the local soil. Accelerograms carry distinct information regarding ground shaking, peak amplitude, duration of strong shaking, frequency content (amplitude of shaking associated with each frequency), and energy content (energy carried by ground shaking at each frequency).

The *peak ground acceleration* (PGA), also termed *peak horizontal acceleration* (PHA), is the most commonly used measure of the intensity of shaking at a site. For instance, a PGA value of 0.6*g* (0.6 times the acceleration due to gravity) suggests that the movement of the ground can cause a maximum horizontal force on a rigid structure equal to 60% of its weight. All points in a rigid structure move with the ground by the same amount and, hence, experience the same PGA. Since inertia forces are related to the horizontal accelerations, the latter are commonly used to describe ground motions. Vertical accelerations are not considered important since the static gravity force of structures counteracts the dynamic forces induced by vertical acceleration during an earthquake.

Ground motions with high peak accelerations are usually the most damaging except for short-duration stray pulses with large amplitude, where little time is available for the system to respond to such excitation. Usually, strong ground motions carry significant energy associated with shaking of frequencies in the range 0.03–30 Hz (i.e., cycles/s). Since peak acceleration does not provide any information on the frequency content and duration of motion, it is supplemented with additional information to characterize ground motion accurately. Pseudo-Horizontal Velocity (PHV) being sensitive to intermediate frequency range provides a better estimate of potential damage for flexible structures such as tall structures, bridges, and so on. Further, peak displacements generally associated with lower frequency components cannot be accurately determined and are therefore not used to measure ground motion.

Generally, the maximum amplitudes of horizontal motions in the two orthogonal directions are about the same. However, the maximum amplitude in the vertical direction is usually less than that in the horizontal direction. In design codes, the vertical design acceleration is taken as one-half to two-thirds of the horizontal design acceleration. In contrast, the maximum horizontal and vertical ground accelerations in the vicinity of the fault rupture do not seem to have such a correlation.

The duration of the strong motion is defined as the time interval in which 90% of the total contribution to the energy of the accelerogram takes place. Usually the time interval between 5% and 95% contributions is taken as the strong motion duration.

1.10 Local Site Effects

Response of a site to an earthquake may influence response of structures as the ground motion gets modified. Local site conditions at the site of interest influence the dynamic characteristics such as amplitude, frequency content, and duration of strong ground motion. The extent of the influence depends upon geometry and material properties of the subsurface materials, topography, and input motion. For important structures, a site-specific analysis is must to account for the influence of local site effects on ground surface motion.

The density and *S*-wave velocity of materials near the surface of earth are smaller than those at greater depths. If the effects of scattering and material damping are neglected, the flow of energy from the depth to the ground surface would be constant. Since density and *S*-wave velocity decrease as waves approach the ground surface, the particle velocity must increase. Thus, there is an increase in the particle velocity.

The effects of topographic irregularities such as ridges, valleys, and slope variations; basin effects; and lateral discontinuities on ground motions may be significant and are discussed as follows.

1.10.1 Topographic Effects

The effects of topographic irregularities are highly complicated. Depending upon the geometry and the type of irregularity, interaction of waves produces complex patterns of amplification and deamplification. In general, the amplification near crest of a ridge is found to increase. The average peak acceleration may be of the order of about 2.5 times the average base acceleration. The topographic acceleration decreases with the increase of angle of incidence of body waves. Moreover, interference between the incident waves and the outgoing diffracted waves causes differential ground motion along the slope of the topography.

1.10.2 Basin Effects

River basins have soft surficial deposits. The curvature of the basin with soft alluvial deposits can trap body waves and cause some incident body waves to propagate through the alluvium as surface waves. These waves can produce stronger shakings of longer durations than those due to only vertically propagating *S*-waves. Since there is potential for significant differential motion across such valleys, the long span structures (bridges, pipelines, etc.) crossing the valleys will suffer heavy damage due to differential settlement. Greater damage will be caused near edges of the basin.

1.10.3 Lateral Discontinuity Effects

Seismic waves travel faster in hard rocks than in the softer deposits. The sites with softer soils are therefore susceptible to stronger shaking. The intensity of damage in narrow zones along the lateral discontinuities is more because of the soft materials in these areas. This is due to the amplitude amplification and local surface wave generation in the softer medium.

In addition to the effects discussed above, following general effects may be noted.

1. With increase in the horizontal extent of the softer soils, the boundary effects of the bedrock on the site response will reduce.
2. The natural vibration of the ground increases with the depth of soil overlying the bedrock. The long-period structures get damaged in areas with greater depth of alluvium.

3. Changes of the soil type horizontally across a site affect the response locally within that site. A structure sited on such ground can get damaged heavily.
4. Liquefaction may occur in stratas with cohesionless soils.

1.11 Classification of Earthquakes

From the viewpoint of ground motion, earthquakes can be classified into the following four groups:

(i) *Practically a single shock.* Motion of this type occurs only at short distances from the epicentre, only on firm ground, and only for shallow earthquakes. When these conditions are not fulfilled, multiple wave reflections change the nature of the motion.

(ii) *A moderately long, extremely irregular motion.* This is associated with moderate distance from the focus and occurs only on firm ground. By analogy with light, it can be said that these motions are nearly white noise. They are usually of almost equal severity in all directions.

(iii) *A long-period ground motion exhibiting pronounced prevailing periods of vibration.* Such motion results from the filtering of earthquakes of the preceding types through layers of soft soil that exhibit linear or almost linear soil behaviour, and from the successive wave reflections at the interfaces of these mantles.

(iv) *A ground motion involving large-scale, permanent deformations of the ground.* At the site of interest, there may be slides or soil liquefaction.

There are ground motions of characteristics intermediate between those that have just been described. For example, the number of significant, prevailing ground periods because of complicated stratification may be so large that a motion of the third group approaches white noise, or soil behaviour may be only moderately non-linear. The nearly white noise type of earthquakes have received the greatest share of attention. This interest in white noise is due to its relatively high incidence, the number of records available, and facilities for simulation in analog and digital computers, or even for analytical treatment of the responses of simple structures. Because of the chaotic nature of these motions, such analytical studies are based necessarily on the theory of probabilities, while the simulation on computers aims at the Monte Carlo studies involving statistical interpretation of the results. The first kind of earthquake can be dealt with deterministically because of its simplicity. The only serious limitation at present is the scarcity of available records.

The third type of motion is obtained from linear filtering of the first or the second type. It is, hence, nearly as amenable to analytical treatment as are the first two types of motion and like the second type of motion, it is also amenable to the Monte Carlo studies.

The fourth kind of earthquake is difficult to study either analytically or through simulation, and it is not especially useful to predict structural responses to this kind of earthquake. Ordinarily, it is impractical to attempt a structural design that would resist a large-scale failure of the ground. Rather, one should normally aim at establishing the conditions under which such phenomena are likely to occur and, where they are likely, either erect the structure in question elsewhere or treat the soil in such a way that the phenomenon becomes unlikely, at least locally.

Some of the other classifications based on earthquake magnitudes and effects of earthquakes are as follows.

Classification Based on Magnitude

Class	*Magnitude*
Great	≥ 8.0
Major	7.0–7.9
Strong	6.0–6.9
Moderate	5.0–5.9
Light	4.0–4.9
Minor	3.0–3.9

Classification Based on Earthquake Effect

Earthquake effect	*Magnitude*
Usually not felt, but can be recorded	< 2.5
Often felt, but only cause minor damage	2.5–5.4
Slight damage to structures	5.5–6.0
May cause lot of damage in populated areas	6.1–6.9
Serious damages	7.0–7.9
Catastrophic, can even destroy communities near the epicentre	≥ 8.0

1.12 Seismic Zoning

The problem of designing economical earthquake-resistant structures rests heavily on the determination of reliable quantitative estimates of expected earthquake intensities in particular regions. However, it is not possible to predict with any certainty when and where earthquakes will occur, how strong they will be, and what characteristics the ground motions will have. Therefore, an engineer must estimate the ground shaking judiciously. A simple method is to use a seismic zone map, wherein the area is subdivided into regions, each associated with a known or assigned seismic probability or risk, to serve as a useful basis for the implementation of code provisions on earthquake-resistant design. Seismic zoning is accomplished with the help of isoseismal maps. An *isoseismal* is a contour bounding areas of equal intensity and different isoseismals when plotted for a particular earthquake constitute an *isoseismal map*. The present seismic zoning map used in India (Appendix II) shows the country divided into four zones (II, III, IV, and V) of approximately equal seismic probability, depending upon the local hazard. Each of these zones is described in terms of the value of its peak ground acceleration, also known as the *design ground acceleration*. The seismic zone map of India has been revised from five to four zones. Zone I from the code has been removed with the consensus between geologists and seismologists that no zone lies below Zone II in India. It is based on the past earthquake data and geological information of the region. However, there is no engineering basis for the same. Associated with each zone is a factor which enters into the expression for determining the total base shear and is known as zone factor (Appendix III). A detailed method of estimating the design of an earthquake-resistant structure is to conduct a site-specific seismic evaluation which takes into account seismic history, active faults in the vicinity of the structure site, and the stress–strain properties of materials through which the seismic waves travel. Normally the zoning is laid down by a code, but outside the area of applicability of this code, the zoning status needs to be based on an assessment of the seismic hazards.

1.13 Response of Structure to Earthquake Motion

The loads or forces which a structure subjected to earthquake motions is called upon to resist, result directly from the distortions induced by the motion of the ground on which it rests. The response (i.e., the magnitude and distribution of the resulting forces and displacements) of a structure to such a base motion is influenced by the properties of the foundations of the structure and surrounding structures, as well as the character of the existing motion.

A simplified behaviour of a building during an earthquake can be seen in Fig. 1.15. As the ground on which the building rests is displaced, the base of the building moves with it. However, the inertia of the building mass resists this motion and causes the building to suffer a distortion (greatly exaggerated in the figure). This distortion wave travels along the height of the structure in much the same manner as a stress wave in a bar with a free end. The continued shaking of the base causes the building to undergo a complex series of oscillations.

When the ground shaking is at a much slower rate than the structure's natural oscillations, the behaviour will be quasi-static; the structure simply moves with the ground with its absolute displacement amplitude, approximately the same as that of the ground. In case ground motion period and natural period of structure are

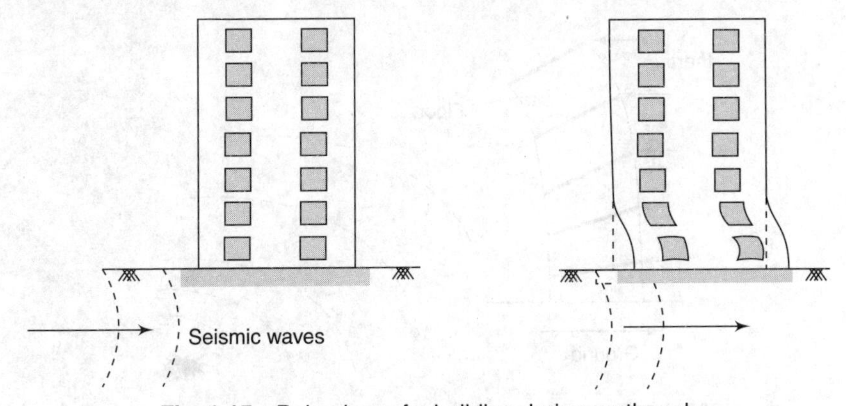

Fig. 1.15 Behaviour of a building during earthquake

similar, resonance occurs and there is large dynamic amplification of the motion. The stiffness and inertial forces at any time in such case are approximately equal and opposite; most resistance to the motion is provided by damping of the system. If ground motion is much faster than the natural oscillations of the structure, then the mass undergoes less motion than the ground. Further, it may be noted that under real earthquake loading, the dynamic amplifications are rather smaller because earthquake time-history, is not a simple sinusoid (does not have constant amplitude), and has a finite duration.

The earthquake motion for the calculation of design seismic actions, at a given point on the surface of the earth, is generally represented by the interaction between the elastic ground acceleration and a structural response called *elastic response spectrum*. At any particular point, the ground acceleration may be described by horizontal components along two perpendicular directions and a vertical component. In most instances only the structural response to the horizontal components of ground motion is considered since buildings are not sensitive to horizontal or lateral distortions. These horizontal forces, equal to mass times acceleration, represent the inertia forces (Fig. 1.16) that occur at the critical instant during the largest cycle of vibration, of maximum deflection and zero velocity, as the structure responds to earthquake motion. The effect of the vertical component of ground motion is generally considered not to be significant and is neglected except in cantilevers. For most structures, experience seems to have justified the viewpoint. In most instances, further simplification of the actual three-dimensional response of a structure is achieved by assuming that the design horizontal acceleration components will act non-concurrently in the direction of each principal plan-axis of a building. It is tacitly assumed that a building designed using this approach will have adequate resistant acceleration acting in any direction.

In addition to ground acceleration, rocking and twisting (rotational) components may be involved. Rocking and torsional effects, due to the horizontal components of ground motion, occur as a result of ground compliance and the non-coincidence

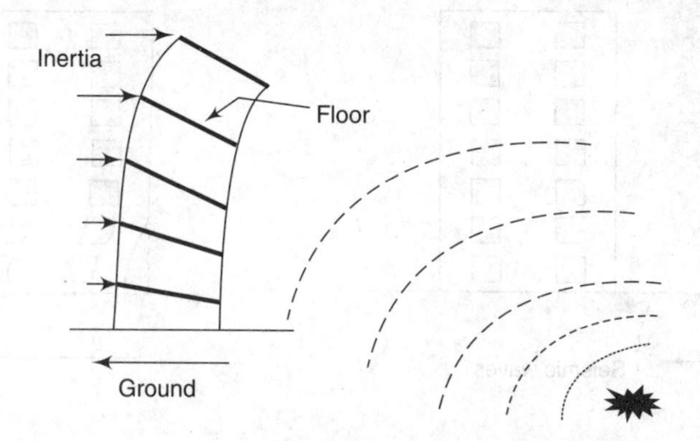

Fig. 1.16 Structure subjected to earthquake excitation

of the centres of mass and rigidity. However, the rotational components of earthquake ground motion are usually negligible.

The principal properties that affect the dynamic response of the acceleration are the mass of the structure, its stiffness, and its damping characteristics. Under certain conditions, the effect of a foundation or a supporting medium may have to be considered. In virtually all earthquake design practice, the structure is analysed as an elastic system; it is acknowledged that the structural response to strong earthquakes involves yielding of the structure, so that the response is inelastic. The effect of yielding in a structure is two-fold. On one hand, stiffness is reduced so that displacements tend to increase. On the other hand, hysteretic yielding absorbs energy from the structure, increasing damping and reducing displacements. The two effects are roughly equal, so yielding does not have a large effect on displacement.

Most of the dynamic analyses of multi-storey structures assume the free-field ground motion to be applied, unaltered, to the building foundation, which is assumed to be rigid. While this approximation may be valid for most structures founded on soil, which is relatively stiff with respect to the structure, cases arise where soil–structure interaction effects produce significantly different results from those based on a rigid foundation assumption. The effect of the soil–structure interaction on earthquake response needs to be considered, as it can vary depending upon the properties of the soil and the structure, as well as the character of the input motion.

1.14 Seismic Design

Severity of ground shaking at a given location during an earthquake may be minor (occurs frequently), moderate (occurs occasionally), or strong (occurs rarely). The probability of a strong earthquake occurring within the expected life of a structure is very low and that of small earthquakes is very high. The annual frequency N of the earthquakes is given by following equation,

$$\log N = a - bM$$

where N is the number of earthquakes of magnitude M during a certain time period, while a and b are determined by fitting the equation to the historical record.

Statistically, about 800 earthquakes of magnitude 5.0–5.9 occur in the world, while only about 18 of magnitude above 7.0 are registered annually. If a building is designed to be earthquake-proof for a rare but strong earthquake, it will be robust but too expensive. The most logical approach to the seismic design problem is to accept the uncertainty of the seismic phenomenon. Consequently, the main elements of the structure are designed to have sufficient ductility, allowing the structure to sway back and forth during a major earthquake, so that it withstands the earthquake with some damage, but without collapse. An earthquake-resistant structure resists the effects of ground shaking; although it may get severely damaged, it does not collapse during a strong earthquake. This implies that the damage should be controlled to acceptable levels, preserving the lives of the occupants of the building at a reasonable cost. Engineers thus tend to make the structures earthquake resistant.

Engineers recognize that damage is unavoidable, but should be allowed to occur at right places and in right amounts. For instance, cracks between columns and masonry filler walls in a frame structure are acceptable, but diagonal cracks through the columns are not. Also the consequences of damage have to be kept in mind during the design process. For example, important structures such as hospitals, schools, public places, dams, bridges, and so on, must sustain very little damage and should be designed for a higher level of earthquake protection.

In the light of the above discussion, seismic design theory must embody the following precepts:

(a) In order to deal effectively with the combination of extreme loading and low probability, the design earthquake is taken as a moderate one; as a test for structural safety, the most severe earthquake, which a structure may be expected to face in its lifetime, is applied. It is reasonable to expect the structure to maintain elastic behaviour.

(b) During a minor earthquake, the load-carrying members of the structure should not be damaged; however, the non-structural parts may sustain repairable damage. During moderate earthquakes, the load-carrying members may sustain repairable damage, while the non-structural parts may even have to be replaced after the earthquake. During a strong earthquake, the load-carrying members may sustain severe damage, but the structure should not collapse. At such times, plastic behaviour of the building is accepted on the premise that the peak forces produced are of short duration and, therefore, can be more readily absorbed by the movement of the structure than a sustained static load can.

(c) One of the concepts involved in seismic design of buildings is to make it ductile. Such a building would ride out an earthquake, not to stand up rigidly but to absorb the earthquake's energy by yielding gently; rigid buildings would attract more earthquake force. This may be visualized as the way a tree moves with strong winds with flexibility. Moreover, ductility gives a warning to the occupants and provides sufficient time to take preventive measures, reducing loss of life in case of strong earthquakes; buildings can undergo large deformations, if ductile, before failure. Satisfactory seismic performance requires, in part, selection of adequate structural strength and ductility. Ductility of reinforced concrete (RC) structures increases with compression steel reinforcement, concrete compression strength, and ultimate compressive strain. On the other hand, it decreases with an increase in tension steel component, yield strength of steel, and axial load. For steel structures, special consideration should be made for joints; the joints should display ductile behaviour.

Summary

Mankind has been bearing the brunt of the furies of nature since time immemorial, and has never given up the quest to develop tools and technologies to meet the challenges. One cannot imagine a greater destructive agent than an earthquake. This chapter is dedicated to first, understanding the cause, nature, effect, and consequences of an earthquake; second, presenting available methodologies to quantify this gigantic force; and third, providing seismic design parameters. Since the response behaviour of the earth/ground is important, strong ground motion and the response of a structure to this motion are described. Seismic zoning is introduced to help the designer implement provisions of the code. The chapter ends by presenting a design philosophy for structures, based on seismic loads.

Exercises

1.1 What is an earthquake? How do human activities induce earthquakes?

1.2 Write short notes on the following:
 (a) Earth's crust
 (b) Earth's mantle
 (c) Causes of volcanic earthquakes
 (d) Seismic waves
 (e) Subduction zone

1.3 Describe the two approaches followed for the prediction of earthquakes. Name the major plates of the earth.

1.4 Explain the plate tectonic theory and its mechanism.

1.5 What are plate tectonics and how are they related to continental drift and sea floor spreading?

1.6 Explain how a subduction zone forms and what occurs at such a plate boundary.

1.7 What is meant by the focus and epicentre of an earthquake? Name the two kinds of body waves and explain how they differ.

1.8 Discuss the main characteristics of seismic waves.

1.9 Distinguish between the following:
 (a) Body waves and surface waves
 (b) Rayleigh waves and love waves
 (c) Lithosphere and asthenosphere

1.10 Discuss briefly the two measures of an earthquake.

1.11 MSK-64 Intensity Scale has an edge over modified MM Intensity Scale. Comment.

1.12 On what is the assignment of an earthquake's magnitude based? Is magnitude the same as intensity? Explain.

1.13 Define saturation. How is moment magnitude a better measure of earthquake size than other magnitudes?

1.14 Describe briefly the direct and indirect effects of an earthquake.

1.15 Write short notes on the following:
  (a) Seismograph
  (b) Modified Mercalli scale
  (c) Seismic design theory
  (d) Strong ground motion
  (e) Tsunamis

1.16 Discuss briefly classification of earthquakes.

1.17 What is strong ground motion? State and discuss their characteristics.

1.18 (a) How is the epicentre of an earthquake located?
  (b) Discuss briefly the need of seismic zoning.

1.19 (a) How is the local magnitude of an earthquake related to the intensity of an earthquake?
  (b) What is the basic design philosophy of seismic design of structures?

1.20 An earthquake causes an average of 2.6 m strike-slip displacement over a 75 km long, 22 km deep portion of a transformed fault. Assuming the average rupture strength along the fault as 180 kPa, estimate the seismic moment and moment magnitude of the earthquake.

Ans: 7.722×10^{21} dyne-cm; 3.925 dyne-cm

CHAPTER 2

Dynamics of Structures and Seismic Response

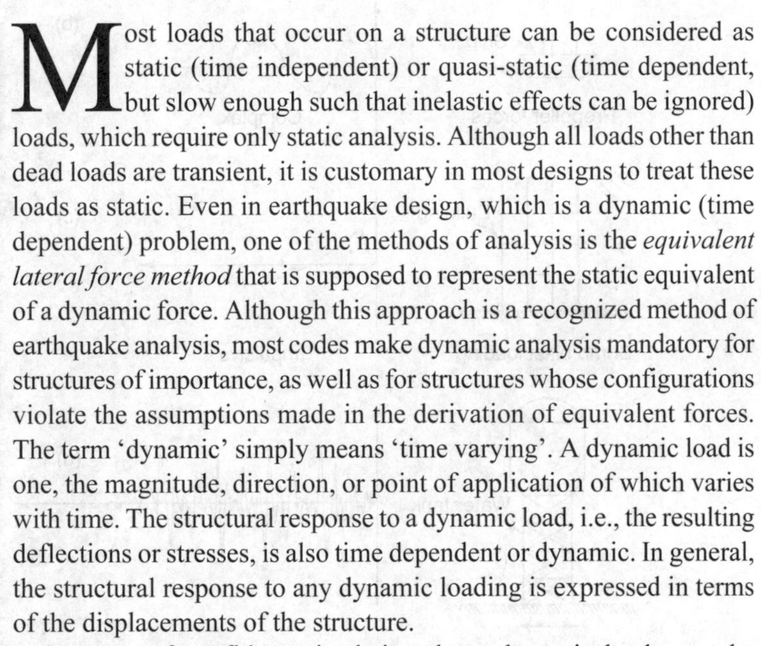

Most loads that occur on a structure can be considered as static (time independent) or quasi-static (time dependent, but slow enough such that inelastic effects can be ignored) loads, which require only static analysis. Although all loads other than dead loads are transient, it is customary in most designs to treat these loads as static. Even in earthquake design, which is a dynamic (time dependent) problem, one of the methods of analysis is the *equivalent lateral force method* that is supposed to represent the static equivalent of a dynamic force. Although this approach is a recognized method of earthquake analysis, most codes make dynamic analysis mandatory for structures of importance, as well as for structures whose configurations violate the assumptions made in the derivation of equivalent forces. The term 'dynamic' simply means 'time varying'. A dynamic load is one, the magnitude, direction, or point of application of which varies with time. The structural response to a dynamic load, i.e., the resulting deflections or stresses, is also time dependent or dynamic. In general, the structural response to any dynamic loading is expressed in terms of the displacements of the structure.

In terms of confidence in their values, dynamic loads may be classified into *deterministic* (or prescribed) and *stochastic* (or random).

If the loading is a known function of time, the loading is said to be *prescribed*, and the analysis of a structural system to a prescribed loading is called *deterministic analysis*. In contrast, the variations of a random force in time may be affected by a number of factors, so its determination always implies a certain probabilistic element. Seismic loads are random in character, though they are usually regarded as deterministic in practical calculations to simplify the design model.

Dynamic loading may also be classified as *periodic loading* [Fig. 2.l(a) and (b)] or *non-periodic loading* [Fig. 2.l(c) and (d)]. Periodic loadings are examples of repetitive loads exhibiting time variation successively for a large number of cycles. The simplest periodic loading is the sinusoidal variation [Fig. 2.l(a)], termed as *simple harmonic*. Non-periodic loadings may be either short-duration impulsive loadings, as shown in Fig. 2.1(c), or long-duration loadings [Fig. 2.l(d)]. Examples of non-periodic loadings are earthquakes, wind, blasts, and explosions.

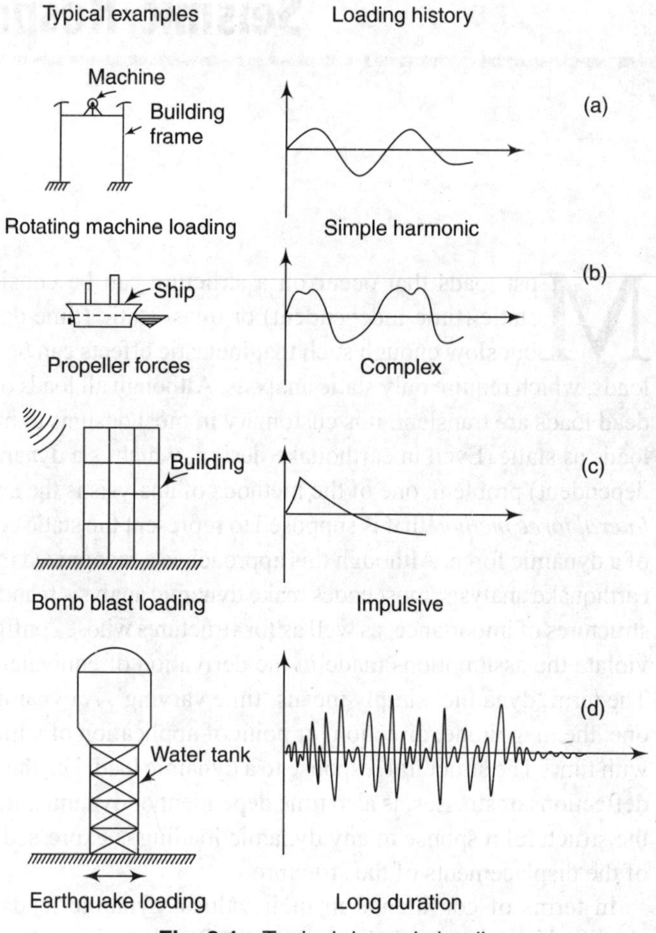

Fig. 2.1 Typical dynamic loadings

Dynamics deals with the motion of nominally rigid bodies. Structural dynamics implies that in addition to having motion, the bodies are non-rigid. In a structural-dynamic problem, the load and response vary with time; hence, a dynamic problem does not have a single solution. Instead, the analyst must establish a succession of solutions corresponding to all items of interest in the response history. Since earthquake forces are considered dynamic, instead of obtaining a single solution as in a static case, a separate solution is required at each instant of time for the entire duration of the earthquake. When a dynamic load $p(t)$ is applied to a structure, e.g., on a simple beam as shown in Fig. 2.2, the resulting displacements are associated with accelerations that produce *inertia forces* resisting the accelerations. Thus the internal moments and shears in the example structure (beam) of Fig. 2.2 must equilibrate not only the externally applied force but also the inertia forces resulting from the acceleration of the beam. These inertia forces cause the system to vibrate. Structural dynamics, however, should not be confused with vibration, whichs implies only oscillatory behaviour.

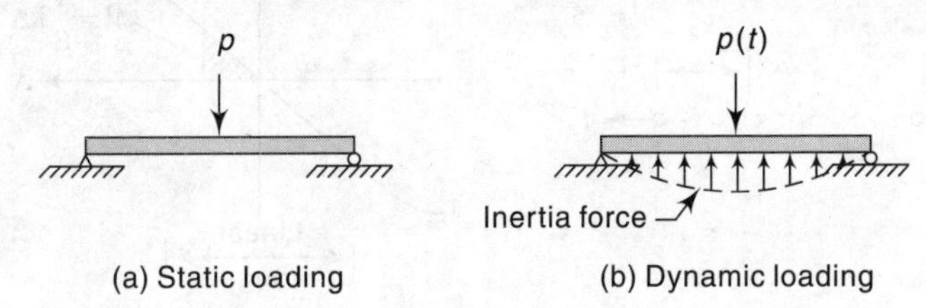

Fig. 2.2 Basic difference between static and dynamic loads

Behaviour of a system subjected to dynamic load is quite complex. However, it can be determined with sufficient accuracy by a simplified mathematical model of the system that may be linear or non-linear. In fact, all physical systems exhibit non-linearity. Accurate modelling will lead to non-linear differential equations and it is difficult to solve and find their solution. Assumptions are made to linearise the system and with application of the principles of dynamics, the differential equations governing the behaviour can be obtained. Newton's second law of motion, d'Alembert's principle, principle of virtual displacement, or the principle of conservation of energy may be used suitably to derive the governing differential equations of motion. A single-degree-of-freedom (SDOF) system leads to one ordinary differential equation of motion and a multi-degrees-of-freedom (MDOF) system leads to a set of ordinary differential equations of motion. The governing differential equations of motion are then solved to find the response of the system. Techniques such as standard (classical) methods for solution of differential equations, time-domain method, frequency-domain method, or numerical methods as appropriate for the particular case may be used. The solution of the governing differential equations of motion gives the displacements, velocities, and accelerations of various masses in dynamic analyses. The ultimate aim of the analysis is to develop a set of curves in the form of response spectrum.

In the following sections an attempt has been made to bring out the essentials of structural dynamics as related to seismic design of buildings. The dynamic analysis consists of defining the analytical model, deriving the mathematical model and solving for the dynamic response. Mathematical modelling of single- and multi-storey structures, with and without damping, is presented briefly.

2.1 Modelling of Structures

To carry out a dynamic analysis, the structure has to be modelled mathematically as a spring–mass–dashpot system. The component that relates force to displacement is usually called a *spring*. Figure 2.3 shows an idealized massless spring and a plot of spring force versus elongation. For most structural materials, for a small value of

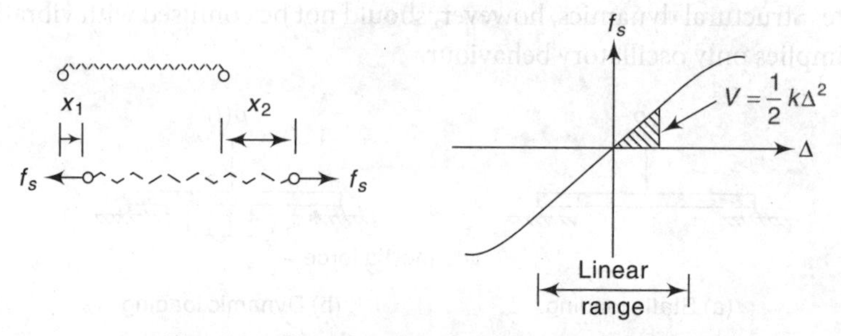

Fig. 2.3 Force–deflection behaviour of a spring

elongation $\Delta = x_2 - x_1$, there is a linear relationship between force and elongation expressed as $f_S = k\Delta$, where k is called the *spring constant*. Within its elastic range, a spring serves as an energy storage device. The energy stored in the spring within the linearly elastic range is called *strain energy*, V, and is expressed as

$$V = \frac{1}{2}k\Delta^2 \tag{2.1}$$

The process by which free vibration steadily diminishes in amplitude is called *damping*. A structure is said to be undergoing free vibration when it is disturbed from its static equilibrium position and then allowed to vibrate without any external dynamic excitation. While a spring serves as an energy storage device, there are also means by which kinetic energy and strain energy are dissipated from a deforming structure. These are called *damping mechanisms*. In a vibrating building these include opening and closing of micro-cracks in concrete, friction at steel connections, and friction between the structure and non-structural elements (e.g., partition walls). It is impossible to identify and describe mathematically each of these energy-dissipating mechanisms in an actual building. As a result, the damping in actual buildings is usually represented in a highly idealized manner. The most commonly used damping element is the viscous damper and, therefore, the most common analytical model of damping employed in structural dynamic analyses is the *linear viscous dashpot* model shown in Fig. 2.4. The damping force f_D is given by

$$f_D = c(\dot{x}_2 - \dot{x}_1) = c\dot{x} \tag{2.2}$$

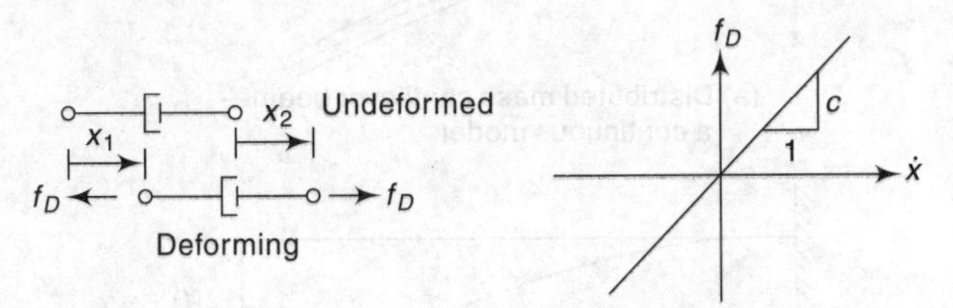

Fig. 2.4 Linear viscous dashpot

where $\dot{x}$ is the velocity across the linear viscous damper. The constant c is called *coefficient of viscous damping.* Its unit is N-s/m. The damping coefficient is selected so that the vibrational energy it dissipates is equivalent to the energy dissipated in all damping mechanisms combined, present in the actual building.

The equivalent viscous damper is intended to model the energy description at deformation amplitudes within the linear elastic limit of the overall structure. Over this range of deformations, the damping coefficient c, determined from experiments may vary with the deformation amplitude. This non-linearity of the damping property is usually not considered explicitly in dynamic analysis.

Additional energy dissipated due to inelastic behaviour of the structure at large deformations is represented by the force-deformation hysteresis loop. This energy dissipation is usually not modelled by a viscous damper. The damping energy dissipation during one deformation cycle is given by the area within the hysteresis loop.

There are, as follows, three approaches of discretization of a structure that may be used depending upon the suitability.

2.1.1 Lumped Mass Approach

The inertia forces resulting from structural displacements are, in turn, influenced by the magnitude of the masses. This makes the analysis complicated and necessitates that the problem is formulated in terms of differential equations. For a complete definition of inertia forces, the displacements and accelerations must be defined for each point. For example, for the cantilever beam shown in Fig. 2.5(a), the displacements and accelerations for each point along the axis of the beam will be required because the mass of the beam is distributed continuously along its length. This reinforces the formulation in terms of partial differential equations because position along the span and time must be taken as independent variables. Such analytical models are called *continuous models*. A continuous model represents an infinite degrees-of-freedom (DOF) system.

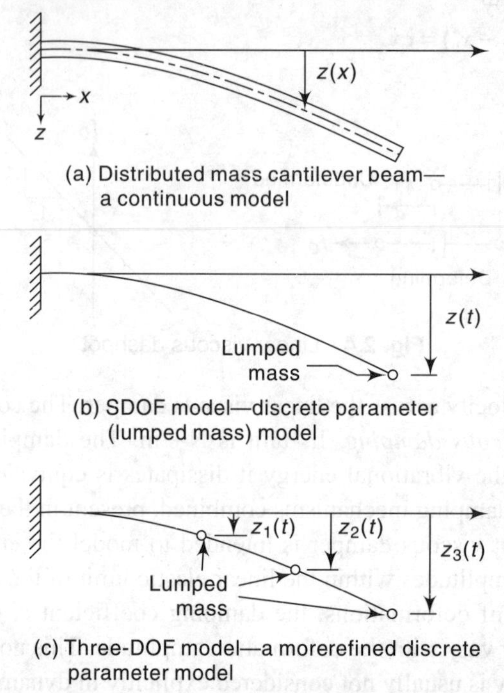

Fig. 2.5 Continuous and discrete parameter (lumped mass) analytical model of cantilever beam

To simplify the analysis, it may be assumed that the mass of the beam is concentrated in a series of discrete points and that the inertia forces will develop only at these mass points. Such discrete points are called *lumps* and the concentrated mass in these points is called *lumped mass* [Fig. 2.5(b) and (c)]. In this case, the displacements and accelerations need to be defined only at these mass points. The lumped mass models depict finite DOF systems.

The number of displacement components to be considered in order to represent the effects of all significant inertia forces of a structure is called the *dynamic degree of freedom* of the structure. For the planar structure shown in Fig. 2.5(c), there would be one degree of freedom for each discrete point if it could move along in a vertical direction only and three degrees of freedom for the whole system. However, if each of the masses was not concentrated in points, but rather had finite rotational inertia, then the degrees of freedom of each discrete point would be two and those of the whole system would be six. If axial deformations are also considered, then the degrees of freedom would be three for each discrete point and nine for the system. Had it been a space member, each mass would have six degrees of freedom and the system would have 18 degrees of freedom.

This approach is most effective in treating a system in which a large proportion of the total mass is actually concentrated in a few discrete points. In order to create a useful analytical model, the analyst should be capable of making a true assessment of the behaviour of the real system. In dynamic analyses of structures, a single-storey structure is modelled as an SDOF system and a multi-storey building as an MDOF system, where it is assumed that the mass is concentrated at the floor level of each storey. Although less accurate than the two following approaches, it is satisfactory for most structural frames.

2.1.2 Generalized Displacement Procedure

The generalized displacement procedure is most effective for a system where the mass is quite uniformly distributed throughout. The underlying assumption is that the deflected shape of the structure can be expressed as the sum of a series of specified displacement patterns; these patterns then become the displacement coordinates of the structure. The example structure of Fig. 2.2 can be assumed to have deflected shapes as shown in Fig. 2.6, and the deflection can be expressed as sum of independent sine-wave contributions given by

$$z_x = \sum_{n=1}^{\infty} b_n \sin \frac{n\pi x}{L} \tag{2.3}$$

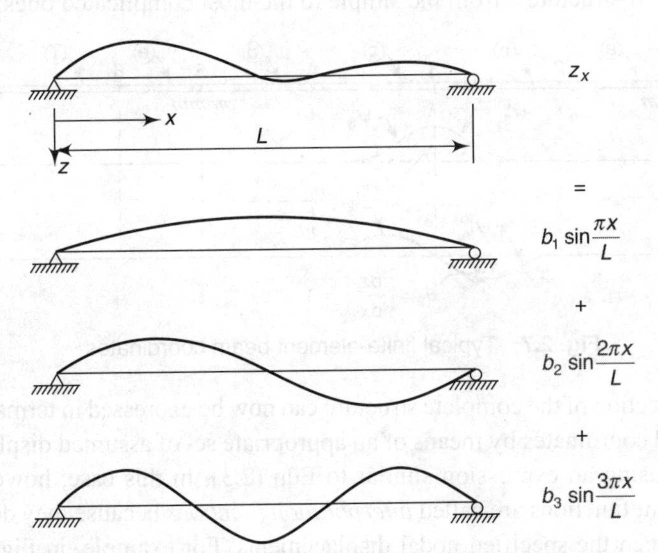

Fig. 2.6 Sine series representation of single beam deflection

The amplitudes of the sine-wave shapes are considered to be the coordinates of the system.

In general, any shapes $\psi_n(x)$, which are compatible with the prescribed geometric-support conditions and which maintain the necessary continuity of internal displacements may be assumed. The generalized expression for the displacement of any one-dimensional structure may be written as

$$z_x = \sum_n \rho_n \psi_n(x) \tag{2.4}$$

For any assumed set of displacement functions $\psi_n(x)$, the resulting shape of the structure depends upon the amplitude terms ρ_n, which will be referred to as *generalized coordinates*. The number of assumed shape patterns represents the degree of freedom considered in this form of idealization.

2.1.3 Finite Element Procedure

The finite element procedure is the most efficient one, especially for expressing the displacements of arbitrary structural configurations. It combines certain features of both, the lumped mass approach and the generalized coordinate approach. It provides a convenient and reliable idealization of the system and is particularly effective in digital computer analyses. For example, the beam shown in Fig. 2.7 is divided into an appropriate number of segments, also called *elements* (a, b, c, ...); they may be equal or unequal in size. The ends of the elements are called *nodal points* (1, 2, 3, ...). The displacements of these nodal points are the generalized coordinates of the structure. The finite element type of idealization is applicable to all types of structures, from the simple to the most complicated ones.

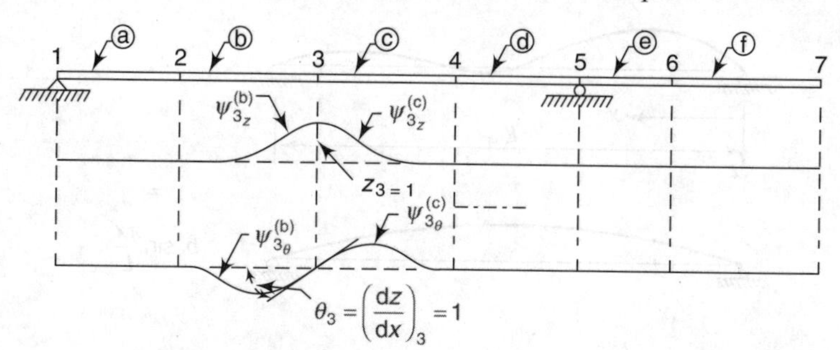

Fig. 2.7 Typical finite-element beam coordinates

The deflection of the complete structure can now be expressed in terms of these generalized coordinates by means of an appropriate set of assumed displacement functions, using an expression similar to Eqn (2.3). In this case, however, the displacement functions are called *interpolation functions* because they define the shape between the specified nodal displacements. For example, in Fig. 2.7 the interpolation functions are associated with the two degrees of freedom of point 3 and they produce transverse displacements in the plane of the figure. In principle, these interpolation functions could be any curve that is internally continuous and that satisfies the geometric displacement conditions imposed by the nodal displacement. For one-dimensional elements it is convenient to use the shapes, which would be produced by these nodal displacements in a uniform beam (these are cubic Hermitian polynomials and are shown in Fig. 2.7). The coordinates used in the finite element method are just a special form of the generalized coordinates. The advantages of this procedure are as follows:

(a) Any desired number of coordinates can be introduced merely by dividing the structure into an appropriate number of segments.
(b) Since the displacement functions chosen for each segment may be identical, computations are simplified.
(c) The equations developed by this approach are largely uncoupled because each nodal displacement affects only the neighbouring elements; thus, the solution procedure is generally simplified.

2.2 Equations of Motion

Dynamic analysis is an approximate analysis with limited degrees of freedom providing sufficient accuracy. The problem is thus reduced to the determination of the time history of these selected displacement components. The mathematical expressions defining the dynamic displacements are called the *equations of motion* of the structure. Any one of the following methods can be used to formulate and solve the equations of motion to provide the required displacement histories.

2.2.1 Direct Equilibration Using d'Alembert's Principle

According to Newton's second law of motion, the rate of change of momentum of any mass m equals the force acting on it. This can be expressed mathematically as the differential equation as follows.

$$p(t) = \frac{d}{dt}\left(m\frac{dx}{dt}\right) \tag{2.5}$$

where $p(t)$ is the applied force vector and $x(t)$ is the position vector of mass m. For most cases in structural dynamics it may be assumed that mass remains constant with time, and Eqn (2.5) may be written as

$$p(t) = m\frac{d^2x}{dt^2} = m\ddot{x}(t) \tag{2.6}$$

or $$p(t) - m\ddot{x}(t) = 0 \tag{2.7}$$

The term $m\ddot{x}(t)$ is called the *inertia force* resisting the acceleration of mass.

The concept that a mass develops an inertia force proportional to its acceleration and opposing it is known as d'Alembert's principle. It permits the equations of motion to be expressed as equations of dynamic equilibrium.

> **Note:** Simple structural dynamics problems can be solved using Newton's second law and d'Alembert's principle. But, for complex systems, methods based on work done (principle of virtual displacement) or based on energy concept (Hamilton's principle) are convenient and more appropriate.

2.2.2 Principle of Virtual Displacements

Complex structures involve a number of interconnected mass points or bodies of finite size, so the equilibration of all the forces acting in the system may be difficult. In such cases, the principle of virtual displacements can be used to formulate the equations of motion as a substitute for the equilibrium relationships.

The principle of virtual displacements states that if a system which is in equilibrium under the action of a set of forces is subjected to a virtual displacement, then the total work done by the forces will be zero.

The principle of virtual work stated as above was stated by Bernoulli and is a static procedure. It was extended to dynamics by d'Alembert who introduced the concept of inertia force.

In this method, the response equations of a system are established by identifying all the forces acting on the mass of the system, including inertia forces. Then the equations of motion are obtained by introducing virtual displacements corresponding to each degree of freedom and equating the work done to zero.

2.2.3 Energy Method—Hamilton's Principle

This principle states that the variation of the kinetic and potential energies plus the variation of work done by the non-conservative forces considered during any time interval t_1 to t_2 must be equal to zero and can be expressed as

$$\int_{t_1}^{t_2} \delta(K - V)dt + \int_{t_1}^{t_2} \delta W_{nc} dt = 0 \tag{2.8}$$

where K is the total kinetic energy of system, V is the potential energy of system (including both strain energy and potential energy of any conservative external forces), W_{nc} is the work done by non-conservative forces acting on system (including damping and any arbitrary external loads), and δ is the variation taken during indicated time interval.

The application of the principle leads directly to the equations of motion for any given system.

Notes: 1. The differential equation such as Eqn (2.6) can be solved by using classical method—solution consisting of the sum of the complementary solution and the particular solution, time-domain method—symbolized by Duhamel's integral, frequency-domain method—using Fourier transform, and numerical methods. Classical method has been used for solving differential equations in this chapter. The complete solution of the linear differential equation of motion by this method consists of the complementary solution $x_c(t)$ and the particular solution $x_p(t)$. Since the differential equation is of the second order, two constants of integration (A and B; Section 2.4.1) are involved which are evaluated from a knowledge of the initial conditions. Duhamel's integral* can be used if the applied force $p(t)$ is defined by a simple function that permits analytical solution for the integral; for complex functions, Duhamel's integral can be evaluated by numerical techniques. The frequency-domain method is the most suitable for dynamic analysis of structures interacting with unbounded media, e.g., soil-structure interaction and fluid-structure interaction (dams interacting with water impounded in reservoirs for great distances upstream). Numerical methods are best suited for complicated ground motions in addition to linear systems of excitation.

2. The first three methods are for linear systems. For strong ground motions involving inelastic behaviour of structures, numerical techniques are the only choice. The scope of the book limits the discussion of the above methods.
3. Since ground acceleration during earthquakes varies randomly, the best results for structural response can be had only by numerical techniques. The numerical methods should, therefore, be preferred over the first three methods for finding the solution of equation of motion for the ground motion.

2.3 Systems with Single Degree of Freedom

The physical properties of any linearly elastic structural system subjected to dynamic loads include its mass m measured in kg (should not be confused with its weight $m \times g$, which is a force measured in N), its elastic properties (flexibility or stiffness), stiffness k, measured in N/m, its energy-loss mechanism (damping) denoted by c, the constant of proportionality between force and velocity measured in N-s/m, and the external source of excitation (loading). A sketch of such a system is shown in Fig. 2.8(a). The entire mass is included in the rigid block.

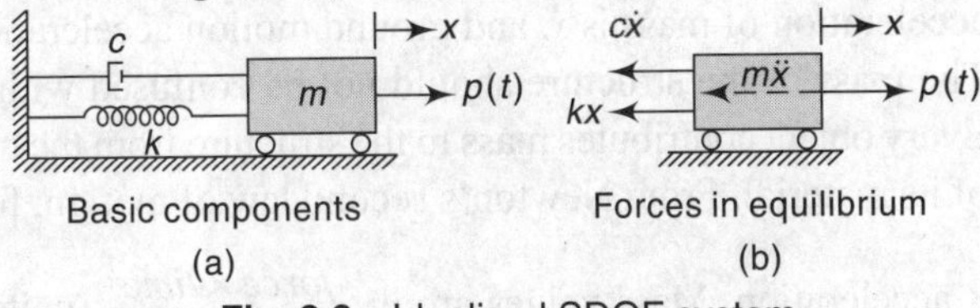

Fig. 2.8 Idealized SDOF system

Rollers constrain this block so that it can move only in simple translation; thus the single displacement coordinate v completely defines its position. The elastic resistance to displacement is provided by the weightless spring of stiffness k, while the energy-loss mechanism is represented by the viscous damper (also known as

---

*The total displacement at time t for an undamped system is given by the equation, (see also Section 2.4.2), $x(t) = \frac{1}{m\omega}\int_0^t p(\tau)\sin[\omega(t-\tau)]d\tau$. The integral in this equation is known as Duhamel's integral.

dashpot) *c*. The external-loading mechanism producing the dynamic response of this system is the time-varying load $p(t)$.

As shown in Fig. 2.8(b), for the applied load $p(t)$ the resulting forces are inertia force f_I, damping force f_D, and the elastic spring force f_S. The equation of motion for this system is given as

$$f_I + f_D + f_S = p(t) \tag{2.9}$$

f_S (elastic force) = spring stiffness × displacement = kx (for a linear system)

f_I (inertia force) = mass × acceleration = $m\ddot{x}$

f_D = (damping force) = damping constant × velocity = $c\dot{x}$

$p(t)$ = the force p varying with time

Equation (2.9) reduces to

$$m\ddot{x} + c\dot{x} + kx = p(t) \tag{2.10}$$

The equation of motion developed above is for a displacement x of the idealized structure of Fig. 2.8(a), assumed to be linearly elastic and subjected to an external dynamic force $p(t)$. In the inelastic range, the force f_S corresponding to deformation x depends on the history of the deformation and on whether the deformation is increasing (positive velocity) or decreasing (negative velocity). Thus the resisting force can be expressed as $f_S(x, \dot{x})$. The derivation of equation of motion for elastic systems can be extended to inelastic systems where the equation of motion becomes

$$m\ddot{x} + c\dot{x} + f_S(x, \dot{x}) = p(t) \tag{2.11}$$

Notes: 1. In the dynamics of structures, motion of the structures is considered as a function of time. The overdot ($\dot{}$) is considered as the differentiation with respect to time. As such, the velocity of mass is $\dot{x}$, the acceleration of mass is $\ddot{x}$, and ground motion acceleration is $\ddot{x}_g$.

2. The mass of the structure should not be confused with its weight. Every object contributes mass to the structure from the mass density of its material. From Newton's second law of motion, force = mass × acceleration. Mass values are in $\frac{force \times time^2}{length}$ units. Since the unit of force is kN, the derived unit of mass is $\frac{\text{kN} \times \text{s}^2}{\text{m}}$. However, the proper SI unit of mass is kg. Therefore, in the solved examples and in the exercises, the unit of mass has been used either N-s^2/m or kg as per convenience. The value of acceleration due to gravity g is taken as 9.81 m/s^2. Therefore, a body with mass of 1 kg weighs 9.81 N. Further, the unit of weight (which is a force) is kN, of stiffness k is N/m, and that of coefficient of damping c is N-s/m

2.4 Dynamic Response of Single-storey Structure

A single-storey structure can be modelled as an SDOF system. Each possible displacement of the structure is known as degree of freedom. For linear dynamic analysis, a structure can be defined by the three key properties—the mass, the stiffness, and the damping. The mass m of the structure is assumed to be concentrated at the floor level of the storey. The horizontal girder in the frame is assumed to be rigid and to include all the moving mass of the structure as shown in Fig. 2.9. In reality, all structures have distributed mass, stiffness, and damping. Assumption of lumped mass, however, is justified as in most cases, it is possible

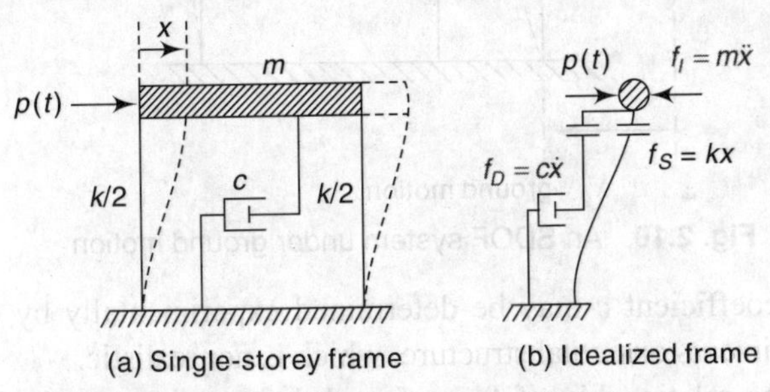

Fig. 2.9 An SDOF system under horizontal force

to obtain reasonably accurate estimates of dynamic behaviour of the structure. The vertical columns are assumed to be weightless and inextensible in the vertical (axial) direction. The resistance to girder displacement provided by each column is represented by its spring constant $k/2$ (since there are two columns). The idealized structure has only one DOF, the lateral displacement x., since it has been idealized with mass concentrated at one location (roof level) for dynamic analysis x is associated with column flexure; the damper c (represented by dashpot) provides a velocity-proportional resistance to this deformation. The system is called SDOF system. Thus, the equation of motion for a single-storey structure will be

$$f_I + f_D + f_S = 0 \tag{2.12}$$

or $$m\ddot{x} + c\dot{x} + kx = 0$$

The dynamic stresses and deflections may be induced in a structure not only by a time-varying applied load but also by motions of its support points and the motions of the building's foundation caused by an earthquake. Figure 2.10 shows a simplified model of the earthquake excitation problem in which the horizontal ground motion displacement caused by the earthquake is indicated by the displacement x_g of the structure base relative to the fixed reference axis. In this case, the structure is subjected to ground acceleration, and total displacement of the mass at any instant can be expressed as the sum of the ground displacement x_g and the column distortion x.

$$f_I = m(\ddot{x} + \ddot{x}_g) \tag{2.13}$$

$$F_D = c\dot{x} \tag{2.14}$$

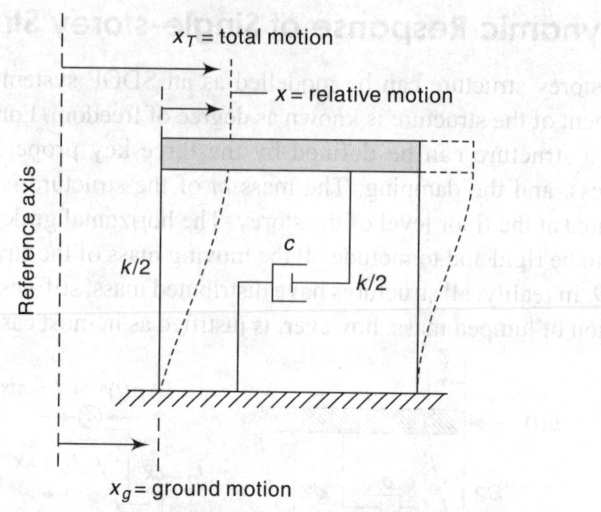

Fig. 2.10 An SDOF system under ground motion

The damping coefficient c may be determined experimentally by conducting vibration experiments on actual structure, which is nonrealistic.

Assuming the relationship of force f_S and deformation x to be linear (i.e., loading and unloading curves to be identical), the system has to be elastic. Hence for linearly elastic system

$$f_S = kx \tag{2.15}$$

From Eqns (2.12) to (2.15)

$$m\ddot{x} + m\ddot{x}_g + c\dot{x} + kx = 0 \tag{2.16}$$

or
$$m\ddot{x} + c\dot{x} + kx = -m\ddot{x}_g \tag{2.17}$$

The equation of motion developed above governs the relative displacement x of the structure of Fig. 2.10 subjected to ground acceleration $\ddot{x}_g$. Till now, the analysis has been done considering elastic system, i.e., the deformations are small. However, for large deformations, the initial loading curve (force-deformation curve) is non-linear and the unloading and reloading curves differ from the initial loading branch of the curve. Such a system is known as inelastic. For inelastic systems, the resulting equation of motion is

$$m\ddot{x} + c\dot{x} + f_S(x, \dot{x}) = -m\ddot{x}_g \tag{2.18}$$

The negative sign in Eqn (2.17) indicates that the effective force opposes the direction of ground acceleration; in practice this has little significance, in as much as the base input must be assumed to act in an arbitrary direction.

A comparison of Eqns (2.10) and (2.17) shows that the equations of motion for the structure are subjected to two separate excitations at each instant of time—ground acceleration $\ddot{x}_g$ and external force $-m\ddot{x}_g$ are one and the same.

Thus the relative displacement or deformation x of the structure due to ground acceleration $\ddot{x}_g$ will be identical to the displacement x of the structure if its base were stationary and if it were subjected to an external force equal to $-m\ddot{x}_g$. The ground motion can, therefore, be replaced by the effective earthquake force as shown in Fig. 2.11.

$$P_{eff} = -m\ddot{x}_g$$

This force is equal to mass times the ground acceleration, acting opposite to the acceleration.

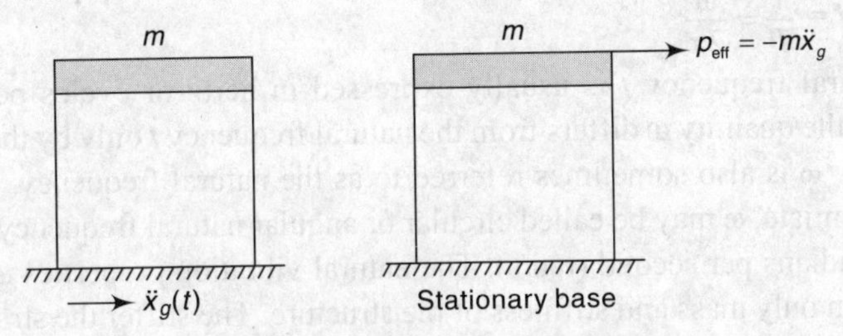

Fig. 2.11 Effective earthquake force

It is important to recognize that the effective earthquake force is proportional to the mass of the structure. Thus the effective earthquake force increases if the structural mass is increased.

The force f_S corresponding to deformation x for such a system depends on the history of the deformations and on whether the deformation is increasing (positive velocity) or decreasing (negative velocity). The resisting force for inelastic system is thus represented as $f_S(x, \dot{x})$.

2.4.1 Free Vibration Response

Motions taking place with the applied force set equal to zero are called *free vibrations.* To establish the free vibration response of the system, let us assume, to begin with, that there is no ground motion and that the SDOF system is without damping. Under these conditions, the system is in motion and is governed only by the influence of the so called *initial conditions;* that is, the given displacement and velocity at time $t = 0$ when the study of the system is initiated. Equation (2.17) can be simplified to

$$m\ddot{x} + kx = 0 \tag{2.19}$$

The solution of Eqn (2.19) is given as

$$x = A \cos \omega t + B \sin \omega t \tag{2.20}$$

Differentiating Eqn (2.20)

$$\dot{x} = -A\omega \sin \omega t + B\omega \cos \omega t \tag{2.21}$$

where ω is circular frequency or angular velocity of the system $= \sqrt{\dfrac{k}{m}}$.

The motion described by Eqn (2.21) is harmonic, and, therefore, periodic. The period T of the motion is determined as

$$T = \frac{2\pi}{\omega}$$

A system executes $1/T$ cycles in 1 s. Thus the period is usually expressed in seconds per cycle or seconds. The value reciprocal to the period is the natural frequency f given by

$$f = \frac{1}{T} = \frac{\omega}{2\pi}$$

The natural frequency f is usually expressed in hertz or cycles per second. Because the quantity ω differs from the natural frequency f only by the constant factor 2π, ω is also sometimes referred to as the natural frequency. However, to differentiate, ω may be called circular or angular natural frequency. The unit of ω is radians per second (rad/s). The natural vibration properties ω, T, and f depend on only mass and stiffness of the structure. The stiffer the structure, the higher the natural frequency of the same mass and the shorter the natural period. Similarly, a heavier (more mass) structure of the same stiffness will have lower natural frequency and longer natural period.

If at time $t = 0$, $x = x(0)$ and $\dot{x} = \dot{x}(0)$, then the constants A and B from Eqns (2.20) and (2.21) will be

$$A = x(0) \quad \text{and} \quad B = \frac{\dot{x}(0)}{\omega} \tag{2.22}$$

Equation (2.20) can be written as

$$x = x(0)\cos\omega t + \frac{\dot{x}(0)}{\omega}\sin\omega t \tag{2.23}$$

This solution represents a simple harmonic motion and is shown in Fig. 2.12(a).

The natural period T defined as the time required for the phase angle ωt to travel from 0 to 2π is given by

$$T = \frac{2\pi}{\omega} = 2\pi\sqrt{\frac{m}{k}} \tag{2.24}$$

$$\text{Amplitude, } X = \sqrt{A^2 + B^2} = \sqrt{[x(0)]^2 + \left[\frac{\dot{x}(0)}{\omega}\right]^2} \tag{2.25}$$

Phase angle is given by θ or θ' ($90° - \theta$) and is shown in Fig. 2.13. The phase angle θ represents the angular distance by which the resultant motion lags behind the cosine term in the response.

$$\theta = \tan^{-1}\left[\frac{\dot{x}(0)}{\omega x(0)}\right] \tag{2.26}$$

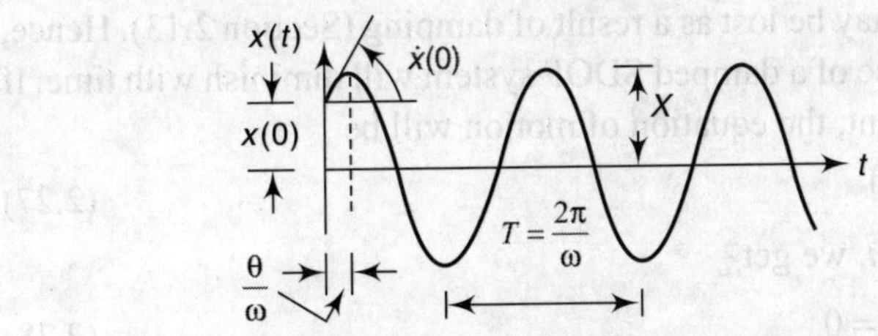

(a) Undamped free vibration response

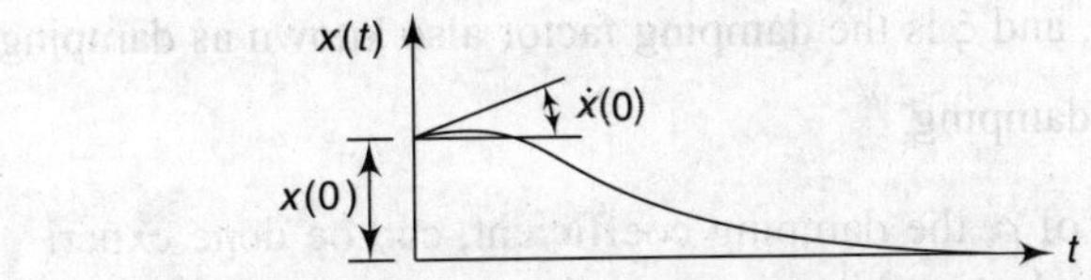

(b) Free vibration response with critical damping

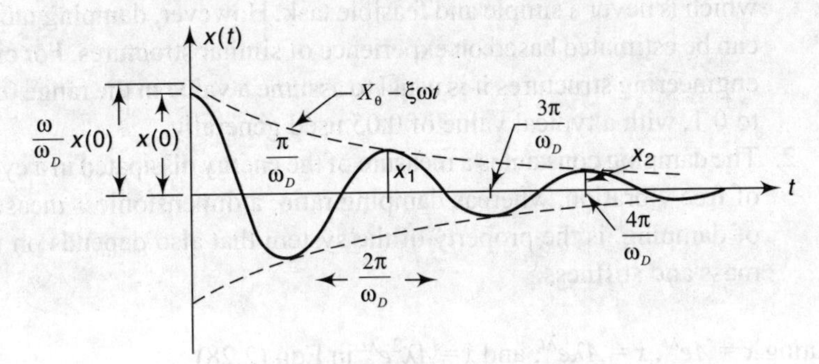

(c) Free vibration response of underdamped system

Fig. 2.12 Response of SDOF system

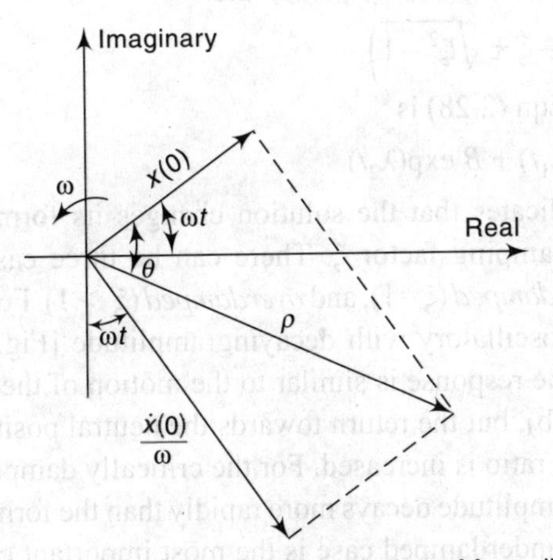

Fig. 2.13 Rotating-vector representation of free vibrations

In real systems, energy may be lost as a result of damping (Section 2.13). Hence, the free vibration response of a damped SDOF system will diminish with time. If viscous damping is present, the equation of motion will be

$$m\ddot{x} + c\dot{x} + kx = 0 \tag{2.27}$$

Dividing Eqn (2.27) by m, we get

$$\ddot{x} + 2\xi\omega\dot{x} + \omega^2 x = 0 \tag{2.28}$$

where $2\xi\omega = \dfrac{c}{m}$; $\omega^2 = \dfrac{k}{m}$, and ξ is the damping factor also known as damping ratio or fraction of critical damping.

Notes: 1. Determination of c, the damping coefficient, can be done experimentally by conducting vibration experiments on actual structure, which is never a simple and feasible task. However, damping ratio ζ can be estimated based on experience of similar structures. For civil engineering structures it is usual to assume a value in the range 0.01 to 0.1, with a typical value of 0.05 used generally.

2. The damping constant is a measure of the energy dissipated in a cycle of free vibration, whereas damping ratio, a dimensionless measure of damping, is the property of the system that also depends on the mass and stiffness.

substituting $x = Ae^{\lambda t}$, $\dot{x} = A\lambda e^{\lambda t}$, and $\ddot{x} = A\lambda^2 e^{\lambda t}$ in Eqn (2.28)

$$Ae^{\lambda t}(\lambda^2 + 2\xi\omega\lambda + \omega^2) = 0$$

or

$$(\lambda^2 + 2\xi\omega\lambda + \omega^2) = 0 \tag{2.29}$$

The roots of above characteristic equation are

$$\lambda_1, \lambda_2 = \omega\left(-\xi \pm \sqrt{\xi^2 - 1}\right) \tag{2.30}$$

The solution of Eqn (2.28) is

$$x = A\exp(\lambda_1 t) + B\exp(\lambda_2 t) \tag{2.31}$$

Equation (2.28) indicates that the solution changes its form according to the magnitude of the damping factor ξ. There can be three cases: *underdamped* ($0 < \xi < 1$), *critically damped* ($\xi = 1$), and *overdamped* ($\xi > 1$). For the underdamped case the motion is oscillatory with decaying amplitude [Fig. 2.12(c)]. For the overdamped case, the response is similar to the motion of the critically damped system of Fig. 2.12(b), but the return towards the neutral position requires more time as the damping ratio is increased. For the critically damped case there is no oscillation, and the amplitude decays more rapidly than the former two cases [Fig. 2.12(b)]. Since the underdamped case is the most important case for structures, it is discussed as follows.

For the underdamped case ($0 < \xi < 1$)

For the underdamped case in which damping is less than the critical, the solution of equation is of the form

$$x = \exp(-\xi\omega t)\left[A\cos\omega\sqrt{1-\xi^2}\,t + B\sin\omega\sqrt{1-\xi^2}\,t\right]$$

or $$x = \exp(-\xi\omega t)[A\cos\omega_D t + B\sin\omega_D t] \tag{2.32}$$

where A and B are constants of integration and ω_D, the damped frequency of the system, is given by

$$\omega_D = \omega\sqrt{1-\xi^2} \tag{2.33}$$

$$\text{damping ratio } \xi = \frac{c}{c_c}$$

where c is the damping coefficient and c_c is the coefficient of critical damping. The value of damping coefficient for real structures is much less than the critical damping coefficient and usually ranges between 2–20% of the critical damping value. The damping coefficient c_c is the smallest value of c that inhibits oscillations completely.

Note: Since for most civil engineering structures the damping ratio is typically less than 0.1, they fall under under-damped system.

A and B can be evaluated from the initial conditions of displacement and velocity, $x(0)$ and $\dot{x}(0)$, and substituted into Eqn (2.32) gives the solution of damped free vibration and can be expressed as

$$x = \exp(-\xi\omega t)\left[x(0)\cos\omega_D t + \frac{\dot{x}(0) + x(0)\xi\omega}{\omega_D}\sin\omega_D t\right] \tag{2.34}$$

Alternatively, this expression can be written as

$$x = \rho\exp(-\xi\omega t)\sin(\omega_D t + \theta) \tag{2.35}$$

$$\text{where amplitude } \rho = \sqrt{[x(0)]^2 + \left[\frac{\dot{x}(0) + x(0)\xi\omega}{\omega_D}\right]^2} \tag{2.36}$$

$$x = \tan^{-1}\left[\frac{x(0)\omega_D}{\dot{x}(0) + x(0)\xi\omega}\right] \tag{2.37}$$

The period of damped vibration is given by

$$T_D = \frac{2\pi}{\omega_D} = \frac{2\pi}{\omega\sqrt{1-\xi^2}} = \frac{T}{\sqrt{1-\xi^2}} \tag{2.38}$$

If the amplitudes at times t_n and $t_n + T_D$ are x_n and x_{n+1}, respectively, the ratio x_n/x_{n+1} can be written as

$$\frac{x_n}{x_{n+1}} = \exp\frac{2\pi\xi}{\sqrt{1-\xi^2}} \tag{2.39}$$

This ratio is called amplitude decay ratio. By taking the natural logarithm on both sides of Eqn (2.39), one can obtain the logarithmic decrement as follows.

$$\delta = \ln\frac{x_n}{x_{n+1}} = \frac{2\pi\xi}{\sqrt{1-\xi^2}} \quad (2.40)$$

Note: For most practical structures, $\xi < 0.2$, and if ξ is small $\sqrt{1-\xi^2} \simeq 1$, the above expression can be rewritten as

$$\delta \simeq 2\pi\xi \quad (2.41)$$

Equation (2.41) can be rewritten as

$$\xi = \delta/2\pi$$

Therefore, a simple way that emerges for the estimation of the damping ratio of an SDOF system is by performing a free vibration test in which the logarithmic decrement is measured when a system is displaced by some initial displacement $x(0)$ and released with initial velocity $\dot{x}(0) = 0$. The above expression can be rewritten in a generalized form to calculate damping ratio as follows.

$$\xi = \frac{1}{2\pi_{(j-i)}}\ln\frac{x_i}{x_j} \quad (2.42)$$

Practically, it is convenient to measure accelerations, and the above expression can be rewritten as

$$\xi = \frac{1}{2\pi j}\ln\frac{\ddot{x}_i}{\ddot{x}_{i+j}} \quad (2.43)$$

The damping ratio ξ can thus be calculated from Eqn (2.42) after determining experimentally the amplitudes of two successive peaks of the system in free vibration.

Damping has an effect of lowering the natural frequency from ω to ω_D and lengthening the natural period from T to T_D. These effects are negligible for damping ratios below 20 per cent, a range that includes most structures. For most structures the damped properties ω_D and T_D are approximately equal to the undamped properties ω and T, respectively. For systems with critical damping $\omega_D = 0$ and $T = \infty$, the system does not oscillate. In this condition, if an SDOF mass is pulled back to its maximum deflection and released, it will come to rest in its undeflected position with no overrun.

Critically damped case

The condition $\xi = 1$ indicates a limiting value of damping at which the system loses its vibratory characteristics; this is called *critical damping*.

The damping coefficient at critical damping is denoted by

$$c = \xi c_c$$

$$\text{where } c_c = 2\omega m = 2\sqrt{mk} \tag{2.44}$$

In a critically damped system, the roots of the characteristic equation are equal and, from Eqn (2.28), they are given as $c_c/2m$.

The general solution of Eqn (2.27) for a critically damped system would be

$$x = (A + Bt)e^{\frac{-c_c}{2m}t} \tag{2.45}$$

For the overdamped case if $\xi > 1$, the system does not oscillate because the effect of damping overshadows the oscillation.

The expression under the radical of Eqn (2.31) is positive, and consequently the solution is given directly by Eqn (2.30).

2.4.2 Forced Vibration Response

When a system is subjected to an exciting force it is forced to vibrate. The force may be harmonic (structures subjected to dynamic action of rotating machinery) or of general type—impulsive, constant, rectangular, triangular, etc. The resulting response of the system to such an excitation is called *forced response*. The exciting forces and the response of the system are briefly presented below.

Forced Harmonic Vibration

The system on excitation will vibrate with the same frequency as that of the excitation. For an SDOF system, the exciting force is assumed to be $p_0 \sin \omega_e t$ or $p_0 \cos \omega_e t$.

For an undamped system, the response is given by

$$x(t) = \frac{f_0/k}{1 - r^2}(\sin \omega_e t - r \sin \omega t)$$

For a damped system, the response is given by

$$x(t) = e^{-\xi\omega t}(A \cos \omega_d t - B \sin \omega_d t) + x_{st}\frac{\sin(\omega_e t - \theta)}{\sqrt{(1 - r^2) + (2r\xi)^2}}$$

where p_0 is the peak amplitude;

ω is the natural frequency of undamped system = $\sqrt{\dfrac{k}{m}}$;

ω_e is the frequency of force or exciting frequency or forcing frequency;

$r = \dfrac{\omega_e}{\omega}$ is the frequency ratio;

$\xi = \dfrac{c}{c_c}$ is the damping ratio;

$x_{st} = \dfrac{p_0}{k}$ is the static deflection of the spring acted upon by the force p_0;

and $\omega_d = \omega\sqrt{1-\xi^2}$ is the frequency of the damped system.

The ratio of steady-state amplitude to the static deflection is called *dynamic magnification factor D* and is given by

$$D = \frac{1}{\sqrt{(1-r^2)+(2r\xi)^2}}$$

Response to General Type Loading

Often real structures are subjected to forces that are not harmonic. These forces and the corresponding responses are described as follows.

(a) ***Impulsive force***

A very large load applied for a very short duration of time that is finite is called an *impulsive force*.

For an undamped system, the response is given by

$$x(t) = \frac{1}{m\omega}\int_0^t p(\tau)\sin[\omega(t-\tau)]d\tau \qquad t \geq \tau$$

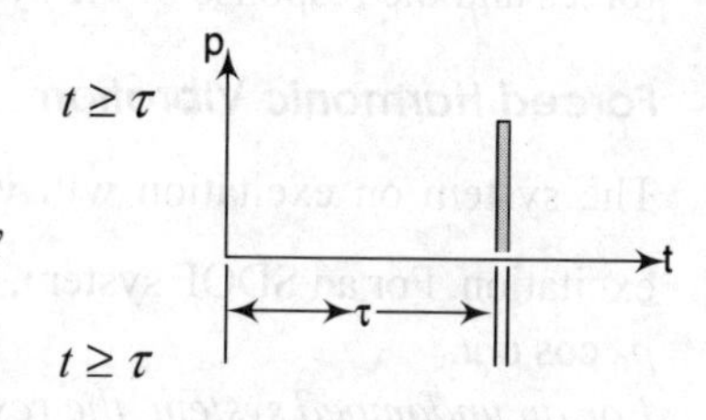

For a damped system, the response is given by

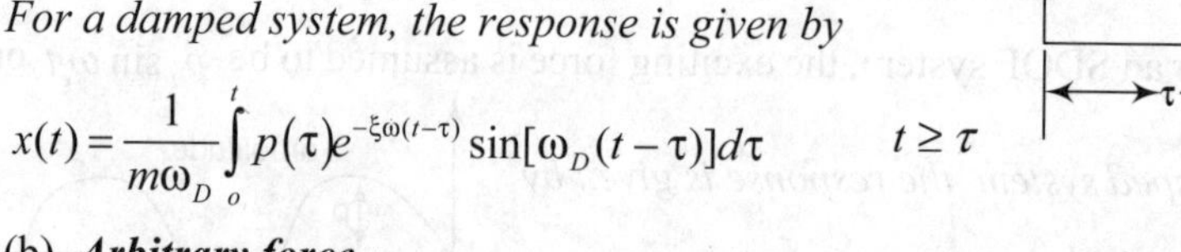

$$x(t) = \frac{1}{m\omega_D}\int_0^t p(\tau)e^{-\xi\omega(t-\tau)}\sin[\omega_D(t-\tau)]d\tau \qquad t \geq \tau$$

(b) ***Arbitrary force***

A force *p(t)* varying arbitrarily with time can be considered as a sequence of infinitesimal short impulses.

For an undamped system, the response is given by

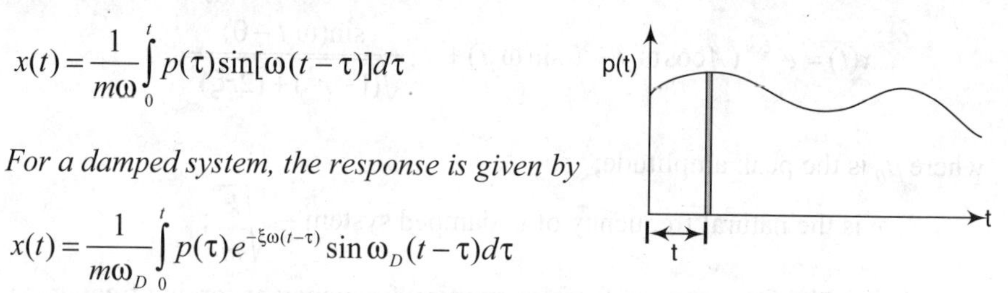

$$x(t) = \frac{1}{m\omega}\int_0^t p(\tau)\sin[\omega(t-\tau)]d\tau$$

For a damped system, the response is given by

$$x(t) = \frac{1}{m\omega_D}\int_0^t p(\tau)e^{-\xi\omega(t-\tau)}\sin\omega_D(t-\tau)d\tau$$

(c) ***Constant force of magnitude p_0 applied suddenly***
For an undamped system, the response is given by

$$x(t) = \frac{p_0}{k}(1 - \cos \omega t)$$

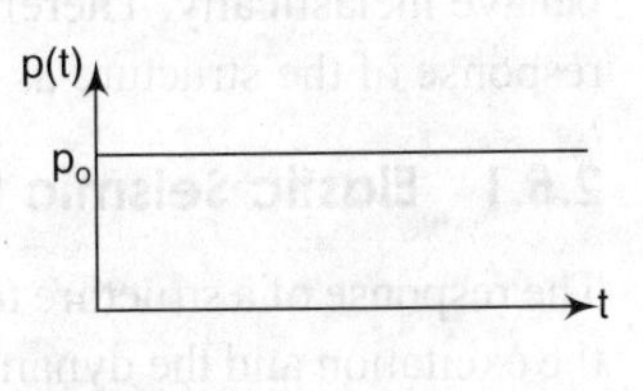

For a damped system, the response is given by

$$x(t) = \frac{p_0}{k}[1 - e^{-\xi\omega t}(\cos \omega_D t + \frac{\xi}{1 - \xi^2}\sin \omega_D t)]$$

(d) ***Rectangular pulse force***
Rectangular pulse force is characterized by a force p_0 applied suddenly but only during a limited time duration t_d.
For an undamped system, the response is given by

$$x(t) = \frac{p_0}{k}[\cos \omega (t - t_d) - \cos \omega t]$$

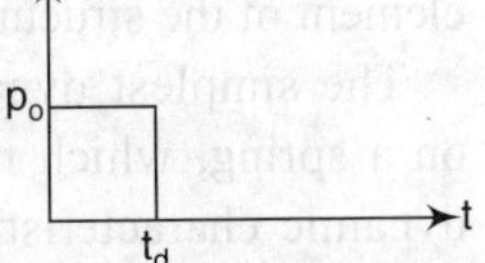

Note: For a damped system, the response is same as that given in (c), since the system is subjected to a constant force.

(e) ***Triangular pulse force***
The system is initially at rest, and subjected to a force *p(t)*, which has initial value p_0 and decreases linearly to zero at time t_d.
For an undamped system, the response is given by

$$x(t) = \frac{p_0}{k\omega t_d}[\sin \omega t - \sin \omega (t - t_d)] - \frac{p_0}{k}\cos \omega t$$

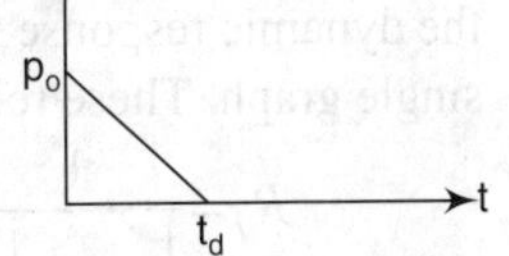

Note: Shaking of the ground during earthquake is usually described by the time variation of ground acceleration $\ddot{x}_g(t)$. The differential equation for response of structures subjected to earthquake excitation is given by Eqn (2.17) and can be rewritten as

$$\ddot{x} + 2\xi\omega\dot{x} + \omega^2 x = -\ddot{x}_g(t)$$

During an earthquake, the ground acceleration varies highly irregularly and classical methods of solutions of differential equations of motion are not realistic. Since the response of a structure to the irregular or transient excitation of an earthquake is quite complex, only numerical methods should be used for structural response.

2.5 Seismic Response of SDOF Structures

The foremost application of structural dynamics is in analysing the response of structures to ground shaking caused by earthquakes. The deformation of the structure may be elastic or inelastic depending on the severity of ground motion. During strong ground motion, large deformation takes place and the structure may behave inelastically. Therefore, it becomes important to understand the inelastic response of the structure as well.

2.5.1 Elastic Seismic Response

The response of a structure to a given dynamic excitation depends on the nature of the excitation and the dynamic characteristics of the structure, i.e., on the manner it stores and dissipates vibrational energy. Seismic excitation is described in terms of displacement, velocity, or acceleration varying with time. When this excitation is applied to the base of a structure, it produces a time-dependent response in each element of the structure and is described in terms of motions or forces.

The simplest dynamic system is the SDOF system consisting of a mass on a spring, which remains in the linear** elastic range when vibrated. The dynamic characteristics of such a system are described by its natural period of vibration T (or frequency ω) and the damping ratio ξ. When subjected to a harmonic base motion described by $x_g = a \sin \omega_e t$, where ω_e is the *exciting frequency* or *forcing frequency,* the response of the mass is fully described in Fig. 2.14. The ratios of response amplitude to input amplitude are shown for displacement response factor R_d, velocity response factor R_v, and acceleration response factor R_a, in terms of the ratio between the frequency of the forcing function ω, and the natural frequency of the system ω_n. The simple relations among the dynamic response factors make it possible to present all the three factors in a single graph. These relation are given as

$$R_d = \frac{1}{\left|1 - \dfrac{\omega_e}{\omega}\right|^2}, \quad R_v = \frac{\omega_e}{\omega} R_d, \quad \text{and} \quad R_a = \left(\frac{\omega_e}{\omega}\right)^2 R_d$$

The significance of the natural period or frequency of the structure is demonstrated by the large amplifications of the input motion at or near the resonance condition, i.e., when $\omega_e/\omega = 1$ (Fig. 2.14). Thus the forcing frequency at which the largest response amplitude occurs is known as *resonant frequency*. However, its value is slightly less than the natural frequency of the system because of damping. Resonant frequency can be determined by setting the first derivatives of R_d, R_v, and R_a with respect to ω_e/ω equal to zero. The importance of damping, particularly near resonance, is also evident. For $\xi = 0.01$, the resonant amplification of the input motion is 50 times for this system, and for $\xi = 0.05$ it reduces to five times the input motion. However, the response of a structure to the irregular or transient excitation of an earthquake is much more complex.

** Linearity implies that acceleration, velocity, and displacement responses bear a straight-line relationship to the force.

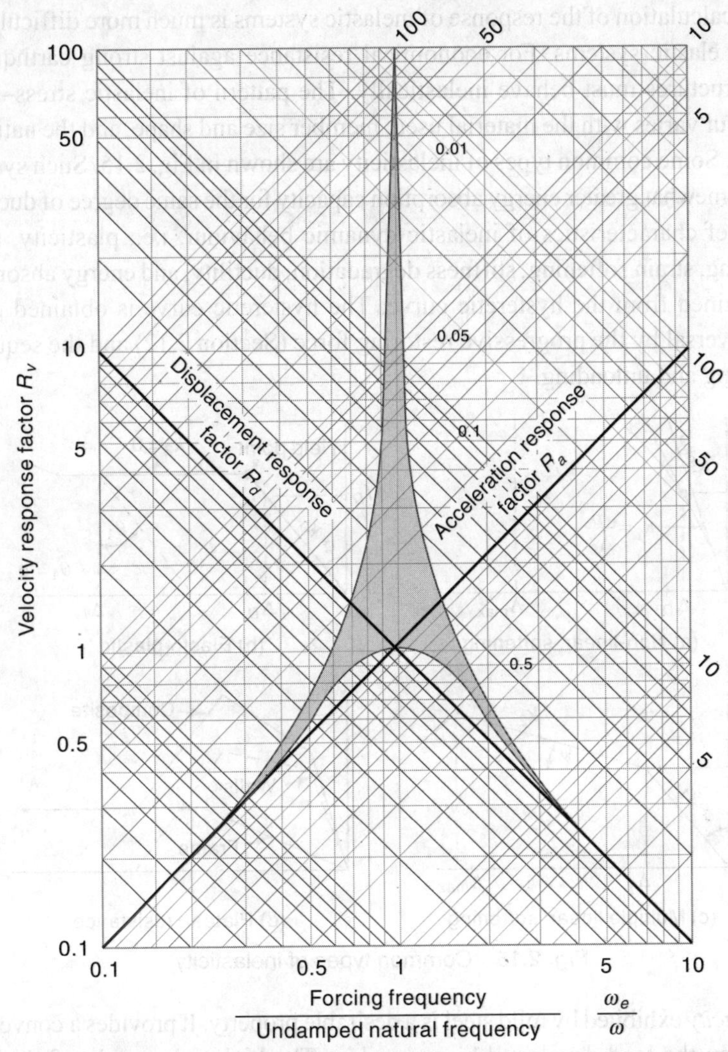

Fig. 2.14 Response of linear elastic SDOF system to a harmonic forcing function

2.5.2 Inelastic Seismic Response

In seismic design for an earthquake of moderate intensity, it is reasonable to assume elastic behaviour for a well-designed and well-constructed structure. However, for very strong motions, this is not a realistic assumption even for a well-designed structure. While structures can be designed to resist severe earthquakes, it is not economically feasible to design buildings to elastically withstand earthquakes of the greatest foreseeable intensity. In order to design structures for strain levels beyond the linear range, the response spectrum has been extended to include the inelastic range.

The calculation of the response of inelastic systems is much more difficult than that for elastic systems. For economical resistance against strong earthquakes most structures must behave inelastically. The pattern of inelastic stress–strain behaviour varies with the material used, member size and shape, and the nature of loading. Some common types of inelasticity are shown in Fig. 2.15. Such systems show somewhat greater energy absorption capacity for the same degree of ductility. The chief characteristics of inelastic dynamic behaviour, i.e., plasticity, strain hardening, strain softening, stiffness degradation, ductility, and energy absorption are obtained from the hysteretic curve. The hysteretic curve is obtained under force reversal by the progressive restoring force (Section 2.12) and the sequence of loading and unloading.

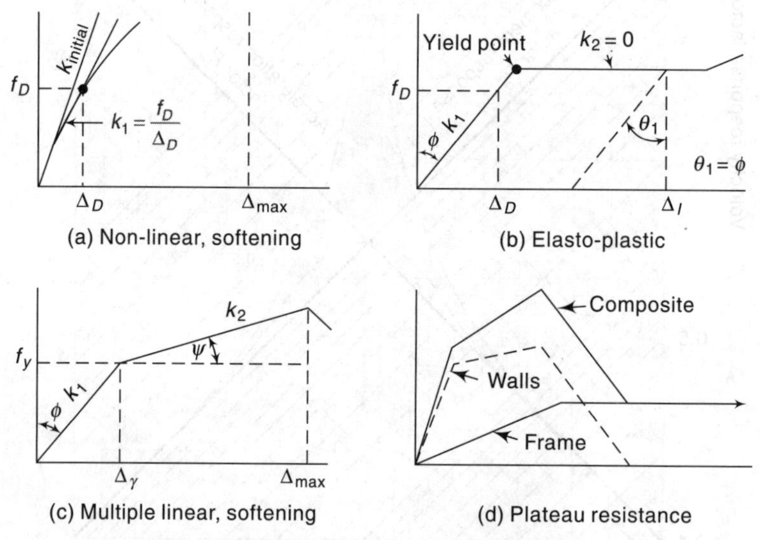

Fig. 2.15 Common types of inelasticity

Plasticity exhibited by mild steel is a desirable property. It provides a convenient control on the load developed by a member. The higher the grade of steel, the shorter the plastic plateau, and the sooner the *strain hardening* effect sets in. *Strain softening* is opposite of strain hardening, involving a loss of stress or strength with increasing strain. *Stiffness degradation* is an important feature of the inelastic cyclic loading of concrete and masonry materials. The stiffness as measured by the overall stress-to-strain ratio of each hysteresis loop reduces with each successive loading cycle. *Ductility* of a member may be defined as the ratio of deformations at failure and at yield. The deformation may be measured in terms of deflection, rotation, or curvature. The numerical value of ductility will also vary depending on the exact combination of applied forces and moments under which the deformations are measured. Ductility is generally desirable in structures because of the relatively gentle onset of failure than that occurring in brittle materials.

Ductility is particularly useful in seismic problems because it is accompanied by an increase in strength in the inelastic range. Steel has the best ductility properties, while concrete can be made moderately ductile with appropriate reinforcement. A *high energy absorption* capacity is often mentioned rather loosely as a desirable property of earthquake-resistant construction. However, a distinction should be made between temporary and permanent absorptions of dissipation of energy. A substantial part of the energy is temporarily stored by the structure as elastic strain energy and kinetic energy. However, when the yield point is exceeded in parts of the structure under strong earthquake motion, permanent energy dissipation in the form of inelastic strain (or hysteretic) energy begins. During the earthquake, the energy is dissipated by damping, which is of course the means by which the elastic energy is dissipated once the forcing ground motion ceases.

2.6 Response Spectrum

The structural response to a particular earthquake can be summarized using a response spectrum, which provides valuable information on the potential effects of ground motion on the structure. A response spectrum shows the peak response of an SDOF structure to a particular earthquake, as a function of the natural period and damping ratio of the structure. The main advantage of response spectrum approach is that earthquakes that look quite different when represented in the time domain may actually contain similar frequency contents, and result in broadly similar response spectra. This uniqueness of response spectra makes it useful for a future earthquake. The El Centro, California, earthquake response spectrum is used worldwide as a reference because of two reasons. One, it contains exhaustive data of ground motion, and two, the data acquisition systems were located very near to the epicentre of the earthquake.

2.6.1 Elastic Systems

The concepts of structural dynamics can be used to analyse the structural response of ground shaking caused by an earthquake. For a linear SDOF system, subjected to ground acceleration, $\ddot{x}_g(t)$, Eqn (2.17) can be rewritten as follows.

$$m\ddot{x} + c\dot{x} + kx = -m\ddot{x}_g(t) \tag{2.46}$$

or
$$\ddot{x} + \frac{c}{m}\dot{x} + \frac{k}{m}x = -\ddot{x}_g(t)$$

or
$$\ddot{x} + 2\xi\omega\dot{x} + \omega^2 x = -\ddot{x}_g(t) \tag{2.47}$$

It is apparent from the above equation that for a given ground motion $\ddot{x}_g(t)$, the deformation response $x(t)$ of the structure depends on the natural frequency ω or natural period T (= $1/\omega$) of the structure and the damping ratio ξ. Assuming different values of T, say 0.1, 0.2, ... for a particular value of ξ Eqn (2.47) is solved for response x. Then a curve is plotted amongst natural period T, deformation x, pseudo-velocity and pseudo-acceleration. The value of deformation when multiplied with ω^2 will give the pseudo-acceleration and that when multiplied with ω will give pseudo-velocity. The above process is repeated with different values of ξ, say 0.01, 0.02, ... and a set of response curves is thus obtained for the particular selected values of ξ.

It is pertinent to note that the magnitude of the mass and spring stiffness of the structure do not independently affect the response to ground motion. However, because the structure is subjected to a base motion and not to a force, the maximum stress that the structure experiences is a function of its stiffness as well as of its period of vibration. In general, the stiffer the spring in the modelled structure, the greater will be the stress in the spring and the smaller its relative deflection or displacement for a given ground motion.

For a specific excitation of a simple system having a particular percentage of critical damping, the maximum response is a function of the natural period of vibration of the system. A plot of the maximum response (e.g., of relative displacement, absolute displacement, acceleration, or spring force) against the period of vibration, or against the natural frequency of vibration or the circular frequency of vibration, is called a *response spectrum.*

Note: Elastic response spectra assume linear structural force-displacement relationship. Further, it is pertinent to note that the response spectra represent only the maximum responses.

The response spectrum for structures is represented by spectral displacement S_d, spectral velocity S_v, and spectral acceleration S_a. With the maximum value of quantity defined as S_v, we have spectral displacement given as

$$S_d = \frac{1}{\omega} S_v = \frac{T}{2\pi} S_v \cong x_{max} \tag{2.48}$$

The spectral velocity S_v is also known as *pseudo-spectral velocity*. The prefix pseudo is used because S_v is not equal to peak velocity.

$$S_v = \omega S_d = \frac{2\pi}{T} S_d \tag{2.49}$$

In structures with damping, S_v is close to the maximum velocity response in the velocity sensitive zone.

$$S_a = \omega^2 S_d = \left(\frac{2\pi}{T}\right)^2 S_d \tag{2.50}$$

For structures subjected to earthquake loads, the maximum base shear V_{max} is given as

$$V_{max} = mS_a \tag{2.51}$$

S_a is known as *spectral acceleration* or more accurately *pseudo-spectral acceleration* because S_a does not exactly represent the peak acceleration value in most cases.

> **Note:** If the mass of the structure and the spectral acceleration are known, the maximum base shear can be calculated using Eqn (2.51).

The diagrams plotting S_d, S_v, and S_a are called the *displacement response spectrum*, *velocity response spectrum*, and *acceleration response spectrum*, respectively. In general, the velocity spectrum is nearly constant in a range of longer natural periods; the acceleration spectrum decreases as the natural period lengthens; and the displacement spectrum increases in proportion to the natural period. Rough sketches of these three spectra are shown in Fig. 2.16. The response spectra may be plotted individually to arithmetic scales or may be combined in tripartite plot (Fig. 2.17). Usually, the tripartite plot displays relative pseudo-velocity (spectral velocity S_v) on the vertical axis, natural frequency f (or period T) on the horizontal axis, and maximum absolute acceleration (spectral acceleration S_v) and relative displacement (spectral displacement S_d). It is usual to choose period of vibration since it is more familiar and preferred by engineers.

When the data of acceleration record of an earthquake are plotted with abscissa and ordinates on a logarithmic scale, a single plot results for reading spectral responses. In fact, diagonal scales for the displacements (sloping 135° with the abscissa) and accelerations (sloping 45° with abscissa) result in a single plot for reading spectral responses. The acceleration and displacement axes are reversed when the spectral values are plotted against natural period rather than natural frequency.

Equation (2.48) can be rewritten as

$$S_v = \omega\, S_d = 2\pi f\, S_d$$

Taking log of both sides

$$\log S_v = \log f + \log (2\pi\, S_d)$$

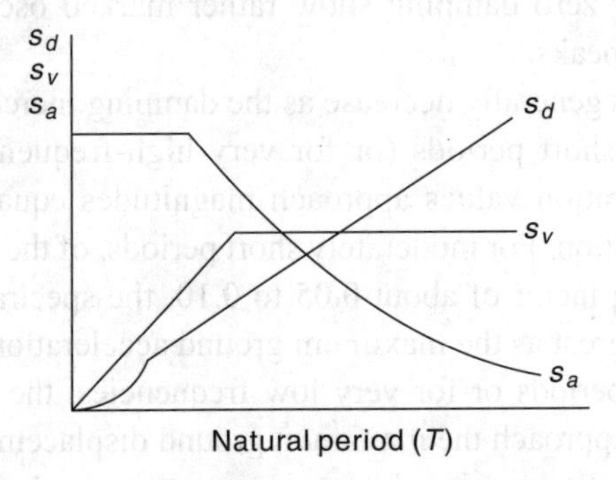

Fig. 2.16 General shapes of response spectra

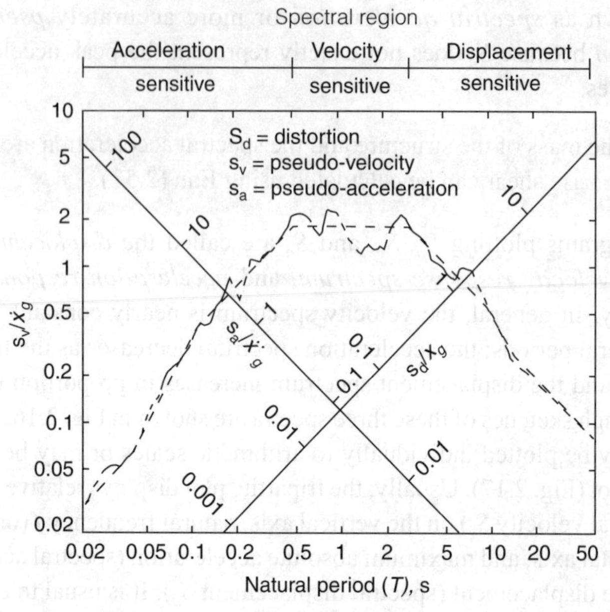

Fig. 2.17 Response spectra for typical ground motion shown by firm line; its idealized version shown by broken line

The plot of log S_v versus log f is a straight line with a slope of 45° for a constant value of S_d.

Equation (2.49) can be rewritten as

$$S_v = S_a/\omega = S_a/2\pi f$$

Taking log of both sides

$$\log S_v = -\log f + \log S_a/2\pi$$

The plot of log S_a versus log f is a straight line with a slope of 135° for a constant value of S_v.

Some of the general characteristics of the response spectrum, plotted for different input data reported by researchers, are as follows:

(a) The spectra for zero damping show rather marked oscillations with very irregular sharp peaks.

(b) The oscillations generally decrease as the damping increases.

(c) For extremely short periods (or for very high-frequency structures), the spectral acceleration values approach magnitudes equal to the maximum ground acceleration. For moderately short periods, of the order of 0.1 to 0.3 s with a damping factor of about 0.05 to 0.10, the spectral accelerations are about twice as great as the maximum ground accelerations.

(d) For very long periods or for very low frequencies, the maximum spectral displacements approach the maximum ground displacements.

(e) For intermediate frequencies, the maximum spectral velocity has a magnitude of several times the input velocity for no damping, ranging down to values that are almost equal to the input maximum ground velocity for about 20 per cent critical damping.

(f) For critical damping in the range of 5 to 10 per cent the maximum spectral acceleration is of the order of twice the maximum ground acceleration, the maximum spectral velocity is of the order of 1.5 times the maximum ground velocity, and the maximum spectral displacement is of the same order as the maximum ground displacement.

Structural frame subjected to the base shear V_{max} as determined for elastic systems can be analysed for member forces and the members designed. The response spectrum concept thus provides a practical approach to apply the knowledge of structural dynamics to the design of structures and development of lateral force requirements in building codes. However, structures are designed for base shear smaller than the maximum base shear—the one that occurs due to the strongest ground motion. This permits structures to suffer damage during intense ground shaking. This is in accordance with the concept that structure should be allowed to damage to an acceptable degree but not collapse; the basic design philosophy of earthquake-resistant design of structures. However, by doing so, the structure deforms beyond the limit of linear behaviour and the structures are allowed to undergo an acceptable degree of damage allowing them to deform into their inelastic range during strong ground motions.

2.6.2 Inelastic Systems

The assumption of elastic behaviour of a well-designed and constructed structure for earthquakes of moderate intensity is reasonable. However, designing structures to remain elastic for strong ground motions will be uneconomical as the force demands will be very large. For economical design use of ductility of the structure may be made to reduce the force demands. Therefore, for strain levels beyond the linear range, the response spectrum is extended to include inelastic range. The spectra for inelastic systems consist of a series of curves corresponding to definite values of ductility ratio μ. From the tripartite logarithmic plot, the spectral velocity and the spectral acceleration are read directly, while the read spectral displacement is multiplied by the ductility ratio μ to obtain the correct spectral displacement.

During earthquakes, structures undergo oscillatory motion with reversal of deformation. The force-deformation plots of Fig. 2.15 show hysteresis loops under cyclic deformations because of inelastic behaviour on initial loading for force less than f_y; the idealized system will be linearly elastic with stiffness k. Yielding begins, when the force reaches f_y and continues at constant force f_y and stiffness $k = 0$.

During initial loading, the system will have the same stiffness as that of the elastoplastic system. Since both the systems have same mass and damping, the natural vibration of the corresponding linear system will be the same as that of an elastoplastic system undergoing small oscillations, i.e., $x \le x_y$.

It is most convenient to explain the concept of response of an inelastic system by the model of Fig. 2.15 (b). Figure 2.18 shows the actual force-deformation curve during initial loading and its elastoplastic idealization. It may be noted that the two curves of Fig. 2.18 should have the same area under them for maximum deformation x_m. Figure 2.19 shows a typical cycle of loading, unloading, and reloading for an elastoplastic system. It may be observed that for deformation x

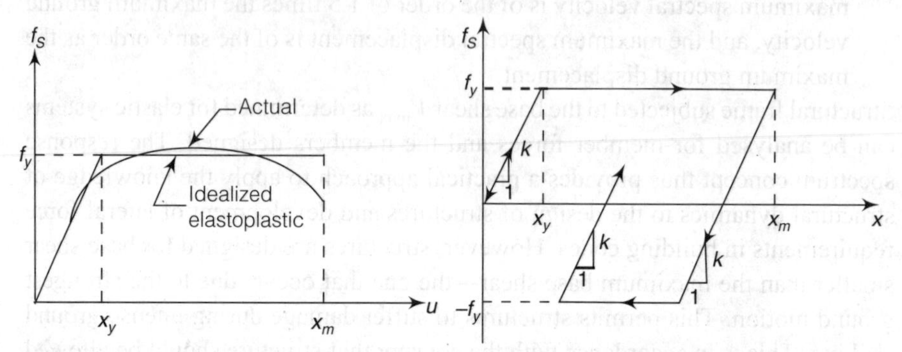

Fig. 2.18 Force-deformation curve during initial loading

Fig. 2.19 Elastoplastic force-deformation relation

at time t, the force f_S depends upon the history of motion of the system.

Figure 2.20 shows elastoplastic system and its corresponding linear system. The normalized yield strength of an elastoplastic system is given by

$$\overline{f_S} = \frac{f_y}{f_S} = \frac{kx_y}{kx_o} = \frac{x_y}{x_o} \tag{2.52}$$

where, f_S is the peak resisting force in corresponding linear system and x_o is the peak deformation corresponding to f_S.

Yielding of a structure also has the effect of limiting the peak force that it must sustain. This force reduction is quantified by a factor known as *yield-reduction factor*.

$$\text{Yield-reduction factor } R_y = \frac{f_S}{f_y} = \frac{x_o}{x_y} \tag{2.53}$$

Note: For a linear system, $\overline{f_y}$ as well as R_y will be unity.

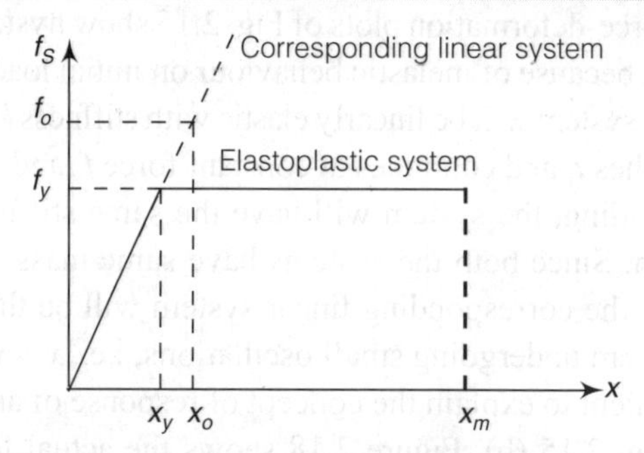

Fig. 2.20 Elastoplastic system and its corresponding linear system

The ratio of maximum deformation of elastoplastic system to the yield deformation of the system is known as ductility factor μ (sometimes referred to as displacement ductility) and is given by

$$\mu = \frac{x_m}{x_y} \tag{2.54}$$

The ductility demand depends on the natural period and the normalized yield strength.

> **Note:** The ductility factor for non-linear system is greater than one and that for corresponding linear system it is equal to one.

The ratio of peak deformation of the elastoplastic system and that of the corresponding linear system can be expressed as follows.

$$\frac{x_m}{x_o} = \frac{x_m}{x_o}.\frac{x_y}{x_y} = \frac{x_m}{x_y}.\frac{x_y}{x_o} = \mu\overline{f_y} = \frac{\mu}{R_y} \tag{2.55}$$

Equation (2.18) for an inelastic system can be rewritten for an elastoplastic system as,

$$m\ddot{x} + c\dot{x} + f_S(x,\dot{x}) = -m\ddot{x}_g(t)$$

or $$\ddot{x} + 2\xi\omega\dot{x} + \omega^2 x_y \overline{f}_S(x,\dot{x}) = -\ddot{x}_g(t) \tag{2.56}$$

where, $\omega = \sqrt{\frac{k}{m}}$, $\xi = \frac{c}{2m\omega}$, $\overline{f_S}(x,\dot{x}) = \frac{f_S(x,\dot{x})}{f_y}$, and $f_y = kx_y = m\omega^2 x_y$

where, $\overline{f}_S(x,\dot{x}) = f_S(x,\dot{x})/f_y$ is a non-dimensional term.

Since $\mu(t) = \frac{x}{x_y}$

$$\Rightarrow x = \mu(t)x_y$$
$$\dot{x} = \dot{\mu}(t)x_y$$
$$\ddot{x} = \ddot{\mu}(t)x_y$$

Equation (2.56) can be rewritten as

$$\ddot{\mu}(t)x_y + 2\xi\omega\dot{\mu}(t)x_y + \omega^2 x_y \overline{f_S}(\mu(t),\dot{\mu}(t)) = -\ddot{x}_g(t) \tag{2.57}$$

$$\mu(t) + 2\xi\omega\dot{\mu}(t) + \omega^2 x_y \overline{f_S}(\mu(t),\dot{\mu}(t)) = \frac{-\ddot{x}_g(t)}{x_y} \tag{2.58}$$

Equation (2.58) is the response in inelastic range. The deformation *x(t)* depends on ω—the natural frequency of the inelastic system vibrating within its linear elastic range, which is the same as the natural frequency of the corresponding linear system, ξ—the damping ratio of the system based on the critical damping of the inelastic system vibrating within its linear elastic range, and x_y—the yield of deformation ($x_y = f_y/k$).

The response spectrum for an inelastic system can be represented by spectral displacement S_{dI}, spectral velocity S_{vI}, and spectral acceleration S_{aI},

$$S_{dI} = x_y,\ S_{vI} = \omega x_y, \text{ and } S_{aI} = \omega^2 x_y$$

A plot of x_y versus T for a particular value of ductility factor μ is called yield deformation response spectrum and that for velocity/acceleration is called velocity/acceleration response spectrum.

A tripartite plot can be had to present quantities x_y, $\dot{x}_y$, and $\ddot{x}_y$, since they are related as follows.

$$\frac{\ddot{x}_y}{\omega} = \dot{x}_y = \omega x_y$$

or $$\frac{T}{2\pi}\ddot{x}_y = \dot{x}_y = \frac{2\pi}{T}x_y$$

The yield strength of an elastoplastic system is given by

$$f_y = kx_y = m(\omega^2 x_y) = m\ddot{x}_y = \frac{\ddot{x}_y}{g}w$$

or $$f_y = \frac{\ddot{x}_y}{g}w \tag{2.59}$$

where w is the weight of the system and $\ddot{x}_y$ is acceleration response for a linear elastic system.

2.7 Design Spectrum

Seismic design of structures is mostly carried out by equivalent static forces evaluated from the maximum acceleration response of the structure under the expected ground shaking. Since, at present, there is no accurate method to predict the expected ground motion for a future earthquake, the design response spectrum, therefore, is constructed incorporating the spectra for several earthquakes whose records are available.

To construct a design spectrum, the spectra for different earthquakes are computed. These are enveloped and smoothened to result in a single curve that encapsulates dynamic characteristics of a large number of possible earthquake accelerograms.

Structures are usually designed with an assumption that they remain linearly elastic when subjected to earthquake-induced ground motion. For moderate ground motions, the assumption is fairly justified, but for a strong ground motion, the structure will be uneconomical. It is because, elastic spectra do not account for the inelasticity that may occur during strong motion earthquakes. In such a case, it becomes necessary to design structures to withstand deformation beyond the elastic limit. A significant reduction of design forces can be had by accounting for the energy absorption and plastic redistribution. To achieve this objective, the elastic design spectrum is modified to account for inelastic deformation of structure.

2.7.1 Elastic Systems

The design spectrum for elastic system provides a basis for calculating the design force and deformation for SDOF systems to be designed to remain elastic. It is constructed in lieu of future earthquakes for the following reasons.

1. New structures are to be designed.
2. Seismic safety evaluation is required for existing structures.

The response spectrums of the site, for recorded earthquakes of different periods, in general, do not display similar peaks and valleys. Also their jaggedness makes them unsuitable for design purposes. Further, for a ground motion in future, it is not possible to predict the jagged response spectrum. To overcome this, the responses in the spectrum are idealized by smooth curves or straight lines. Therefore, the design spectrum consists of a set of smooth curves or a series of straight lines with one curve for each level of damping. To accomplish this, statistical analysis is made for the set of recorded ground motions for a site. Thus, a design spectrum represents the average characteristics of many ground motions normalized so as to have same peak ground acceleration for all the ground motions under consideration. The procedure for generating the design spectrum is not described as it is beyond the scope of the book.

The response spectrum should not be confused with design spectrum. Although the former is a description of a particular ground motion, the design spectrum is a specification of the level of seismic design force or deformation.

The spectrum to be used for designing structures should be representative of ground motions recorded at the site during past earthquakes. If the ground motion records for the site are not available, there can be two alternatives. First, the ground motion records of other sites with similar conditions—magnitude of earthquake, site distance from the epicentre, fault mechanism, geology of the travel path of seismic waves, and soil conditions at site—may be used. Second, the available records may be used judicially making suitable adjustments.

Note: Elastic response spectrum shows the peak response of an SDOF structure to a particular earthquake as a function of the natural period and damping ratio of the structure. The codal spectra of IS 1893: 2002 are given for 5 per cent damping and for three different soil types: Type I—rock or hard soil, Type II—medium soil, and Type III—soft soils. A table of factors corresponding to different damping ratios is also given with the help of which spectral values for various other percentages of damping can be determined. The design spectrum is specifically prepared for a structure at a particular project site.

The same may be used for design at the discretion of the project authorities and by using various parameters such as zone factor, importance factor, and response reduction factor.

2.7.2 Inelastic Systems

A reasonable design spectrum for an elasto-plastic system can be derived merely by taking account of the fact that the spectral displacement of the elasto-plastic system is practically the same as that for an elastic system having the same period of vibration. Consequently, one can obtain a design spectrum for the elasto-plastic system by dividing the ordinates of the spectrum response for the elastic system, at each period, by the ductility ratio for which it is desired to be designed.

The tripartite logarithmic scales used to plot these spectra give simultaneously, for any SDOF system of natural period T and a specified ductility ratio, the spectral values of displacement, velocity and acceleration. These spectra are usually plotted as a series of curves corresponding to definite values of the ductility ratio. The ductility ratio is defined as the ratio of the maximum displacement of the structure in the inelastic range to the displacement corresponding to the yield point. The spectra for the elasto-plastic system have the same general characteristics as spectra for elastic systems, but in general the spectrum plots appear to be displaced downward, at each frequency by an amount that is dependent on the ductility factor. Also, the two sources of energy absorption, viscous damping and plastic behaviour, affect the response in about the same way and are roughly additive in their effects. However, the influence of viscous damping diminishes as the ductility ratio increases or as the energy absorption increases.

The plot with firm lines *abcd* of Fig. 2.21 shows a design response spectrum of an elastic system. The plots with broken lines are derived design spectra for inelastic systems to obtain maximum displacements (the outer one) and maximum accelerations (the inner one). For very low frequencies, the maximum structural displacement approaches the maximum ground displacement; for very flexible structures, the displacements in the low-frequency region are conserved. In this region of frequencies, the acceleration of the structure is reduced, since the force for an elastoplastic structure does not increase when yielding occurs.

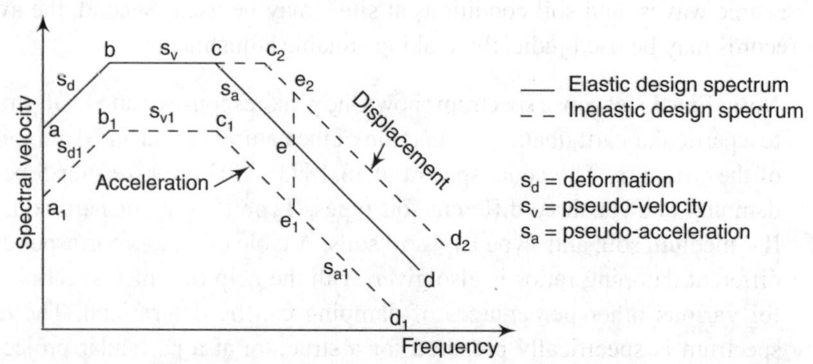

Fig. 2.21 Typical design spectrum for plastic system

Further, acceleration also reaches a maximum value when yielding occurs. Thus the acceleration is reduced by factor $1/\mu$. For very high frequencies, the spectral acceleration values approach maximum ground acceleration; for very stiff structures, the ground acceleration is directly transferred to the structure and is therefore conserved. Here the deformations are greater than the elastic deformations and the energy must be conserved. The lines S_a and S_{al} differ by a factor of $\sqrt{2(\mu-1)}$.

To construct inelastic design spectrum, the elastic design spectrum *abcd* is plotted for specified damping. Lines a_1b_1 and b_1c_1 are drawn parallel to lines *ab* and *bc,* respectively, by dividing the ordinates of S_d and S_v by the specified ductility ratio μ. To locate point e_1, the ordinate at *e* of the elastic spectrum is divided by $\sqrt{2(\mu-1)}$. From point e_1, a line *cd* is drawn at 45° until it intersects the line *bc* at point *b*. The desired spectrum for acceleration will be $a_1b_1c_1d_1$. The maximum inelastic displacement response will be abc_2d_2. To obtain ordinates of segment c_2d_2, the ordinates in the segment c_1d_1 are multiplied with ductility ratio μ.

Note: The procedure laid above gives a simple method to use yield-reduction factor and ductility factor for designing regular structures, where inelasticity can be assumed to be reasonably uniformly distributed, taking account of non-linear approach. However, for irregular structures, Pushover technique or non-linear time-history analysis must be performed.

2.8 Systems with Multiple Degrees of Freedom

Structures cannot always be modelled as SDOF systems. In fact, structures are continuous systems and possess infinite degrees of freedom. Multi-storey buildings are the most suitable example. A thorough knowledge and understanding of the concepts discussed for SDOF system are of prime importance. It is because, a structure modelled with MDOF system is transformed to consist of a number of SDOF independent systems and then each one is solved as an SDOF system. The response spectrum of SDOF system is extended to solve MDOF system as well. The MDOF system may be divided into two groups according to their deformation characteristics. In one group the floor moves only in the horizontal direction and there is no rotation of a horizontal section at the level of floors. Such buildings are referred to as *shear buildings.* In the other group of structures, the floors move in both rotational and horizontal directions and are referred to as *moment-shear buildings*.

For the present, we shall consider one of the most instructive and practical type of structure, which involves many degrees of freedom, the multi-storey shear building. The following assumptions are made about the structure:

(a) The total mass of the structure is concentrated at the levels of the floors, although it is distributed throughout the building. This assumption is justified in case of multi-storey buildings where most of the building mass is indeed

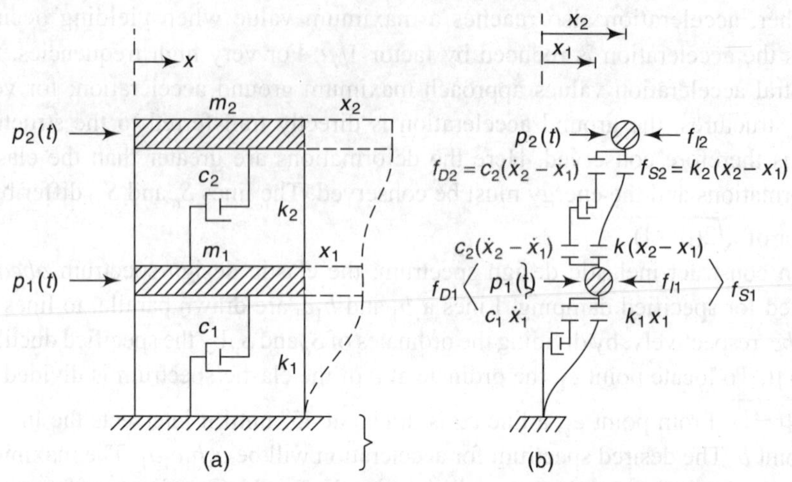

Fig. 2.22 Two-degrees-of-freedom system under horizontal forces

at the floor levels. This assumption transforms the problem from a structure with infinite degrees of freedom (due to distributed mass), to a structure that has only as many degrees as it has lumped masses at the floor levels. For example, the structure shown in Fig. 2.22 has two degrees of freedom.

(b) The girders on the floors are infinitely rigid as compared to the columns and the deformation of the structure is independent of the axial forces present in the columns. This assumption introduces the requirements that the joints between girders and columns are fixed against rotation and the girders remain horizontal during motion.

It must be noted that a building may have any number of bays and that it is only as a matter of convenience that we represent the shear buildings solely in terms of a single bay. Further, a shear building can be idealized as a single column [Fig. 2.22 (b)] having concentrated masses at floor levels, and the columns as massless springs. The *stiffness coefficient* or *spring constant* k_j is the force required to produce a unit displacement of the two adjacent floor levels. For a uniform column with the two ends fixed against rotation, the spring constant is $12EI/h^3$, and for a column with one end fixed and the other pinned it is $3EI/h^3$, where E is the modulus of elasticity of the material, I the moment of inertia, and h the height of the storey.

2.8.1 Equations of Motion

The equations of motion are developed for a simple MDOF system; a two-storey shear frame is selected to permit easy visualization of elastic, damping, and inertial forces. The following equations of motion are obtained for a two-storey shear building [Fig. 2.22(a)].

$$f_{I1} + f_{D1} + f_{S1} = p_1(t)$$

$$f_{I2} + f_{D2} + f_{S2} = p_2(t) \tag{2.60}$$

The inertial forces in the equations are

$$f_{I1} = m_1\ddot{x}_1 \tag{2.61}$$

$$f_{I2} = m_2\ddot{x}_2$$

or in matrix form

$$\begin{Bmatrix} f_{I1} \\ f_{I2} \end{Bmatrix} = \begin{bmatrix} m_1 & 0 \\ 0 & m_2 \end{bmatrix} \begin{Bmatrix} \ddot{x}_1 \\ \ddot{x}_2 \end{Bmatrix} \tag{2.62}$$

or $\quad f_I = m\ddot{x} \quad (2.63)$

where f_I, $\ddot{x}$, and m are the *inertial-force vector*, *acceleration vector*, and *mass matrix*, respectively.

As shown in Fig. 2.22(b), the lumped masses are concentrated at floor levels, and the mass matrix is therefore a diagonal matrix. The restoring forces and displacements are related as follows.

$$f_{S1} = k_1x_1 - k_2(x_2 - x_1) = (k_1 + k_2)x_1 - k_2x_2$$

$$f_{S2} = k_2(x_2 - x_1) = -k_2x_1 + k_2x_2 \tag{2.64}$$

By introducing k_{11}, k_{12}, k_{21}, and k_{22}

$$k_{11} = k_1 + k_2 \text{ and } k_{12} = -k_2$$

$$k_{21} = -k_2 \text{ and } k_{22} = k_2 \tag{2.65}$$

and substituting into Eqn (2.64), the following equations are obtained (Fig. 2.23)

$$f_{S1} = k_{11}x_1 + k_{12}x_2$$

$$f_{S2} = k_{21}x_1 + k_{22}x_2 \tag{2.66}$$

If k_{ij} is the force applied to the ith storey when the jth storey is subjected to a unit displacement, while all other stories remain undisplaced, then by the

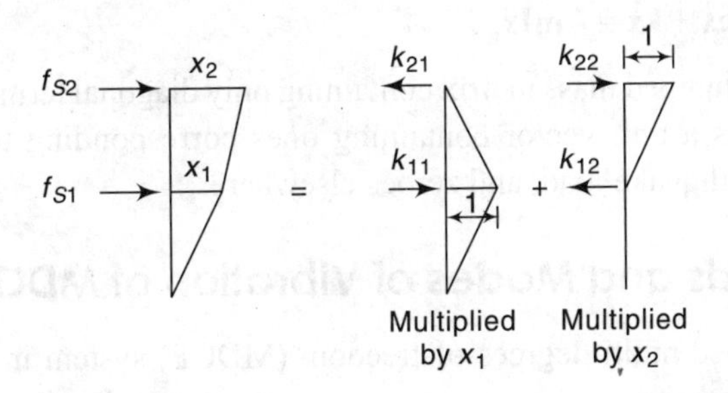

(a) Total deflection (b) Decomposition of deflection

Fig. 2.23 Load and deflection of two-DOF system

Maxwell–Betti reciprocal theorem

$$k_{ij} = k_{ji} \tag{2.67}$$

Equation (2.66) can be written as

$$\begin{Bmatrix} f_{S1} \\ f_{S2} \end{Bmatrix} = \begin{bmatrix} k_{11} & k_{12} \\ k_{21} & k_{22} \end{bmatrix} \begin{Bmatrix} x_1 \\ x_2 \end{Bmatrix} \tag{2.68}$$

or $$\mathbf{f_S} = \mathbf{kx} \tag{2.69}$$

where $\mathbf{f_S}$, $\mathbf{x}$, and $\mathbf{k}$ are the elastic-force vector, displacement vector, and stiffness matrix, respectively. Equation (2.67) indicates that $\mathbf{k}$ is a symmetrical matrix.

If damping forces induced by viscous damping are assumed to be proportional to relative velocities then

$$\begin{Bmatrix} f_{D1} \\ f_{D2} \end{Bmatrix} = \begin{bmatrix} c_{11} & c_{12} \\ c_{21} & c_{22} \end{bmatrix} \begin{Bmatrix} \dot{x}_1 \\ \dot{x}_2 \end{Bmatrix} \tag{2.70}$$

or $$\mathbf{f_D} = \mathbf{c\dot{x}} \tag{2.71}$$

where $\mathbf{f_D}$, $\dot{\mathbf{x}}$, and $\mathbf{c}$ are the viscous damping-force vector, velocity vector, and viscous damping matrix, respectively. The applied load vector is

$$\mathbf{p(t)} = \begin{Bmatrix} p_1(t) \\ p_2(t) \end{Bmatrix} \tag{2.72}$$

Using Eqns (2.63), (2.69), (2.71), and (2.72), the equations of motion for the two-degrees-of-freedom system can be written as

$$\mathbf{f_I} + \mathbf{f_D} + \mathbf{f_S} = \mathbf{p(t)} \tag{2.73}$$

or

$$\mathbf{m\ddot{x}} + \mathbf{c\dot{x}} + \mathbf{kx} = \mathbf{p(t)} \tag{2.74}$$

This expression is essentially in the same form as the equation for an SDOF system [Eqn (2.10)].

If ground acceleration, $\ddot{x}_g$, is applied to the structure, then

$$\mathbf{m\ddot{x}} + \mathbf{c\dot{x}} + \mathbf{kx} = -\mathbf{mI\ddot{x}_g} \tag{2.75}$$

where $\mathbf{m}$ is the lumped mass matrix containing only diagonal terms, $\mathbf{k}$ is a banded matrix, and $\mathbf{I}$ is a unit vector containing ones corresponding to DOFs in the direction of earthquake load, and zeroes elsewhere.

2.9 Periods and Modes of Vibration of MDOF Systems

For an undamped multi-degrees-of-freedom (MDOF) system in free vibration, Eqn (2.75) reduces to

$$\mathbf{m\ddot{x} + kx = 0} \tag{2.76}$$

The solution of Eqn (2.76) is assumed to be

$$\boldsymbol{x} = \hat{\boldsymbol{x}} \sin \omega t \tag{2.77}$$

where $\hat{\boldsymbol{x}}$ represents the vibrational shape (mode shape) of the system. Differentiating Eqn (2.77) twice

$$\ddot{\boldsymbol{x}} = -\omega^2 \hat{\boldsymbol{x}} \sin \omega t \tag{2.78}$$

Substituting Eqns (2.77) and (2.78) into Eqn (2.76), we get

$$\mathbf{k\hat{x} - \omega^2 m\hat{x} = 0} \tag{2.79}$$

Equation (2.79) is called the *frequency equation* with respect to the circular frequency ω. For a system with *n* degrees of freedom, there will be *n* natural circular frequencies from Eqn (2.79). An *N*-DOF system can thus vibrate in *N* different modes, each having a different mode shape and each occurring at a particular natural frequency. The lowest value of ω is called the *first natural circular frequency* ω_1. The ω are numbered sequentially so that the *n*th lowest value of ω is the *n*th natural *circular frequency; by* substituting it into Eqn (2.79), the relative displacements *x* of the system, which represent the shape of vibration or the *modal* shape, can be determined. For a two-DOF system, Eqn (2.79) would become

$$\begin{aligned}(k_{11} - \omega^2 m_1)\hat{x}_1 + k_{12}\hat{x}_2 &= 0 \\ k_{21}\hat{x}_1 + (k_{22} - \omega^2 m_2)\hat{x}_2 &= 0\end{aligned} \tag{2.80}$$

For $\hat{\boldsymbol{x}}$ to have a nontrivial solution, the determinant of Eqn (2.80) must be zero.

$$\begin{vmatrix} k_{11} - \omega^2 m_1 & k_{12} \\ k_{21} & k_{22} - \omega^2 m_2 \end{vmatrix} = 0 \tag{2.81}$$

or $$(m_1\omega^2 - k_{11})(m_2\omega^2 - k_{22}) - k_{12}k_{21} = 0 \tag{2.82}$$

Four roots can be derived from Eqn (2.82)

$$\omega_{12}^2 = \frac{1}{2}\left[\left(\frac{k_{11}}{m_1} + \frac{k_{22}}{m_2}\right) \pm \sqrt{\left(\frac{k_{11}}{m_1} - \frac{k_{22}}{m_2}\right)^2 + \frac{4k_{12}k_{21}}{m_1 m_2}}\right]$$

The positive roots ω_1 and ω_2, respectively, correspond to the first and second natural circular frequencies. By substituting them into Eqn (2.80), the ratio of displacements, $\hat{x}_2/\hat{x}_1$, is uniquely determined for each ω_1 and ω_2 as shown in Fig. 2.23. The modal shapes corresponding to ω_1 and ω_2 are called the *first* and *second mode,* respectively. As evident from the condition specified by Eqn (2.80),

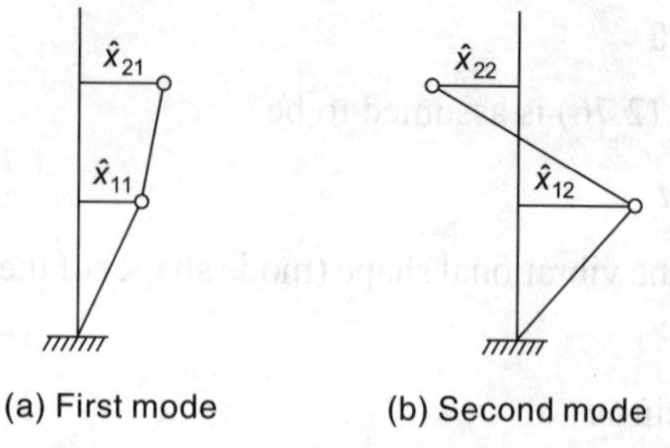

Fig. 2.24 Modal shapes of a two-DOF system

only displacement ratios of $\hat{x}$ can be obtained. In usual practice, the maximum displacement corresponding to the top or the lowest storey is taken to be unity. If a system has N degrees of freedom, then the nth modal shape ϕ_n is written as

$$\phi_n = \begin{bmatrix} \phi_{1n} \\ \phi_{2n} \\ \vdots \\ \phi_{Nn} \end{bmatrix} = \frac{1}{\hat{x}_{kn}} \begin{bmatrix} \hat{x}_{1n} \\ \hat{x}_{1n} \\ \vdots \\ \hat{x}_{1n} \end{bmatrix} \tag{2.83}$$

Here, $\hat{x}$ represents the reference component. The square matrix, consisting of n-modal-shape vectors, is called the *modal-shape matrix* and is expressed as

$$\phi = [\phi_1 \quad \phi_2 \quad \cdots \quad \phi_N] \begin{bmatrix} \phi_{11} & \phi_{12} & \cdots & \phi_{1N} \\ \phi_{21} & \phi_{22} & \cdots & \phi_{2N} \\ \vdots & \vdots & \vdots & \vdots \\ \phi_{N1} & \phi_{N2} & \cdots & \phi_{NN} \end{bmatrix} \tag{2.84}$$

Modal-shape vectors possess an orthogonality relationship for elastic systems.

2.10 Elastic Response of MDOF Systems

A multi-storey building behaves in a much more complex manner than the simple system considered in Fig. 2.22. A multi-storey building has one degree of freedom for each storey and it may vibrate with as many different mode shapes and periods as it has degrees of freedom. The response history of any element of such a structure is a function of all the modes of vibration, as well as of its position within the overall structural configuration. Such a building can oscillate in any of these modes at the particular frequency of that mode. The fundamental frequency of the system corresponds, in general, to a motion that involves displacement of all of the masses towards the same side. However, the higher modes correspond to reversals in the directions of motion of the various masses, with inflection points in the system between the base and the top (Fig. 2.25).

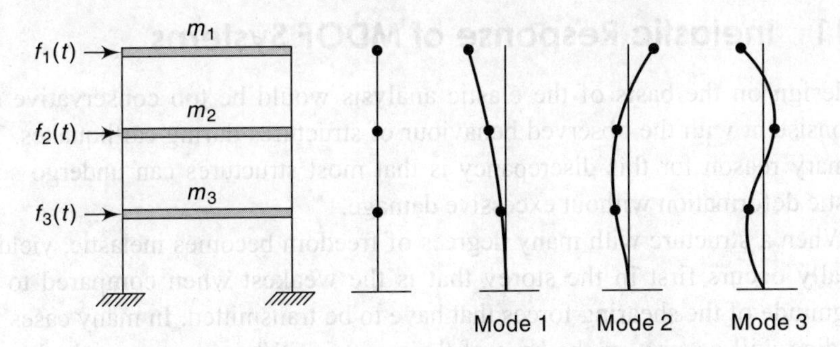

Fig. 2.25 Typical mode shapes for a 3-storey frame

So long as the structure remains elastic and is undamped, or when the damping forces satisfy certain requirements, it is possible to analyse the structure as if it were a system of simple SDOF elements. Each element is considered to have its particular frequency, and to be excited by the ground motion in a manner determined by a *participation coefficient* and the acceleration, displacement, or velocity desired for spectrum response. An example of such an analysis is given in Section 5.10. The procedure outlined gives the maximum response for each of the modes. The actual responses are nearly independent functions of time, and the maxima in the different modes do not necessarily occur at the same time. Although it is possible to obtain the time history of the motion in each of the modes, this is an extremely complex and tedious calculation and has been done only rarely.

For the most general types of damping, including viscous damping, modal vibration in independent uncoupled modes cannot exist. The types of viscous damping for which modal analysis is possible are linear combinations of (a) damping proportional to the relative velocity between the masses, where the damping coefficient is proportional to the spring constant coupling the same masses, and (b) damping proportional to the velocity of each mass related to the ground, with each damping coefficient proportional to the magnitude of the attached mass. For these kinds of damping, and for certain other restricted damping arrangements, modal vibrations are possible; where they are possible, the modes have the same shape as for the undamped system. In these cases, the maximum possible response of the system (stress, deformation, displacement, velocity, etc.) is given by the sum of maximum modal responses without taking into account the sign. This is an absolute upper limit for the response of the system, and is excessively conservative. However, it is suggested that the probable value of the maxima response is approximately equal to the square root of the sum of the squares of the modal maxima. This concept arises from the consideration of equal probability of modal responses in any mode, and is in accord with perfectly random distributions of the expected values for each of the modal components. The accuracy of this approach increases with the number of degrees of freedom.

2.11 Inelastic Response of MDOF Systems

A design on the basis of the elastic analysis would be too conservative and inconsistent with the observed behaviour of structures during earthquakes. The primary reason for this discrepancy is that most structures can undergo some plastic deformation without excessive damage.

When a structure with many degrees of freedom becomes inelastic, yielding usually occurs first in the storey that is the weakest when compared to the magnitude of the shearing forces that have to be transmitted. In many cases this yielding will occur near the base of the structure. When an area at the base or within the structure yields, the forces that can be transmitted through the yielded region cannot exceed the value of the yield shear for that storey, provided the system is essentially elasto-plastic. Consequently, the shears and the accompanying acceleration and relative deflection for the upper region of the structure are reduced in magnitude, when compared to the values for an elastic structure subjected to the same base motion.

In other words, since the region above the part of the structure that yields behaves essentially in an elastic manner, the effect of yielding near the base of the structure is to reduce the shear. The upper parts of the structure must, therefore, be designed by limiting the base shear magnitude. As a consequence of this, if the total base shear for which the structure is designed is some fraction of the maximum computed value for an elastic system, yielding will occur in the lowest storey and the shears in the remaining part of the structure will have magnitudes appropriate to the revised value of the base shear. If provision is made for the absorption of energy in the lower storeys, the structure should be adequately strong, provided that the shearing forces for which it is designed in the upper storeys are consequently related to the base shear design, even though the structure may yield near the base.

When a structure deforms inelastically to a major extent, its higher modes of oscillation are inhibited and its major deformation takes place in the one mode in which the inelastic deformation is most prominent, which is generally the fundamental mode. However, there are situations in which principal plastic deformation occurs at a higher mode than the fundamental mode. In effect, when the lower portion of the structure becomes inelastic, the period of vibration is effectively increased. In any event where large amounts of plastic behaviour occur, the modal analysis concept is no longer applicable and the structure behaves in many respects like an SDOF system, corresponding to the entire mass of the structure supported by the elements that become plastic.The base shear can, therefore, be computed for the modified structure, with its fundamental period defining the modified spectrum for which the design should be made. However, the fundamental period of this modified structure will not differ materially from the fundamental period of the original elastic frame structure. In the case of a shear-wall structure, the fundamental period is longer. As a result, it is usually appropriate to use the frequency of the fundamental mode in design recommendations, without taking the higher-mode frequencies into account directly.

However, it is desirable to consider a distribution of shearing stresses in the structure, which take into account the higher-mode excitation of the part above the region that becomes plastic. This is done implicitly in code recommendations, by providing for a variation of the lateral force coefficient with the height of the structure. In other words, the distribution of local seismic force over the height of the building, corresponding to a uniformly varying acceleration ranging from a zero value at the base to a maximum at the top, accounts quite well for the moments and shears in the structure. The distribution also takes into account the fact that the local acceleration at higher elevations in the structure is greater than at lower elevations, because of the greater magnitudes of motion at higher elevations.

2.12 Restoring Force

The restoring force in the structure is proportional to the deformation induced in the structure during seismic excitation. The constant of proportionality is referred to as stiffness of the structure. To study inelastic response of a discrete mass system, a mathematical model of restoring force is set up. This defines the relationship between the storey shear and storey deflection.

The simplest models of hysteresis are shown in Fig. 2.26. The bilinear hysteresis models are shown in Fig. 2.26(a) and (b). The model is *positive bilinear* when the line *AB* has a positive slope and *negative bilinear* when the line *AB* has a negative slope. In case the slope is zero the result is an *elasto-plastic model* [Fig. 2.26(b)]. An elasto-plastic or positive bilinear model is often used to represent the restoring force characteristics of steel frames. The *trilinear* model [Fig. 2.26(c)] is usually used for RCC and composite (steel and RCC) frames. Here, lines *OABC* constitute the spectral curve. Line *CD* is parallel to and twice as long as line *OA* and line *DE* is parallel to and twice as long as line *AB*. Points *A* and *B* correspond to the points of cracking and yielding, respectively. However, for the frame subjected to high axial force, a negative bilinear model is sometimes used for RCC frames.

The curve shown in Fig. 2.26(d) is the most realistic curve, as it represents both the Bauchinger effect as well as the effect of sequential yielding of members. However, the models shown in Fig. 2.26(a), (b), and (c) are the choice of designers for their simplicity. Figure 2.27 shows a *degrading-type* model. The model allows for the effect of load reversals in inelastic ranges, in an RCC frame that yields by flexure. The *slip-type* model is shown in Fig. 2.28. This model is very useful for a bolt connection.

2.13 Damping

The energy produced in the structure by ground motion is basically dissipated through internal friction within the structural and non-structural members.

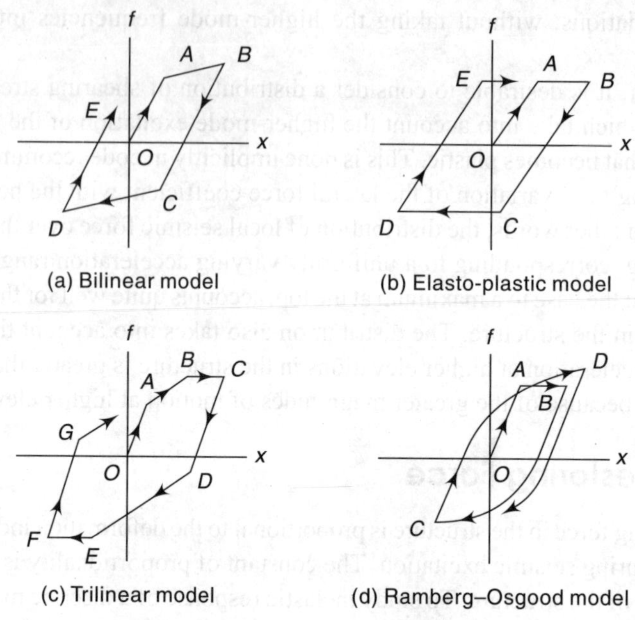

Fig. 2.26 Massing-type models

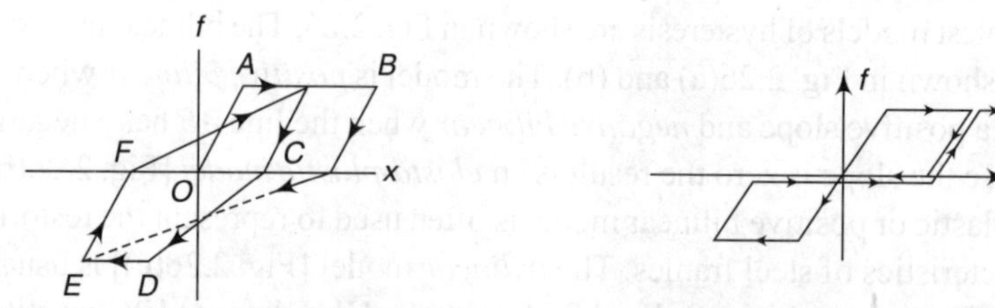

Fig. 2.27 Degrading-type model

Fig. 2.28 Slip-type model

Damping may be defined as the process by which free vibration steadily diminishes in amplitude. A vibrating structure may be simulated with a tuning fork. However, the structures do not resonate with the purity of the tuning fork because of damping. The extent of damping depends upon the construction materials used, the type of construction, and the presence of non-structural elements. Damping is measured as a percentage of critical damping. In a dynamic system, critical damping is the minimum amount of damping necessary to prevent oscillation altogether. In damping, the energy of the vibrating system is dissipated by various mechanisms, and often more than one mechanism may be present at the same time. Given below are the types of damping mechanisms of structures under earthquake disturbances.

External viscous damping External viscous damping is caused by the air or water surrounding a structure and is insignificant, because of lower viscosity of air or water, as compared to other types of damping.

Body-friction damping Body-friction damping, also called *Coulomb damping*, occurs because of friction at connections or support points. It is constant regardless of the velocity or amount of displacement. It is usually treated either as *internal viscous damping* when the level of displacement is small or as *hysteresis damping* when it is large. Body friction is large in infilled masonry walls when the walls crack, and provides very effective seismic resistance. Structures with bolted connections have more friction damping than structures with welded connections. An RCC structure has more frictional damping than a prestressed structure, since the latter shows relatively lesser cracking.

Internal viscous damping This is the damping associated with material viscosity and is also known as material damping. It is proportional to velocity, so that the damping ratio increases in proportion to the natural frequency of the structure. Internal viscous damping is readily included in dynamic analysis by introducing a dashpot. It is frequently used to represent all kinds of damping.

Hysteresis damping Hysteresis damping takes place when a structure is subjected to load reversals in the inelastic range. In Fig. 2.29, a one-cycle hysteresis loop is shown in terms of the force-deflection relationship. The one-cycle hysteresis loop swells outwards. Energy corresponding to the area of the loop is dissipated in the cycle. The dissipation in energy is defined as hysteresis damping. It is unaffected by the velocity of the structure but increases with the level of displacement. Analysis with such a spring model is usually very complicated. Instead, hysteresis damping is often replaced by equivalent viscous damping and an elastic analysis is performed. An inelastic spring system vibrating under stationary sinusoidal base motion is replaced by an elastic damped spring system, which is subjected to the same motion and has the same natural frequency and energy-dissipation capacity as the inelastic system. Let us consider a system having restoring-force characteristics as shown in Fig. 2.29. In this system, the spring constant varies with the force. In the equivalent damped mass system, the spring constant is assumed to be the one represented by the line *AOC* and the equivalent viscous damping is given by

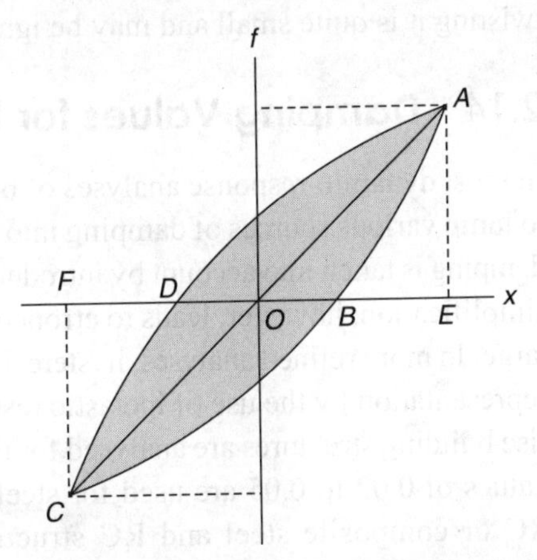

Fig. 2.29 Load–deflection hysteresis loop

$$\xi_{eq} = \frac{1}{2\pi}\frac{\text{area of loop } ABCDEA}{\Delta OAE + \Delta OCF} = \frac{1}{2\pi}\frac{\Delta V}{V} \tag{2.85}$$

where ΔV is the energy loss and V is the maximum strain energy.

This equation is derived by equating the area of *ABCDA* with the energy dissipation by viscous damping.

Radiation damping Radiation damping is also known as material damping or internal viscous damping. The radiation damping increases in proportion to the natural frequency of the structure. When a building structure vibrates, elastic waves propagate through the semi-infinitely extended ground on which the structure is built. The energy input into the structure is dissipated by this wave propagation. Radiation damping is a measure of the energy loss from the structure through radiation of waves away from the footing; it is a purely geometrical effect. The dissipation in energy, defined as radiation damping, is a function of the elastic constant E, the density ρ, Poisson's ratio of the ground, the mass of the structure per unit area m/A, and the ratio of the spring constant of the structure to the mass k/m. Radiation damping increases and eventually the structural response decreases as the structure becomes stiffer, the ground becomes more flexible, and embedment becomes deeper. Radiation damping is smaller for higher-mode vibration, which is the reverse in the case of internal viscous damping. The radiation and ground hysteresis damping are not additive to structural damping. For horizontal and vertical translations, radiation damping may be quite large, while for rocking or twisting it is quite small and may be ignored in most practical design problems.

2.14 Damping Values for Buildings

In most dynamic-response analyses of building structures, it is common practice to lump various sources of damping into viscous damping. In this case, hysteresis damping is taken into account by introducing an equivalent viscous damping. This simplification, however, leads to erroneous results when the level of deflection is large. In more refined analyses, hysteresis damping is often considered in stiffness representation by the use of inelastic restoring-force characteristics. When high-rise building structures are analysed for their earthquake response, damping-ratio values of 0.02 to 0.05 are used for steel and RC or composite steel and RC structures. Damping ratios corresponding to higher modes are assumed to increase in proportion to natural frequencies. The damping ratios for various building materials are given in Table 2.1. Values for damping for a range of constructions are indicated in Table 2.2. These values are suitable for normal response spectrum or modal analysis, in which viscous damping, equal in all modes, is assumed.

Table 2.1 Damping ratio for various building materials

Material	Damping ratio (ξ)
Concrete	5%
Steel	<2%
Wood	12%
Clay	8–10%
Brick	5–7%

Table 2.2 Typical damping ratio for structures

Type of construction	Damping ratio (ξ)
Steel frame, welded, with all the walls of flexible construction	2%
Steel frame, welded or bolted with stiff cladding and all internal walls flexible	5%
Steel frame, welded or bolted with concrete shear walls	7%
Concrete frame, with all walls of flexible construction	5%
Concrete frame, with stiff cladding, and all internal walls flexible	7%
Concrete frame, with concrete or masonry shear walls	10%
Concrete or masonry shear wall buildings	10%
Timber shear wall construction	15%

Notes:

1. The term 'frame' indicates beam and column bending structures as distinct from shear structures.
2. The term 'concrete' includes both reinforced and prestressed concrete in buildings. For isolated prestressed concrete members such as in bridge decks, damping values less than 5% are appropriate, e.g., 1–2% if the structure remains substantially uncracked.

Before yield, the base shear decreases with increased critical damping. However, it increases after the yield point. Inelastic deformations and hysteretic damping increase the earthquake resistance of a structure beyond that provided by their elastic strength. The effectiveness of damping in reducing the response is smaller for inelastic systems and decreases as the yielding deformation increases. This implies that the effect of damping is more before yielding of the structure and decreases with increase in yielding. Further, the effect is more in velocity sensitive region. Because of this reason yielding cannot be considered in terms of a fixed amount of equivalent viscous damping.

2.15 Uncertainties of Dynamic Analysis

The problems involved in adequately representing seismic behaviour in structural analysis are numerous and many compromises have to be made even in sophisticated analysis. Any dynamic analysis starts with an assumed base movement. This base movement is intended to simulate the earth movement that would actually occur at the building site during an earthquake. With the current state of scientific advancement in this field it is not possible to predict the characteristics of this movement at any given site. The distribution patterns of ground accelerations have shown that a close estimate of the ground motions of a particular site is rather difficult, even if an accelerograph is located close by. There may be differences between the ground motions of sites only a fraction of a kilometre apart. These differences are caused by variations in the propagation paths of the seismic waves, by surface and subsurface topography, and by details of local geological and soil conditions.

The amount of damping in a building structure is uncertain but this has a very important effect on its dynamic response. Determination of the damping coefficients to be used (in dynamic analysis) is one of the most important and difficult steps in seismic analysis. There are relatively few applicable test data to support an accurate estimate of the true damping of a structure. Most available test results are based on very small-amplitude distortions or on component tests and the results probably do not accurately reflect the damping that might be expected for the large-amplitude motions associated with a severe earthquake. And yet, small changes in assumed damping may significantly change the calculated response of a structural system.

Another serious uncertainty is the reduction that must be made in the elastic or linear response of a building, as calculated by a dynamic analysis, in order to allow for the ductility of the structure. One method is to divide the calculated elastic response by the ductility factor to obtain the response of the actual structure. Since the ductility factor may range between three and six or more, determination of the ductility factor is largely a matter of judgement. It is evident that the choice of a ductility factor for any given structure will have a very large influence on the final result of a dynamic analysis.

The mathematical modelling of a structure for the purpose of dynamic analysis is subject to other important uncertainties. Shear walls, or shear walls in conjunction with moment-resisting frames, are commonly used for lateral bracing of multi-storey buildings. Determination of the stiffness of such bracing systems is, however, almost impossible. Non-structural partitions and filler walls can have an important effect on the dynamic response of a building, and their stiffness is a source of uncertainties. Also their stiffness will change during an earthquake due to progressive damage to these elements. Prefabricated outside-wall panels, which are often used in high-rise buildings, unless properly mounted so as to allow free movement of the panel relative to the building structure, may greatly increase the stiffness of the building. This free movement must not be subject to any impairment due to improper design, poor installation, or deterioration of the mount details.

The uncertainties involved in calculating the deflection, and consequently the dynamic action, of a frame, also affect the dynamic analysis greatly.

Summary

The chapter deals with dynamics as related to seismic design of buildings. The process for development of mathematical models of single-degree-of-freedom (SDOF) system and multi-degrees-of-freedom (MDOF) system are described. SDOF system is idealized by a mathematical model of mass-spring-dashpot system for its dynamic analysis. The differential equation of SDOF system is derived for free vibration with and without damping and the dynamic properties required for the analysis derived. The differential equation of motion for the damped SDOF system is given by $m\ddot{x} + c\dot{x} + kx = 0$.

The methods for solutions of differential equations of motion are discussed. The classical method has been used to derive solutions of the differential equation of motion. The expression for the solution of this differential equation of motion depends on the magnitude of the damping ratio ξ. The three cases based on damping ratio, the underdamped ($\xi < 1$), critically damped ($\xi = 1$), and overdamped systems ($\xi > 1$) are discussed. For real structures, the damping ratios are usually less than 20% of critical damping ($\xi < 0.2$) and are, therefore, underdamped.

The concept of response spectrum is introduced. It is a plot of maximum response—the maximum displacement, pseudo-velocity, and pseudo-acceleration—as ordinate and the natural frequency as abscissa. Once these curves are constructed, the analysis for design of structures is reduced to calculation of the natural frequency of the system and the use of design spectrum. Earthquakes often induce non-linear response in structures. Both the elastic and inelastic responses of SDOF and MDOF systems are presented. The importance of the response of an SDOF system is emphasized as this forms the basis for the analysis of an MDOF system.

The peak force induced in a linearly elastic system by ground motion is $V_{max} = mS_a$, where m is the mass of the system and S_a is the pseudo-spectral acceleration corresponding to natural vibration period and damping ratio ξ of the system. In case of strong ground motions, structures do not behave linearly elastically and are subjected to inelastic deformations. The system behaviour is assumed to be elastoplastic, being the simplest model, and analysis is performed. The concept of normalized yield strength, reduction factor, and ductility ratio are discussed. The effect of non-linearity, in general, reduces the seismic demand of structures and is normally accounted for by a simple modification to the linear analysis procedure. The method to derive inelastic design response spectrum from the elastic response spectrum is detailed.

The modelling of a multi-storey shear building and the assumptions made therein are presented. The shear building idealization simplifies the dynamic analysis as it permits the representation of structure by lumped rigid masses interconnected by elastic springs. The governing differential equations of motion for ground acceleration are of the form

$$\mathbf{m}\ddot{\mathbf{x}} + \mathbf{c}\dot{\mathbf{x}} + \mathbf{k}\mathbf{x} = -\mathbf{m}\mathbf{I}\ddot{\mathbf{x}}_\mathbf{g}$$

For a dynamic system with few degrees of freedom, the natural frequencies and the modal shapes may be determined by calculating the roots of the resulting characteristic equation. For large degrees of freedom numerical methods should be used. The elastic and inelastic response of MDOF systems is discussed. The types of damping and restoring forces, and uncertainties of dynamic analysis associated with mathematical modelling and idealization of structure are described.

Solved Problems

2.1 A vibrating system consisting of a mass of 50 kg and a spring of stiffness 4×40^4 N/m is viscously damped. The ratio of two consecutive amplitudes is 20:18. Determine the natural frequency of undamped system. Also determine the damping ratio and damped natural frequency.

Solution

Natural circular frequency $\omega = \sqrt{k/m} = \sqrt{4\times10^4/50} = 28.284$ rad/s

Natural frequency $f = \dfrac{\omega}{2\pi} = \dfrac{28.28}{2\pi} = 4.5$ cps

Damping ratio $\xi \approx \dfrac{\delta}{2\pi}$

$$\delta = \ln\frac{x_1}{x_2} = \ln\frac{20}{18} = 0.105$$

$$\xi \approx \frac{0.105}{2\pi} = 0.017$$

Natural frequency of damped system,

$$\omega_D = \omega\sqrt{1-\xi^2} = 28.28\sqrt{1-0.017^2} = 28.275\,\text{rad/s}$$

$$f_D = \frac{\omega_D}{2\pi} = 4.5\,\text{cps}$$

2.2 An SDOF system consists of a mass with weight of 175 kg and a spring constant, $k = 530$ kN/m. While testing the system a relative velocity of 30 cm/s was observed on application of a force of 450 N. Determine the damping ratio, damped frequency of vibration, logarithmic decrement, and the ratio of two consecutive amplitudes.

Solution

Mass $m = 175$ kg

Stiffness $k = 530$ kN/m $= 530 \times 10^3$ N/m

Velocity $\dot{x} = 30$ cm/s $= 0.3$ m/s

Damping force $f_D = c\dot{x}$

$$\Rightarrow c = \frac{f_D}{\dot{x}} = \frac{450}{0.3} = 1500 \text{ N-s/m}$$

Coefficient of critical damping $c_c = 2\sqrt{km} = 2\sqrt{530 \times 10^3 \times 175} = 19261.4$ N-s/m

Damping ratio $\xi = \dfrac{c}{c_c} = \dfrac{1500}{19,261.4} = 0.07788 = 7.79\%$

Natural circular frequency $\omega = \sqrt{\dfrac{k}{m}} = \sqrt{\dfrac{530 \times 10^3}{175}} = 55.03$ rad/s

Damped frequency $\omega_D = \omega\sqrt{1-\xi^2} = 55.03\sqrt{1-0.07788^2} = 54.865$ rad/s

Logarithmic decrement $\delta = \dfrac{2\pi\xi}{\sqrt{1-\xi^2}} = \dfrac{2\pi \times 0.07788}{\sqrt{1-0.07788^2}} = 0.49$

Let the consecutive amplitudes be x_1 and x_2

$$\delta = \ln\frac{x_1}{x_2}$$

or $$0.49 = \ln\frac{x_1}{x_2}$$

or $$\frac{x_1}{x_2} = e^{0.49} = 1.63$$

2.3 In an experiment on a certain structure modelled as an SDOF system, the amplitude of free vibration decreased from 10 mm to 4 mm. If the logarithmic decrement was 0.1018 and undamped natural frequency is 40 rad/s, determine the damping ratio, damped period, and number of cycles completed.

Solution

Undamped natural frequency $\omega = 40$ rad/s
Logarithmic decrement $\delta = 0.1018$

The ratio between the first amplitude x_1 and the nth amplitude can be expressed as

$$\frac{x_1}{x_2} = \frac{x_1}{x_2} \times \frac{x_2}{x_3} \times \frac{x_3}{x_4} \times \cdots \times \frac{x_{n-1}}{x_n}$$

Taking natural logarithm of the above expression

$$\ln\frac{x_1}{x_n} = \ln\frac{x_1}{x_2} + \ln\frac{x_2}{x_2} + \ln\frac{x_3}{x_4} + \cdots + \ln\frac{x_{n-1}}{x_n}$$

$$\ln\frac{x_1}{x_n} = \delta + \delta + \delta + \cdots + \delta$$

$$= (n-1)\delta$$

or $$n-1=\frac{1}{\delta}\ln\frac{x_1}{x_n}$$

or $$n=1+\frac{1}{\delta}\ln\frac{x_1}{x_n}$$

$$x_1=10\text{ mm},\ x_2=4\text{ mm}$$

$$n=1+\frac{1}{0.1018}\ln\frac{10}{4}=10$$

Number of cycles = 10

Damping ratio $\xi\simeq\frac{\delta}{2\pi}=\frac{0.1018}{2\pi}=0.016$

Damped frequency

$$\omega_D=\omega\sqrt{1-\xi^2}$$
$$=40\sqrt{1-0.016^2}=39.995\text{ rad/s}$$

Damped period

$$T_D=\frac{2\pi}{\omega_D}=\frac{2\pi}{39.995}=0.157\text{ s}$$

The time for 10 cycles is = 10 × 0.157 = 1.57 s

2.4 The properties mass *m*, stiffness *k*, and natural frequency ω of an undamped SDOF system are to be determined by a harmonic excitation test. At an excitation frequency of 4 Hz the response tends to increase without bound. Then, a weight *W* of 22 N is attached to the mass *m*, the resonance occurred at 3 Hz. Determine the dynamic properties of the system.

Solution

Let the natural circular frequency of the system be ω.

Resonance frequency ω_{e1} = 4 Hz, mass $m_1=m$, resonance $\omega=\omega_{e1}$

Natural circular frequency $\omega=\sqrt{\frac{k}{m}}$

$$\Rightarrow\frac{k}{m}=\omega^2=4^2=16 \tag{1}$$

Resonance frequency,

$$\omega_{e_2}=3\text{ Hz},\qquad \text{mass, } m_2=m+\frac{W}{g}=m+\frac{22}{9.81}=m+2.243\text{ kg}$$

At resonance, $\omega_2=\omega_{e2}$

$$\text{Natural circular frequency } \omega=\sqrt{\frac{k}{m+2.243}} \tag{2}$$

From Eqns (1) and (2), $\dfrac{k}{m+2.243} = 3^2 = 9$

$$\frac{16}{9} = \frac{m+2.243}{m}$$

$$\Rightarrow m = 2.884 \text{ kg}$$

From Eqn (1)

$$\frac{k}{m} = 16$$

stiffness, $k = 16 \times m = 16 \times 2.884 = 46.14$ N/mm

Natural frequency of space $f = \dfrac{1}{2\pi}\sqrt{\dfrac{k}{m}}$

$$= \frac{1}{2\pi}\sqrt{\frac{46.14}{2.884}}$$

$$= 0.636 \text{ cps}$$

2.5 An empty elevated water tank is pulled by a steel cable by applying a 30 kN force. The tank is pulled horizontally by 5 cm. The cable is suddenly cut and the resulting free vibration is recorded. At the end of five complete cycles, the time is 2.0 s and the amplitude is 2 cm. Determine the damping ratio, natural period of undamped vibration, effective stiffness, effective weight, and damping coefficient for the given data.

Solution

Damping ratio $= \dfrac{1}{2\pi j}\ln\dfrac{x_i}{x_{i+j}} = \dfrac{1}{2\times 3.14\times 5}\ln\dfrac{5}{2} = 0.0292$

Therefore damping factor $\xi = 0.0292 \times 100 = 2.92\%$

Natural period of undamped vibration

$$T = T_D\sqrt{1-\xi^2}$$

Dumped period $T_D = \dfrac{2.0}{5} = 0.4$ s

$$T = 0.4\sqrt{1-0.0292^2} = 0.4 \text{ s}$$

Effective stiffness $k = \dfrac{\text{force}}{\text{displacement}} = \dfrac{30}{0.05} = 600$ kN/m

Circular natural frequency $\omega = \dfrac{2\pi}{T} = \dfrac{2\pi}{0.4} = 15.7$ rad/s

$$\text{Mass } m = \frac{k}{\omega^2} = \frac{600}{15.7^2} = 2.43 \text{ kN-s}^2/\text{m}$$

Effective weight $W = mg = 2.43 \times 9.81 = 23.84$ kN

$$\text{Damping coefficient } c = \xi(2\sqrt{km})$$

$$= 0.0292 \times (2 \times \sqrt{600 \times 2.43})$$

$$= 2.23 \text{ kN-s/m}$$

2.6 An SDOF system is modelled as shown in Fig. 2.30. It has the following properties.

Mass, $m = 2$ kg

Stiffness, $k = 15{,}000$ N/m

Coefficient of damping, $c = 45$ N/m/s

Determine the natural circular frequency, damping factor, and damped frequency of the system shown in Fig. 2.30 Write the equation of free response for determining the time history response of the system.

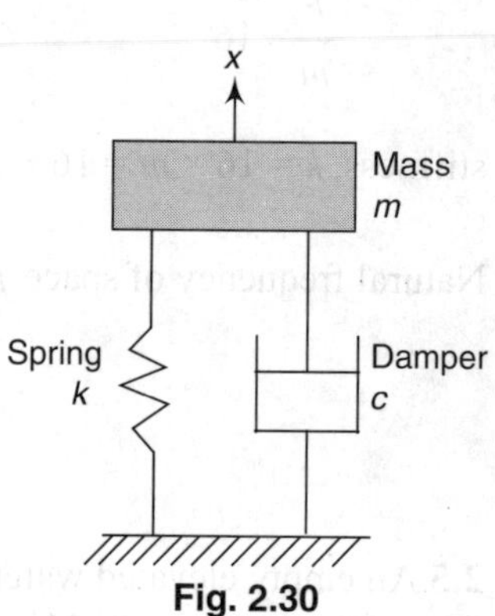

Fig. 2.30

Solution

$$\text{Natural circular frequency } \omega = \sqrt{\frac{k}{m}} = \sqrt{\frac{15{,}000}{2}} = 86.60 \text{ rad/s}$$

Coefficient of critical damping $C_c = 2\,m\omega = 2 \times 2 \times 86.60 = 346.4$ rad/s

$$\text{Damping ratio } \xi = \frac{c}{c_c} = \frac{86.60}{346.4} = 0.25$$

Damped frequency of the system

$$\omega_D = \omega\sqrt{1-\xi^2} = 86.6 \times \sqrt{1-0.25^2} = 83.85 \text{ rad/s}$$

Since the damping ratio is less than one, the system is underdamped. For underdamped case,

displacement $x = \exp(-\xi\omega t)\,[A \cos \omega_D t + B \sin \omega_D t]$ (1)

or $x = \exp(-0.25 \times 86.60\,xt)\,[A \cos 83.85t + B \sin 83.85t]$

or $x = \exp(-21.65t)\,[A \cos 83.85t + B \sin 83.85t]$ (2)

Expression for velocity can be obtained by differentiating Eqn (1) with respect to time t.

$$\text{Velocity } \dot{x} = \exp(-\xi\omega t)\left[(B\omega_D - A\xi\omega) \cos \omega_D t - (A\omega_D + B\xi\omega) \sin \omega_D t\right] \quad (3)$$

or $x = \exp(-21.65t)\,[B \times 0.25 - A \times 21.65) \cos 0.25t -$
$[A \times 0.25 + B \times 21.65t) \sin 0.25t]$ (4)

Note: Equations (2) and (4) can be solved simultaneously to determine constants A and B provided the initial conditions, i.e., the value of x at time $t = 0$, are specified, and time history can be plotted.

2.7 Determine the free vibration response of an SDOF system shown in Fig. 2.31 at time $t = 0.20$ s for the following data:

Natural circular frequency $\omega = 12$ rad/s
Damping factor $\xi = 0.15$
Initial velocity $\dot{x}(0) = 10$ cm/s
Initial displacement $x(0) = 5$ cm

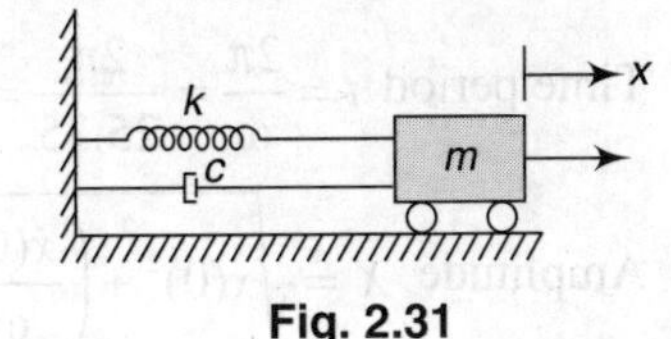

Fig. 2.31

Solution

Displacement at any time t is given by

$$x = \exp(-\xi\omega t)\left(x(0)\cos\omega_D t + \frac{\dot{x}(0) + x(0)\xi\omega}{\omega_D}\sin\omega_D t \right)$$

$$\omega_D = \omega\sqrt{(1-\xi^2)} = 12 \times \sqrt{(1-0.15^2)} = 11.86 \text{ rad/s}$$

$$\xi\omega = 0.15 \times 12 = 1.8 \text{ rad/s}$$

Displacement at time $t = 0.20$ s

$$x = \exp(-1.8 \times .20)\left[5 \times \cos(11.86 \times 0.2) + \frac{(10 + 5 \times 1.8)}{11.86}\sin(11.86 \times 0.2) \right]$$

$$= 0.697676\ (5 \times (-0.718194) + 1.602 \times 0.6958)$$

$$= -1.7276 \text{ cm}$$

In order to get velocity at 0.20 s, differentiating displacement equation with respect to t

$$\dot{x} = -\xi\omega\exp(-\xi\omega t)\left[(x(0)\cos\omega_D t + \frac{\dot{x}(0) + x(0)\xi\omega}{\omega_D}\sin\omega_D t) \right]$$

$$+ \exp(-\xi\omega t)\left[(-x(0)\omega_D \sin\omega_D t + \frac{\dot{x}(0) + x(0)\xi\omega}{\omega_D}\omega_D \cos\omega_D t \right]$$

$$= -1.8 \times 0.697676 \times [5 \times (-0.718194) + 1.062 \times 0.6958] + 0.697676$$
$$\times [-5 \times 11.86 \times 0.6958 + 1.602 \times 11.860 \times (-0.718194)]$$

$$= -34.72 \text{ cm/s}$$

2.8 A mass of 0.07 kg is suspended from a spring of stiffness 45 N/m. The mass is pulled downwards by 15 mm from its equilibrium position and then released. The upward velocity observed was 25 mm/s. Determine the maximum velocity, maximum acceleration, and the phase angle.

Solution

$x(0) = 15$ mm = 1.5 cm; $\dot{x}(0) = -25$ mm/s = -2.5 cm/s

Mass $m = 0.07$ kg; stiffness $k = 45$ N/m

Natural circular frequency $\omega = \sqrt{\frac{k}{m}} = \sqrt{\frac{45}{0.07}} = 25.35$ rad/s

Time period $t = \frac{2\pi}{\omega} = \frac{2\pi}{25.35} = 0.248$ s

Amplitude $X = \sqrt{x(0)^2 + \left(\frac{\dot{x}(0)}{\omega}\right)^2}$

Maximum velocity $\dot{x}_{\max} = X\omega = \sqrt{x(0)^2 + \left(\frac{\dot{x}(0)}{\omega}\right)^2} \times \omega$

$$= \sqrt{1.5^2 + \left(\frac{2.5}{25.35}\right)^2} \times 25.35 = 38.11 \text{ cm/s}$$

Maximum acceleration $\ddot{x}_{\max} = x\omega^2 = \dot{x}_{\max}\,\omega$

$$= 38.11 \times 25.35 = 996.08 \text{ cm/s}^2$$

Phase angle $\theta = \tan^{-1}\left(\frac{\dot{x}(0)}{x(0)\,\omega}\right) = \tan^{-1}\left(\frac{2.5}{1.5 \times 25.35}\right) = 3.76$

2.9 Determine the lateral stiffness of rigid steel frame braced as shown in Fig. 2.32.

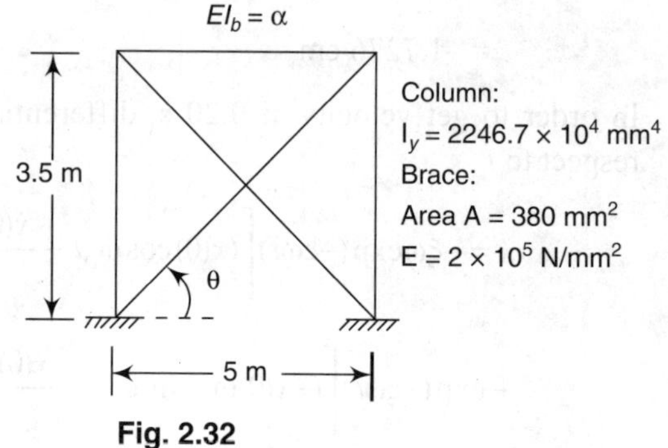

Fig. 2.32

Solution

The lateral stiffness of the braced frame will be the sum of the lateral stiffness of the columns and the braces.

Lateral stiffness of each column

$$k_{col} = \frac{12Ely}{h^3}$$

$$= \frac{12 \times 2 \times 10^5 \times 2246.7 \times 10^4}{3500^3} = 1257.62 \text{ N/mm}$$

Lateral stiffness of brace

The frame has two cross braces. Of these one will be in tension, providing lateral stiffness and the other will be in compression, which will buckle under axial compression and therefore will contribute little to the lateral stiffness; this may be neglected. Therefore, only one brace will contribute to lateral stiffness.

$$\cos\theta = \frac{5}{\sqrt{5^2 + 3.5^2}} = 0.819$$

$$k_{\text{brace}} = \frac{AE}{L}\cos^2\theta$$

$$= \frac{380 \times 2 \times 10^5}{5000} \times 0.819^2$$

$$= 51004.66 \text{ N/mm}$$

Total lateral stiffness of frame

$$k_{\text{frame}} = 2 \times k_{\text{col.}} + k_{\text{brace}}$$

$$= 2 \times 1257.62 + 51004.66$$

$$= 53519.9 \text{ N/mm}$$

Notes: 1. It may be observed that in a braced frame the stiffness of columns is quite small as compared to that of brace and is therefore usually ignored in the designs.

2. The stiffness of brace is formulated as follows.
Refer to Fig. 2.32,
Elongation $e_l = x\cos\theta$
Axial tension in brace = P
Elastic force $f_S = P\cos\theta$

Now, $P = \frac{AE}{L}(x\cos\theta)$

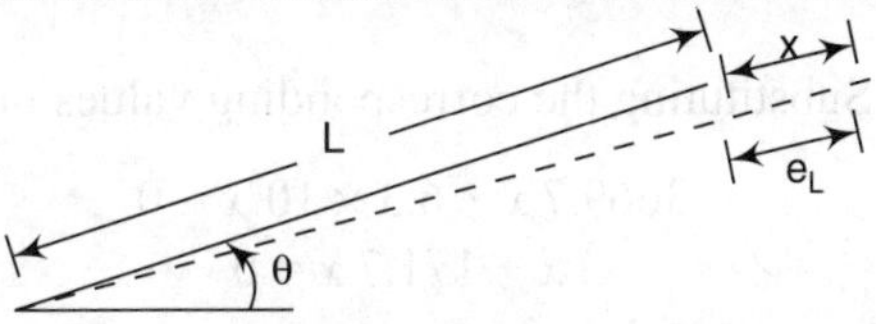

or $\frac{f_S}{\cos\theta} = \frac{AE}{L}(x\cos\theta)$

or $f_S = \frac{AE}{L}(x\cos^2\theta)$

or $\frac{f_S}{x} = k_{\text{brace}}$

$$= \frac{AE}{L}\cos^2\theta$$

Fig. 2.33

2.10 A 120-m long prestressed box girder is supported on four supports—two abutments and two symmetrically located piers—as shown in Fig. 2.34. The cross-sectional area of the deck is 12 m^2. The weight of the deck is idealized as lumped; the unit weight of concrete is 25 kN/m^3. The weight of the piers can be neglected. Each pier consists of four 8-m tall columns of square cross-section, with $I_x = I_y = 0.12$ m^4. Formulate the equation of motion governing free vibration in the longitudinal direction. Also find the natural circular frequency, natural period, and natural frequency of the free vibration of the deck slab. Modulus of elasticity of concrete $E = 28{,}000$ MPa.

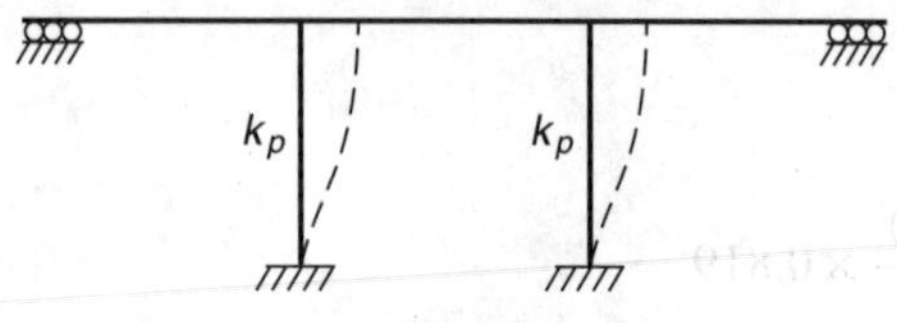

Fig. 2.34

Solution

The weight per unit length lumped at the deck level = 12 × 25 = 300 kN/m

The total lumped weight at the deck level, W = 120 × 300 = 36,000 kN

$$\text{Therefore, mass} = \frac{W}{g} = \frac{36{,}000}{9.81} = 3669.7 \text{ kN-s}^2/\text{m}$$

Assume that the girder is supported on abutments and piers as shown in Fig. 2.34. The longitudinal stiffness provided by each pier is

$$k_p = 4\left(\frac{12EI_x}{h^3}\right) = 4 \times \left(\frac{12 \times 28{,}000 \times 10^3 \times 0.12}{8^3}\right) = 3.15 \times 10^5 \text{ kN/m}$$

Two piers provide total stiffness of $k = 2 \times 3.15 \times 10^5$

$= 6.3 \times 10^5$ kN/m

The equation governing longitudinal displacement x is given by

$$m\ddot{x} + kx = 0$$

Substituting the corresponding values in the above equation

$$3669.7\,\ddot{x} + 6.3 \times 10^5 x = 0$$

$$\ddot{x} + 171.7\,x = 0$$

$$\text{Natural circular frequency } \omega = \sqrt{k/m} = \sqrt{6.3 \times 10^5/3669.7}$$

$$= 13.10 \text{ rad/s}$$

$$\text{Natural period, } T = 2\pi\sqrt{\frac{m}{k}} = 2 \times \pi \times \sqrt{\frac{3669.7}{6.3 \times 10^5}} = 0.48 \text{ s}$$

$$\text{Natural frequency} = \frac{1}{T} = \frac{1}{0.48} = 2.08 \text{ cps}$$

2.11 Derive the equation of motion for a cantilever concrete beam carrying a weight w sustained from a spring at its free end as shown in Fig. 2.35(a), given the following data.

Modulus of elasticity of concrete $E = 22{,}000$ MPa
Moment of inertia of beam $= 1.2 \times 10^{-4}$ m^4
Length of beam $= 3.6$ m
Coefficient of stiffness of spring $k = 40$ kN/m
Neglect the mass of the beam and spring.
Determine also the natural circular frequency, natural period of vibration, and natural frequency of the system if the suspended weight is 30 kN.

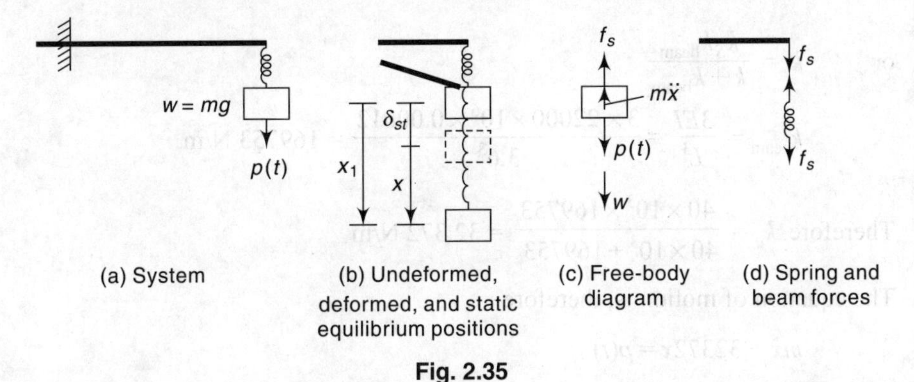

Fig. 2.35

Solution

Figure 2.35(b) shows the deformed position of the free end of the beam, spring, and mass. The displacement of the mass x is measured from its initial position, with respect to the beam and spring in their original undeformed position.
Equilibrium of the forces shown in Fig. 2.35(c) gives

$$m\ddot{x} + k_e x_1 = w + p(t) \tag{2.11.1}$$

where k_e is the effective stiffness of the system.
And the displacement x_1 can be expressed as

$$x_1 = \delta_{\text{st}} + x$$

Since δ_{st} does not vary with time, $\ddot{x}_1 = \ddot{x}$

δ_{st} is the static displacement due to weight w
Therefore, $k_e \delta_{st} = w$
Therefore, the equation of motion reduces to

$$m\ddot{x} + k_e x = p(t)$$

In order to determine the effective stiffness k_e of the system, equating the displacements

$$x_1 = \delta_{\text{spring}} + \delta_{\text{beam}} \tag{2.11.2}$$

where δ_{beam} is the deflection of right end of the beam and δ_{spring} is the deformation in the spring. With reference to Fig. 2.35(d)

$$f_S = k\delta_{spring} = k_{beam}\delta_{beam}$$

Now Eqn (2.11.2) can be rewritten as

$$\frac{f_S}{k_e} = \frac{f}{k} + \frac{f_S}{k_{beam}}$$

or $$k_e = \frac{k_S k_{beam}}{k + k_{beam}}$$

$$k_{beam} = \frac{3EI}{L^3} = \frac{3 \times 22000 \times 10^6 \times 0.00012}{3.6^3} = 169753 \text{ N/m}$$

Therefore, $k_e = \dfrac{40 \times 10^3 \times 169753}{40 \times 10^3 + 169753} = 32{,}372$ N/m

The equation of motion is, therefore

$$m\ddot{x} + 32372x = p(t)$$

Natural frequency $= \dfrac{1}{2\pi}\sqrt{\dfrac{k}{m}}$

$$k_e = 32372 \text{ N/m}$$

$$m = W/g = 30 \times 10^3/9.81 = 3058 \text{ N-s}^2\text{/m}$$

Therefore, natural frequency $= \dfrac{1}{2 \times 3.14}\sqrt{\dfrac{32372}{3058}} = 0.518$ cps

Time period $T = \dfrac{1}{0.518} = 1.93$ s

Natural circular frequency $\omega = \dfrac{2\pi}{T} = \dfrac{2 \times \pi}{1.93} = 3.26$ rad/s

2.12 For the system shown in Fig. 2.36(a), formulate the equation of motion governing the undamped free vibration.

Solution

The structural system of Fig. 2.36(a) has been modelled as shown in Fig. 2.36(b). For mass m_1 [Fig. 2.36(c)]

$$m_1\ddot{x}_1 + k_1x_1 - k_2(x_2 - x_1) = 0$$

or $$m_1\ddot{x}_1 + (k_1 + k_2)x_1 - k_2x_2 = 0 \tag{2.12.1}$$

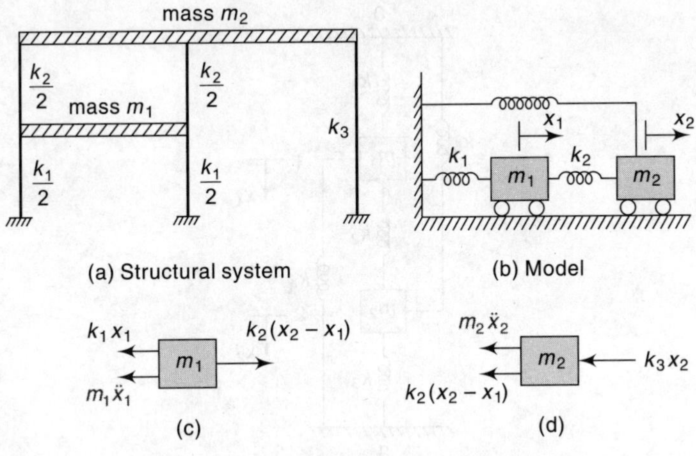

Fig. 2.36

For mass m_2 [Fig. 2.36(d)]

$$m_2\ddot{x}_2 + k_2(x_2 - x_1) + k_3x_2 = 0$$

or
$$m_2\ddot{x}_2 - k_2x_1 + (k_2 + k_3)x_2 = 0 \tag{2.12.2}$$

Combining Eqns (2.12.1) and (2.12.2) in vector form, the required equation of motion for undamped free vibration is

$$\begin{bmatrix} m_1 & \\ & m_2 \end{bmatrix}\begin{bmatrix} \ddot{x}_1 \\ \ddot{x}_2 \end{bmatrix} + \begin{bmatrix} k_1 + k_2 & -k_2 \\ -k_2 & k_2 + k_3 \end{bmatrix}\begin{bmatrix} x_1 \\ x_2 \end{bmatrix} = 0$$

2.13 A simply supported massless beam is shown in Fig. 2.37. The beam carries two masses of equal magnitude at every third point of beam span. Model the system and formulate the equation of motion governing the undamped free vibration in the vertical direction.

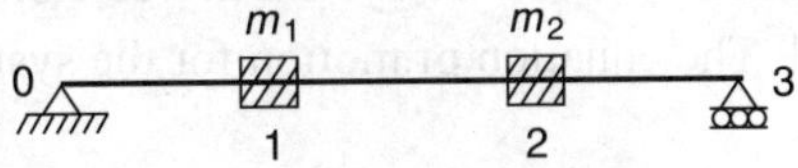

Fig. 2.37

Solution

The modelling of the system consists of five springs as shown in Fig. 2.38. Springs k_1, k_2, and k_3 are for the portions of the beam 0–1, 1–2, and 2–3, respectively. Since deflection at mass 1 (but not at mass 2) will cause a reaction at support 3, this is accounted for by spring k_4. Similarly, deflection at mass 2 (but not at mass 1) will cause a reaction at support 0, which is accounted for by spring k_5.

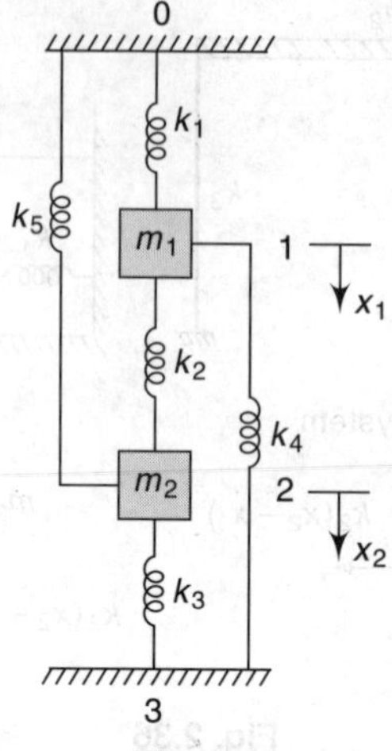

Fig. 2.38

Formulation of equations of motion for mass m_1

$$m_1\ddot{x}_1 + k_1x_1 - k_2(x_2 - x_1) + k_4x_1 = 0$$

or $$m_1\ddot{x}_1 + (k_1 + k_2 + k_4)\, x_1 - k_2x_2 = 0 \tag{2.13.1}$$

Formulation of equations of motion for mass m_2

$$m_2\ddot{x}_2 + k_5x_2 + k_2(x_2 - x_1) + k_3x_2 = 0$$

or $$m_2\ddot{x}_2 - k_2x_1 + (k_2 + k_3 + k_5)x_2 = 0 \tag{2.13.2}$$

Combining Eqns (2.13.1) and (2.13.2) in vector form, the required equation of motion for undamped free vibration is

$$\begin{bmatrix} m_1 & \\ & m_2 \end{bmatrix}\begin{bmatrix} \ddot{x}_1 \\ \ddot{x}_2 \end{bmatrix} + \begin{bmatrix} k_1 + k_2 + k_4 & -k_2 \\ -k_2 & k_2 + k_3 + k_5 \end{bmatrix}\begin{bmatrix} x_1 \\ x_2 \end{bmatrix} = 0$$

2.14 Show that for an undamped MDOF system in free vibration, the modal shape vectors are orthogonal. The equation of motion for the system is $\mathbf{m}\ddot{\mathbf{x}} + \mathbf{k}\mathbf{x} = \mathbf{0}$.

Solution

The equation of the motion for the MDOF system given is

$$\mathbf{m}\ddot{\mathbf{x}} + \mathbf{k}\mathbf{x} = \mathbf{0} \tag{2.14.1}$$

The solution of Eqn (2.14.1) is assumed to be

$$\mathbf{x} = \hat{\mathbf{x}} \sin \omega t \tag{2.14.2}$$

where $\hat{\mathbf{x}}$ represents the vibration shape of the system. By differentiating Eqn (2.14.2) twice, the acceleration experienced by the mass during its oscillating motion is given by

$$\ddot{\mathbf{x}} = -\omega^2\hat{\mathbf{x}} \sin \omega t \tag{2.14.3}$$

Substituting Eqns (2.14.2) and (2.14.3) into Eqn (2.14.1), we get

$$\mathbf{k}\hat{\mathbf{x}} - \omega^2\mathbf{m}\hat{\mathbf{x}} = \mathbf{0} \tag{2.14.4}$$

Equation (2.14.4) is called the frequency equation with respect to the circular frequency ω [see Eqn (2.79)].

Modal-shape vectors possess an orthogonality relationship. To demonstrate this relationship, let us consider $\hat{\mathbf{x}}_n$, the *n*th modal-shape vector. From Eqn (2.14.4), one can obtain

$$\mathbf{k}\hat{\mathbf{x}}_\mathbf{n} - \omega_n^2\mathbf{m}\hat{\mathbf{x}}_\mathbf{n} = \mathbf{0} \tag{2.14.5}$$

By premultiplying Eqn (2.14.5) by the transpose of the *m*th modal-shape vector $\hat{\mathbf{x}}_m$ we obtain

$$\hat{\mathbf{x}}_\mathbf{m}^\mathrm{T}\mathbf{k}\hat{\mathbf{x}}_\mathbf{n} - \omega_n^2\hat{\mathbf{x}}_\mathbf{m}^\mathrm{T}\mathbf{m}\hat{\mathbf{x}}_\mathbf{n} = \mathbf{0} \tag{2.14.6}$$

Interchanging *m* and *n* in Eqn (2.14.6)

$$\hat{\mathbf{x}}_\mathbf{n}^\mathrm{T}\mathbf{k}\hat{\mathbf{x}}_\mathbf{m} - \omega_m^2\hat{\mathbf{x}}_\mathbf{n}^\mathrm{T}\mathbf{m}\hat{\mathbf{x}}_\mathbf{m} = \mathbf{0} \tag{2.14.7}$$

Considering the symmetrical characteristics of matrices **m** and **k**

$$\hat{\mathbf{x}}_\mathbf{m}^\mathrm{T}\mathbf{k}\hat{\mathbf{x}}_\mathbf{n} = \hat{\mathbf{x}}_\mathbf{n}\mathbf{k}\hat{\mathbf{x}}_\mathbf{m}$$

$$\hat{\mathbf{x}}_\mathbf{m}^\mathrm{T}\mathbf{m}\hat{\mathbf{x}}_\mathbf{n} = \hat{\mathbf{x}}_\mathbf{n}^\mathrm{T}\mathbf{m}\hat{\mathbf{x}}_\mathbf{m}$$

Subtracting Eqn (2.14.6) from Eqn (2.14.7) leads to

$$(\omega_n^2 - \omega_m^2)\hat{\mathbf{x}}_\mathbf{n}\mathbf{m}\hat{\mathbf{x}}_\mathbf{m} = \mathbf{0}$$

With the condition that $\omega_\mathbf{n}^2 - \omega_\mathbf{m}^2 \neq 0\ (\mathbf{m} \neq \mathbf{n})$ then

$$\omega_\mathbf{n}^2 - \omega_\mathbf{m}^2 \neq 0\ (\mathbf{m} \neq \mathbf{n}) \tag{2.14.8}$$

Another form of expression for Eqn (2.14.8) is

$$\sum \mathbf{m_i}\hat{\mathbf{x}}_\mathbf{in}\hat{\mathbf{x}}_\mathbf{im} = \mathbf{0} \tag{2.14.9}$$

This equation indicates that the two modal-shape vectors, $\hat{\mathbf{x}}_\mathbf{n}$ and $\hat{\mathbf{x}}_\mathbf{m}$, are orthogonal with respect to the mass matrix **m**. Substituting Eqn (2.14.8) into Eqn (2.14.7) gives

$$\hat{\mathbf{x}}_\mathbf{n}^\mathrm{T}\mathbf{k}\hat{\mathbf{x}}_\mathbf{m} = \mathbf{0} \text{ with } \mathbf{n} \neq \mathbf{m} \tag{2.14.10}$$

Thus the modal-shape vectors are orthogonal to each other also with respect to the stiffness matrix **k**.

2.15 Derive the amplitude of the *n*th modal shape and the earthquake participation factor of the *n*th mode using the equation $\mathbf{x} = \phi\mathbf{Y}$. The equation of motion for the system is $\mathbf{m}\ddot{\mathbf{x}} + \mathbf{kx} = \mathbf{0}$.

Solution

The modal-shape matrix given is

$$\mathbf{x} = \phi \mathbf{Y} \tag{2.15.1}$$

An NDOF system contains n individual modal shapes. Arbitrary displacements, **x**, of the system can be expressed as the sum of the nth-modal-shape vector ϕ_n multiplied by the amplitude $\mathbf{Y}_n$ (Fig. 2.39).

$$\mathbf{x} = \sum_{n=1}^{N} \phi_n \mathbf{Y}_n \tag{2.15.2}$$

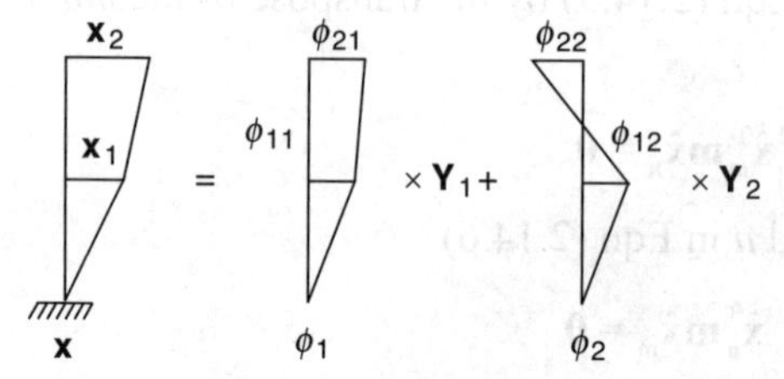

Fig. 2.39 Deflection as the sum of modal components

The vector $\boldsymbol{Y}$ is called the *general coordinate vector* or *the normal coordinate* of the system. By pre-multiplying Eqn (2.15.1) by $\phi_n{}^{\mathrm{T}}\boldsymbol{m}$ and considering the orthogonality condition, the amplitude corresponding to the nth modal shape, $\mathbf{Y}_\mathbf{n}$, can be derived as

$$\phi_\mathbf{n}^{\mathrm{T}} \mathbf{m} \mathbf{x} = \phi_\mathbf{n}^{\mathrm{T}} \mathbf{m} \phi_\mathbf{n} \mathbf{Y}_\mathbf{n}$$

$$\mathbf{Y}_\mathbf{n} = \frac{\phi_\mathbf{n}^{\mathrm{T}} \mathbf{m} \mathbf{x}}{\phi_\mathbf{n}^{\mathrm{T}} \mathbf{m} \phi_\mathbf{n}} \tag{2.15.3}$$

or
$$\mathbf{Y}_\mathbf{n} = \frac{\sum_{i=1}^{N} m_i \phi_{in}^{\mathrm{T}} x_i}{\sum_{i=1}^{N} m_i \phi_{in}^2} \tag{2.15.4}$$

In case of a two-DOF system

$$Y_1 = \frac{m_1\phi_{11}x_1 + m_2\phi_{21}x_2}{m_1\phi_{11}^2 + m_2\phi_{21}^2} \qquad Y_2 = \frac{m_1\phi_{12}x_1 + m_2\phi_{22}x_2}{m_1\phi_{12}^2 + m_2\phi_{22}^2} \tag{2.15.5}$$

$\dot{x}$ and $\ddot{x}$ can also be expressed by using the normal coordinates, since x is now expressed as in Eqn (2.15.2). When the equation for forced vibration [Eqn (2.74)] is to be solved with respect to normal coordinates, the right side of the equation also must be expressed with respect to these coordinates. First, a unit vector **1** is decomposed to

$$1 = \sum_{n=1}^{N} \phi_n \beta_n \tag{2.15.6}$$

or $$1 = \phi\beta \tag{2.15.7}$$

In the case of the two-DOF system, the expression of Eqn (2.15.7) is as represented in Fig. 2.40.

$$\mathbf{1} = \phi_1\beta_1 + \phi_2\beta_2 \tag{2.15.8}$$

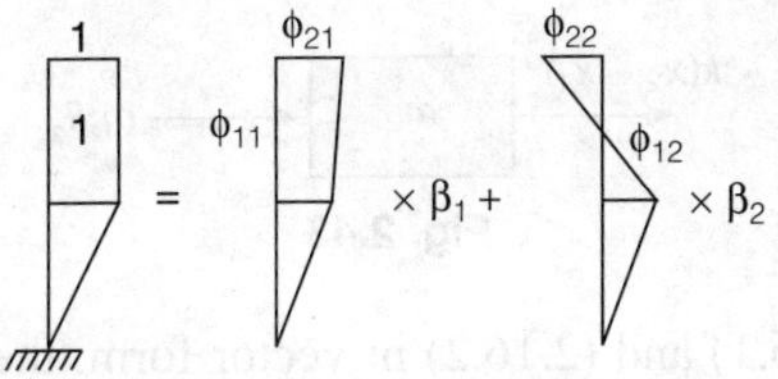

Fig. 2.40 Unit deflection as the sum of modal components

To find β_n, Eqn (2.15.7) is premultiplied by $\phi_n^T m$.

$$\phi_n^T m1 = \phi_n^T m\phi\beta \tag{2.15.9}$$

From the orthogonality condition

$$\beta_n = \frac{\phi_n^T m1}{\phi_n^T m\phi_n} = \frac{\sum_{i=1}^{N} m_i \phi_{in}}{\sum_{i=1}^{N} m_i \theta_{in}^2} \tag{2.15.10}$$

β_n represents the relative participation of the nth modal shape in the entire vibration of the system. It is also called the *earthquake-participation factor* for the nth mode.

2.16 Determine the natural frequency and mode shapes for different modes for the system shown in Fig. 2.41 ($m_1 = m_2 = m$).

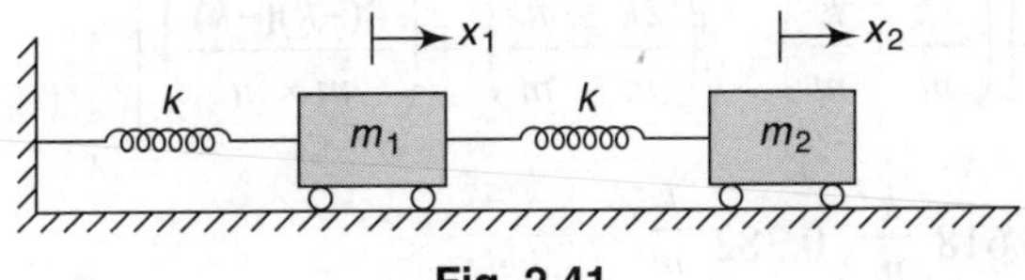

Fig. 2.41

Solution

The system shown above has two degrees of freedom.

Formulation of equation of motion for mass m_1 (Fig. 2.42)

$$m_1\ddot{x}_1 + kx_1 - k(x_2 - x_1) = 0$$

or $$m_1\ddot{x}_1 + 2kx_1 - kx_2 = 0 \tag{2.16.1}$$

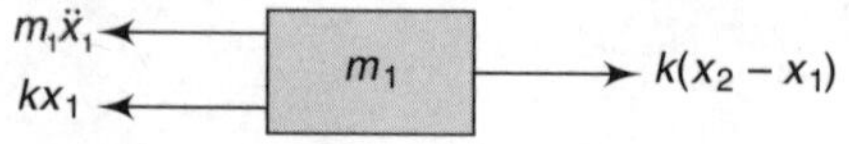

Fig. 2.42

Formulation of equation of motion for mass m_2 (Fig. 2.43)

$$m_2\ddot{x}_2 + k(x_2 - x_1) = 0$$

or $$m_2\ddot{x}_2 - kx_1 + kx_2 = 0 \tag{2.16.2}$$

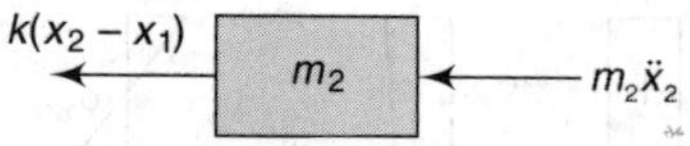

Fig. 2.43

Combining Eqns (2.16.1) and (2.16.2) in vector form, the required equation of motion for undamped free vibration is

$$\begin{bmatrix} m_1 & \\ & m_2 \end{bmatrix}\begin{bmatrix} \ddot{x}_1 \\ \ddot{x}_2 \end{bmatrix} + \begin{bmatrix} 2k & -k \\ -k & k \end{bmatrix}\begin{bmatrix} x_1 \\ x_2 \end{bmatrix} = 0$$

Since $m_1 = m_2 = m$, the required equation of motion is

$$\begin{bmatrix} m & \\ & m \end{bmatrix}\begin{bmatrix} \ddot{x}_1 \\ \ddot{x}_2 \end{bmatrix} + \begin{bmatrix} 2k & -k \\ -k & k \end{bmatrix}\begin{bmatrix} x_1 \\ x_2 \end{bmatrix} = 0$$

The natural frequency and mode shape for different modes can be given by

$$\omega_{12}^2 = \frac{1}{2}\left[\left(\frac{k_{11}}{m_1} + \frac{k_{22}}{m_2}\right) \pm \sqrt{\left(\frac{k_{11}}{m_1} - \frac{k_{22}}{m_2}\right)^2 + \left(\frac{4k_{12}k_{21}}{m_1 m_2}\right)}\right]$$

Putting $k_{11} = 2k$, $k_{12} = -k$, and $k_{22} = k$ in the above equation

$$\omega_{12}^2 = \frac{1}{2}\left[\left(\frac{2k}{m} + \frac{k}{m}\right) \pm \sqrt{\left(\frac{2k}{m} - \frac{k}{m}\right)^2 + \left(\frac{4(-k)(-k)}{m \times m}\right)}\right]$$

$$= 2.618\ \frac{k}{m},\ 0.382\ \frac{k}{m}$$

$$\omega_1^2 = 0.382\ \frac{k}{m}$$

$$\omega_2^2 = 2.618\ \frac{k}{m}$$

Lowest natural frequency will be the fundamental frequency, i.e., ω_1 is the fundamental frequency.

For the first mode

$$(k_{11} - m_1\omega_1^2)\hat{x}_1 + k_{12}\hat{x}_2 = 0$$

$$\left(2k - 0.382\frac{k}{m}m\right)\hat{x}_1 + (-k)\hat{x}_2 = 0$$

or $\quad 1.618\hat{x}_1 = \hat{x}_2$

or $\quad \dfrac{\hat{x}_1}{\hat{x}_2} = \dfrac{1}{1.618}$

For the second mode

$$(k_{11} - m_1\omega_2^2)\hat{x}_1 + k_{12}\hat{x}_2 = 0$$

$$\left(2k - 2.618\frac{k}{m}m\right)\hat{x}_1 + (-k)\hat{x}_2 = 0$$

or $\quad -0.618\hat{x}_1 = \hat{x}_2$

or $\quad \dfrac{\hat{x}_1}{\hat{x}_2} = -\dfrac{1}{0.618}$

The first and second mode shapes are shown in Fig. 2.44.

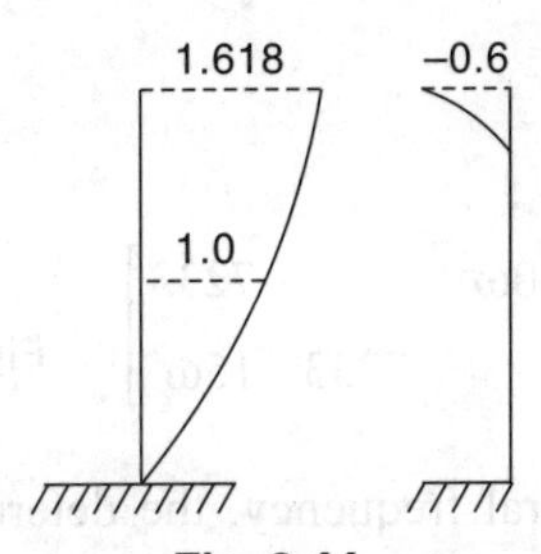

Fig. 2.44

2.17 A model of two-storey RCC frame is shown in Fig. 2.45(a). Determine the natural frequency, assuming the beam–column joints to be rigid, for the following data.

Dimensions of columns 250 × 250 mm

Storey height 3 m

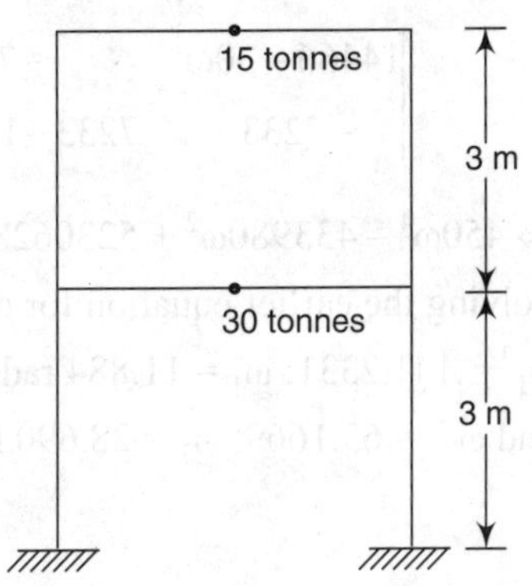

Fig. 2.45(a) Two-storey RCC frame

Solution

Stiffness of column $k = \dfrac{12EI}{h^3}$

Modulus of elasticity,

$$E = 5000\sqrt{f_{ck}} = 5000 \times \sqrt{25} \quad = 25000 \text{ N/nm}^2$$

$$= 2.5 \times 10^6 \text{ N/m}^2$$

Moment of inertia of column, $I = \dfrac{0.25 \times 0.25^3}{12} = 3.255 \times 10^{-4} \text{ m}^4$

Height of column, $h = 3\text{m}$

Stiffness of each storey $= k + k = 2k = \dfrac{2 \times 12EI}{h^3} = \dfrac{24EI}{h^3}$

$$= \frac{24 \times 2.5 \times 10^6 \times 3.255 \times 10^{-4}}{3^3} = 7233 \times 10^3 \text{ N/m}$$

Refer to Fig. 2.45(b).
Mass of the bottom storey, $m_1 = 30 \times 1000 = 30{,}000$ kg
Mass of the top storey, $m_2 = 15 \times 1000 = 15{,}000$ kg
Stiffness matrix:

$$k = \begin{bmatrix} k_1 + k_2 & -k_2 \\ -k_2 & k_2 \end{bmatrix} = \begin{bmatrix} 7233 + 7233 & -7233 \\ -7233 & 7233 \end{bmatrix}$$

m_2 = 15 tonnes

k

m_1 = 30 tonnes

k

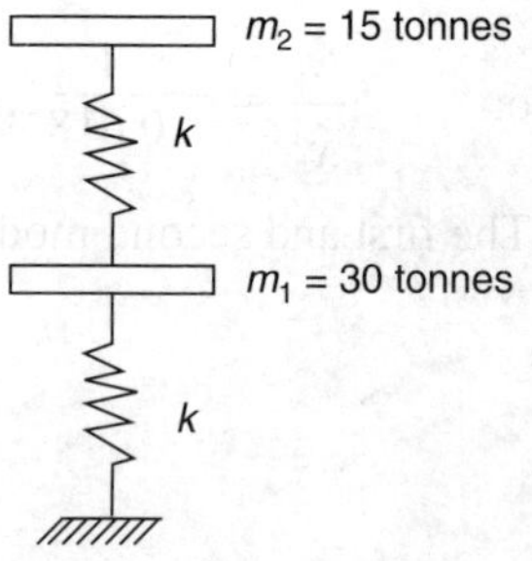

Fig. 2.45(b) Spring-mass model

Mass matrix:

$$m = \begin{bmatrix} 30 & 0 \\ 0 & 15 \end{bmatrix}$$

$$[k] - \omega^2[m]: \begin{bmatrix} 14466 - 30\omega^2 & -7233 \\ -7233 & 7233 - 15\omega^2 \end{bmatrix}$$

For determination of natural frequency, the determinant of the above matrix should be zero.

$$\begin{vmatrix} 14466 - 30\omega^2 & -7233 \\ -7233 & 7233 - 15\omega^2 \end{vmatrix} = 0$$

$\Rightarrow 450\omega^4 - 433980\omega^2 + 52306289 = 0$

Solving the earlier equation for ω;

$\omega_1^{\,2} = 141.2331$; $\omega_1 = 11.884$ rad/s

and $\omega_2^{\,2} = 63.1669$; $\omega_2 = 28.690$ rad/s

2.18 A 3-tonne hammer (Fig. 2.46) is mounted on a 15-tonne concrete block. The hammer drops from a height of 1.5 m. The static deflection in the spring k_1 is found to be 2 cm, whereas for k_2 it is 0.2 cm. Analyse the resulting motion of the hammer and concrete block.

Solution

The given system is modelled as two-DOF system as shown in Fig. 2.46(b).

$$m_1 = 1.5t = 15{,}000 \text{ kg}$$
$$m_2 = 3.0t = 3{,}000 \text{ kg}$$

$$k_1 = \frac{(15{,}000 + 3{,}000) \times 9.81}{2 \times 10^{-2}} = 8.829 \times 10^6 \text{ N/m}$$

$$k_2 = \frac{3{,}000 \times 9.81}{0.2 \times 10^{-2}} = 14.715 \times 10^6 \text{ N/m}$$

Combined stiffness matrix

$$k = \begin{bmatrix} k_1 + k_2 & -k_2 \\ -k_2 & k_2 \end{bmatrix} = \begin{bmatrix} 23544 & -14715 \\ -14715 & 14715 \end{bmatrix} \times 10^3 \text{ N/m}$$

Mass matrix

$$m = \begin{bmatrix} m_1 & 0 \\ 0 & m_2 \end{bmatrix} = \begin{bmatrix} 15 & 0 \\ 0 & 3 \end{bmatrix} \times 10^3 \text{kg}$$

Now, $[k] - \omega^2 [m]$

$$= \begin{bmatrix} 23544 - 15\omega^2 & -14715 \\ -14715 & 14715 - 3\omega^2 \end{bmatrix}$$

For determination of natural frequency, the determinant of the above matrix should be zero.

$$45\omega^4 - 291375\omega^2 + 130007025 = 0$$

Solving the earlier equation for ω

$$\omega_1^2 = 5992.9; \quad \omega_1 = 21.956 \text{ rad/s}$$

and $\omega_2^2 = 482.1; \quad \omega_2 = 77.414$ rad/s

Equation of motion

$$x = A\cos(21.956t - \phi_1) + B\cos(77.414t - \phi_2)$$

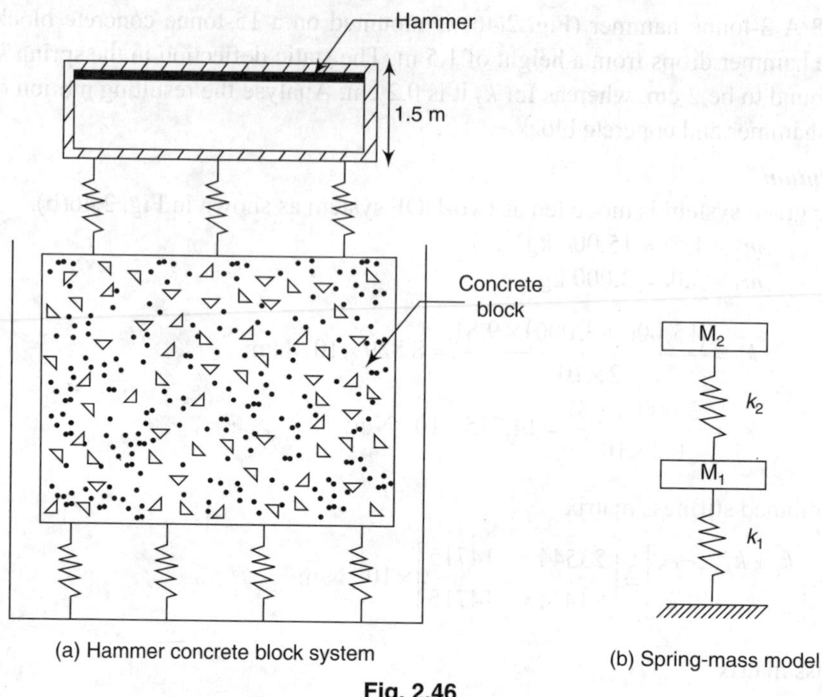

(a) Hammer concrete block system

(b) Spring-mass model

Fig. 2.46

Exercises

2.1 What are the various types of dynamic loads? State some of the characteristics of seismic loads.

2.2 Name the various modelling techniques of structures. Discuss lumped mass approach in detail.

2.3 Give the merits and demerits of the three techniques of modelling structures discussed in the chapter.

2.4 Write short notes on the following:
   (a) d'Alembert's principle
   (b) Inertia force
   (c) Hamilton's principle
   (d) Uncertainties of dynamic analysis

2.5 Derive a mathematical expression defining the dynamic displacements using d'Alembert's principle.

2.6 What are the various methods for solving the differential equation of motion? State the conditions under which these are preferred.

2.7 Define the following:
   (a) Natural frequency
   (b) Damped frequency
   (c) Time period

(d) Critical damping
(e) Damping ratio
(f) Ductility factor

2.8 Discuss the following:
(a) Response factors
(b) Response spectra
(c) Resonance
(d) Restoring force
(e) Damping
(f) Yield-reduction factor

2.9 Discuss the following:
(a) Response spectrum for elastic and inelastic systems
(b) Design response spectrum for elastic and inelastic systems

2.10 How is the response spectrum constructed for inelastic systems?

2.11 Write the equation of motion and determine the effective stiffness using the basic definition of stiffness for the spring–mass systems shown in Fig. 2.47.

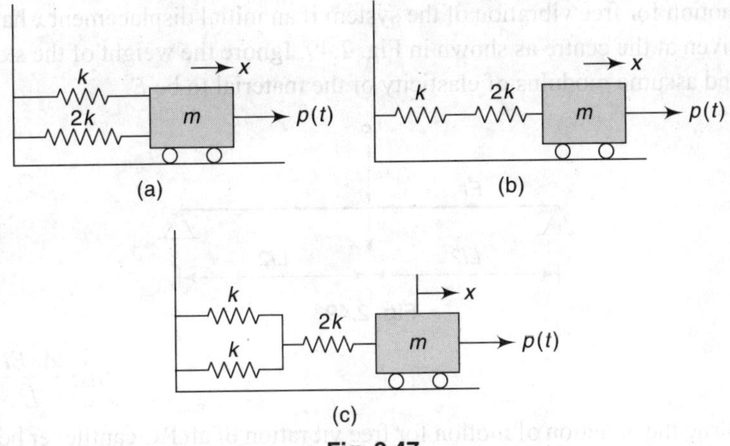

Fig. 2.47

Ans: (a) $m\ddot{x} + 3kx = p(t)$

(b) $m\ddot{x} + \frac{2}{3}kx = p(t)$

(c) $m\ddot{x} + kx = p(t)$

2.12 A rod made of an elastic material with modulus of elasticity E, having cross-sectional area A and length L is fixed on top, carrying a mass m at its lower end, as shown in Fig. 2.48. Derive the equation governing longitudinal motion of the system. Ignore mass of the rod and measure displacement x from the static equilibrium position.

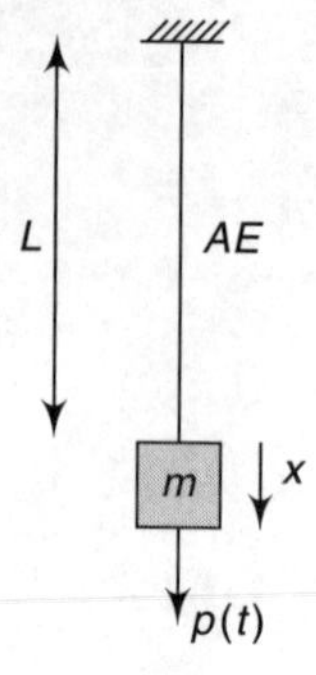

Fig. 2.48

Ans: $m\ddot{x} + kx = p(t)$

$$k = \frac{AE}{L}$$

2.13 A steel rod of uniform cross-sectional area with radius r is simply supported at its ends and carries a point load P at the mid-point. Write the equation of motion for free vibration of the system if an initial displacement x has been given at the centre as shown in Fig. 2.49. Ignore the weight of the steel rod and assume modulus of elasticity of the material to be E.

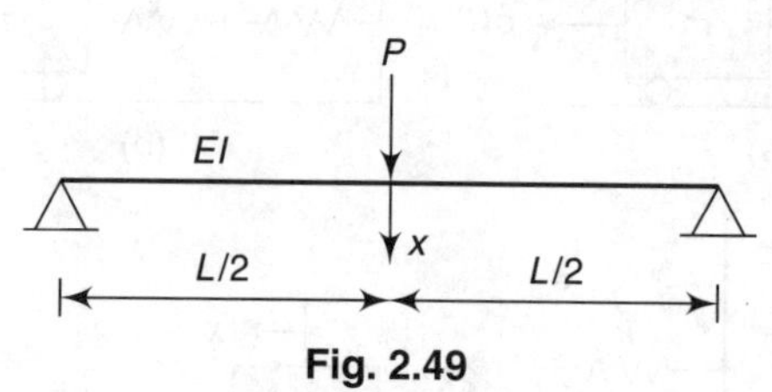

Fig. 2.49

Ans: $\frac{48EI}{L^3} x = 0$

2.14 Write the equation of motion for free vibration of an RC cantilever beam of length 2.00 m, cross-sectional dimensions 200 × 450 mm, carrying a point load of 20 kN at its free end. An initial displacement x is applied downwards at the free end of the beam as shown in Fig. 2.50. Ignore the weight of the beam and assume modulus of elasticity of the material to be 25000 MPa.

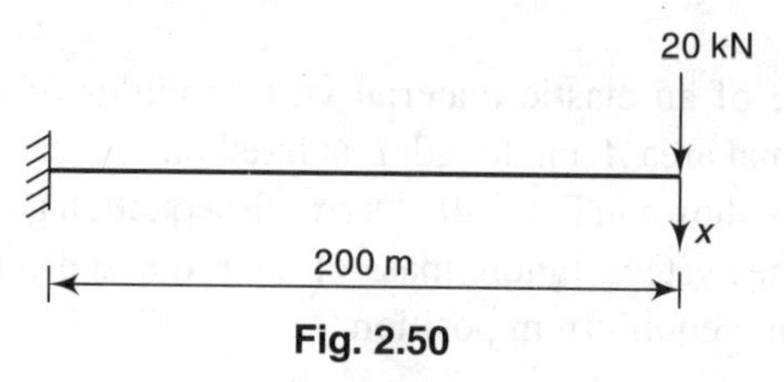

Fig. 2.50

Ans: $kx = 20 \times 10^3$

$k = 14.24 \times 10^3$ N/mm

2.15 A beam with uniform flexural rigidity EI is shown in Fig. 2.51. Write the equation of motion for free vibration of the system, assuming the beam to be massless.

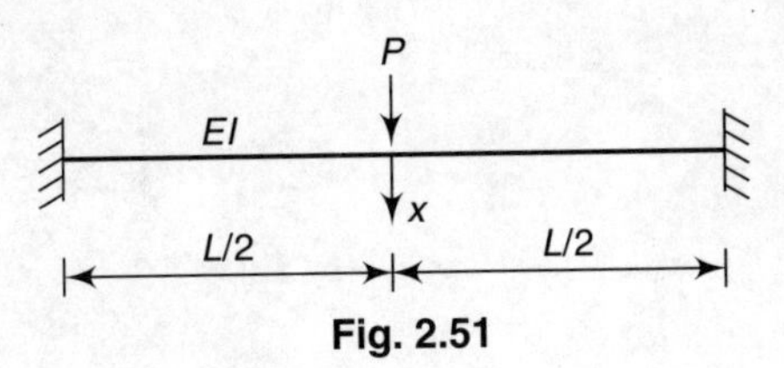

Fig. 2.51

$$Ans: \frac{192EIx}{L^3} = 0$$

2.16 Determine the natural circular frequency and natural period of vibration of a system of weight 5×10^4 N. The lateral stiffness of the system is 3×10^4 N/m. What is the mass of the system?

Ans: 2.43 rad/s; 0.386 cps; 5096.8 kg

2.17 It is observed experimentally that the amplitude of free vibration of a certain structure modelled as an SDOF system decreases from 10 to 4 mm in 10 cycles. Determine damping of the system in terms of percentage critical damping. *Ans:* 1.62%

2.18 An SDOF system is excited by a sinusoidal force. At resonance the amplitude of displacement was measured to be 50 mm. At an exciting frequency of one-tenth of the natural frequency of the system, the displacement amplitude was measured to be 5 mm. Estimate the damping ratio of the system.

Ans: 0.0495

2.19 A diver weighing 70 kg standing at the end of a 1.50-m cantilever diving board, oscillates at a frequency of 2 Hz. Calculate the flexural rigidity EI of the diving board. *Ans:* 1266.4 kg/m^2

2.20 For a system with damping ratio ξ, determine the number of free vibration cycles required to reduce the displacement amplitude to 10% of the initial amplitude, if the initial velocity is zero. *Ans:* $1 + 0.37/\xi$

2.21 Calculate the ratio of successive amplitudes of vibration if the viscous damping ratio is known to be (a) $\xi = 0.01$ and (b) $\xi = 0.25$.

Ans: (a) 1.0648 (b) 4.8066

CHAPTER

3

Dynamics of Soils and Seismic Response

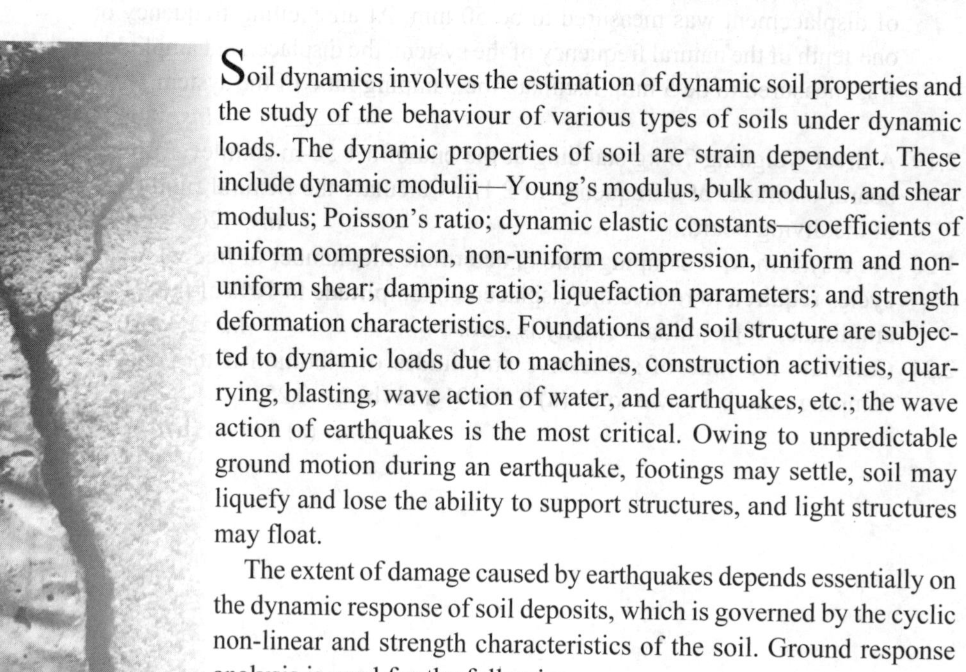

Soil dynamics involves the estimation of dynamic soil properties and the study of the behaviour of various types of soils under dynamic loads. The dynamic properties of soil are strain dependent. These include dynamic modulii—Young's modulus, bulk modulus, and shear modulus; Poisson's ratio; dynamic elastic constants—coefficients of uniform compression, non-uniform compression, uniform and non-uniform shear; damping ratio; liquefaction parameters; and strength deformation characteristics. Foundations and soil structure are subjected to dynamic loads due to machines, construction activities, quarrying, blasting, wave action of water, and earthquakes, etc.; the wave action of earthquakes is the most critical. Owing to unpredictable ground motion during an earthquake, footings may settle, soil may liquefy and lose the ability to support structures, and light structures may float.

The extent of damage caused by earthquakes depends essentially on the dynamic response of soil deposits, which is governed by the cyclic non-linear and strength characteristics of the soil. Ground response analysis is used for the following.

(a) To predict ground surface motion for the development of design response spectra
(b) To evaluate dynamic stresses and strains for the evaluation of liquefaction hazards
(c) To determine the earthquake-induced forces

To ensure the safety of a structure susceptible to earthquakes, it is important to ascertain the subsoil condition and the bearing of the soil–structure relationship on the seismic behaviour of the structures. Owing to the lack of reliable data on the wide range of soil and soil–structure problems encountered in earthquake engineering, codes of practice offer little (and conflicting) guidance on seismic foundation design. Consideration needed for the seismic design of foundations is discussed also.

3.1 Stress Conditions of Soil Element

Soil dynamics problems are generally divided into either small strain amplitude (amplitudes of the order of 0.0001–0.01%, e.g., in the case of machine foundation) or large strain amplitude (amplitudes of the order of 0.01–0.1%, e.g., in the case of strong motion earthquakes, blasts, nuclear explosions) responses. To solve many important problems, particularly those dominated by wave propagation effects, only strains of small amplitude are induced in the soil. However, for other problems, such as those involving the stability of masses of soil, large-amplitude strains are induced. Under the influence of earthquake loading, the soil element may be in any one of the following conditions.

- The initial static stress is large and the additional stresses induced by the earthquake are small [Fig. 3.1(a)]. Here, a symmetrical pulsating stress system is superimposed on an initial sustained stress.
- The sustained stress is small and the pulsating strain is large. The combined effect is shown in Fig. 3.1(b).

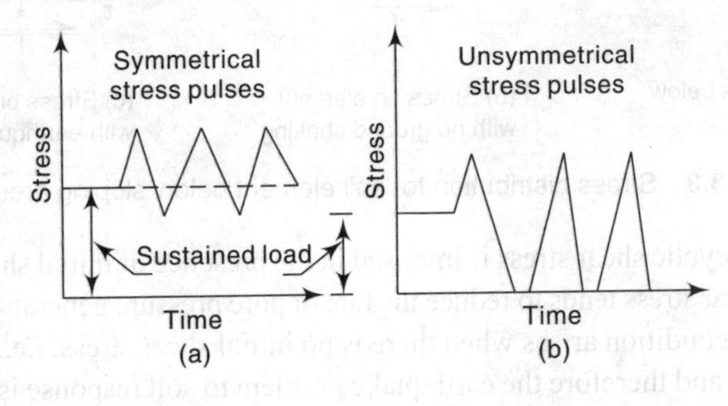

Fig. 3.1 Loading condition on soil elements during earthquakes

Figure 3.2 (a) shows an element of soil located at some depth below a horizontal ground surface, e.g., a soil element below a footing simply resting on soil. When there is no ground motion, the soil element will be subjected to vertical effective stress $\overline{\sigma}$ and horizontal effective stress $K_0\overline{\sigma}$, where K_0 is the coefficient of earth pressure at rest [Fig. 3.2 (b)]. Initially the soil element will not be subjected to any shear stresses. However, due to ground shaking, cyclic shear stress τ_h will be imposed on the soil element as shown in Fig. 3.2 (c).

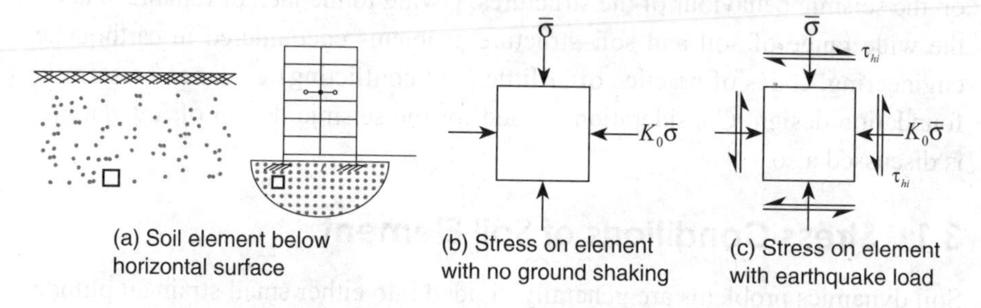

Fig. 3.2 Stress conditions for soil element below horizontal surface

In case of soil element below sloping ground, e.g., a soil element in an embankment, as shown in Fig. 3.3 (a), the soil element will also have initial shear stress τ_{hi} in addition to vertical effective stress $\overline{\sigma}$ as shown in Fig. 3.3 (b), when there is no ground shaking. During earthquakes, additional cyclic shear stress τ_h will also be there (Fig. 3.3 (c)).

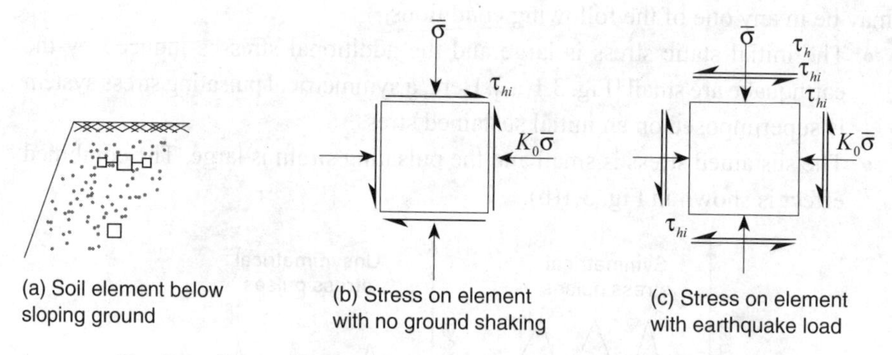

Fig. 3.3 Stress distribution for soil element below sloping ground

When the cyclic shear stress is imposed in the presence of initial shear stresses, the cyclic shear stress tends to reduce the rate of pore pressure generation. Thus the most critical condition arises when there is no initial shear stress, i.e., the ground is horizontal, and therefore the earthquake problem to soil response is considered under essentially level ground.

3.2 Dynamic Behaviour of Soil

Soil behaviour under dynamic loading depends on the strain magnitude, the strain rate, and the number of loading cycles. The strength of certain soils increases under rapid cyclic loading, while saturated sand or sensitive clay may lose strength with vibration. The behaviour of various types of soil under earthquake loading is discussed in the following sections.

3.2.1 Settlement of Dry Sands

Loose sands can get compacted under vibration. With earthquake loading, such compaction causes settlement. It is therefore important to assess the degree of vulnerability of a given loose sand deposit to settlement. Dry sands densify very quickly. Settlement of a dry sand deposit is usually complete by the end of an earthquake. This densification subjected to earthquake loading depends on the density of sand, the amplitude of the cyclic shear strain induced in the sand, and the number of cycles of shear strain applied during earthquake. Though it is difficult to predict settlement with accuracy, it appears that sands with relative density (D_r) below 60% or standard penetration resistance below 15 are susceptible to significant settlement. The amount of compaction achieved by any given earthquake will obviously depend on the magnitude and duration of the vibration as well as on the relative density as shown in Fig. 3.4.

A simple method to predict settlement consists of determining the critical void ratio, e_{cr}, above which a granular soil deposit will compact when vibrated and comparing it with the void ratio e of the stratum. If the void ratio of the stratum

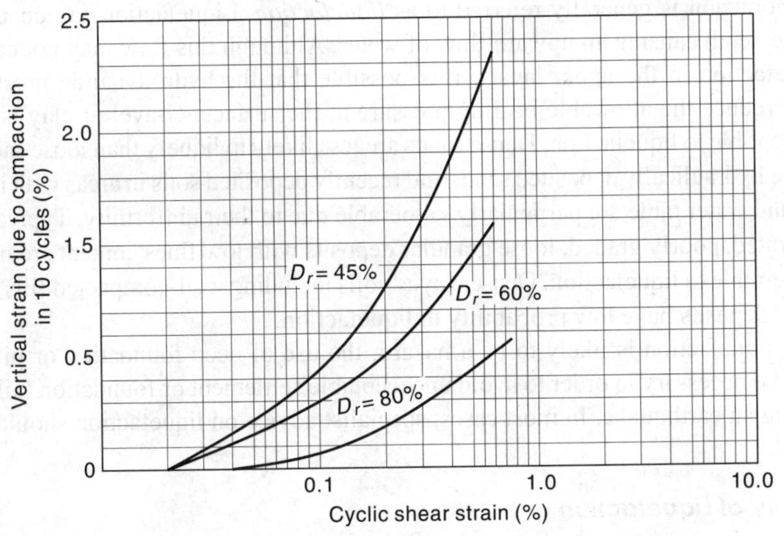

Fig. 3.4 Effect of relative density on settlement

is greater than the critical void ratio, the maximum amount of possible settlement can be determined by

$$\Delta H = \frac{e_{cr} - e}{1 - e} H \tag{3.1}$$

where H is the depth of the stratum. The critical void ratio is given by

$$e_{cr} = e_{min} + (e_{max} - e_{min}) \exp[-0.75a/g] \tag{3.2}$$

where e_{min} is the minimum possible void ratio, e_{max} is the maximum possible void ratio, a is the amplitude of applied acceleration, and g is the acceleration due to gravity.

The limitation of this method is that the effect of factors such as confining pressure and number of cycles is ignored.

3.2.2 Liquefaction of Saturated Cohesionless Soils

One of the causes of destruction during an earthquake is the failure of the ground structure. Ground failure may occur due to the presence of fissures, abnormal or unequal movements, or loss of strength.

Saturated soil deposits take more time to settle than dry sands; settlement of the former can take place only after earthquake-induced pore pressure dissipates. The time required for this settlement to take place depends on the permeability and compressibility of the soil, and on the length of drainage path. Under earthquake loading, some soils may compact, increasing the pore water pressure and causing a loss in shear strength, and behave like liquid mud (a viscous liquid). This phenomenon is generally referred to as *liquefaction*. Liquefaction can occur at some depth causing an upward flow of water. Although this flow may not cause liquefaction in the upper layers, it is possible that the hydrodynamic pressure may reduce the allowable bearing pressure at the surface. Gravel or clay is not susceptible to liquefaction. Dense sands are less likely to liquefy than loose sands, while hydraulically deposited sands and recently deposited soils in areas with high ground water table are particularly vulnerable due to their uniformity. Typically, saturated, poorly graded, loose granular deposits with low fines content are most susceptible to liquefaction. Dense clayey soils including well-compacted fills and older deposits have low probability to liquefaction.

If liquefaction is likely to be a hazard, the use of deep foundation or piling may be necessary in order to avoid unacceptable settlement or foundation failure during an earthquake. In most cases, specialist advice on liquefaction should be sought.

Theory of Liquefaction

The strength of sand is entirely due to internal friction. In a saturated state, the strength may be expressed as

$$S = (\sigma_n - u)\tan\phi \tag{3.3}$$

where S is the shear strength, σ_n is the normal stress on any plane at depth Z (Fig. 3.5), $u = \gamma_w Z$, u is the pore pressure, γ_w is the unit weight of water, and ϕ is the angle of internal friction.

Considering stress on a horizontal plane at depth Z

$$\sigma_n = \gamma_{sat} Z \tag{3.4}$$

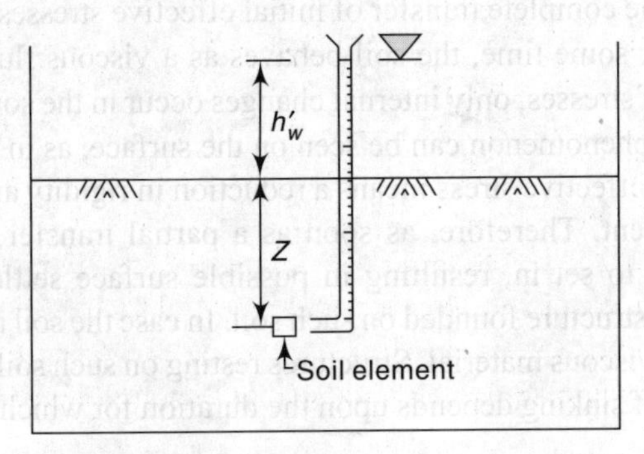

Fig. 3.5 Increase of pore-water pressure

Equation (3.3) can now be expressed as

$$S = \bar{\sigma}_n \tan\phi \qquad (\bar{\sigma}_n = \sigma_n - u) \tag{3.5a}$$

or

$$S = \gamma_b Z \tan\phi \tag{3.5b}$$

where $\bar{\sigma}_n$ is the effective normal stress, γ_{sat} is the unit weight of saturated soil, and γ_b is the submerged unit weight of the soil.

If there is an increase in the pore pressure $+\Delta u = \gamma_w h'_w$ due to the vibration of the ground, the strength may be expressed as

$$S = (\gamma_b Z - \Delta u)\tan\phi$$

$$= (\gamma_b Z - \gamma_w h'_w)\tan\phi$$

$$= \bar{\sigma}_{dyn}\tan\phi \tag{3.6}$$

where $\bar{\sigma}_{dyn}$ is the effective dynamic stress and h'_w is the height of water rise in stand pipe.

Therefore, with the development of additional positive pore pressure, the strength of the sand reduces. For a complete loss of soil strength

$$\gamma_b Z = \gamma_w h'_w$$

or

$$\frac{h'_w}{z} = \frac{\gamma_b}{\gamma_w} = \frac{S_s - 1}{1 + e} = i_{cr} \tag{3.7}$$

where S_s is the specific gravity of soil solids, e is the void ratio, and i_{cr} is the critical hydraulic gradient.

Loss of soil strength occurs due to the transfer of intergranular stress from the grains to the pore water. Thus, if this transfer is complete, there is a complete loss of strength. However, if stress is only partially transferred from the grains to the pore water, only a partial loss of strength occurs. Since the stress condition is cyclic, a momentary transfer of all the initial effective confining pressure to the pore water may not be of great engineering significance if the subsequent behaviour of sand is satisfactory with respect to load-carrying capacity considerations, particularly in dense sands. If the complete transfer of initial effective stresses to the pore water is maintained for some time, the soil behaves as a viscous fluid. In the case of partial transfer of stresses, only internal changes occur in the soil and no apparent evidence of this phenomenon can be seen on the surface, as in the former case.

A decrease in effective stress means a reduction in rigidity and, hence, greater strain or settlement. Therefore, as soon as a partial transfer of stress occurs, settlement tends to set in, resulting in possible surface settlement as well as settlement of the structure founded on such soil. In case the soil remains liquefied, it behaves like a viscous material. Structures resting on such soil will start sinking into it. The rate of sinking depends upon the duration for which the sand remains liquefied.

The structure may settle if the stress transfer from the soil grains to the pore water is partial and will sink if the soil is completely liquefied. Thus, as soon as liquefaction occurs, the process of consolidation starts, followed by surface settlement, which results in closer packing of the sand particles. During this process, the pore pressure starts dissipating and, in the field, water flows only upwards. The upward seepage forces as a result of the flow may further reduce the effective stresses. This may cause liquefaction in layers that were not liquefied initially.

During and following earthquake, seismically induced excess pore pressures are dissipated by the upward flow of pore water. Consequently, the soil particles are subjected to upward acting forces. If the hydraulic gradient driving the flow reaches a critical value, the vertical effective stress drops to zero and the soil is in *quick condition*. In such a condition the soil particles are carried to the surface. Through the localized cracks and channels formed, sand particles may eject on the ground surface; the phenomenon is known as *sand boils*.

Research has shown that because of the large number of factors responsible for the collapse of a sand structure, the criteria considered should not be limited just to the critical void ratio or the density of sand. The critical values of the intensity of dynamic disturbance, stress condition of the soil, and weight of surcharge and hydraulic gradient of the water passing through the soil structure (i.e., geometric arrangement of soil particles) should also be considered. Even if the factors listed above are considered, no ready index for the liquefaction of sands can be evolved. However, liquefaction is likely to occur under the following soil conditions.

(a) The sandy layer is within 15–20 m of the ground level and is not subjected to high overburden pressure.

(b) The layer consists of uniform medium-size particles (Fig. 3.6).
(c) The layer is saturated, i.e., it is below the ground water level.
(d) The standard penetration test value is below a certain level (Fig. 3.7).

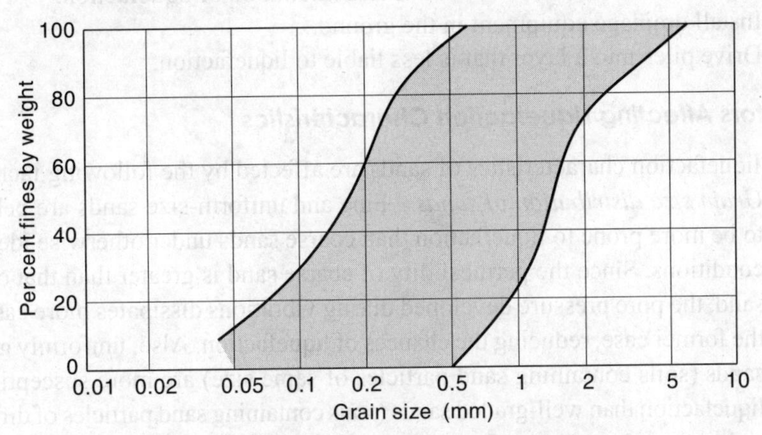

Fig. 3.6 Critical zone for grain size distribution curves (Ohsaki 1970)

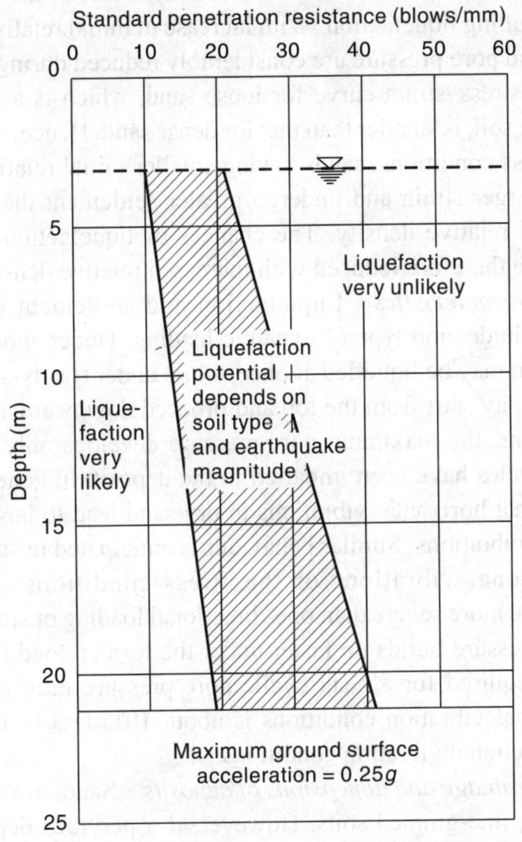

Fig. 3.7 Liquefaction potential evaluation chart (Seed et al. 1971)

To reduce the possibility of liquefaction, the following measures can be considered.

1. Increase the relative density of the sands by compaction.
2. Replace the soil with soil that has less likelihood of liquefaction.
3. Install drainage equipment in the ground.
4. Drive piles into a layer that is less liable to liquefaction.

Factors Affecting Liquefaction Characteristics

The liquefaction characteristics of sands are affected by the following factors.

(a) *Grain size distribution of sands* Fine and uniform-size sands are believed to be more prone to liquefaction than coarse sands under otherwise identical conditions. Since the permeability of coarse sand is greater than that of fine sand, the pore pressure developed during vibrations dissipates more easily in the former case, reducing the chances of liquefaction. Also, uniformly graded sands (soils containing sand particles of same size) are more susceptible to liquefaction than well-graded sands (soils containing sand particles of different sizes in good proportion).

(b) *Initial relative density* Initial relative density is one of the most important factors controlling liquefaction. With increase in initial relative density, both settlement and pore pressure are considerably reduced during vibration. The slope of the stress–strain curve for loose sand, which is a measure of the rigidity of the soil, is smaller than that for dense sand. Hence, under otherwise identical stress conditions, sands having smaller initial relative density will experience larger strain and undergo greater settlement than those having higher initial relative density. The chances of liquefaction and excessive settlement are therefore reduced with increased relative density.

(c) *Vibration characteristics* Liquefaction and settlement depend on the nature, magnitude, and type of dynamic loading. Under shock loading, the whole stratum may be liquefied at once, while under steady-state vibrations, liquefaction may start from the top and proceed downwards. Under steady-state vibrations, the maximum pore pressure develops only after a certain number of cycles have been imparted to the deposit. In general, it has also been found that horizontal vibrations in dry sand lead to larger settlements than vertical vibrations. Similar behaviour is anticipated in saturated sands. Multidirectional vibrations or the stress conditions created by an earthquake are more severe than one-directional loading or stress conditions; pore water pressure builds up faster under the former conditions. Also, the stress ratio required for a peak cyclic pore pressure ratio of 100% under multidirectional vibration conditions is about 10% less than that required under unidirectional vibration conditions.

(d) *Location of drainage and dimensions of deposits* Sands are generally more pervious than fine-grained soils. However, if a pervious deposit has large dimensions, then the drainage path, i.e., the drainage length of water from large soil deposits, increases and under rapid loading during an earthquake the deposit may behave as if it were undrained. Therefore, the chances of liquefaction increase in such deposits. Gravel drains may be introduced to stabilize a potentially liquefiable sand deposit. The drains are considered fully effective if the material used to construct them is about 200 times more permeable than the soil that holds them. The drainage path is reduced by the introduction of drains within the large deposits.

(e) *Magnitude and nature of superimposed loads* The initial effective stress on a sample is isotropic. To transfer large initial effective stress to the pore water, either the intensity of vibrations or the number of particular stress cycles must be large. Hence, large initial effective stress reduces the possibility of liquefaction.

(f) *Method of soil formation* Sands are generally not known to display a characteristic structure as do fine-grained soils such as clays. However, recent investigations have demonstrated that the liquefaction characteristics of saturated sand under cyclic loading are significantly influenced by the method of sample preparation and the soil structure.

(g) *Period under sustained load* The age of a sand deposit may influence its liquefaction characteristics. A study of the liquefaction of undisturbed sand and its freshly prepared sample indicates that the liquefaction resistance may increase by 75%. This strength increase might be due to some form of cementation or welding, which may occur at the contact points between sand particles, and might be associated with secondary compression of soil.

(h) ***Trapped air*** A part of the developed pore pressure might get dissipated due to the compression of air trapped in the water. Hence, trapped air helps reduce the possibility of liquefaction.

(i) *Previous strain history* Sands may have been subjected to certain strains during previous earthquakes. This may influence the liquefaction characteristics.

Effects of Soil Liquefaction on Structures

Liquefaction has pronounced destructive effects on structures and a proper assessment must be made.

1. For moderate and steep slopes, the completely liquefied soil layer may flow (ride) on a layer of liquefied soil. In case of gentle slopes, the superficial block of soils laterally displace because of liquefaction of a subsurface layer known as *lateral spread.*
2. For flat and almost flat grounds the liquefied soil at depth may disconnect the overlying soils from the underlying ground. This may cause the upper soil to oscillate back and forth in the form of ground waves and consequently end up in ground fissures and fracture of rigid pavements, pipe lines, etc.
3. The increased pore water pressure during earthquake causes the soil to lose its strength and hence the bearing capacity.

4. Since in the process of liquefaction, the pore water pressure dissipates and the soil gets densified, after the liquefaction is over, the settlement of structures may take place.
5. The liquefied soil behaves as a heavy fluid with no internal friction consequently increasing the lateral pressure on retaining walls.

3.3 Dynamic Design Parameters of Soils

The basic parametric data used for the dynamic response analyses of soil or soil–structure systems are those of shear modulus, damping, Poisson's ratio, and density; stiffness and damping are the most important parameters. These are discussed in the following subsections.

3.3.1 Shear Modulus

Soil stiffness is influenced by cyclic strain amplitude, number of loading cycles, void ratio, etc. The secant shear modulus of soil element varies with cyclic shear strain amplitude. For small strains, the shear modulus of a soil can be taken as the mean slope of the stress–strain curve. At large strains, the stress–strain curve becomes markedly non-linear, so that the shear modulus is far from constant but is dependent on the magnitude of the shear strain (Fig. 3.8). For both sand and clay, the shear modulus reduces as the strain level increases. The shear modulus measured from the wave velocity corresponds to 10^{-5} to 10^{-4} of strain.

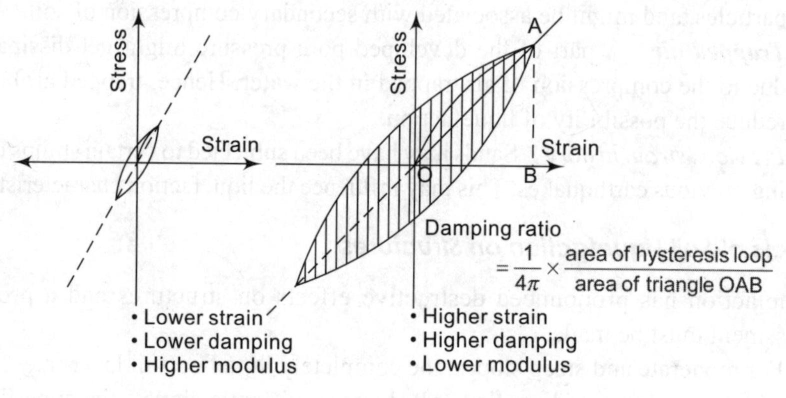

Fig. 3.8 Effect of shear strain on damping and shear modulus of soils

Field tests concentrate on finding the shear wave velocity v_S and calculating the shear modulus G from the relationship as follows.

$$G = \rho v_S^2 \tag{3.8}$$

where ρ is the mass density of the soil. Laboratory methods generally measure G more directly from stress–strain tests. It is clear from Fig. 3.9 that the level of strain, i.e., low or high, at which G is measured must be known. Figure 3.8 shows the average relationship of shear modulus and strain for clay and sand. Shear

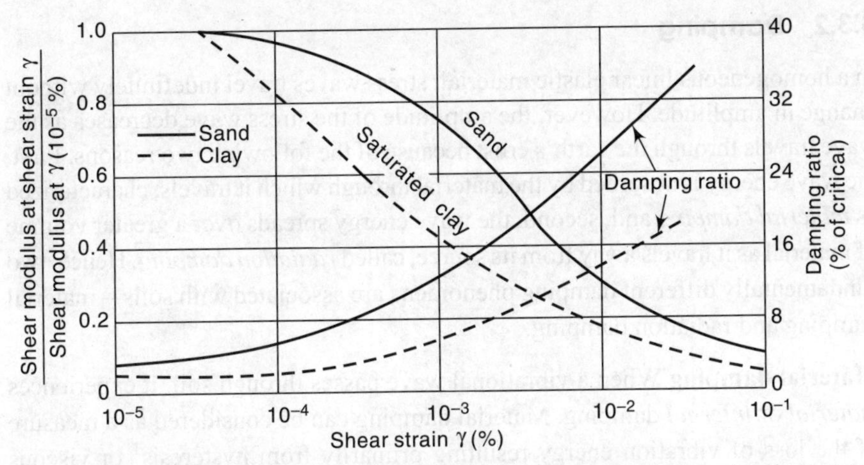

Fig. 3.9 Average relationship of shear modulus to shear strain for sand and saturated clays

strains developed during earthquakes may increase by about 10^{-3}% during small earthquakes to 10^{-1}% during large-scale vibrations, and the maximum strain in each cycle will be different. For earthquake-resistant design purposes, a value of two-thirds the G measured at the maximum strain developed may be used. Alternatively, an appropriate value of G can be calculated from the following relationship

$$G = \frac{E}{2(1+\mu)} \tag{3.9}$$

where E is the Young's modulus and μ is the Poisson's ratio. In the absence of specific data, the low-strain values of E and μ may be used (Tables 3.1 and 3.2).

Table 3.1 Typical modulus of elasticity values for soils and rocks

Soil type	E (MN/m^2)	Soil type	E (MN/m^2)
Soft clay	Up to 15	Dense sand and gravel	90–200
Firm, stiff clay	10–50		
Very stiff, hard clay	25–200	Sandstone	Up to 50,000
Silty sand	7–10	Chalk	5000–20,000
Loose sand	15–50	Limestone	25,000–100,000
Dense sand	50–120	Basalt	15,000–100,000

Table 3.2 Typical values of Poisson's ratio for soils

Soil type	Poisson's ratio (μ)
Sand	0.35
Saturated clay	0.50
Most other soils	0.40

3.3.2 Damping

In a homogeneous linear elastic material, stress waves travel indefinitely without change in amplitude. However, the amplitude of the stress wave decreases as the wave travels through the earth's crust because of the following two reasons. First, the wave energy is absorbed by the material through which it travels, characterized as *material damping* and, second, the wave energy spreads over a greater volume of material as it travels away from its source, called *radiation damping*. Hence, two fundamentally different damping phenomena are associated with soils—material damping and radiation damping.

Material damping When a vibrational wave passes through soil, it experiences *material* or *internal* damping. Material damping can be considered as a measure of the loss of vibration energy resulting primarily from hysteresis[1] or viscous mechanism in the soil. Material damping is therefore because of the inelastic behaviour of the soil supporting the foundation. Damping is conveniently expressed as a fraction of critical damping, and hence referred to as a damping ratio. A physical definition of damping ratio is shown graphically in Fig. 3.7. Since in most cases the strain level experienced during earthquakes ranges from 10^{-5} to 10^{-1}, damping as high as 10% and 16% for clay and sand, respectively, can be expected during an earthquake.

Radiation damping Radiation damping is also known as *geometric damping* or *geometric attenuation*. It occurs as the dynamic forces in the structure cause foundation to deform the soil producing shear waves that travel away from the foundation and reduction of wave amplitude by spreading of energy over greater volume of material. This is a purely geometrical effect. There is no convenient method to measure radiation damping of the soil in the field. Typically, radiation damping is much greater than material damping.

3.4 Soil–Structure Interaction

The process, wherein the response of the soil influences the motion of the structure and the response of the structure influences the motion of the soil is referred to as *soil–structure interaction*. A detailed analytical treatment of the soil–structure interaction problem is beyond the scope of the book. However, the basic soil–structure interaction effect and the dynamic response on structures are introduced. Further, it may be taken note of that the soil–structure interaction effect may be very little on dynamic response of many structures, whereas for some other structures it may be significant. A study of recent earthquakes has indicated that

[1] Hysteresis is a phenomenon wherein the energy loss per cycle is related to the internal friction under repeated loading and unloading. Hysteretic action has the effect of increasing the overall damping of the system and reducing the deformation of the structure.

understanding the relationship between the period of vibration of structures and the period of vibration of the supporting soil is profoundly important for determining the seismic response of the structure. The pattern of structural damage is directly related to the depth of the soil alluvium overlying the bedrock, which in turn is directly related to the period of vibration of the soil. Considering shear waves travelling vertically through a soil layer of depth H, the period of horizontal vibration of the soil is given by

$$T_H = \frac{4H}{(2n-1)v_S} \tag{3.10}$$

where n is an integer and v_s is the velocity of the shear wave.

Therefore, to evaluate the seismic response of a structure at a given site the dynamic properties of the combined soil–structure system must be examined. The nature of the subsoil may influence the response of the structure in the following three ways.

1. The phenomenon of soil amplification may occur, in which the seismic excitation at the bedrock is modified during transmission through the overlying soils to the foundation. This may cause *attenuation*.
2. The fixed-base dynamic properties of the structure may be significantly modified by the presence of soils overlying the bedrock. This will include changes in the mode shape and period of vibration.
3. A significant part of the vibration of a flexibly supported structure may be dissipated by material damping and radiation damping in the supporting medium.

Points 2 and 3 are investigated under soil–structure interaction, which may be defined as the interdependent response relationship between a structure and the supporting soil. The behaviour of the structure depends partly on the nature of the supporting soil, and similarly the behaviour of the stratum is modified by the presence of the structure. It follows that *soil amplification* (point 1) will also be influenced by the presence of the structure, as soil–structure interaction effects a difference between the motion at the base of the structure and the free-field motion which would have occurred at the same point in the absence of the structure. In practice, however, this factor is seldom taken into account while determining the soil amplification. The general free-field motion that is applied to the soil–structure model is discussed in the following section.

As discussed earlier, for a structure founded on rock, the large stiffness of latter constraints the rock motion close to free-field motion and therefore the structure is considered to be fixed-base structure. The same is assumed for structures on hard and firm soils. The assumption is well justified for structures flexible compared to the soil strata over which the foundation rests. However, for structures stiffer than the soil strata, the structural response can get influenced significantly by the flexibility of the soil. The effects of soil–structure interaction are two fold—the kinematic interaction, which is due to change of material at soil–structure interface, and the inertial effect, which is due to vibration of structure inducing deformation of the supporting soil. These affects can be explained as follows.

Soil is considered to be elastic medium through which seismic waves travel. These waves will encounter a discontinuity at the soil–foundation interface. Because of change of material properties scattering, diffraction, reflection, and refraction of seismic waves will take place modifying the nature of ground motion. Since foundation is rigid compared to the adjacent soil mass, the deformation of the latter is constrained at the soil–foundation interface. This further leads to slippage across the soil–foundation interface, which is a non-linear phenomenon. This is known as *kinematic interaction effect*. Its analysis accounts for geometry and stiffness properties of foundation. Further, because of earthquake, structure vibrates inducing base shears and moments. Consequently, the supporting soil is deformed, gets stressed and modifies the response. Since this effect is due to dynamic response of structural system, it is known as *inertial interaction effect*.

The kinematic interaction can induce rocking and torsion modes of deformation which are not present in free-field motion, whereas inertial interaction produces foundation movements that would not occur in a fixed-base structure. In general, soil–structure interaction will cause the natural frequency of a soil–structure system to be lower than the natural frequency of structure itself. Also due to the radiation damping, the soil–structure system exhibits greater damping than that of structure alone.

Note: Owing to the difficulties involved in making dynamic analytical models of the soil system, it has been a common practice to ignore soil–structure interaction effects, simply by treating structures as if they were rigidly based, regardless of the soil condition.

3.5 Dynamic Analysis of Soil–Structure Systems

A comprehensive dynamic analysis of the soil–structure system is the most demanding analytical task in earthquake engineering. The cost, complexity, and validity of such exercises are major considerations. There are two main problems to be overcome. First, the large computational effort generally required for foundation analysis makes thc appropriate choice of the foundation model very important. Second, there are great uncertainties in defining a ground motion design that represents not only the nature of earthquake vibration appropriate for a site but also a suitable level of risk.

3.5.1 Soil Models

In an ideal model, the earthquake motion should be applied to the bedrock rather than to the entire soil–structure system. This is not a very realistic method at the present time, because much less is known about bedrock motion than surface motion and there is a great scatter in the possible results for the soil amplification effect. At present the most realistic methods of analysis seem to be those that apply *free-field motion* to the base of the structure. This may be achieved easily using a simple spring at the base of the structure or a substructuring technique in which the foundation dynamic characteristics are predetermined and the soil and structure responses are superimposed on each other.

Specific models used for analyses normally consist of structural and soil components. Appropriate modelling depends upon the physical properties and configuration of the structural components and their interaction with the soil components. Since soil is not perfectly rigid, the ground near a building deforms in response to the vibrations of the building; that is, the building and the ground interact with each other under disturbances caused by earthquakes. A realistic dynamic model of soil requires a representation of soil stiffness and material and radiation damping, allowing for strain dependence (non-linearity) and variation of soil properties in three dimensions. The non-linear properties of the foundation element should reflect the possibility of soil yielding, sliding or uplift, as well as inelastic structural behaviour where appropriate. While various analytical techniques exist for handling the different aspects of soil behaviour, they all have drawbacks—varying combinations of expensiveness and inaccuracy. Many soil models have been proposed; some are relatively simple while others need rigorous formulation. There are four methods of modelling soil.

1. Spring model—Equivalent static springs and viscous damping at base level only
2. Lumped mass model—Shear beam analogy using continua or lumped masses and springs distributed vertically through the soil profile
3. Semi-infinite model—Elastic or viscoelastic half-space
4. Finite-element model

Spring Model

Figure 3.10 (a) shows a foundation element model representing a spread footing or a pile group with cap. Vertical force F_y, lateral force F_x and moment M_z act upon the element causing it to translate in x and y directions and rotate about z-axis. An uncoupled, single node of the element is shown in Fig. 3.10 (b). Here, the soil components k_x, k_y, and k_r represent stiffness and strength in each of the independent degrees of freedom. The single node representation is appropriate when the structural components are rigid and do not interact significantly with soil. However, when structural components are relatively flexible and interact significantly with soil material, the Winkler model [Fig. 3.10 (c)] is preferred.

The simplest method of modelling soil is to use springs at the base level to represent the horizontal, rocking, vertical, and torsional stiffness of the soil (Fig. 3.11). Probably this is the simplest model for analysing the rocking motion of

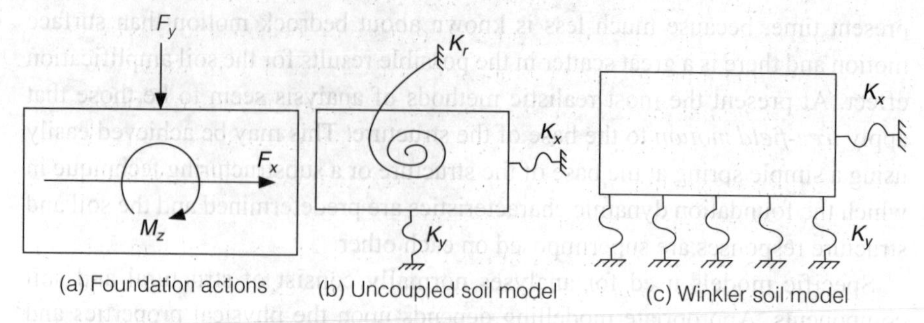

Fig. 3.10 General foundation models

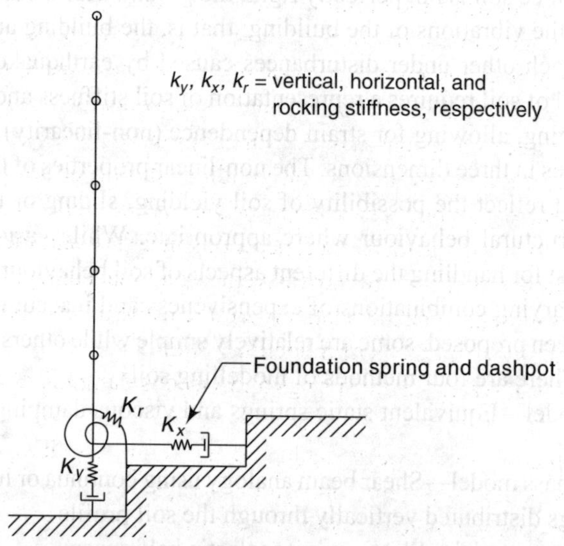

Fig. 3.11 Soil–structure analytical model representing soil flexibility with a simple spring at the base of the structure

a building due to ground disturbance. In this model, the building is assumed to be supported by springs, which represent the characteristics of the ground as shown in Fig. 3.11. The spring, which resists the rotation of the building, is identified as the *rocking spring*. A dashpot can be included if some viscous damping is expected in the ground.

The spring stiffness or constant can be estimated experimentally as well as theoretically. For experimental estimation, the ground is excited using a vibration generator. In the theoretical approach, it is assumed that the ground is a semi-infinite body and that a dynamic force is applied to the foundation. The spring stiffness depends on the shear modulus, which in turn varies with the level of shear strain. Hence, for the linear elastic calculation, the spring stiffness should be calculated corresponding to a value of shear strain less than the maximum expected value. For instance, if the spring stiffness at low strain is k_o, then a value of k equal to $0.67k_0$ may be used in the analysis. Alternatively, a series of comparative analyses may be done using a range of values of k, particularly if in situ tests have not been conducted; in this case it may be appropriate to select k from the following ranges.

For translation: $0.5k_0 \leq k \leq k_0$

For rocking: $0.33k_0 \leq k \leq k_0$

The spring model has the following limitations:

1. It is difficult to accurately represent the effects of material damping and radiation damping in the foundation with the spring model shown in Fig. 3.11. The total amount of viscous damping for the foundation is likely to increase considerably than that for the superstructure. A conservative compromise between the structural and soil damping values will generally be necessary.
2. The software for dynamic analyses uses equal damping in all modes of vibration. But the damping in soil in different modes of vibration varies considerably. This requires some intermediate value of damping to be used to arrive at a realistic result.

Lumped Mass Model

In the lumped mass model, the ground is represented by vertically linked lumped masses as shown in Fig. 3.12. Each lumped mass, with its spring constant and damping coefficient, represents one ground layer. These properties are difficult to determine, however, and the model does not take energy dissipation into account. Furthermore, the assumption that the surrounding soil is perfectly rigid is questionable. Non-linearity may be assumed for using iterative linear analysis.

Semi-infinite Model

In the semi-infinite model, the ground is assumed to be a uniform elastic or viscoelastic semi-infinite body. Stiffness and damping are treated as frequency dependent. Radiation damping can be included and the damping effect of the soil can also be incorporated into the analysis by assuming that the ground is viscoelastic.

Finite-element Model

The use of finite elements for modelling the foundation of a soil–structure system is the most comprehensive method available. The ground is discretized into finite elements (Fig. 3.13). This allows the incorporation of radiation damping and three-dimensionality. The non-uniformity of soil properties (vertical and horizontal variation of soil stiffness) is modelled by assigning different material properties to each finite element. Inelastic soil behaviour can be simulated by

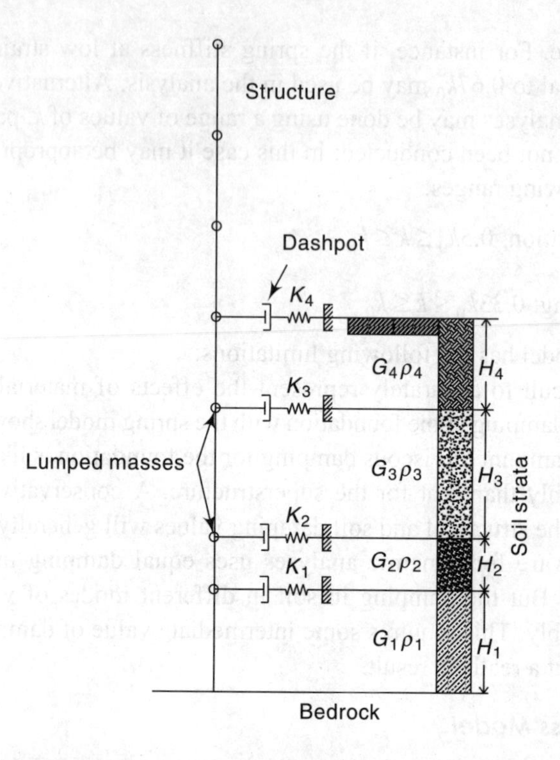

Fig. 3.12 Analytical model of soil structure representing the vertical soil profile by the lumped parameter system of springs and dashpots

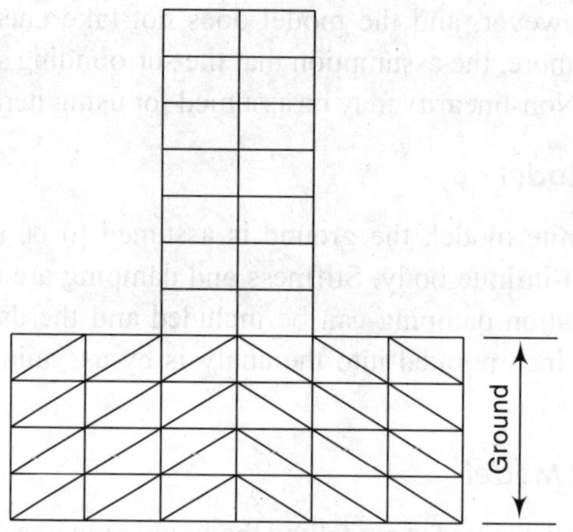

Fig. 3.13 Finite element model of soil and structure

means of non-linear finite-element computation. The embedment of footings can also be readily dealt with. One disadvantage of this model is the high cost of analysis. Although a full three-dimensional model is generally too expensive, three dimensions should be simulated.

Since finite-element analysis is expensive, the discretization should be carefully selected. If the ground consists of layers with uniform material properties in the horizontal plane, one-dimensional discretization is suitable. If the soil is confined in a long, narrow valley, a two-dimensional model is useful. If the stratum or the contact area between the building and the ground is symmetrical about the vertical axis of revolution, axisymmetrical analysis is useful. In any case, a rigid boundary confining the energy dissipation of the ground must be defined in the discretization.

Three methods are available for simulating radiation of energy through boundaries in the finite-element model.

(i) ***Elementary boundaries*** These boundaries do not absorb energy and rely on the distance to the boundary to minimize the effect of reflected waves.

(ii) ***Viscous boundaries*** These boundaries attempt to absorb the radiating waves, modelling the far field by a series of dashpots and springs.

(iii) ***Consistent boundaries*** These boundaries are the best absorptive boundaries available at present, reproducing the far field in a way consistent with the finite-element expansion used to model the core region.

Non-linearity of soil behaviour can be modelled with non-linear finite elements; however, the necessary time-domain analysis is very expensive. Alternatively, non-linearity could theoretically be simulated in repetitive linear model analyses with the adjustment of the modulus and damping in each cycle as a function of strain level. In frequency-domain solutions (e.g., when using consistent boundaries), non-linearity can be approximately simulated using the iterative approach again. Material damping may also be accounted for using a viscoelastic finite-element model.

3.5.2 Methods of Analysis

When a structure is provided with a rigid foundation (assumed to be massless) supported on elastic soil deposit, the base of the structure will be treated as fixed. The material supporting the fixed-base system should depend only on the mass m and stiffness k of the structure. The natural frequency ω_{sy} and the hysteretic damping ratio ξ_{sy} of such a system will be given by

$$\omega_{sy} = \sqrt{\frac{k}{m}}$$

$$\xi_{sy} = \frac{c\omega_{sy}}{2k}$$

However, in general, these conditions are not met and the soil beneath the foundation may be compliant; the foundation can translate and rotate as well.

Thus, the soil–foundation system can be idealized by the translational spring, rotational spring, and dashpot as shown in Fig. 3.11. The dashpot will represent both material and radiation damping. The natural frequency ω_e and the hysteretic damping ratio ξ_e of the resultant model will be

$$\omega_e = \frac{\omega_{sy}}{\sqrt{1 + \dfrac{k}{k_x} + \dfrac{kh^2}{k_r}}}$$

$$\xi_e = \frac{\omega_e^2}{\omega_{sy}^2}\xi_{sy} + \left(1 - \frac{\omega_e^2}{\omega_{sy}^2}\right)\xi_g + \frac{\omega_e^2}{\omega_x^2}\xi_x + \frac{\omega_e^2}{\omega_r^2}\xi_r$$

where ω_x, ω_r are translational and rocking natural frequencies, respectively.

ξ_g, ξ_x, ξ_r are material, translational, and rotational damping ratios, respectively.

The above equations for response analysis are not suitable for detailed dynamic analysis of important structures. Methods of analysis in use for practical soil–structure interaction problems—the direct method and the multistep method—are as follows. Formulation of a detailed dynamic analysis procedure is beyond the scope of the book.

Direct method

This method performs finite element analysis over soil–structure foundation system modelled as one unit. The free field input motion is specified along the base and sides of the model and resulting response, for a finite element model, is computed from the equation of motion

$$[m]\{\ddot{x}\} + [k^*]\{x\} = -[m]\{\ddot{x}_{ff}(t)\} \tag{3.11}$$

where $[m]$ is the global mass matrix;

$\{x\}$ is the nodal points displacement vector;

$\{\ddot{x}_{ff}(t)\}$ is the specified free-field accelerations at the boundary nodal points.

Note: The free field motion assumed to be constant at the base of structure is not true for large structures such as bridges and arch dams.

Multistep method

This method uses the principle of superposition. However, the assumption is valid for linear analysis only and thus limits the scope. The process consists of first performing kinematic analysis resulting in foundation input motion. This input motion is then applied as inertial loads on the structure.

The deformation due to kinematic interaction is computed with the assumptions that the structure is massless and, the foundation is also massless but has stiffness by

$$[m_{so}]\{\ddot{x}_{ki}\}+[k^*]\{x_{ki}\}=-[m_{so}]\ddot{x}_b(t) \tag{3.12}$$

where $[\mathbf{m}_{so}]$ is the mass matrix of soil;

$\{\ddot{x}_{ki}\}$ is the foundation input motion for kinematic interaction;

$[k^*]$ is the complex stiffness matrix; and

$\ddot{x}_b(t)$ is the time history of base acceleration.

It may be noted that the structure and foundations do have mass and respond dynamically producing movement of foundation.

Notes: 1. The complex stiffness, $k^* = k(1-2i\xi)$.

2. By using the complex stiffness matrix, a viscously damped system can be represented as an undamped system. In actual, the soil system is not a viscously damped system.

The deformation due to inertial interaction are computed from the equation of motion,

$$[m]\{\ddot{x}_{ii}\}+[k^*]\{x_{ii}\}=-[m_{st}]\{\ddot{x}_{ki}(t)+\ddot{x}_b(t)\} \tag{3.13}$$

where $[m_{st}]$ is mass matrix of structure assuming soil as massless

$\{\ddot{x}_{ii}\}$ is displacement vector for internal interaction.

The right hand side of Eqn (3.13) represents inertial loading on the structure-foundation system. The two equations of motion, Eqns (3.11) and (3.12), when added give the overall response which is same as that given by Eqn (3.11).

3.6 Seismic Considerations for Foundations

Foundation systems consist typically of either shallow or deep elements or sometimes a combination of both. Shallow foundations are normally isolated or continuous spread footings or rafts. Resistance to uplift of these foundations is restricted to superimposed loads. Most deep foundation elements are piles or piers. These rely on friction and on end bearing vertical support. These provide resistance to uplift provided they are adequately tied to the structure. Compared with shallow foundations deep foundations are relatively stiff. Combined system of shallow and deep elements is not encouraged because of the inherent differences in strength and stiffness, particularly in the inelastic range.

The forces acting on a foundation are vertical force, overturning moment, and horizontal base shear. Since seismic forces are usually applied for a short duration, it is assumed that the soil remains elastic under these forces. While designing foundations that can withstand seismic forces, the deformation of the adjacent soil and the effect of seismic forces acting upon the superstructure are considered. Thus, there are two interrelated problems: (i) The effect of a structure's foundation upon the motions and forces induced in the structure during an earthquake (soil–structure interaction) and (ii) the requirement of the foundations to resist the base shear and overturning moment produced by the superstructure. The current knowledge and ability to estimate the seismic force transferred from the superstructure to the substructure is quite advanced; however, it is meagre with respect to the effects of deformation of adjacent soil. Both seismic vertical stresses due to the overturning effect and seismic horizontal stresses due to base shear should be duly considered. The main problem of foundation design is providing for the transfer of base shear to the ground. This aspect is discussed briefly for various types of foundations in the following sections.

3.6.1 Shallow Foundations (Spread Footing)

Shallow foundations are proportioned so as to keep the maximum bearing pressure due to overturning moments and gravity loads within the safe bearing pressure limits. These bear vertically directly against the underlying soil to resist vertical, horizontal, and rotational loads. In general, spread footing properly designed for static loadings perform, adequately under seismic loading. However, significant settlements may cause problems. The ground properties of interest for seismic design of foundations are its stiffness, damping, and strength. These three properties of the soil column affect the transmission of seismic motion of the bed rock to the ground surface. The density and degree of over-consolidation of the soil column are the indicators of the likely behaviour under seismic excitation.

Stiffness of granular and cohesive soils is highly non-linear with increasing strains beyond threshold strain. Although the stiffness of soils reduces with increase of strain, the material damping increases. The stiffness and strength of soils depend on the effective stress in the ground. The effective stresses are influenced by the pore water pressures, which vary with fluctuations in ground water table and stress changes on the ground. For cohesive soils undrained shear strength is the soil strength parameter adjusted for the rapid rate of loading and cyclic degradation effects under the earthquake loads, if required. The cyclic undrained shear strength taking account of possible built-up pore water pressure is the soil strength parameter for cohesionless soil. For rocks, unconfined compressive strength may be used as soil strength parameter.

It is assumed that most of the resistance (shear resistance) to the base shear is provided by the friction between the soil and the bottom surfaces of the foundation. The total resistance R_f to the lateral movement of the structure may be taken as $R_f = D_f\Phi_s$, where D_f is the dead load of the element under consideration and Φ_s is the coefficient of sliding friction.

In case there is a possibility of development of passive soil pressure against the subsurface elements, the horizontal resistance should be accounted for. Appropriate measures must be taken on site in such a case, e.g., adequate compaction of backfill against the sides of the footing. Since shallow foundations are most vulnerable to damage, it is common practice to tie column pads (independent footings) to the beams, even for very low structures founded on soft soils.

Notes: 1. The horizontal seismic stress due to soil–structure interaction is problematic, and since little is known about these, it is customary to assume more arbitrary distributions for horizontal stress between the soil and foundations than for vertical stress.

2. Assessing the foundation response for non-linear behaviour of soils is not a simple task. Near the footing, the strains may be high and the ground response soft. Whereas at remote locations, the strains will be small and the ground behaviour stiff. The non-linear response of soils results in different amplifications of the bed rock motion through the soil column. The greatest amplification of bed rock acceleration occurs at low peak acceleration levels.

Ductile detailing

For earthquake-resistant design, the lateral strength, the deformability, and ductility capacity of the structural elements are of concern. Ductility in the structural elements arises from inelastic material behaviour and the manner in which the reinforcement is detailed. To ensure that the structural elements possess enough ductility and do not fail in a brittle manner, minimum reinforcement in structural elements is specified by the codes.

Column bases

1. The minimum percentage of steel required is 0.15% of cross-sectional area, each way.
2. The special confining reinforcement of columns should extend at least 300 mm in the column base.

Beams

1. Minimum percentage of longitudinal steel = 1%
2. Maximum percentage of longitudinal steel = 6%
3. Minimum diameter of longitudinal steel = 12 mm
4. Minimum diameter of the links = 8 mm

Special considerations

A structure may fail primarily due to ground shaking. However, damages caused due to fault rupture, topographic amplifications, slope instability, liquefaction, and shakedown settlements are no way less important.

As far as possible, important structures should be away from active fault zone. Structures built on sloping grounds are affected by slope instability and topographic amplifications. In case of flow sliding, the foundation of a structure may displace unless it is constructed below the plane above which slipping may occur. Where deep landslides are expected, the structures may translate down-hill or get undermined by the slip.

Because of the topographical effects, the amplitudes at the ridges may be as high as 2.5 times the amplitude at the base of the hill. For average slope, angles smaller than 15°, the topographic effects are negligible. The basin effect causes amplification of ground motion, increase in the duration, and a shift to lower frequencies that are highly damaging to taller structures during strong ground shaking.

Liquefaction is the process by which non-cohesive and granular sediments below water table temporarily lose strength and behave as a viscous liquid during strong ground shaking. Typically, saturated, poorly graded, loose, granular deposits with low fines content are highly susceptible to liquefaction. If soils are prone to liquefaction, ground improvement and piling may be the choice.

3.6.2 Deep Foundations (Pile Foundation)

Not much literature is available on seismic design of deep foundations. The practice is to rely on normal structural and geotechnical static design techniques, taking into account seismically enhanced soil pressures.

The design of pile foundations is based on the consideration of vertical and horizontal stresses and the structural integrity of the foundation. The base shear is resisted by the lateral bearings on pile foundations. The vertical seismic loads on individual piles may vary greatly depending on their position in relation to the rest of the pile group and the superstructure. Some piles, particularly those at the edges or corners of pile systems, may have to bear large tensile as well as compression forces during earthquakes. It must be carefully ensured that the strata contiguous to and below the piles have sufficient adhesive, shear, and bearing strength under seismic conditions. Since little is known of the stress deformations involved in the soil–pile interaction during earthquakes, the most difficult aspect of seismic design of piles is lateral strength. However, a brief about considerations for the pile foundations follows.

Piles are treated as slender columns supported by surrounding soils and therefore do not suffer buckling. However, when placed in very soft soils, they do buckle. When their slenderness ratio is less than 50, they behave like a short column. In case of liquefaction, since the lateral support may be lost, the strength of the soil layer susceptible to liquefaction should be ignored; piles with large axial load may buckle. Apart from large axial loads piles may be called for to support loads due to inertial, kinematic, and lateral spreading effects. These are briefly described in the following paragraphs.

The load-bearing soil strata into which piles are driven may change their character under strong cyclic loading; piles have to bear inertial as well as kinematic loads. Therefore, liquefaction potential of all the soil layers through which the pile passes should be taken into consideration. The response of a pile shaft under seismic loading can be considered as follows.

1. The zone of approximately eight times the pile diameter beneath the soil surface is dominated by inertial loading effects. This zone is known as *near surface zone*.
2. The zone between 12 to 15 times the pile diameter beneath the soil surface is dominated by kinematic effects. This zone is known as *deep zone*.
3. The zone between near surface zone and deep surface zone is influenced by both inertial and kinematic effects. This zone is called as *intermediate zone.*

The structural requirements of piles in the intermediate and deep zones are influenced by the kinematic response. In deep zones, the presence of piles has little effect on the ground motion or natural frequency of stratum. The pile and soil motion are likely to be practically coincident for frequency up to at least 1.5 times the natural frequency of the stratum.

If the piles are treated as flexible, they will move with the surrounding soil and therefore will attract the inertial shear load imposed by the superstructure under earthquake loading. Rigid piles, on the other hand, attract significant soil load, as the piles stay in position and will be subjected to passive pressures on their either side in alternative load cycles. This additional load will have to be considered in the pile design.

Further, additional loading on the piles and pile caps that may arise due to lateral spreading of soils must also be taken care of. Piles are often used to stabilize the u/s and d/s of earthen dams. Such piles are subjected to large lateral forces created by the soil passive pressures, once the whole slope is subjected to lateral spreading. The laterally spreading ground layers can impose additional loading on pile foundations. Since liquefied soil has very little shear resistance, even the ground with gentle slope may also spread laterally. In case of pile foundation, since the lateral spreading will be in the order of metres, sufficient soil strains will build up to generate full passive earth pressure, especially if there are non-liquefied layers above the liquefied layers. Further, if the upper layers are of clayey nature with low hydraulic conductivity, then they exacerbate the problem by helping the liquefied layer to retain the excess pore water pressure for longer duration.

Sufficient continuity reinforcement must be provided between the piles and the pile cap, and the piles themselves must be able to develop the required tensile, compression, and bending strengths. Suitable confinement reinforcement must be provided (as for columns, Chapter 8) when plastic hinges are likely to form in the tops or bottoms of reinforced concrete piles.

3.7 Test of Soil Characteristics

In designing the foundations of a building structure, the designer must first determine the soil characteristics of the construction site. Borehole drilling and penetration tests are the two standard site tests. The data needed for the seismic design of foundations can be obtained from the following tests.

3.7.1 Field Tests

Measurements of in situ dynamic soil properties have great advantage as these do not require sampling which otherwise may change the soil structure, the stress condition, and the chemical and thermal conditions. Since most tests are performed on large volume of soil, the property of interest is not based on small size of specimen which may not actually reflect the properties of the entire mass. However, indirect determination of both the properties and the uncontrolled pore-water drainage are some of the drawbacks.

Bore hole test Bore hole drilling is usually done from the ground level to a depth of about 2 to 4 times the width of the footings. The depth of drilling, however, depends very much upon the size and importance of the building. The various bore-hole tests are *seismic down-hole test* and *up-hole test*, and *cross-hole test*. In the down-hole and up-hole tests, only one bore-hole is drilled. In the down-hole test, impulse source is located on the ground surface and receivers are placed inside the hole. The arrangement of the impulse source and receivers is made reverse of the above arrangement in the up-hole test. The source is triggered and the waves generated are received at the receiver. The travel time of *P*- and *S*-waves are recorded. A number of receivers are located along the depth of the bore-holes and a plot of travel time is generated. The slope of the travel time curve at any depth represents the wave propagation velocity at that depth. It may be noted that the down-hole test is more commonly used.

Cross-hole test In this test, two parallel bore-holes are drilled. In one of the bore-hole, the impulse energy source is placed. In the other bore-hole, a receiver is placed at the same depth. The objective is to measure the wave propagation velocity along the horizontal path.

Sometimes three bore holes are drilled, with two having receivers and the third with impulse energy source. This arrangement facilitates computation of material damping ratio by the measurement of amplitude attenuation.

The shear modulus of soil can be estimated from a shear wave velocity test. An explosive charge or a hammer is used to produce waves in the soil. The velocity is measured by applying the excitation at one bore hole and measuring the velocity at another bore hole or by applying an excitation on the ground and measuring the velocity at a bore hole.

Penetration-resistance test In principle, the test is similar to bore-hole test as discussed above, but no bore-hole is made. The dynamic cone penetrometre consists of a cone penetrometre outfitted with a geophone or accelerometre. The test is performed primarily for measuring the relative density and degree of compaction of the soil. *Hollow-tube samplers* and *cone penetrometres* are used for these tests. The tests are usually of two types: dynamic and static. The standard penetration-resistance test can also provide data that are used for judging potential liquefaction and calculating the allowable bearing capacity of sandy ground.

Seismic reflection/refraction tests These tests are conducted to ascertain the thickness of surficial layers. A source produces a pulse or waves, whose times of arrival are measured at different locations. The source generally produces *P*-waves, *S*-waves, and surface waves; the relative amplitude of each of them depends on how the impulse is generated.

The fundamental period *T* of soil is an important property to be considered for the earthquake-resistant design of structures. It can be estimated by a *micro-tremor test* or from a measurement of small earthquake disturbances.

3.7.2 Laboratory Tests

Soil samples collected from a construction site can be tested in a laboratory to determine soil characteristics such as weight per unit volume, cohesion, internal friction angle, water content, liquid limit, plastic limit, void ratio, compression index, preconsolidated load, and sensitivities. Further, particle-size distribution and relative-density tests are useful for checking potential soil liquefaction.

Resonant column test, ultrasonic pulse test, cyclic triaxial test are some of the important laboratory tests of which the last one is the most popular and reliable test.

The *cyclic triaxial test* is a useful means of estimating the damping ratio of the soil. In this test, hydraulic pressure is applied to a cylindrical sample, and then reversed loading is applied to the cylinder in a longitudinal direction. The elastic modulus *E* is estimated from the stress–strain curve. The shear modulus *G* can then be computed from *E* and the measured Poisson's ratio. The moduli *E* and *G* can also be determined by applying axial and torsional vibrations to the cylindrical sample.

Summary

This chapter discusses the response of soil deposits subjected to dynamic loading, governed by dynamic soil properties. The response depends upon the state of stress in the soil prior to the dynamic loading and the stresses imposed by the dynamic loading.

Dynamic loading leads to the settlement and liquefaction of cohesionless soils. The factors that affect the liquefaction phenomenon are grain size distribution, vibration characteristics, drainage conditions, magnitude of load, soil formation, period of loading, etc.

The dynamic shear modulus and the damping of soil are the two basic parameters that influence the dynamic response analysis and soil–structure interaction. The soil–structure interaction relates the period of vibration of the structure and the soil. The analysis of dynamic soil–structure interaction is based on the free-field motion applied to the base of structure. For such an analysis, it is important to choose the appropriate soil model, which depends upon the type of soil encountered and the type of structure to be analysed. The soil models available for analysis are the spring model, the lumped mass model, the semi-infinite model, and the finite-element model. Equations of motion by direct method and multistep method for the analyses of soil–structure interaction are presented.

In the design of foundations, the interaction between the foundation and the superstructure is considered. Shallow foundations are vulnerable to damage during earthquakes, and hence for isolated footings, column pads are tied with beams. Special considerations for shallow foundations and their ductile detailing are discussed.

Soil layers in the strata may become unstable due to cyclic shear stresses induced by earthquake loading. Under such circumstances, pile foundations are the choice as these transfer loads to deeper firm strata. The base shear resistance of piles is by a combination of passive pressure, lateral load resistance of piles, and friction between the footing and the ground. For piles not driven deep, the passive pressure is not accounted for. Friction effects should be limited unless justified. Consideration for loads due to lateral spreading of soils is emphasized.

A variety of field and laboratory techniques have been introduced in order to determine the dynamic characteristics of soil.

Solved Problems

3.1 If the in situ unit weight of a soil deposit is 15 kN/m^3, determine the settlement of the 3-m thick soil layer under vibration. (Assume $G = 2.70$, $e_{cr} = 0.81$.)

Solution

$$\text{Void ratio } e = \frac{G\gamma_w}{\gamma} - 1$$

$$= \frac{2.70 \times 10}{15} - 1$$

$$= 0.80$$

From the correlation given in Eqn (3.1), the settlement

$$\Delta H = \frac{e_{cr} - e}{1 - e} H$$

$$= \frac{0.81 - 0.8}{1 - 0.8} \times 3$$

$$= 0.15 \text{ m}$$

3.2 In a 10-m thick sand deposit, the ground water table is 3 m below the ground surface. The unit weight of the soil is 17 kN/m^3 up to 3 m from the ground surface and 15 kN/m^3 below that. During an earthquake, the water table rises up to the ground level. Determine the effective dynamic stress at the bottom of the sand layer.

Solution

The effective normal stress at the bottom of the sand layer

$$= (\gamma_1 Z_1 + \gamma_2 Z_2) - \gamma_w Z$$

$$= (17.00 \times 3 + 19.00 \times 7) - 7 \times 10$$

$$= 184 - 70$$

$$= 114 \text{ kN/m}^2$$

The effective dynamic stress at the bottom of the sand layer

$$\overline{\sigma}_{\text{dyn}} = \overline{\sigma}_n - \gamma_w h'_w$$

$$= 114 - 10 \times 3$$

$$= 84 \text{ kN/m}^2$$

Exercises

3.1 To ensure the safety of a structure during an earthquake, what are the important considerations from the viewpoint of soil?

3.2 Write notes on the following regarding soil selection:
   (a) Dynamic soil parameters
   (b) Settlement of dry sands
   (c) Liquefaction
   (d) Soil amplification

3.3 Define liquefaction. What are the factors that affect liquefaction?

3.4 Derive an expression for the condition under which a structure will sink during an earthquake.

3.5 State the soil conditions under which liquefaction can occur.

3.6 What are the measures taken to reduce the possibility of liquefaction?

3.7 Discuss how soil and structure interact during an earthquake.

3.8 Discuss briefly the various soil models used for the dynamic analysis of soil–structure systems. Which of these is the most favourable? State the advantages of this model.

3.9 Write notes on the following:
   (a) Seismic bore hole tests
   (b) Effects of liquefaction on structures
   (c) Sand boils
   (d) Damping in soil

3.10 Discuss briefly seismic considerations for:
   (a) Shallow foundations
   (b) Deep foundations

3.11 Describe briefly the methods of dynamic analysis of soil–structure interaction system.

3.12 In a cross bore hole test the shear wave velocity was observed to be 115 m/s. Determine the value of dynamic shear modulus, if the density of soil is 18.2 kN/m^3.

Ans: 2.41×10^4 kN/m^2

3.13 A sand deposit consists of two layers. The top layer is 3 m thick and bottom layer is 5 m thick. Determine the total settlement of the soil deposit under earthquake-caused vibrations. For the top layer, $e = 0.63$ and $e_{cr} = 0.70$; for the bottom layer, $e = 0.75$ and $e_{cr} = 0.76$.

Ans: $e = 0.77$ m

3.14 A soil profile consists of a 4-m thick surface layer of sand ($\gamma = 18.5$ kN/m^3) overlying a 2-m thick layer of sand ($\gamma = 16.5$ kN/m^3). The water table is at the ground surface. During an earthquake, water in a driven stand pipe rises 2 m above the ground surface. Determine the effective dynamic stress at depths of 4 m and 6 m from the ground surface.

Ans: at 4 m = 10 kN/m^2; at 6 m = 27 kN/m^2

CHAPTER 4

Conceptual Design

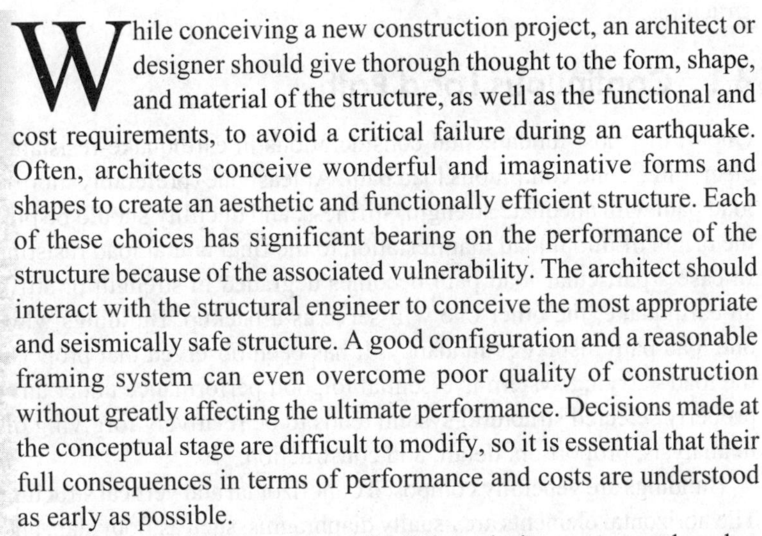

While conceiving a new construction project, an architect or designer should give thorough thought to the form, shape, and material of the structure, as well as the functional and cost requirements, to avoid a critical failure during an earthquake. Often, architects conceive wonderful and imaginative forms and shapes to create an aesthetic and functionally efficient structure. Each of these choices has significant bearing on the performance of the structure because of the associated vulnerability. The architect should interact with the structural engineer to conceive the most appropriate and seismically safe structure. A good configuration and a reasonable framing system can even overcome poor quality of construction without greatly affecting the ultimate performance. Decisions made at the conceptual stage are difficult to modify, so it is essential that their full consequences in terms of performance and costs are understood as early as possible.

Observing the performance of buildings during strong earthquakes has been an age-old means of educating engineers and builders on proper and improper construction of earthquake-load-resisting systems. A structural engineer can draw upon a broad database of engineering observations that have been reported following earthquakes. An observation of performance that suggests incipient collapse is certainly noteworthy. Learning from such observations, faulty construction can be avoided. Observations on good performances of buildings are also noteworthy, as they serve as examples of desirable structural systems.

The basic factors contributing to the proper seismic behaviour of a building, in a rational conceptual design of the structural system, are simplicity, symmetry of the building, ductility, and transfer of the lateral loads to the ground without excessive rotation. Complex structural systems that introduce uncertainties in the analysis and detailing, or that rely on effectively non-redundant load paths, can lead to unanticipated and potentially undesirable structural behaviour. The behaviour of a structure during an earthquake depends largely on the form of the superstructure and on how the earthquake forces are carried to the ground. For this reason the overall form, regular configuration, flow of loads, and the framing system of building may be of serious concern if not taken care of in the first stage of planning.

The first nine sections of this chapter deal with the fundamental principles that should be observed strictly at the conceptual planning stage of a building since these have great impact on the overall performance of the building during strong ground motion. Next, functional planning and the framing systems are discussed. The type of framing system opted will govern the load transfer mechanism for gravity and lateral loads and may dictate choice of particular materials for construction to be used for good performance and economy. The chapter ends with the desirable properties of construction materials for earthquake-resistant structures.

4.1 Continuous Load Path

One of the most fundamental considerations in earthquake-resistant design is a clear, direct, and continuous load path. At least one (preferably more) continuous load path with adequate strength, stiffness, and ductility should be provided from the origin of initial load manifestation to the final lateral load resisting elements. In case a particular load path becomes degraded in strength or stiffness during an earthquake, the other one will serve as a backup. Buildings with more than one load path display redundancy. It has been observed that proper selection of the load-carrying system is essential to good performance under any loading. A properly selected structural system tends to be relatively forgiving of oversights in analysis, proportion, detail, and construction.

Buildings are generally composed of horizontal and vertical structural elements. The horizontal elements are usually diaphragms, such as floor slab, and horizontal bracing in special floors; and the vertical elements are the shear walls, braced frame, and moment-resisting frames. Horizontal forces produced by seismic motion are directly proportional to the masses of building elements and are considered to act at the centres of the mass of these elements. The earthquake forces developed at different floor levels in a building are brought down along the height to the ground through the shortest path. The general path for load transfer, in a conceptual sense, is opposite to the direction in which seismic loads are delivered to the structural elements. Thus the path for load transfer is as follows—inertia forces generated in an element, such as a segment of exterior curtain wall, are delivered through structural connections to a horizontal diaphragm; the diaphragm distributes these forces to vertical components; and finally the vertical elements transfer the forces into the foundations and eventually to the ground (Fig. 4.1).

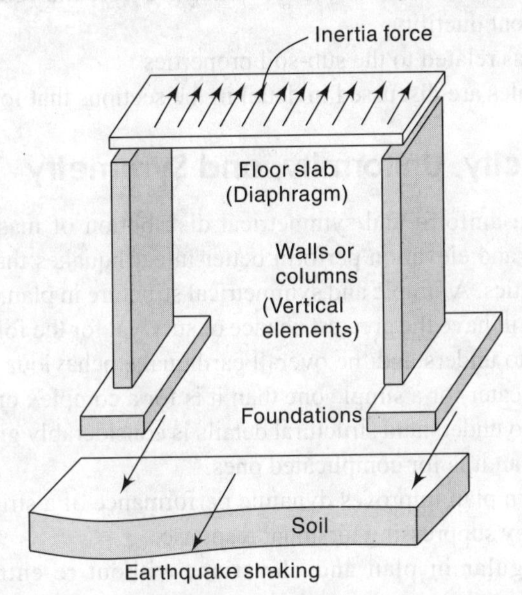

Fig. 4.1 Flow of seismic inertia forces through the structural components

A deviation or discontinuity in this load-transfer path results in poor performance of the building. Failure to provide adequate strength, stiffness, and ductility of individual elements in the system or failure to tie individual elements together can result in distress or complete collapse of the system. One of the earliest lessons from earthquakes was the realization that structural and non-structural elements must be adequately tied to the structural system. Concrete diaphragms, with appropriate struts, ties, and boundary elements, should be provided with adequate reinforcement to transmit the seismic forces.

4.2 Overall Form

A structure is conceived and designed to transfer the seismic forces to the ground safely. However well the structure may have been designed, it is said to be acceptable only if it meets all the established configuration-related requirements from the observed failures during past earthquakes. Buildings having simple, regular, and compact layouts, incorporating a continuous and redundant lateral force-resisting system, tend to perform well during earthquakes and, thus, are

desirable. While planning a particular structure, the guiding principles to be borne in mind are as follows. The structure should have the following characteristies:
(a) have a direct and continuous load path
(b) be simple and symmetrical
(c) not be too elongated in plan or elevation, i.e., the size should be moderate
(d) have uniform and continuous distribution of strength, mass, and stiffness
(e) have horizontal members which form hinges before the vertical members
(f) have sufficient ductility
(g) have stiffness related to the sub-soil properties

These principles are discussed in detail in the sections that follow.

4.3 Simplicity, Uniformity, and Symmetry

Buildings with a uniform and symmetrical distribution of mass, strength, and stiffness in plan and elevation perform better in earthquakes than those lacking these characteristics. A simple and symmetrical structure in plan, e.g., a square or circular shape, will have the greatest chance of survival for the following reasons:

1. The ability to understand the overall earthquake behaviour of a structure is markedly greater for a simple one than it is for a complex one.
2. The ability to understand structural details is considerably greater for simple structures than it is for complicated ones.
3. Uniformity in plan improves dynamic performance of a structure during an earthquake by suppressing torsional response.

Buildings regular in plan and elevation, without re-entrant corners or discontinuities in transferring the vertical loads to the ground, display good seismic behaviour as well. It is important that the plan of a structure is symmetrical in both directions. In general, buildings with simple geometry in plan as shown in Fig. 4.2(b) perform well during earthquakes. Buildings with re-entrant corners, such as U, V, T, and + shapes in plan [Fig. 4.2(a)], may sustain significant damage during earthquakes and should be avoided. H-shapes, although symmetrical, should not be encouraged either. The probable reason for the damage is the lack of proper detailing at the corners, which is complex. To check the bad effects of these interior corners in the plan, the building can be broken into parts using a separation joint at the junction. There must be enough clearance at the separation joints so that the adjoining portions do not pound each other. Figure 4.3 shows such cases of elongated, L-shaped and H-shaped buildings.

A building having a simple plan but a lack of symmetry in the columns or walls (more partition walls in the shorter direction than that in the longer direction), or an irregularity in the elevation (Fig. 4.5), produces torsional effects, which are difficult to assess properly and can be destructive. External lifts and stairwells provide similar dangers; they tend to act on their own in earthquakes, making it difficult to predict force concentrations, torsions, and out-of-balance forces. To avoid torsional deformation, the centre of stiffness of a building should coincide

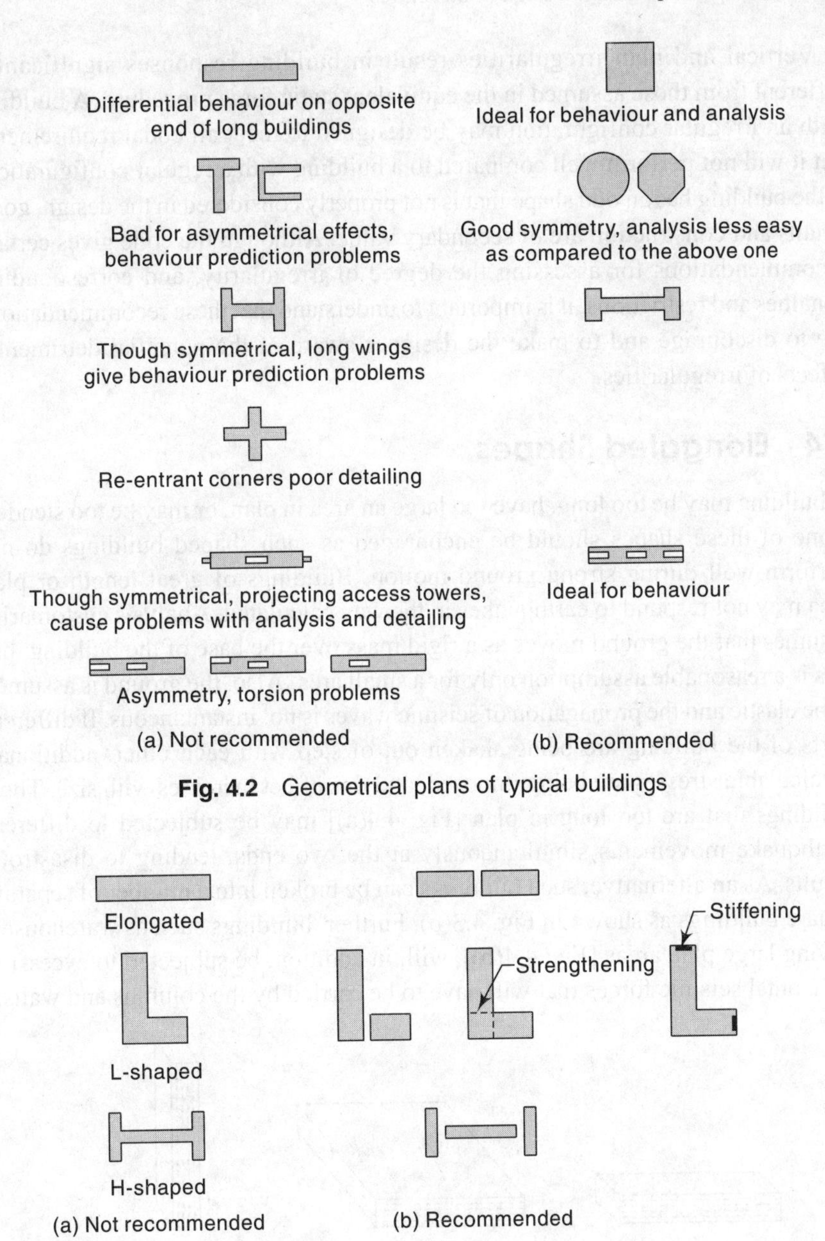

Fig. 4.2 Geometrical plans of typical buildings

Fig. 4.3 Broken layout concept

with the centre of mass. It is desirable to have symmetry both in the building configuration, as well as in the structure, in order to satisfy this condition. The torsions of unsymmetrical structures can lead to a failure of corner columns and walls at the perimeter of the building. The twisting effect of buildings is discussed in Section 4.7.

Vertical and plan irregularities result in building responses significantly different from those assumed in the equivalent static force procedure. A building with an irregular configuration may be designed to meet all codal requirements but it will not perform well compared to a building with a regular configuration. If the building has an odd shape that is not properly considered in the design, good details and construction are of secondary value. Although the code gives certain recommendations for assessing the degree of irregularity, and corresponding penalties and restrictions, it is important to understand that these recommendations are to discourage and to make the designer aware of the potential detrimental effects of irregularities.

4.4 Elongated Shapes

A building may be too long, have too large an area in plan, or may be too slender. None of these shapes should be encouraged as such shaped buildings do not perform well during strong ground motion. Buildings of great length or plan area may not respond to earthquakes in the way calculated. Analysis customarily assumes that the ground moves as a rigid mass over the base of the building, but this is a reasonable assumption only for a small area. Also, the ground is assumed to be elastic and the propagation of seismic waves is not instantaneous. If different parts of the building are being shaken out of step with each other, additional, incalculable stresses are being imposed, and this effect increases with size. Thus, buildings that are too long in plan [Fig. 4.4(a)] may be subjected to different earthquake movements simultaneously at the two ends, leading to disastrous results. As an alternative, such buildings can be broken into a number of separate square buildings as shown in Fig. 4.3(b). Further, buildings such as warehouses, having large plan areas [Fig. 4.4(b)], will, in addition, be subjected to excessive horizontal seismic forces that will have to be carried by the columns and walls.

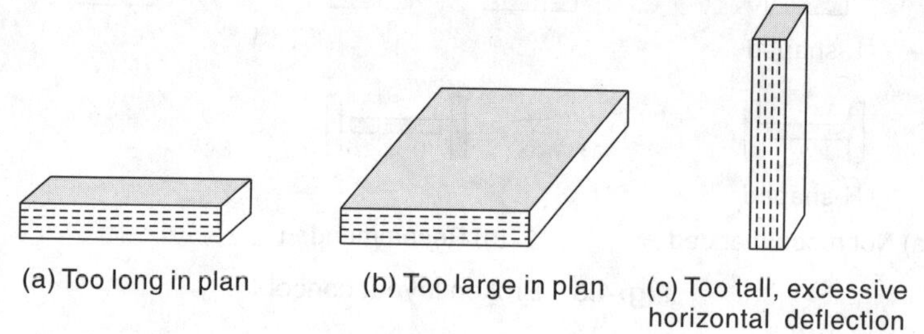

Fig. 4.4 Elongated shapes—one of the overall dimensions much lesser or much smaller than the other (adapted from Murty, 2005)

In tall buildings [Fig. 4.4(c)] with large height-to-base ratio (slenderness ratio > 4), the horizontal movement of the floors during ground shaking is large. For buildings with slenderness ratio less than 4, the movement is reasonable. The more slender a building, the worse the overturning effects of an earthquake. The axial column force due to the overturning moment in such buildings tends to become unmanageably large. Also, the compressive and pull-out forces acting on the foundation increase tremendously.

4.5 Stiffness and Strength

Building with a uniform distribution of stiffness and strength in plan and elevation generally performs well during earthquakes. Strength is the property of an element to resist force. Stiffness is the property of an element to resist displacement. When two elements of different stiffnesses are forced to deflect the same amount, the stiffer element will carry more of the total force because it takes more force to deflect it. Stiffness greatly affects the structure's uptake of earthquake-generated forces. On the basis of stiffness, the structure may be classified as *brittle* or *ductile*. A brittle structure, having greater stiffness, proves to be less durable during an earthquake, while a ductile structure performs well in earthquakes.

Sudden changes in stiffness and strength between adjacent storeys are very common. Such changes are associated with setbacks (in penthouses and other small appendages), changes over the height of a structural system (e.g., discontinuous shear walls), changes in storey height, changes in materials, and unanticipated participation of non-structural components. A common problem with such discontinuities is that inelastic deformations tend to concentrate in or around the discontinuity. These sudden changes in stiffness, strength, or mass in either vertical or horizontal planes of a building can result in distribution of lateral loads and deformations different from those that are anticipated for a uniform structure. A sudden change of lateral stiffness up a building is not advised for the following reasons:

(a) Even with most sophisticated and expensive computerized analysis, the earthquake stress cannot be determined adequately.
(b) The structural detailing poses practical problems.

Drastic changes in the vertical configuration as shown in Fig. 4.5 cause changes in stiffness and strength between adjacent storeys of a building and should be avoided. Such discontinuity in the vertical configuration of a building is not recommended. Failures due to discontinuity of vertical elements of the lateral load-resisting system have been among the most notable and spectacular.

Buildings with vertical setback as shown in Fig. 4.5(a) cause a sudden jump of earthquake forces at the level of discontinuity. A large vibrational motion takes place in some portions and a large diaphragm action is required at the border to transmit forces from the top to the base. It may be noted that the effects of setbacks cannot be predicted by normal code equivalent static analysis.

Uniformity of strength and stiffness in elevation helps to avoid the formation of weak or soft storeys. Buildings that have fewer columns or walls in a particular storey, or that have an unusually tall storey [Fig. 4.5(b)] are prone to damage or collapse. One of the most common forms of discontinuity of vertical elements occurs when shear walls that are present in upper floors are discontinued in the lower floors. The result is frequent formation of a soft storey that concentrates damage. Figure 4.5(c) shows a building having shear walls (RCC walls for carrying earthquake forces) that do not go all the way to the ground, but terminate at an intermediate storey level. It is advocated that the stiffness of the lower storey, the so-called soft storey, be reduced, so that a reduced dynamic force is transmitted to the superstructure. However, this argument is based on simple elastic analysis. When realistic inelastic and geometrical non-linear effects are taken into account, the plastic deformations tend to concentrate in the soft storey, and may cause the entire building to collapse.

The unequal height of the columns [Fig. 4.5(d)] causes twisting and damage to the short columns of the building. It is because shear force is concentrated in the relatively stiff short columns that fail before the long columns. In a structural frame, long columns can be turned into short columns by the introduction of spandrels. Buildings with columns that hang or float on beams at an intermediate storey [Fig. 4.5(e)] have discontinuities in the load transfer path.

The most common form of vertical discontinuity arises because of unintended effects of nonstructural elements. The problem is most severe in structures having relatively flexible lateral load-resisting systems, because in such cases the non-structural component can comprise a significant portion of the total stiffness. A common cause of failure is the infilled frames. If properly designed, the infill can improve the performance of the frame due to its stiffening and strengthening action. However, soft storeys may result if infills are omitted in a single storey (often the first storey). Even if infills are placed continuously and symmetrically throughout the structure, a soft storey may be formed if one or more infill panels should fail.

Partial height frame infills are also common. In this form of construction an infill extends between columns, from the floor level to the bottom of the window line, leaving a relatively short portion of the column exposed in the upper portion of the storey. The shear required to develop flexural yield in the shortened column can be substantially higher than for the full-length column. If the designer has not considered this effect of the infill, shear failure of this so-called *captive column* can result before flexural yield. Complete collapse of the column (and building) can occur if it is not well equipped with transverse steel. This form of distress is a common cause of building damage and collapse during earthquakes.

Apparent vertical irregularities can occur due to the interaction between adjacent structures having inadequate separation. A tall building adjacent to a shorter building may experience irregular response due to the effects of impact between the two structures. This effect can be exacerbated by local column damage due

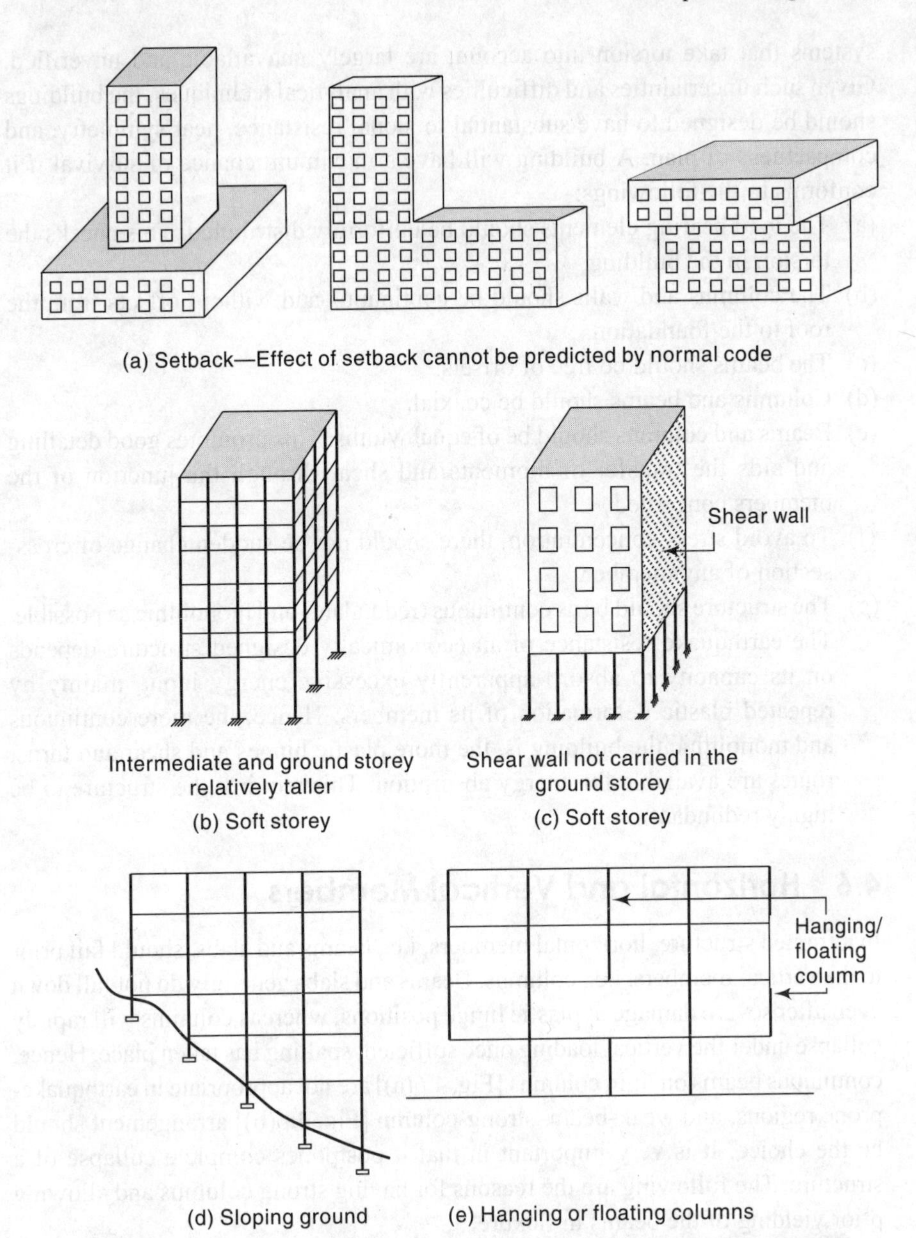

(a) Setback—Effect of setback cannot be predicted by normal code

(b) Soft storey

(c) Soft storey

(d) Sloping ground

(e) Hanging or floating columns

Fig. 4.5 Discontinuation in vertical configuration of buildings (adapted from Murty, 2005)

to the pounding of the roof of the small building against the columns of the taller one.

Mass, stiffness, and strength plan irregularities can result in significant torsional response. Inelastic torsional response cannot, at present, be rectified with the results of elastic analysis. Techniques for inelastic analysis of complete building systems that take torsion into account are largely unavailable and unverified. Given such uncertainties and difficulties with analytical techniques, the buildings should be designed to have substantial torsional resistance, near symmetry, and compactness of plan. A building will have a maximum chance of survival if it conforms to the followings:

(a) The load bearing elements should be uniformly distributed. This checks the torsion in the building.
(b) The columns and walls should be continuous and without offsets from the roof to the foundation.
(c) The beams should be free of offsets.
(d) Columns and beams should be coaxial.
(e) Beams and columns should be of equal widths. This promotes good detailing and aids the transfer of moments and shear through the junction of the members concerned.
(f) To avoid stress concentration, there should not be sudden change of cross-section of any member.
(g) The structure should be as continuous (redundant) and monolithic as possible. The earthquake resistance of an economically designed structure depends on its capacity to absorb apparently excessive energy input, mainly by repeated plastic deformation of its members. Hence, the more continuous and monolithic the building is, the more plastic hinges and shear and thrust routes are available for energy absorption. This requires the structure to be highly redundant.

4.6 Horizontal and Vertical Members

In a framed structure, horizontal members, i.e., beams and slabs, should fail prior to the vertical members, i.e., columns. Beams and slabs generally do not fall down even after severe damage at plastic hinge positions, whereas columns will rapidly collapse under the vertical loading once sufficient spalling has taken place. Hence, continuous beams on light columns [Fig. 4.6(a)] are not appropriate in earthquake-prone regions, and weak-beam–strong-column [Fig. 4.6(b)] arrangement should be the choice. It is very important in that it postpones complete collapse of a structure. The following are the reasons for having strong columns and allowing prior yielding of the beams in flexure.

(a) Failure of a column means the collapse of the entire building.
(b) In a weak-column structure, plastic deformation is concentrated in a particular storey, as shown in Fig. 4.6(c), and a relatively large ductility factor is required.
(c) In both shear and flexural failures of columns, degradations are greater than those in the yielding of beams.

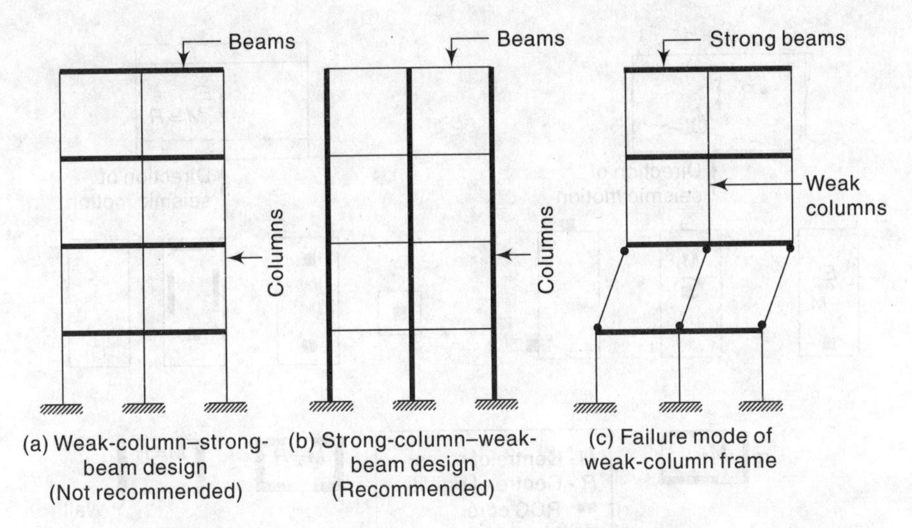

Fig. 4.6 Weak-beam–strong-column concept

4.7 Twisting of Buildings

Torsional forces from ground motion are not usually of great concern unless the building has an inherently low torsional strength. Twist in buildings causes different portions at the same floor level to move horizontally by different amounts. Irregularities of mass, stiffness, and strength in a building can result in significant torsional response. However, torsion arises from eccentricity in the building layout—when the centre of mass of the building does not coincide with its centre of rigidity. If there is torsion, the building will rotate about its centre of rigidity due to the torsional moment about the centre of structural resistance. The torsional response may be inadequately represented by a linear dynamic analysis, because yielding caused by lateral-torsional response can reduce the stiffness on one side of a building and further increase the eccentricity between mass and stiffness centres.

The recommended plan configurations of buildings to avoid torsional moments due to distribution of mass and stiffness of elements is illustrated in Fig. 4.7. This additional torsion will have to be dealt alongwith the torsional component of ground motion. This may cause a large increase in the lateral forces acting on bracing elements and on other parts of the structure, in proportion to their distances from the centre of rotation.

Torsion in buildings during earthquakes can be most simply explained by analogy with a rope swing. Consider a wooden cradle tied with coir ropes to a sturdy branch of a tree. Buildings behave like this swing, except that they are anchored at the bottom rather than at the top. That is to say that buildings are essentially inverted swings. The walls and columns are like ropes, the ground is like the branch of tree to which the ropes are tied, and the upper floors or storeys are like the wooden cradle. In a single-storey building, the roof acts as the wooden cradle [Fig. 4.8(a)] of a swing and in multi-storey buildings, the upper

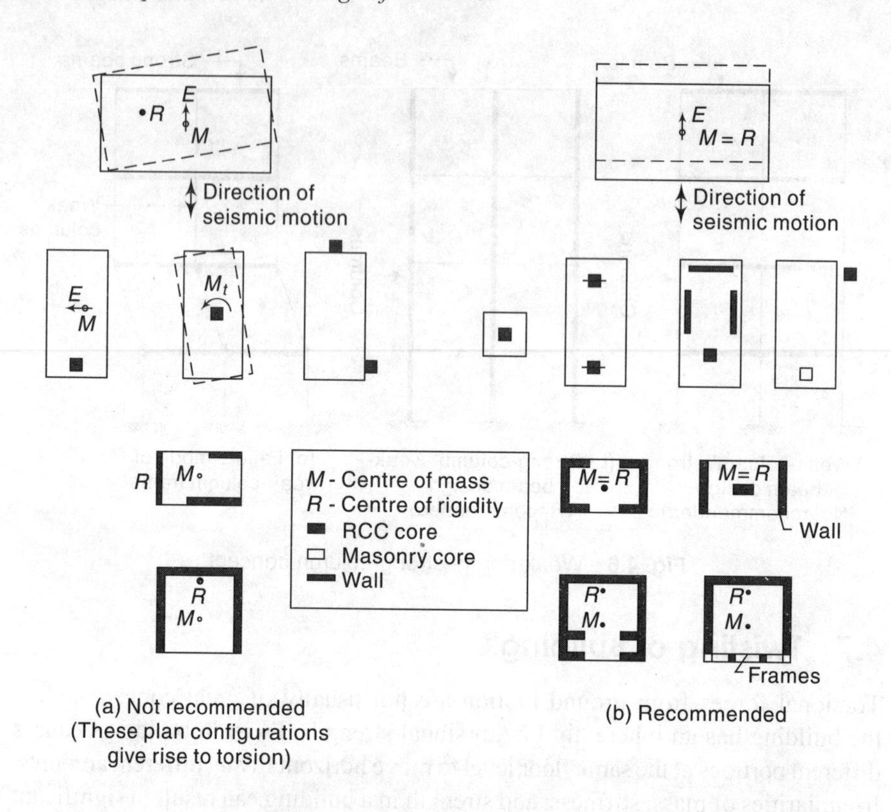

(a) Not recommended (These plan configurations give rise to torsion)

(b) Recommended

Fig. 4.7 Distribution of mass and stiffness of elements in plan

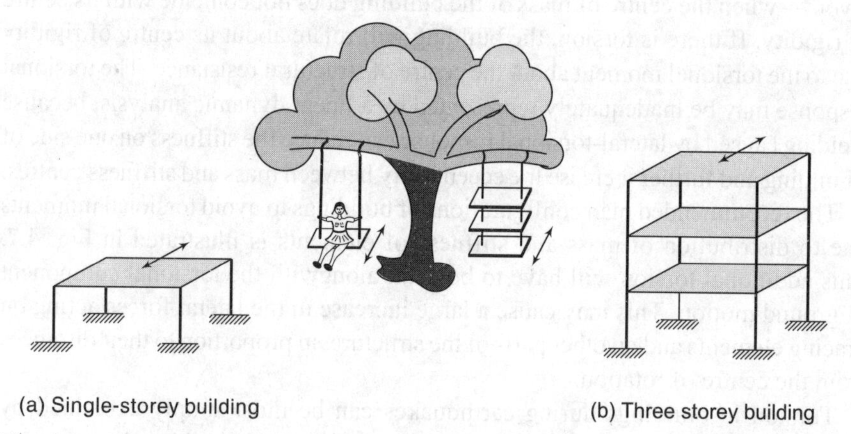

(a) Single-storey building

(b) Three storey building

Fig. 4.8 Horizontal shaking of single- and three-storey buildings and their simulation with rope and swing (adapted from Murty, 2005)

floors act as a stack of wooden cradles suspended by the ropes at regular intervals [Fig. 4.8(b)].

Now consider a rope swing tied symmetrically with two equal ropes. If one sits in the middle of the cradle, it will swing back and forth in a symmetric fashion without any sideways swinging or tilting. Similarly, when a symmetric building, loaded uniformly, is shaken by an earthquake, it swings back and forth such that all points on a floor move horizontally in the same direction and by the same amount at any given time. However, if one sits on the cradle of the swing on any one side, it tilts, causing the ropes and, thus the swing to twist [Fig. 4.9(a)]. Similarly, if the mass on the structure of a building is more on one side than on the other [Fig. 4.9(b)], then the lighter side is displaced by a greater amount when the building is subjected to ground movement. This is to say that the building undergoes horizontal displacement as well as rotational motion. To understand this sort of motion better, try sitting on a park swing on one side of the cradle and swinging fast. Instead of swinging back and forth, the swing will turn about its centre of mass, causing the rope to twist. This is the kind of motion an unequally loaded structure undergoes. And although the ropes of a swing are flexible enough to twist under torsion and then come back to their original position, the walls and columns of a building are not.

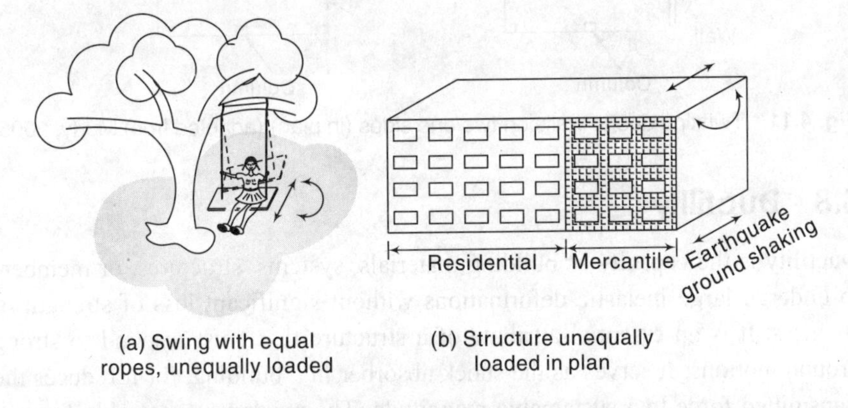

(a) Swing with equal ropes, unequally loaded

(b) Structure unequally loaded in plan

Fig. 4.9 Torsional vibration of a structure with even vertical members (adapted from Murty, 2005)

A rope swing with unequal rope lengths [Fig. 4.10(a)] on either side will undergo motion similar to that described above. The structural counterparts of this are buildings on slopes. These have unequal columns as shown in Fig. 4.10(b) and the floors experience twisting about a vertical axis because of the varying stiffness of columns. Buildings having walls only on two sides and thin columns along the other also experience twist as shown in Fig. 4.11. The twist induces more damage in the columns and walls on the side that moves more. The best way to minimize twist is by ensuring that buildings have symmetry in plan with respect to vertical members and loads.

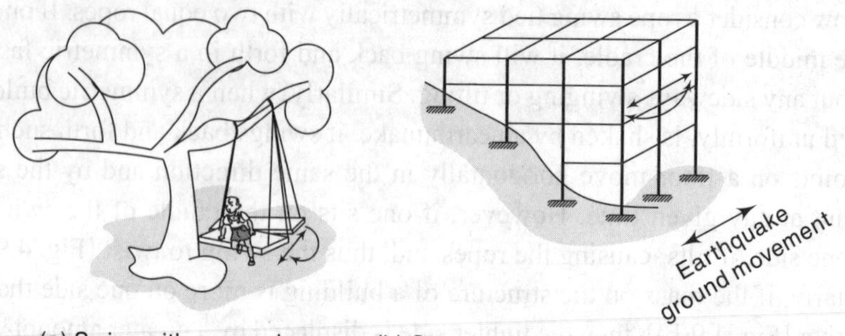

Fig. 4.10 Torsional vibration of a structure with uneven vertical members loaded unequally in plan (adapted from Murty, 2005)

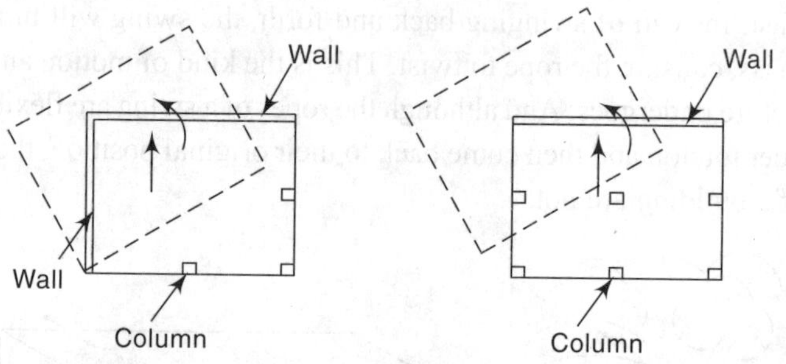

Fig. 4.11 Twisting due to walls on two/one sides (in plan) (adapted from Murty, 2005)

4.8 Ductility

Ductility is the capacity of building materials, systems, structures, or members to undergo large inelastic deformations without significant loss of strength or stiffness. It is an essential attribute of a structure that must respond to strong ground motions. It serves as the shock absorber in a building, for it reduces the transmitted force to a sustainable magnitude. The resultant sustainable force is traditionally used to design a hypothetically elastic representation of the building. Therefore, the survivability of a structure under strong seismic action relies on the capacity to deform beyond the elastic range, and to dissipate seismic energy through plastic deformation.

Formally, ductility refers to the ratio of the displacement just prior to ultimate displacement or collapse, to the displacement at first damage or yield. This is a very important characteristic of a building since it greatly reduces the effect or *response* that is produced in the structure by an earthquake. This is because the building is set in vibration by the energy of an earthquake. This vibration, as well as the accompanying deflection, is reduced by the energy that is absorbed by the large inelastic deflections of a ductile structure. Some materials, such as steel and wood, are inherently ductile, while others, such as masonry and concrete, are brittle and fail suddenly. Building elements constructed with ductile materials have a *reserve capacity* to resist earthquake overloads. Therefore, buildings constructed with ductile elements, such as steel and adequately reinforced concrete, tend to withstand earthquakes much better than those constructed with brittle materials such as unreinforced masonry.

One way of achieving ductility in structural members is by designing elements, with known limits, which deform in a ductile manner. For example, in RCC members, the amount and location of steel should be such that the failure of the member occurs by steel reaching its strength in tension before concrete reaches its strength in compression. This is referred to as *ductile failure*. In RCC buildings the seismic inertia forces generated at floor levels are transferred through the various beams and columns to the ground. The correct building components need to be made ductile. The failure of a beam causes localized effects. However, the failure of a column can affect the stability of the whole building. Therefore, it is better to make the beams ductile rather than the columns. Such a design method is known as strong-column–weak-beam design method.

Table 4.1 Types of brittle failure

Structure	Overturning
Foundation	Rotational shear failure
Structural steel	Bolt shear or tension failure
	Member buckling
	Member tension failure
	Member shear failure
	Connection tearing
Reinforced concrete	Bond or anchorage failure
	Member tension failure
	Member shear failure
Masonry	Out-of-plane bending failure
	Toppling

Ductility can also be achieved by avoiding any possibility of brittle failure (Table 4.1). As an example, a tension bolt in a steel beam–column connection should be at a safe stress level when the beam has reached its ultimate moment. For the entire structural system to be ductile, the following requirements must be met.

(a) Any mode of failure should involve the maximum possible redundancy.

(b) Brittle-type failure modes, such as overturning, should be adequately safeguarded so that ductile failure occurs first.

Ductility is often measured by hysteretic behaviour of critical components, such as a column–beam assembly of a moment frame. The hysteretic behaviour is usually examined by observing the cyclic moment–rotation (or force–deflection) behaviour of the assembly. The slopes of the curves represent the stiffness of the structure, and the enclosed areas are sometimes full and flat, or they may be lean and pinched. Structural assemblies with curves enclosing a large area representing large dissipated energy are regarded as superior systems for resisting seismic loading.

Brundtland, UNEP 1987

BuroHappold

B. Happold 2012

4.9 Flexible Building

Whether a structure should be stiff or flexible has always been a point of discussion. The ground shaking during an earthquake contains a group of many sinusoidal waves of different frequencies having periods in the range of 0.03 to 33 s. The base of the building swings back and forth when the ground shakes. The building oscillates back and forth horizontally and after some time comes back to the original position. The time taken (in seconds) for one complete back and forth motion is called the *fundamental natural period T* of the building; the higher the flexibility, the greater the value of T. The fundamental time periods of some structures are given in Table 4.2. Depending upon the value of T for the building and the characteristics of earthquake ground motion, some buildings are shaken more than the others. In a stiff (rigid) building, every part moves by the same amount as the ground; for a flexible building, different parts move by different amounts.

Table 4.2 Fundamental time period of some of the structures

Type of structure	Fundamental time period (T)
Moment resisting RC frame building without brick infill walls	$0.075h^{0.75}$
Moment resisting steel frame building without brick infill walls	$0.085h^{0.75}$
All other buildings including moment-resisting RC frame with brick infill walls	$0.09h/\sqrt{d}$

Ideally, a flexible structure with moment-resisting frames (beam and column type) is built so that the non-structural elements such as partitions and infill walls are isolated from the frame movements. This is necessary because a flexible structure tends to exhibit large lateral deflections, which induce damage in non-structural members. Even the lifts and shaft walls are completely separated. There is no extra safety margin provided by the non-structure as in traditional construction. In tall buildings, oscillations due to wind gusts can cause discomfort to the occupants, and a stiff structure is desirable. For a flexible structure, materials such as masonry are not suitable and steel work is the usual choice. For greater stiffness, diagonal braces or RCC shear wall panels may be incorporated in steel frames. Concrete can be readily used to achieve almost any degree of stiffness. The non-structures such as partitions may greatly stiffen a flexible structure and, hence, must be allowed in the structural analysis.

Depending on the situation, either a flexible or a stiff structure can be made to work, but the advantages and disadvantages of the two forms need careful consideration. These advantages and disadvantages are given in Table 4.3.

Table 4.3 Flexible structures vs stiff structures

	Flexible structures	Stiff structures
Advantages	1. Especially suitable for short period sites, and for buildings with long periods	1. Suitable for long period sites
	2. Ductility arguably easier to achieve	2. Easier to reinforce stiff reinforced concrete (i.e., shear walls)
	3. More amenable to analysis	3. Non-structure easier to detail
Disadvantages	1. High response on long period sites	1. High response on short period sites
	2. Flexible framed reinforced concrete is difficult to reinforce	2. Approximate ductility not easy to knowingly achieve
	3. Non-structure may invalidate analysis	3. Less amenable to analysis
	4. Non-structure difficult to detail	

4.10 Functional Planning

The functional planning of a building affects the way in which it can accommodate its structural skeleton. The principal categories of buildings from the point of view of a lateral load-resisting system are as given in Table 4.4 and are discussed in the section to follow. The vertical divisions of the building create problems, making it

Table 4.4 Lateral load-resisting systems

Framing system	Description
Bearing-wall system	The walls are load-bearing walls. Some of the bearing walls may be shear walls. The system is designed for gravity as well as for lateral loads. Under lateral loads the walls act like cantilevers. The shear distribution is proportional to the moments of inertia of the cross-sections of the walls. The relative displacements of the floors result from bending deformation of the walls.
Moment-resisting frames	These are the frames in which the beams, columns, and joints resist earthquake forces, primarily by flexure. These frames, when subjected to lateral forces, exhibit zero moments at mid-height of the columns, shear distribution proportional to the moments of inertia of the columns, and relative displacements (or inter-story drifts) proportional to the shear forces. This is the reason why sometimes these frames are referred to as shear systems. The continuity of the frame also assists in resisting gravity loading more efficiently by reducing positive moments in the centre span of girders. These are preferred because of least obstruction to access. However, this system

(*Contd*)

Table 4.4 (*Contd*)

	is recommended only up to 30 storeys due to a limitation on the drift.
Dual systems	These consist of moment-resisting frames either braced or with shear walls. The coupling of the above two systems completely alters the moment and shear diagrams of both the walls and the frame. The characteristic of this combination is that in the lower floors the wall restrains the frame, while in the upper floors the frame inhibits the large displacements of the wall. As a result, the frame exhibits a small variation in storey shear between the first and the last floors. The two systems may be designed to resist the total design force in proportion to their lateral stiffness.
Tube systems	It is a fully three-dimensional system that utilizes the entire building perimeter to resist lateral loads. For taller buildings, the relatively recent framed-tube, trussed-tube, tube-in-tube, and bundled-tube systems are used.

difficult to avoid irregularities in mass or stiffness. However, the service cores and exterior cladding provide an opportunity to incorporate shear walls or braced panels. One of the main objectives in preliminary planning is to establish the optimum locations for service cores and for stiff structural elements that should be continuous to the foundation. The initial structural and architectural plans may be in conflict, but it is essential to arrive at a satisfactory compromise at the concept planning stage itself.

4.11 Framing Systems

The load-bearing wall system is the most common building system for low-rise structures where gravity loads are the dominant loads. However, this system is inherently weak in resisting lateral loads, and is seldom recommended for multi-storey buildings. The framework of a multi-storey building consists of a number of beams and columns built monolithically, forming a network. The ability of a multi-storey building to resist the lateral forces depends on the rigidity of the connections between the beams and the columns. When the connections are fully rigid, the structure as a whole is capable of resisting the lateral forces. The moment-resisting frame is thus the fundamental structural system. However, if the strength and stiffness of a frame are not adequate, the frame may be strengthened by incorporating load-bearing walls, shear walls, and/or bracings (Fig. 4.12). Shear walls and bracings are also useful in preventing the failure of non-structural components by reducing drift. Shear walls are walls situated in advantageous positions in a building that can effectively resist lateral loads originating from earthquakes or winds. These may be made of RCC, steel, composite, and masonry. RCC shear walls are most commonly used in multi-storey structures and are described in detail in Chapter 8.

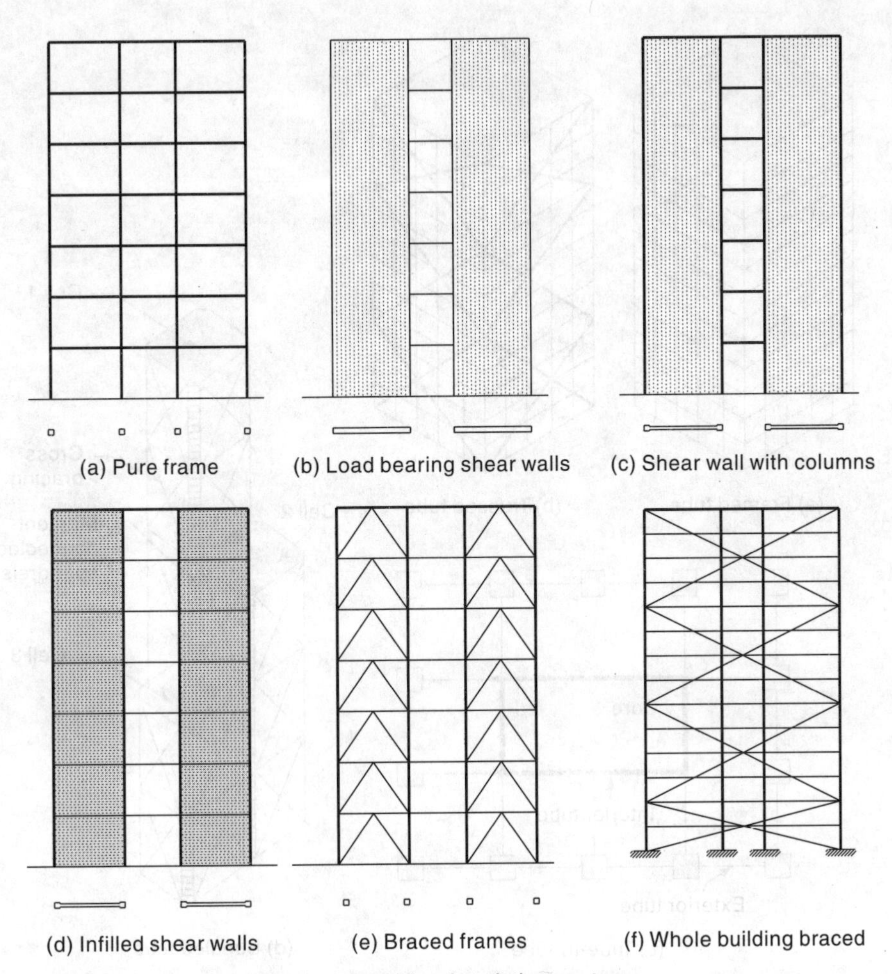

Fig. 4.12 Lateral load-resisting systems

The flat slab system is one of the most favourite reinforced concrete structural forms with the architects. It provides architectural flexibility, maximum usage of space, easier formwork, and shorter construction period. However, the flat slab systems need special attention as these perform poorly under earthquake loading and are less efficient. This is primarily due to the absence of deep beams or shear walls in this form of construction. When subjected to even moderate earthquakes, the excessive deformations cause damage to the non-structural members creating panic.

For buildings taller than about forty storeys, the effect of lateral forces becomes increasingly intense, and *tube systems* become economical. Tube systems may be classified as *framed-tube*, *trussed-tube*, *tube-in-tube*, and *bundled-tube* systems. In the framed-tube system [Fig. 4.13(a)], closely spaced columns are tied at each floor level by deep spandrel beams, thereby creating the effect of a hollow tube, perforated by openings for windows. This system represents a logical evolution

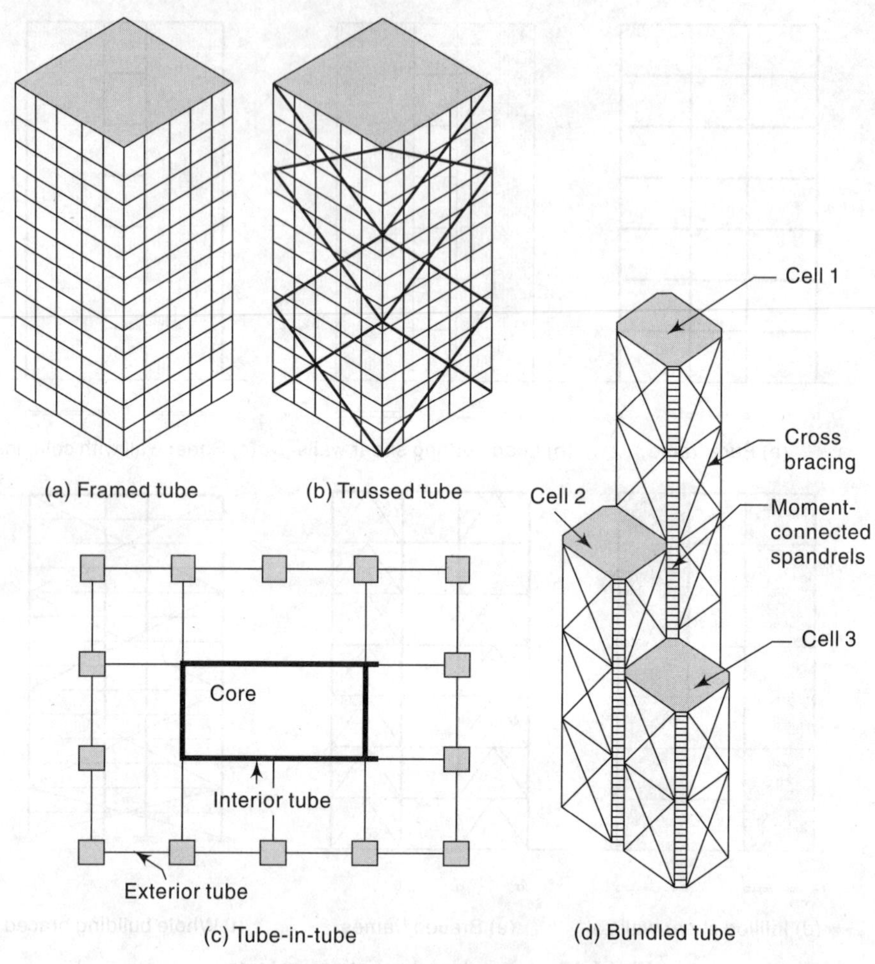

Fig. 4.13 Types of tube structures

of the conventional framed structure, possessing the necessary lateral stiffness with excellent torsional qualities, while retaining the flexibility of planning. The trussed-tube system shown in Fig. 4.13(b) is an advancement over the framed-tube system. The diagonal members, along with girders and columns, form a truss system that imparts a great deal of stiffness to the building. The tube-in-tube system [Fig. 4.13(c)] consists of an exterior tube that resists the bending moment due to lateral forces and an interior slender tube, which resists the shear produced by the lateral forces. The bundled-tube system [Fig. 4.13(d)] is made up of a number of tubes separated by shear walls; the tubes rise to various heights and each tube is designed independently.

In a multi-storey building, the moment-resisting frames, along with shear walls [Fig. 4.14(a)] or the bracing, work to resist lateral forces. Frames deform in a predominantly shear mode [Fig. 4.14(b)], where the relative storey deflection depends on the shear applied at the storey level. The walls deform in an essentially bending mode [Fig. 4.14(c)]. A structural framework with load-bearing walls hence exhibits an intermediate form of behaviour as shown in Fig. 4.14(d). In the lower part of the building, the walls resist the greater part of the shear force, but the shear gradually decreases in higher storeys. If flexural deformation occurs in a load-bearing wall, the adjacent boundary beam undergoes a large deformation and should have adequate ductility. Also, adjacent columns are subjected to large axial force, so the difficulties arise, both in designing the column cross-section and in dealing with the pull-out force on the foundation. To overcome this situation, the building may be braced or shear walls may be provided, as shown in Fig. 4.12.

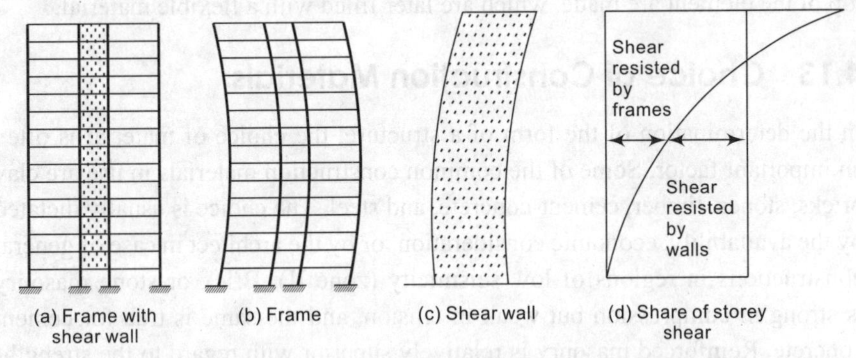

Fig. 4.14 Contribution of frames and shear walls to storey shear

The framed-tube system combines the behaviour of a true cantilever, such as a shear wall, with that of a beam–column frame. Overturning under the lateral load is resisted by the tube form, causing compression and tension in the columns. The shear from the lateral load is resisted by bending in columns and beams, primarily in the two sides of the building parallel to the direction of the lateral load.

4.12 Effect of Non-structural Elements

Non-structural elements such as claddings, infill walls, partition walls, etc., interfere with the free deformation of the structure and thus become structurally very responsive in earthquakes. If the material used in construction is flexible, the non-structures will not affect the structure significantly. However, these are often made with brittle materials like bricks, concrete blocks, etc., and so affect the overall behaviour of the structure in the following ways.

(a) The natural period of vibration of the structure may be reduced and may cause a change in the intake of seismic energy and, consequently, a change in the seismic stresses of the structure.
(b) The lateral stiffness of the structure may redistribute, changing the stress distribution.
(c) The structure may suffer pre-mature failure, usually in shear or by pounding.
(d) Non-structures may suffer excessive damage due to shear forces or pounding.

The more flexible the basic structure, the worse the above effects will be. The structure will suffer pronounced effects if the non-structural elements are asymmetric or not the same on successive floors. There are two approaches to deal with such problems in structures and to create low seismic response. One way is to include these shear elements into the official structure, as analysed, and to detail accordingly. This approach is suitable for stiff buildings. The other way is to prevent the non-structural elements from contributing their shear stiffness to the structure. This approach is appropriate particularly for a flexible structure. To achieve this objective, gaps against the structure, up the sides, and along the top of the element are made, which are later filled with a flexible material.

4.13 Choice of Construction Materials

In the determination of the form of a structure, the choice of material is often an important factor. Some of the common construction materials in use are clay bricks, stones, timber, cement-concrete, and steel. The choice is usually dictated by the availability, economic consideration, or by the architect in case of general constructions in regions of low seismicity (zone II). Brick or stone masonry is strong in compression but weak in tension, and the same is true for cement concrete. Reinforced masonry is relatively superior with regard to the strength-to-weight ratio, degradation, and deformability, and it is also less expensive. Concrete, although the most favourable building material, produces beams that are brittle in shear, and columns brittle both in compression and shear. However, proper ductile detailing improves the behaviour and performance of the member. Introduction of concrete shear walls improves to a large extent the behaviour of RCC buildings under strong ground motions. RCC structures are inferior to steel structures with respect to strength-to-weight ratio, degradation, and deformability. In the prestressed concrete structures, the introduction of prestressing adversely affects the deformability and hence the seismic characteristics of the building. Prestressed concrete is used for medium- and low-rise buildings. For tall buildings, steel is generally preferable. There is little to choose between RCC and steel for medium-rise buildings as long as the structures are well designed and detailed. Steel has an edge over concrete, since it has high strength-to-mass ratio and steel members are ductile both in flexure and shear. Steel, though expensive, is the ultimate choice to make a building ductile. Steel is most suitable for high-rise structures. However, it is not often used for low- to medium-rise buildings because of high cost. Timber, because of its high strength-to-weight ratio performs well for low-rise buildings. However, wooden structures are inferior in fire resistance.

The order of suitability of various construction materials, recommended for various types of buildings is given in Table 4.5. However, it is far from fixed, as it will depend on qualities of locally available materials, skill of the labour available, construction method, and the quality control exercised.

Table 4.5 Structural materials in appropriate order of suitability

Order of suitability	Type of building High-rise	Medium-rise	Low-rise
1.	Steel	Steel	Steel
2.	In situ reinforced concrete	In situ reinforced concrete	In situ reinforced concrete
3.		Good precast concrete	Steel
4.		Prestressed concrete	Prestressed concrete
5.		Good reinforced masonry	Good reinforced masonry
6.			Precast concrete
7.			Primitive reinforced masonry

For the purpose of earthquake resistance (for strucutres in zones III, IV, and V), construction materials should have the following desirable properties.

(a) High ductility: High plastic deformation capacity can enhance the load carrying capacity of the members.

(b) High strength-to-weight ratio: Since the inertial force is a function of mass of the structure, it is advantageous to use light and strong materials or structural systems.

(c) Orthotropy and homogeneity: The basic physical model in seismology is that of a perfectly elastic medium in which the infinitesimal strain approximation of elastic theory is adopted. Anisotropy imperfections in elasticity and inhomogeneities modify the responses predicted by simpler theories and thus are undersirable.

(d) Ease in making full strength connections: Since both ductile and brittle members can result from a combination of, e.g., brittle concrete and ductile steel, performance of structural elements cannot be evaluated by materials alone. Further, the structural continuity at connections is of great importance in evaluating the behaviour of an entire structural system.

(e) Cost: A building plan is often discarded beacuse of high cost despite its superior physical quality. The cost of the overall structure should be reasonable.

Summary

This chapter explains how the quality of a structure depends on the form of the structure and how careful planning at the conceptual stage can improve its seismic performance. There are two main fundamental requirements for seismic performance levels of buildings. First, the no-collapse condition that requires the structure to retain its full vertical load-bearing capacity. Moreover, the structure must be left with sufficient lateral strength and stiffness to protect life even during strong aftershocks. Second, the damage limitation performance level including the cost of damage and the total cost of the structure should not be disproportionately high. The factors such as redundancy, simplicity, and symmetry with a minimum of changes in the section, stiffness, strength, flexibility, and ductility of structures, when accounted for in the conceptual design assist in meeting the no-collapse and the damage-limitation requirements. Although, these are qualitative in nature, yet are sound principles to be observed during planning stage for good performance of a building. Suitable framing systems conforming to the principles of earthquake-resistant configuration, which eliminate vulnerabilities in the structural system, are described. The chapter ends with an introduction to the desirable properties of the materials to be used for earthquake-resistant construction.

Solved Problems

4.1 A building having non-uniform distribution of mass is shown in Fig 4.15. Locate its centre of mass.

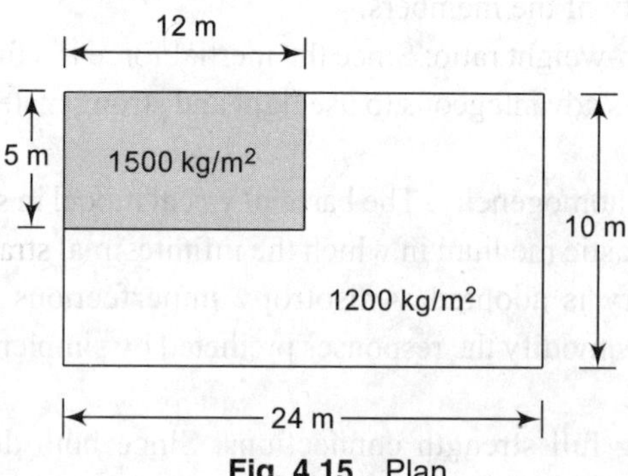

Fig. 4.15 Plan

Solution

Let us divide the roof slab into three rectangular parts as shown in Fig 4.16. Mass of part I is 1500 kg/m^2, while that of other two parts is together 1200 kg/m^2. Let origin be at point A, and the coordinates of the centre of mass be at (X,Y).

$$X = \frac{(12\times5\times1500)\times6+(12\times5\times1200)\times18+(24\times5\times1200)\times12}{(12\times5\times1500)+(12\times5\times1200)+(24\times5\times1200)}$$

$$= 11.65 \text{ m}$$

$$Y = \frac{(12\times5\times1500)\times7.5+(12\times5\times1200)\times7.5+(24\times5\times1200)\times2.5}{(12\times5\times1500)+(12\times5\times1200)+(24\times5\times1200)}$$

$$= 5.15 \text{ m}$$

Hence, coordinates of centre of mass are (11.65, 5.15).

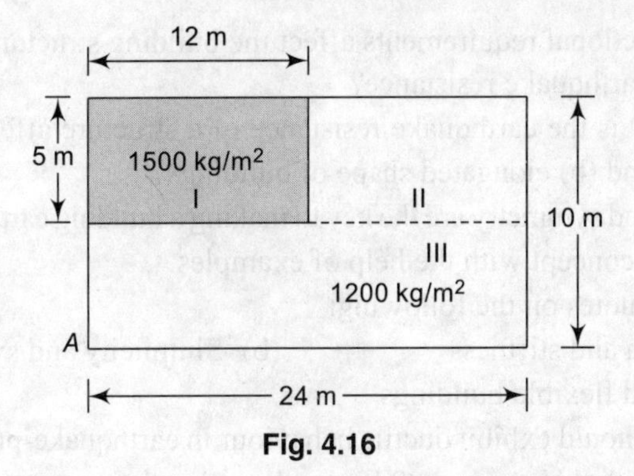

Fig. 4.16

4.2 The plan of a simple one-storey building is shown in Fig. 4.17. All the columns and the beams have same cross-sections. Obtain its centre of stiffness.

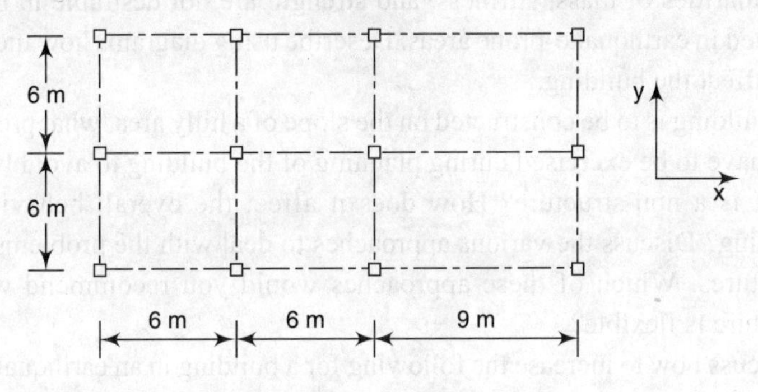

Fig. 4.17 Plan

Solution

In the *y*-direction there are three identical frames located at uniform spacing. Hence, the coordinate of centre of stiffness is located symmetrically, i.e., at $y = 6.0$ m from the left bottom corner.

In the *x*-direction, there are four identical frames having equal stiffness. However, the spacing is not uniform. Let the lateral stiffness of each transverse frame be k, and the coordinates of the centre of stiffness be (X, Y)

$$X = \frac{k \times 0 + k \times 6 + k \times 12 + k \times 21}{k + k + k + k} = 9.75$$

Hence, the coordinates of centre of stiffness are (9.75, 6.0).

Exercises

4.1 The architect and the structural engineer must coordinate at the planning stage of a building structure. Comment.

4.2 How do functional requirements affect the building structure from the point of view of earthquake resistance?

4.3 In what way is the earthquake resistance of a structure affected by (a) non-symmetry and (b) elongated shape of buildings?

4.4 Simplicity and symmetry are the key to making a building earthquake resistant. Explain the concept with the help of examples.

4.5 Write short notes on the following:
(a) Strength and stiffness (b) Simplicity and symmetry
(c) Stiff and flexible buildings

4.6. A building should exhibit ductile behaviour in earthquake-prone regions. Do you agree with this statement? If yes, then give the measures and provisions you would make at the conceptual stage to make a building stiff.

4.7 Irregularities of mass, stiffness, and strength are not desirable in buildings situated in earthquake-prone areas. Describe using diagrams how these occur and affect the building.

4.8 If a building is to be constructed on the slope of a hilly area, what precautions will have to be exercised during planning of the building to avoid twisting?

4.9 What is a non-structure? How does it affect the overall behaviour of a building? Discuss the various approaches to deal with the problems of non-structures. Which of these approaches would you recommend when the structure is flexible?

4.10 Discuss how to increase the following for a building in an earthquake-prone area:
(a) Period of vibration (b) Energy-dissipation capacity
(c) Ductility

4.11 What type of problems will an engineer face with the design of building frames shown in Fig. 4.18?

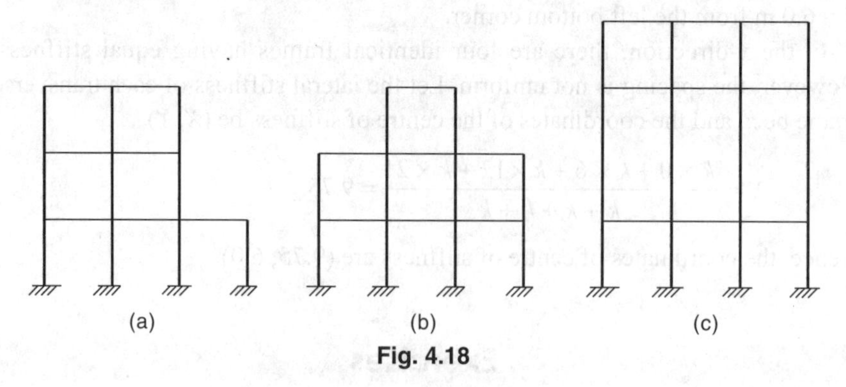

Fig. 4.18

4.12 For the building shown in Fig. 4.19, locate the centre of mass. The building has non-uniform distribution of mass as shown in the figure.

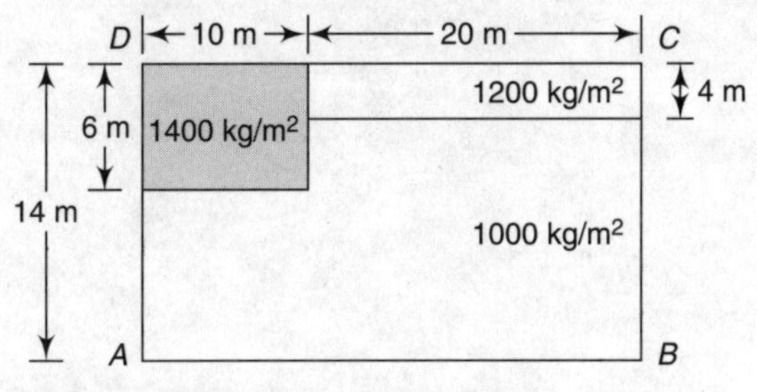

Fig. 4.19 Plan of building

Ans: 14.652 m, 7.383 m

CHAPTER 5

Code-based Analysis Method and Design Approaches

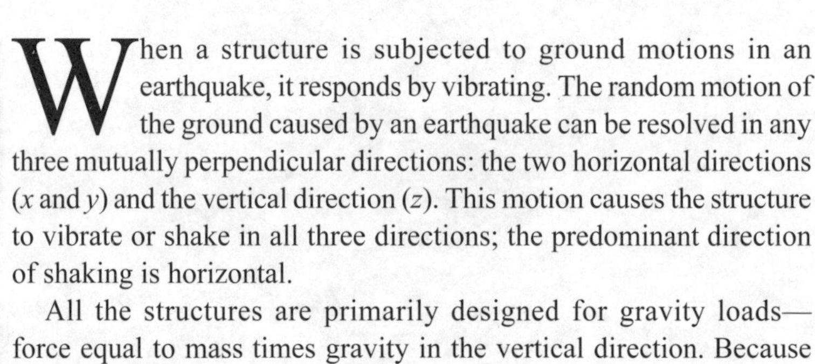

When a structure is subjected to ground motions in an earthquake, it responds by vibrating. The random motion of the ground caused by an earthquake can be resolved in any three mutually perpendicular directions: the two horizontal directions (x and y) and the vertical direction (z). This motion causes the structure to vibrate or shake in all three directions; the predominant direction of shaking is horizontal.

All the structures are primarily designed for gravity loads—force equal to mass times gravity in the vertical direction. Because of the inherent factor of safety used in the design specifications, most structures tend to be adequately protected against vertical shaking. However, earthquake-generated vertical inertia forces must be considered in the design unless checked and proved to be insignificant. In general, buildings are not particularly susceptible to vertical ground motion, but its effect should be borne in mind in the design of RCC columns, steel column connections, and prestressed beams. Vertical acceleration should also be considered in structures with large spans, those in which stability is a criterion for design, or for overall stability analysis of structures. Structures designed only for vertical shaking, in general, may not be able to safely sustain the effect of horizontal shaking. Generally, however, the inertia forces generated by the horizontal components of ground motion require greater consideration in seismic design. Hence, it is necessary to ensure that the structure is adequately resistant to horizontal earthquake shaking too.

As the ground on which a building rests is displaced, the base of the building moves suddenly with it, but the roof has a tendency to stay in its original position. The tendency to continue to remain in its original position is known as *inertia.* So the upper part of the structure will not respond instantaneously but will lag because of inertial resistance and flexibility of the structure. Since the roofs and foundations are connected with the walls and columns, the roofs are dragged along with the walls/columns. The building is thrown backwards and the roof experiences a force called the inertia force (Fig. 5.1). The maximum inertia force acting on a simple structure during an earthquake may be obtained by multiplying the roof mass m by the acceleration a. When designing a building according to the codes, the lateral force is considered in each of the two orthogonal horizontal directions of the structure. For structures having lateral force-resisting elements (e.g., braced frames, shear walls) in both directions, the design lateral force is considered along one direction at a time, and not in both the directions simultaneously. Structures with lateral force-resisting elements in directions other than the two orthogonal directions are analysed for the load combinations given in Section 5.2.

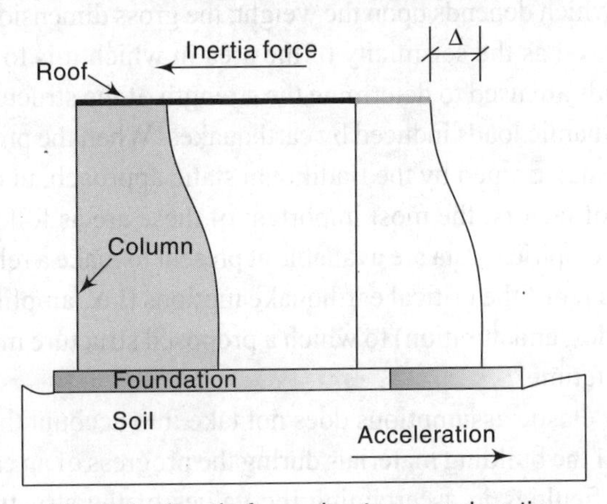

Fig. 5.1 Inertia force due to ground motion

The steps involved in an adequate earthquake-resistant design for a structure include the following:

1. Selecting a workable overall structural concept as discussed in Chapter 4
2. Establishing preliminary trial member sizes
3. Performing a structural analysis of the members to verify that stress and displacement requirements are satisfied

4. Providing structural and non-structural details so that the building will accommodate the distortions and stresses. Elements which cannot accommodate these stresses and distortions, such as rigid stairs, partitions, and irregular wings, should be isolated to reduce detrimental effects to the lateral force-resisting system.

5.1 Seismic Design Requirements

The two most important elements of concern to a structural engineer are calculation of seismic design forces and the means for providing sufficient ductility. In most structural engineering problems, dead loads (DLs), live loads, and wind loads can be evaluated with a fair degree of accuracy. However, the situation with regard to earthquake forces is entirely different. The loads or forces that a structure sustains during an earthquake, result directly from the distortion induced in the structure by the motion of the ground on which it rests. Base motion is characterized by displacements, velocities, and accelerations, which are erratic in direction, magnitude, duration, and sequence.

Earthquake loads (ELs) are inertia forces related to the mass, stiffness, and energy-absorbing (e.g., damping and ductility) characteristics of the structure. The design seismic loading recommended by building codes is in the form of static lateral loading, which depends upon the weight, the gross dimensions, and the type of structure, as well as the seismicity of the area in which it is to be built. These static design loads are used to determine the strength of the structure necessary to withstand the dynamic loads induced by earthquakes. When the proper earthquake design loads are determined by the traditional static approach, uncertainties arise from a number of factors; the most important of these are as follows:

(a) Not enough empirical data are available at present to make a reliable prediction of the character of the critical earthquake motions (i.e., amplitude, frequency characteristics, and duration) to which a proposed structure may be subjected during its lifetime.

(b) Analysis by elastic assumptions does not take into account the change in the properties of the building materials during the progress of an earthquake. This presents difficulties in ascertaining the values of the structural parameters affecting the dynamic response (e.g., stiffness and damping), as well as the dynamic properties of the soil or supporting medium.

(c) Soil–structure interaction and geological conditions have a profound effect on structural performance. At present there is no clear-cut method to correctly incorporate these effects.

Despite these uncertainties, the structure should perform satisfactorily beyond the elastic-code-stipulated stress. Ductility, the foremost important property in the inelastic range thus becomes a necessity for an earthquake-resistant design of a structure. It is generally accepted that sufficient ductility will be achieved by following the codes. However, design codes are prepared for regular structures. For structures requiring high ductility, e.g., a light flexible structure attached to a large structure, careful analysis may be required.

Importance of Ductility

The seismic forces specified in the code are quite small in comparison to the actual forces (4–6 times) expected at least once in the lifetime of the building. In spite of the large difference, the structures designed for the lateral loads specified by the code have survived severe earthquakes because of the following reasons.

1. Due to the ductility of the structure, energy is dissipated by post-elastic deformations.
2. It is the reduced response due to increased damping and soil–structure interaction.

Figure 5.2 shows the relationship between lateral design forces for an elastic structure and for a yielding ductile structure. Much larger design forces are required for an elastic structure without ductility.

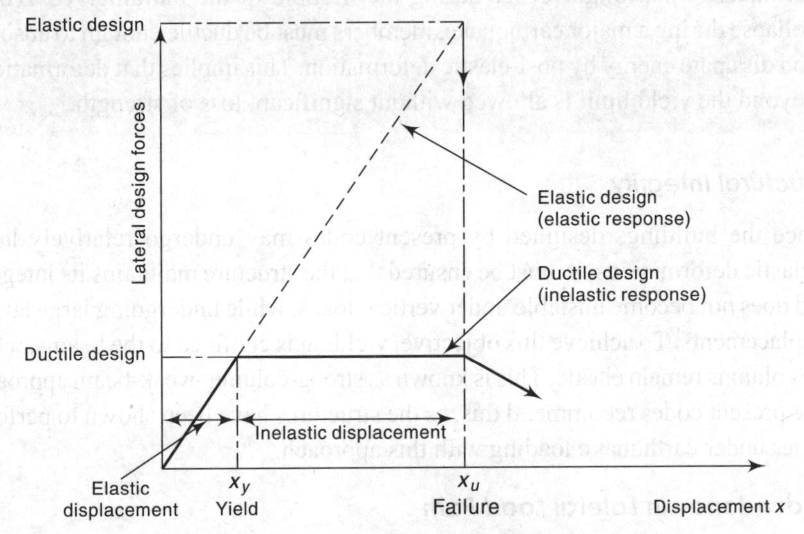

Fig. 5.2 Lateral forces and ductilty (ductily factor x/x_y, ductile capacity x_u/x_y)

During the life of a structure located in a seismically active zone, it is generally expected that the structure will be subjected to many small earthquakes, a few moderate earthquakes, one or more large earthquakes, and, possibly, a very severe earthquake. If the earthquake motion is severe, most structures will yield in some of their elements. The energy absorption capacity of the yielding structure will limit the damage. Thus, buildings that are properly designed and detailed can survive earthquake forces, which are substantially greater than the design forces that are associated with allowable stresses in the elastic range. It is evident that it would be uneconomical to design a structure to withstand the greatest likely earthquake, without damage within the elastic range. Hence, the structure is allowed to be damaged in case of severe shaking. The cost of securing the structure against strong shaking must be weighed against the importance of the structure and the probability of earthquakes. Seismic design concepts must consider the building's proportions, and details of its ductility (capacity to yield) and reserve energy absorption capacity, to ensure that it survives the inelastic deformations that would result from a maximum expected earthquake. Special attention must be given to connections that hold the lateral force-resisting elements together.

The basic intent of design theory for earthquake-resistant structures is that buildings should be able to resist minor earthquakes without damage, resist moderate earthquakes without structural damage but with some non-structural damage, and resist major earthquakes without collapse but with some structural and non-structural damage. This indicates that damage during earthquakes is acceptable as long as loss of life is avoided. The objective is to have structures that will behave elastically and survive without collapse under major earthquakes that might occur during the lifetime of the building. To avoid collapse during a major earthquake, members must be ductile enough to absorb and dissipate energy by post-elastic deformation. This implies that deformation beyond the yield limit is allowed without significant loss of strength.

Structural Integrity

Since the buildings designed by present codes may undergo relatively large inelastic deformations, it must be ensured that the structure maintains its integrity and does not become unstable under vertical loads, while undergoing large lateral displacements. To achieve this objective, yielding is confined to the beams, while the columns remain elastic. This is known as strong-column–weak-beam approach. The present codes recommend this, as the structures have been shown to perform better under earthquake loading with this approach.

Redundancy in Lateral Load Path

Structural systems that combine several lateral load-resisting elements or sub-systems generally have been observed to perform well during earthquakes. Redundancy in the structural system permits redistribution of internal forces in the event of the failure of key elements. When the primary element or system yields or fails, the lateral force can be redistributed to a secondary system to prevent progressive failure. Without capacity for redistribution, global structural collapse can result from failure of individual members or connections. Redundancy can be provided by several means—a dual system, a system of interconnected frames that enable redistribution among frames after yield has initiated in individual frames, and multiple shear walls. Redundancy combined with adequate strength, stiffness, and continuity can alleviate the need for excesses in ductile detailing.

Non-structural Elements

Elements attached to the floors of buildings (e.g., mechanical equipment, ornamentation, piping, non-structural partitions) respond to floor motion in much the same manner as the building responds to ground motion. However, the floor motion may vary substantially from the ground motion. The high frequency components of the ground motion tend to be filtered out at the higher levels in the building, while the components of ground motion that correspond to the natural periods of vibrations of the building tend to be magnified. If the elements are rigid and are rigidly attached to the structure, the forces on the elements will be in the same proportion to the mass as the forces on the structure. But elements that are flexible, and have periods of vibration close to any of the predominant modes of the building's vibration, will experience forces in a proportion substantially greater than the forces on the structure. The analysis of non-structural elements is described in Chapter 10.

Regular and Irregular Configurations

The importance of the rational conceptual design of structures is discussed in Chapter 4. It is ensured that lateral loads are transferred to the ground without excessive rotations in ductile manner. Reasonable strength and ductility can be achieved by the mandatory requirements of the code. A building with irregular configuration may be designed to meet all code requirements, but it will not perform well as a building with a regular configuration. If the building has an odd shape that is not properly considered in the design, then good design and construction are of secondary value.

However, to perform well in an earthquake, a building should possess four main attributes—simple and regular configuration, adequate lateral strength, stiffness, and ductility. Buildings having simple regular geometry and uniformly distributed mass and stiffness in plan as well as in elevation suffer much less damage than buildings with irregular configurations. When a building has an irregular feature, such as asymmetry in plan or vertical discontinuity, the assumptions used in analysis of buildings with regular features may not apply. These irregularities result in building responses significantly different from those assumed in the equivalent static force procedure. Therefore, it is best to avoid creating buildings with irregular features. A building is considered as irregular if at least one of the conditions given in Appendices IV and V is applicable.

A building is analysed for its response to ground motion by representing the structural properties in an idealized mathematical model as an assembly of masses interconnected by springs and dampers. The tributary weight to each floor level is lumped into a single mass, and the force-deformation characteristics of the lateral force-resisting walls or frames between floor levels are transformed into equivalent storey stiffness. Because of the complexity of the calculations, the use of a computer program is necessary, even when the equivalent static force procedure is used in design.

5.2 Design Earthquake Loads

The random motion of the ground caused by an earthquake causes inertia forces in a structure both in the horizontal (x and y) and vertical directions (z). These design earthquake loads and their combinations are discussed in the following subsections.

5.2.1 Design Horizontal Earthquake Load

When the lateral load-resisting elements are oriented along orthogonal horizontal directions, the structure should be designed so that the effects due to a full design earthquake load act in one horizontal direction at a time. When the lateral load-resisting elements are not oriented along the orthogonal horizontal directions, the structure should be designed for the effects due to a full design earthquake load in one horizontal direction, and 30% of the design earthquake load in the other direction. For instance, the building should be designed for ($\pm EL_x \pm 0.3EL_y$) as well as ($\pm 0.3EL_x \pm EL_y$), where x and y are two orthogonal horizontal directions and EL is the value of EL adopted for design.

5.2.2 Design Vertical Earthquake Load

Due to the random earthquake ground motions, which cause the structure to vibrate, all structures experience a constant vertical acceleration (downward) that may be additive or subtractive to the gravity depending on the direction of ground motion at that instant. Factor of safety for gravity loads is usually sufficient to cover the earthquake-induced vertical acceleration. When effects due to vertical ELs are to be considered, the design vertical force is calculated by considering the vertical acceleration as two-thirds of the horizontal acceleration. However, this is a possible median value of vertical acceleration and certainly should not be used for sensitive structures.

5.2.3 Combination for Two- or Three-component Motion

When responses from the three earthquake components are to be considered, they may be combined using the assumption that when the maximum response from one component occurs, the response from the other two components is 30% of their maximum. All possible combinations of the three components (EL_x, EL_y, and EL_z) including variations in sign (plus or minus) should be considered. Thus, the response due to earthquake force (EL) is the maximum for the following three cases.

(a) $\pm EL_x \pm 0.3EL_y \pm 0.3EL_z$
(b) $\pm EL_y \pm 0.3EL_z \pm 0.3EL_x$
(c) $\pm EL_z \pm 0.3EL_x \pm 0.3EL_y$

where x and y are the two orthogonal directions and z is the vertical direction.

As an alternative to the above procedure, the response (EL) due to the combined effect of the three components can be obtained by computing the square root of the sum of the squares (SRSS), that is

$$EL = \sqrt{(EL_x)^2 + (EL_y)^2 + (EL_z)^2}$$

The above combination procedure applies to the same response quantity (say, moment in a column about its major axis, or storey shear in a frame) due to different components of the ground motion.

5.2.4 Basic Load Combinations

The following different load combinations of gravity and lateral loads with appropriate load factors as given by the codes are worked out. The structure is then analysed and designed for the combination that yields the most critical value. Here, the terms DL, IL, EL stand for response quantities due to DL, IL, and designated EL, respectively.

For plastic design of steel structures, the following load combinations should be accounted for. Here 1.7 and 1.3 are the partial safety factors.

(a) 1.7(DL + IL)
(b) 1.7(DL ± EL)
(c) 1.3(DL + IL ± EL)

For the limit state design of reinforced and prestressed concrete structures, the following load combinations should be accounted for. Here, 1.5, 1.2, and 0.9 are the partial safety factors.

(a) 1.5(DL + IL)
(b) 1.2(DL + IL ± EL)
(c) 1.5 (DL ± EL)
(d) 0.9DL ± 1.5EL

Masonry and timber structures are usually designed using allowable stress design method. However, provisions in IS code to achieve ductility must be incorporated.

Notes: 1. In the load combination 0.9DL ± 1.5EL, since the horizontal loads are reversed and since in some situations the design is governed by effect of horizontal load minus effect of gravity load, a load factor higher than 1 for gravity loads will be unconservative. Hence, a load factor of 0.9 is specified for gravity loads.

2. In limit state design method, partial safety factors for the load combinations are the same for RCC and steel structures. However, these are different for the two types of construction materials, concrete and steel, because there is less control on the quality of concrete than the steel; steel is rolled in the rolling mills under controlled conditions.

5.3 Permissible Stresses

The permissible stresses are specified by the codes for design of structures subjected to static loading. These are increased considering safety factors to account for the transient effect.

When earthquake forces are considered along with other normal design forces, the permissible stresses in material, in the elastic method of design, may be increased by one-third. However, for steels with a definite yield point, the stress may be limited to yield stress. For steels without a definite yield point, the stress will be limited to 80% of the ultimate strength or 0.2% of proof stress, whichever is smaller. For prestressed concrete members, the tensile stress in the extreme fibres of the concrete may be permitted so as not to exceed two-thirds of the modulus of rupture of concrete. The allowable bearing pressure in soils is increased as given in Table 5.1, depending upon the type of foundation of the structure and the type of soil.

Table 5.1 Percentage of permissible increase in allowable bearing pressure or resistance of soils

Foundation	Type of soil mainly constituting the foundation		
	Type I—rock soil or hard soil Well-graded gravel and sand gravel mixtures, with or without clay binder, and clayey sands poorly graded or sand clay mixtures having N above 30, where N is the standard penetration value.	*Type II—medium soil* with N between 10 and 30 and poorly graded sands (SP) or gravelly sands with little or no fines (SP) with $N > 15$.	*Type III—soft soil* All soils other than SP with $N < 10$.
Piles passing through any soil but resting on soil type I	50	50	50
Piles not covered under conditions stated above	–	25	25
Raft foundations	50	50	50
Combined isolated RCC footing with tie beams	50	25	25
Isolated RCC footing without tie beams, or unreinforced strip foundations.	50	25	–
Well foundations	50	25	25

Notes:

1. The allowable bearing pressure should be determined in accordance with IS 6403 or IS 1888.

(Contd)

Table 5.1 (*Contd*)

2. If any increase in bearing pressure has already been permitted for forces other than seismic forces, the total increase in allowable bearing pressure when the seismic force is also included should not exceed the limits specified above.
3. Desirable minimum field values of *N*: If soils of smaller *N* values are met, either compacting should be adopted to achieve these values or deep pile foundations going to stronger strata should be used.
4. The values of *N* (corrected values) are at the founding level and the allowable bearing pressure shall be determined in accordance with IS 6403 or IS 1888.

Seismic zone level (in metres)	*Depth below ground*	*N values*	*Remark*
III, IV, and V	≤ 5	15	For values of depths between 5 m and 10 m, linear interpolation is recommended
	≥ 10	20	
II (for important structures only)	≤ 5	15	
	≥ 10	20	

5. The piles should be designed for lateral loads, neglecting the lateral resistance of soil layers liable to liquefy.
6. IS 1498 and IS 2131 may also be referred to.
7. Isolated RCC footing without tie beams or unreinforced strip foundation should not be permitted in soft soils with $N < 10$.

(i) See IS 1498 (ii) See IS 2131

5.4 Seismic Methods of Analysis

After selecting the structural model, it is possible to perform analysis to determine the seismically induced forces in the structures. The analysis can be performed on the basis of the external action, the behaviour of the structure or structural materials, and the type of structural model selected. The analysis process can be classified as shown in Fig. 5.3. Depending on the nature of the considered variables, the method of analysis can be classified as shown in Fig. 5.4. Based on the type of external action and behaviour of structure, the analysis can be further classified as *linear static analysis*, *linear dynamic analysis*, *non-linear static analysis*, or *non-linear dynamic analysis*.

Linear static analysis or equivalent static analysis can be used for regular structures with limited height. Linear dynamic analysis can be performed in two ways, either by the response spectrum method or by the elastic time-history method. The significant difference between linear static and linear dynamic analyses is the level of the forces and their distribution along the height of the structure.

Non-linear static analysis is an improvement over linear static or dynamic analysis in the sense that it allows inelastic behaviour of the structure. The method is simple to implement and provides information on the strength, deformation, and ductility of the structure, as well as the distribution of demands. This permits

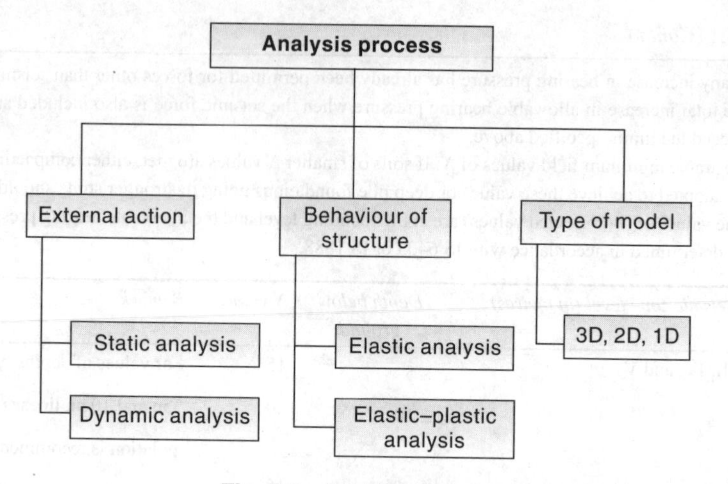

Fig. 5.3 Analysis process

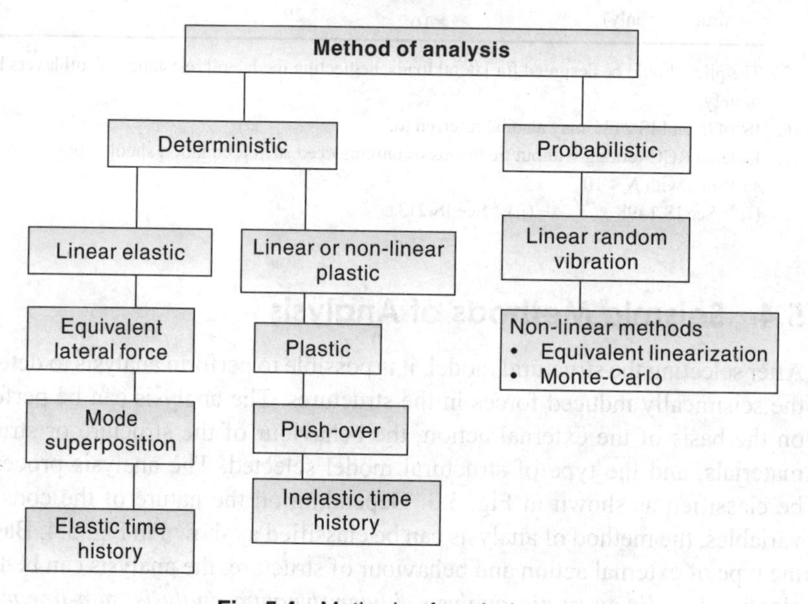

Fig. 5.4 Methods of analysis

the identification of the critical members that are likely to reach limit states during the earthquake, to which attention should be paid during the design and detailing process. But the non-linear static method is based on many assumptions, which neglect the variation of loading patterns, the influence of higher modes of vibration, and the effect of resonance. In spite of the deficiencies, this method, known as *push-over analysis*, provides a reasonable estimation of the global deformation capacity, especially for structures that primarily respond according to the first mode.

A *non-linear dynamic analysis* or *inelastic time-history analysis* is the only method to describe the actual behaviour of a structure during an earthquake.

The method is based on the direct numerical integration of the differential equations of motion by considering the elasto-plastic deformation of the structural element.

The scope of this book limits the discussion to only methods of elastic analysis; namely, the seismic coefficient method, dynamic analysis, and a brief description of the time-history method. These are explained in the sections that follow.

5.4.1 Basic Assumptions

The following assumptions are made in the analysis of earthquake-resistant design of structures.

(a) An earthquake causes impulsive ground motions, which are complex and irregular in character, with each change in period and amplitude lasting for a small duration. Therefore, resonance of the type visualized under steady-state sinusoidal excitations will not occur, as it would need time to build up such amplitudes. However, there are exceptions where resonance-like conditions have been seen to occur between long-distance waves and tall structures founded on deep soft soils.

(b) An earthquake is not likely to occur simultaneously with winds or powerful floods and sea waves.

The probability of occurrences of strong earthquake motion along with strong winds and/or maximum sea waves is low. Therefore, it is justified to assume that these hazardous events are not occurring at the same time.

(c) The value of elastic modulus of materials, wherever required, may be taken as the one used for static analysis, unless a more definite value is available for use in such a condition.

It may be noted that the values of modulus of elasticity for various construction materials display large variations.

5.4.2 Methods of Elastic Analysis

The most commonly used methods of analysis are based on the approximation that the effects of yielding can be accounted for by linear analysis of the building using the design spectrum for inelastic systems. Forces and displacements due to each horizontal component of ground motion are separately determined by analysis of an idealized building having one lateral degree of freedom per floor in the direction of the ground motion component being considered. Such analysis may be carried out by the equivalent lateral force procedure (static method) or response spectrum analysis procedure (dynamic method). Another refined method of dynamic analysis is the elastic time-history method.

Both the equivalent lateral force and response spectrum analysis procedures lead directly to lateral forces in the direction of the ground motion component. The main differences between the two methods are in the magnitude and distribution of the lateral forces over the height of the building. The equivalent lateral force method is mainly suited for preliminary design of the building. The preliminary design of the building is then used for response spectrum analysis or any other refined method such as the elastic time-history method.

Equivalent Lateral Force Method (Seismic Coefficient Method)

Seismic analysis of most structures is still carried out on the assumption that the lateral force is equivalent to the actual (dynamic) loading. This method requires less effort because, except for the fundamental period, the periods and shapes of higher natural modes of vibration are not required. The base shear, which is the total horizontal force on the structure, is calculated on the basis of the structure's mass, its fundamental period of vibration, and corresponding shape. The base shear is distributed along the height of the structure, in terms of lateral forces, according to the code formula. Planar models appropriate for each of the two orthogonal lateral directions are analysed separately; the results of the two analyses and the various effects, including those due to torsional motions of the structure, are combined. This method is usually conservative for low- to medium-height buildings with a regular configuration.

Response Spectrum Analysis

This method is also known as *modal method* or *mode superposition method*. The method is applicable to those structures where modes other than the fundamental one significantly affect the response of the structure. This method is based on the fact that, for certain forms of damping—which are reasonable models for many buildings—the response in each natural mode of vibration can be computed independently of the others, and the modal responses can be combined to determine the total response. Each mode responds with its own particular pattern of deformation (mode shape), with its own frequency (the modal frequency), and with its own modal damping. The time history of each modal response can be computed by analysis of an SDOF oscillator with properties chosen to be representative of the particular mode and the degree to which it is excited by the earthquake motion. In general, the responses need to be determined only in the first few modes because response to earthquake is primarily due to lower modes of vibration.

A complete modal analysis provides the history of response—forces, displacements, and deformations—of a structure to a specified ground acceleration history. However, the complete response history is rarely needed for design; the maximum values of response over the duration of the earthquake usually suffice. Because the response in each vibration mode can be modelled by the response of an SDOF oscillator, the maximum response in the mode can be directly computed from the earthquake response spectrum. Procedures for combining the modal maxima to obtain estimates (but not the exact value) of the maximum of total response are available.

In its most general form, the modal method for linear response analysis is applicable to arbitrary three-dimensional structural systems. However, for the purpose of design of buildings, it can often be simplified from the general case by restricting its application to the lateral motion in a plane. Planar models appropriate for each of two orthogonal lateral directions are analysed separately and the results of the two analyses and the effects of torsional motions of the structures are combined.

Generally, the method is applicable to analysis of the dynamic response of structures, which are assymmetrical or have areas of discontinuity or irregularity, in their linear range of behaviour. In particular, it is applicable to analysis of forces and deformations in multi-storey buildings due to medium-intensity ground shaking, which causes a moderately large but essentially linear response in the structure.

Elastic Time-history Method

A linear time-history analysis (THA) overcomes all the disadvantages of a modal response spectrum analysis provided non-linear behaviour is not involved. This method requires greater computational efforts for calculating the response at discrete times. One interesting advantage of such a procedure is that the relative signs of response quantities are preserved in the response histories. This is important when interaction effects are considered among stress resultants.

5.4.3 Limitations of Equivalent Lateral Force and Response Spectrum Analysis Procedures

The assumptions common to the equivalent lateral force procedure and the response spectrum analysis procedure are as follows—(a) forces and deformations can be determined by combining the results of independent analyses of a planar idealization of the building for each horizontal component of ground motion, and by including torsional moments determined on an indirect, empirical basis and (b) non-linear structural response can be determined to an acceptable degree of accuracy, by linear analysis of the building using the design spectrum for inelastic systems. Both analysis procedures are likely to be inadequate if the dynamic response behaviour of the building is quite different from what is implied by these assumptions, and also if the lateral motions in two orthogonal directions and the torsional motions are strongly coupled.

Buildings with large eccentricities at the centres of storey resistance relative to the centres of floor mass, or buildings with close values of natural frequencies of the lower modes and essentially coincident centres of mass and resistance, exhibit coupled lateral-torsional motions. For such buildings independent analyses for the two lateral directions may not suffice, and at least three degrees of freedom per floor—two translational motions and one torsional—should be included in the idealized model. The modal method, with appropriate generalizations of the concept involved, can be applied to analyses of the model. Because natural modes of vibration will show a combination of translational and torsional motions, it is necessary while determining the modal maxima to account for two facts: that a given mode might be excited by both horizontal components of ground motion; and modes that are primarily torsional can be excited by translational components of ground motion. Because natural frequencies of a building with coupled lateral torsional motions can be rather close to each other, the modal maxima should not be combined in accordance with the SRSS formula; instead a more general formula should be employed.

5.4.4 Equivalent Lateral Force versus Response Spectrum Analysis Procedures

Both, the equivalent lateral force procedure and the response spectrum analysis procedure, are based on the same basic assumptions and are applicable to buildings that exhibit a dynamic response behaviour in reasonable conformity with the implications of the assumptions made in the analysis. The main difference between the two procedures lies in the magnitude of the base shear and distribution of the lateral forces. Although in the modal method the force calculations are based on compound periods and mode shapes of several modes of vibration, in the equivalent lateral force method, they are based on an estimate of the fundamental period and simple formulae for distribution of forces which are appropriate for buildings with regular distribution of mass and stiffness over height.

It would be adequate to use the equivalent lateral force procedure for buildings with the following properties—seismic force-resisting system has the same configuration in all storeys and in all floors; floor masses do not differ by more than, say, 30% in adjacent floors; and cross-sectional areas and moments of inertia of structural members do not differ by more than about 30% in adjacent storeys. For other buildings, the following sequence of steps may be employed to decide whether the modal analysis procedure ought to be used.

1. Compute lateral forces and storey shears using the equivalent lateral force procedure.
2. Approximate the dimensions of structural members.
3. Compute lateral displacements of the structure as designed in step 2 due to lateral forces in step 1.
4. Compute new sets of lateral forces and storey shears with the displacements computed in step 3.
5. If at any storey the recomputed storey shear (step 4) differs from the corresponding original value (step 1) by more than 30%, the structure should be analysed by the modal analysis procedure. If the difference is less than this value the modal analysis procedure is unnecessary, and the structure should be designed using the storey shears obtained in step 4; they represent an improvement over the results of step 1.

This method for determining modal analysis is efficient as well as effective. It requires far less computational effort than the use of the modal analysis procedure.

The seismicity of the area and the potential hazard due to failure of the building should also be considered in deciding whether the equivalent lateral force procedure is adequate. For example, even irregular buildings that may require modal analysis according to the criterion described may be analysed by the equivalent lateral force procedure if they are not located in higher seismic zones and do not house the critical facilities necessary for post-disaster recovery or a large number of people.

5.5 Factors in Seismic Analysis

The factors taken into account in assessing lateral design forces and the design response spectrum are described as follows.

5.5.1 Zone Factor [IS 1893 (Part 1): 2002, Clause 6.4]

Seismic zoning assesses the maximum severity of shaking that is anticipated in a particular region. The zone factor (Z), thus, is defined as a factor to obtain the design spectrum depending on the perceived seismic hazard in the zone in which the structure is located. The basic zone factors included in the code are reasonable estimate of effective peak ground acceleration. Zone factors as per IS 1893 (Part 1): 2002 are given in Table 5.2.

Table 5.2 Zone factor (Z)

Seismic zone	II	III	IV	V
Seismic intensity	Low	Moderate	Severe	Very severe
Z	0.10	0.16	0.24	0.36

5.5.2 Importance Factor [IS 1893 (Part 1): 2002, Clause 7.2]

The importance factor is a factor used to obtain the design seismic force depending upon the functional use of the structure. It is customary to recognize that certain categories of buildings should be designed for greater levels of safety than the others, and this is achieved by specifying higher lateral design forces. Such categories are

(a) Buildings which are essential after an earthquake—hospitals, fire stations, etc.
(b) Places of assembly—schools, theatres, etc.
(c) Structures the collapse of which may endanger lives—nuclear plants, dams, etc.

The importance factors are given in Table 5.3.

Table 5.3 Importance factor (I)

Structure	Importance factor (I)
Important service and community buildings, such as hospitals; schools; monumental structures; emergency buildings like telephone exchanges, television stations, radio stations, railway stations, fire station buildings; large community halls like cinemas, assembly halls; and subway stations, power stations	1.5
All other buildings	1.0

Notes:

1. The design engineer may choose values of importance factor I greater than those mentioned above.
2. Buildings not covered in the table above may be designed for a higher value of I, depending on economy and strategy considerations. These could be buildings such as multi-storey buildings having several residential units.
3. This table does not apply to temporary structures like excavations, scaffolding, etc.

To achieve higher level of seismic performance, structures are designed for importance factor (*I*) of 1.5. When $I > 1$, the effective response reduction factor *R* value reduces the inelastic behaviour. Consequently, there is a reduction in the potential damage to the structure. As per IS 1893 (Part 4): 2005, higher values of importance factors may be used in situations to follow. An importance factor of 2 should be considered for industrial structures that can lead directly or indirectly to extensive loss of life/property to population at large in the areas adjacent to the plant complex, and 1.75 should be used in case when considerations for serious fire hazards/extensive damage within the plant are anticipated.

5.5.3 Response Reduction Factor [IS 1893 (Part 1): 2002, Clause 6.4]

The basic principle of designing a structure for strong ground motion is that the structure should not collapse but damage to the structural elements is permitted. Since a structure is allowed to be damaged in case of severe shaking, the structure should be designed for seismic forces much less than what is expected under strong shaking, if the structures were to remain linearly elastic. Response reduction factor (*R*) is the factor by which the actual base shear force should be reduced, to obtain the design lateral force. Base shear force is the force that would be generated

Table 5.4 Response reduction factor (*R*) for building systems

Lateral load-resisting system	Response reduction factor (*R*)
Building frame systems	
Ordinary RCC moment-resisting frame (OMRF)	3.0
Special RCC moment-resisting frame (SMRF)	5.0
Steel frame with	
(a) Concentric braces	4.0
(b) Eccentric braces	5.0
Steel moment-resisting frame designed as per SP 6(6)	5.0
Load bearing masonry wall buildings	
(a) Unreinforced	1.5
(b) Reinforced with horizontal RCC bands	2.5
(c) Reinforced with horizontal RCC bands and vertical bars at corners of rooms and jambs of openings	3.0
Buildings with shear walls	
Ordinary RCC shear walls	3.0
Ductile shear walls	4.0
Buildings with dual systems	
Ordinary shear wall with OMRF	3.0
Ordinary shear wall with SMRF	4.0
Ductile shear wall with OMRF	4.5
Ductile shear wall with SMRF	5.0

Notes:

1. The values of response reduction factors are to be used for buildings with lateral load-resisting elements, and not just for the lateral load-resisting elements built in isolation.

(*Contd*)

Table 5.4 (*Contd*)

2. OMRF are those designed and detailed as per IS 456 or IS 800, but not meeting ductile detailing requirement as per IS 13920 or SP 6(6).
3. SMRF are those designed as OMRF and meeting the ductile detailing requirement.
4. Buildings with shear walls also include buildings having shear walls and frames, but where
   - frames are not designed to carry lateral loads, or
   - frames are designed to carry lateral loads but do not fulfil the requirements of 'dual systems'.
5. Reinforcement should be as per IS 4326.
6. Prohibited in Zones IV and V.
7. Ductile shear walls are those designed and detailed as per IS 13920.
8. Buildings with dual systems consist of shear walls (or braced frames) and moment-resisting frames such that (i) the two systems are designed to resist the total design force in proportion to their lateral stiffness, considering the interaction of the dual system at all floor levels; and (ii) the moment-resisting frames are designed to independently resist at least 25% of the design seismic base shear.

if the structure were to remain elastic during its response to the design basis earthquake (DBE) shaking. The values of response reduction factors arrived at empirically based on engineering judgement are given in Table 5.4. IS 1893 (Part I): 2002 uses R in the design to account for ductility. For example, a high value such as five, for special RCC moment-resisting frames, reflects their high ductility. Overstrength, redundancy, and ductility together contribute to the fact that an earthquake-resistant structure can be designed for a much lower force than that imparted by strong shaking of the structure (Fig. 5.5).

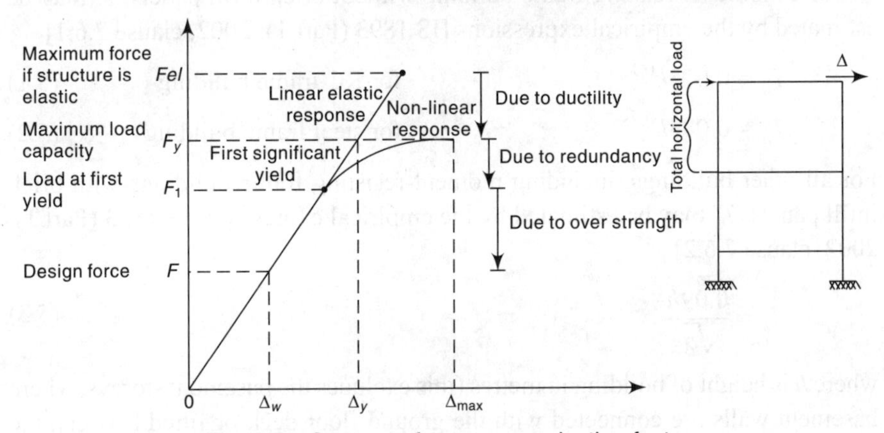

Fig. 5.5 Concept of response reduction factor

Overstrength The factors that account for the yielding of a structure at loads higher than the design loads are

(a) Partial safety factors on seismic loads, gravity loads, and materials
(b) Material properties such as over-sized member, strain hardening, confinement of concrete, and higher material strength under cyclic loads
(c) Strength contributions of non-structural elements
(d) Special ductile detailing

Redundancy Redundancy is a fundamental characteristic for good performance in earthquakes. It is a good practice to provide a building with redundant system such that failure of a single connection or element does not adversely affect the lateral stability of the structure. Yielding at one location in the structure does not imply yielding of the structure as a whole. Load redistribution in the members of redundant structures provides additional safety margin. Sometimes, the additional margin due to redundancy is considered within the overstrength term itself.

Ductility Higher ductility indicates that a structure can withstand stronger shaking without collapse. When a structure yields there is more energy dissipation in the structure due to hysteresis. Also, the structure becomes softer and its natural period increases, which implies that the structure has to now resist a lower seismic force.

5.5.4 Fundamental Natural Period

The fundamental natural period T_a is the first (longest) modal time period of vibration of the structure. Because the design loading depends on the building period, and the period cannot be calculated until a design has been prepared, IS 1893 (Part 1): 2002, clause 7.6, provides formulae from which T_a may be calculated.

For a moment-resisting frame building without brick infill panels, T_a may be estimated by the empirical expressions [IS 1893 (Part 1): 2002, clause 7.6.1]

$$T_a = 0.075h^{0.75} \quad \text{for RC frame building} \tag{5.1}$$

$$T_a = 0.085h^{0.75} \quad \text{for steel frame building} \tag{5.2}$$

For all other buildings, including moment-resisting frame buildings with brick infill panels, T_a may be estimated by the empirical expression [IS 1893 (Part 1): 2002, clause 7.6.2]

$$T_a = \frac{0.09h}{\sqrt{d}} \tag{5.3}$$

where h is height of building in metres (this excludes the basement storeys, where basement walls are connected with the ground floor deck or fitted between the building columns. But it includes the basement storeys, when they are not so connected), and d is the base dimension of the building at the plinth level, in metres, along the considered direction of the lateral force.

5.5.5 Design Response Spectrum [IS 1893 (Part 1): 2002, Clause 6.4.5]

The design response spectrum is a smooth response spectrum specifying the level of seismic resistance required for a design. Seismic analysis requires that the design spectrum be specified. IS 1893 (Part 1): 2002 stipulates a design acceleration spectrum or base shear coefficients as a function of natural period. These coefficients are ordinates of the acceleration spectrum, divided by acceleration due to gravity. This relationship works well in SDOF systems. The spectral ordinates are used for the computation of inertia forces. Figure 5.6 relates to the proposed 5% damping for rocky or hard soil sites and Table 5.5 gives the multiplying factors for obtaining spectral values for various other damping (note that the multiplication is not to be done for zero period acceleration). The design spectrum ordinates are independent of the amounts of damping (multiplication factor of 1.0) and their variations from one material or one structural solution to another.

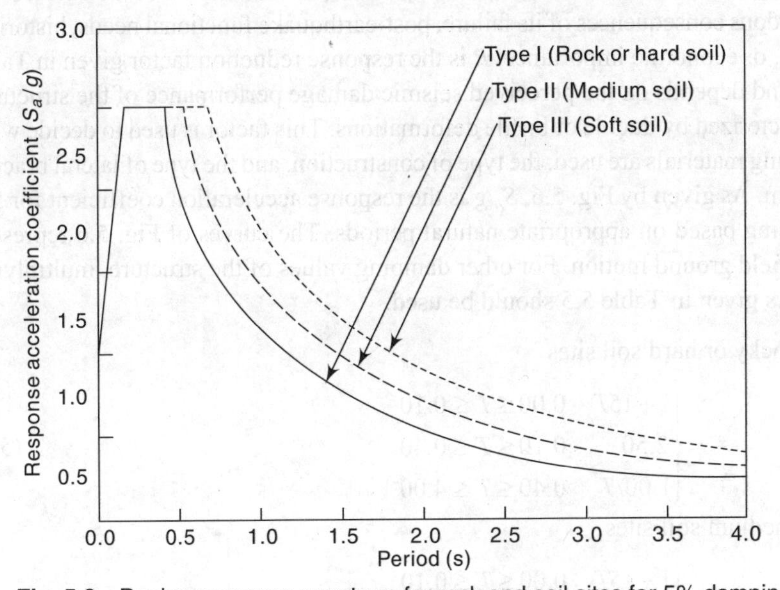

Fig. 5.6 Design response spectrum for rock and soil sites for 5% damping

Table 5.5 Multiplying factor for obtaining spectral values for damping (other than 5% damping)

Damping (%)	0	2	5	7	10	15	20	25	30
Factor	3.20	1.40	1.00	0.90	0.80	0.70	0.60	0.55	0.50

5.6 Seismic Base Shear

The total design lateral force or design seismic base shear (V_B) along any principal direction is determined [IS 1893 (Part 1): 2002, clause 7.5.3] by

$$V_B = A_h W \tag{5.4}$$

where A_h is the design horizontal acceleration spectrum value, using the fundamental natural period T in the considered direction of vibration and W is the seismic weight of the building (Section 5.7). The design horizontal seismic coefficient A_h for a structure is determined by the expression [IS 1893 (Part 1): 2002, clause 6.4.2]

$$A_h = \frac{ZIS_a}{2Rg} \tag{5.5}$$

For any structure with $T \leq 0.1$ s, the value of A_h will not be taken less than $Z/2$ whatever be the value of I/R. In Eqn (5.5), Z is the zone factor given in Table 5.2 for the maximum considered earthquake (MCE). The factor 2 in the denominator is used so as to reduce the MCE zone factor to the factor for DBE. I is the importance factor given in Table 5.3, and depends upon the functional use of the structure, the hazardous consequences of its failure, post-earthquake functional needs, historical value, or economic importance. R is the response reduction factor given in Table 5.4, and depends on the perceived seismic damage performance of the structure, characterized by ductile or brittle deformations. This factor is used to decide what building materials are used, the type of construction, and the type of lateral bracing system. As given by Fig. 5.6, S_a/g is the response acceleration coefficient for 5% damping based on appropriate natural periods. The curves of Fig. 5.6 represent free-field ground motion. For other damping values of the structure, multiplying factors given in Table 5.5 should be used.

For rocky or hard soil sites

$$\frac{S_a}{g} = \begin{cases} 1+15T & 0.00 \leq T \leq 0.10 \\ 2.50 & 0.10 \leq T \leq 0.40 \\ 1.00/T & 0.40 \leq T \leq 4.00 \end{cases} \tag{5.6}$$

For medium soil sites

$$\frac{S_a}{g} = \begin{cases} 1+15T & 0.00 \leq T \leq 0.10 \\ 2.50 & 0.10 \leq T \leq 0.55 \\ 1.36/T & 0.55 \leq T \leq 4.00 \end{cases} \tag{5.7}$$

For soft soil sites

$$\frac{S_a}{g} = \begin{cases} 1+15T & 0.00 \leq T \leq 0.10 \\ 2.50 & 0.10 \leq T \leq 0.67 \\ 1.67/T & 0.67 \leq T \leq 4.00 \end{cases} \tag{5.8}$$

5.7 Seismic Weight [IS 1893 (Part 1): 2002, Clause 7.4]

The seismic weight of the whole building is the sum of the seismic weights of all the floors. The seismic weight of each floor is its full DL plus the appropriate amount of IL, the latter being that part of the ILs that may reasonably be expected to be attached to the structure at the time of earthquake shaking. It includes the weight of permanent and movable partitions, permanent equipment, a part of the live load, etc. While computing the seismic weight of each floor, the weight of columns and walls in any storey should be equally distributed to the floors above and below the storey. Any weight supported in between storeys should be distributed to the floors above and below in inverse proportion to its distance from the floors [IS 1893 (Part 1): 2002, clause 7.3].

As per IS 1893: (Part I), the percentage of IL as given in Table 5.6 should be used. For calculating the design seismic forces of the structure, the IL on the roof need not be considered. A reduction in IL is recommended for the following reasons.

1. All the floors may not be occupied during earthquake.
2. A part of earthquake energy may get absorbed by non-rigid mountings of IL.

Table 5.6 Percentage of IL to be considered in seismic weight calculation

Imposed uniformly distributed floor load (kN/m^2)	Percentage of IL
Upto and including 3.0	25
Above 3.0	50

Notes:

1. The proportions of IL indicated above for calculating the lateral design forces for earthquakes are applicable to average conditions.
2. Where the probable loads at the time of earthquake are more accurately assessed, the designer may alter the proportions indicated or even replace the entire IL proportions by the actual assessed load.
3. Lateral design force for earthquakes should not be calculated on contribution of impact effects from ILs.
4. Other loads apart from those given above (e.g., snow and permanent equipment) should be considered as appropriate.

5.8 Distribution of Design Force [IS 1893 (Part 1): 2002, Clause 7.7]

Buildings and their elements should be designed and constructed to resist the effects of design lateral force. The design lateral force is first computed for the building as a whole and then distributed to the various floor levels. The overall design seismic force thus obtained at each floor level is then distributed to individual lateral load-resisting elements, depending on the floor diaphragm action.

5.8.1 Equivalent Lateral Force Method

This method of finding design lateral forces is also known as the *static method* or the *equivalent static method* or the *seismic coefficient method*. This procedure does not require dynamic analysis, however, it accounts for the dynamics of building in an approximate manner. The static method is the simplest one—it requires less computational effort and is based on formulae given in the code of practice. First, the design base shear is computed for the whole building, and it is then distributed along the height of the building. The lateral forces at each floor level thus obtained are distributed to individual lateral load-resisting elements.

Vertical distribution of base shear to different floor levels [IS 1893 (Part 1): 2002, Clause 7.7.1] The design base shear (V_B) is distributed along the height of the building as per the following expression.

$$Q_i = V_B \frac{W_i h_i^2}{\sum_{i=1}^{n} W_i h_i^2} \tag{5.9}$$

where Q_i is the design lateral force at floor *i*, W_i is the seismic weight of floor *i*, h_i is the height of floor *i* measured from the base, and *n* is the number of storeys in the building, i.e., the number of levels at which the masses are located.

Distribution of horizontal design lateral force to different lateral force-resisting elements [IS 1893 (Part 1): 2002, Clause 7.7.2] In the case of buildings in which floors are capable of providing rigid horizontal diaphragm action, the total shear in any horizontal plane is distributed to the various vertical elements of the lateral force-resisting system, assuming the floors to be infinitely rigid in the horizontal plane. For buildings in which floor diaphragms cannot be treated as infinitely rigid in their own plane, the lateral shear at each floor is distributed to the vertical elements resisting the lateral forces, accounting for the in-plane flexibility of the diaphragms.

- A floor diaphragm is considered to be flexible, if it deforms such that the maximum lateral displacement measured from the chord of the deformed shape at any point of the diaphragm is more than 1.5 times the average displacement of the entire diaphragm.
- RC monolithic slab-beam floors or those consisting of prefabricated/precast elements with topping of reinforced screed can be taken as rigid diaphragms.

5.8.2 Response Spectrum Method

Dynamic analysis may be performed either by response spectrum method or by the time-history method. In the response spectrum method, the peak response of a structure during an earthquake is obtained directly from the earthquake response (or design) spectrum. The design spectrum specified in Section 5.5.5 or a site-specific design spectrum, which is specially prepared for a structure at a particular project site, may be used. This procedure gives an approximate peak response, which is quite accurate for structural design purposes. In this approach, the multiple modes of response of a building to an earthquake are taken into account. For each mode, a response is read from the design spectrum, based on the modal frequency and the modal mass. The responses of different modes are combined to provide an estimate of total response of the structure using modal combination methods as described in Section 5.8.2.2 later. The procedure for determining distribution of lateral forces to each storey, as per IS 1893 (Part 1): 2002, clause 7.8.4.5, is outlined in Section 5.8.2.1.

Modal analysis is an alternative procedure to the equivalent lateral force method performed to obtain the design lateral forces at each floor level along the height of the building and its distribution to individual lateral load-resisting elements. The base shear is thus distributed to different levels along the height of the buildings and to the various lateral load-resisting elements in a way similar to that in the equivalent lateral force method.

The design base shear V_B calculated using the dynamic analysis procedure is compared with a base shear $\bar{V}_B$, calculated using a fundamental period T_a. Where V_B is less than $\bar{V}_B$, all the response quantities, e.g., member forces, displacement, storey forces, storey shears, and base reactions, should be multiplied by $\bar{V}_B/V_B$. The value of damping for buildings may be taken as 2 per cent and 5 per cent of the critical value, for the purposes of dynamic analysis of steel and RCC buildings, respectively.

- The analytical model for dynamic analysis of buildings with unusual configuration should be such that it adequately models the types of irregularities present in the building configuration. Buildings with plan irregularities as defined in Appendix IV and with vertical irregularities defined in Appendix V cann ot be modelled for dynamic analysis by the method explained in this section.
- For irregular buildings, lesser than 40 m in height in Zones II and III, dynamic analysis, though not mandatory, is recommended.

The assumptions made in the expressions used for design load calculation and load distribution with height in this procedure are as follows.

(a) the fundamental mode dominates the response and (b) mass and stiffness are evenly distributed with building height, thus giving a regular mode shape. In tall buildings, higher modes can be quite significant, and in irregular buildings, mode shapes may be somewhat irregular.

Hence, for tall and irregular buildings, though dynamic analysis is generally preferred, the method outlined in Section 5.8.2.2 cannot be applied as per clause 7.8.1 of IS 1893 (Part 1): 2002.

Undamped free-vibration analysis of the entire building is performed as per established methods of mechanics, using the appropriate masses and elastic stiffness of the structural system, to obtain natural periods T and mode shapes ϕ of those of its modes of vibration that need to be considered. Determination of natural frequencies and mode shapes is described in Appendix VI.

As per IS 1893 (Part 1): 2002, clause 7.8.4.2, the number of modes to be used in the analysis should be such that the total sum of modal masses of all modes considered is at least 90% of the total seismic mass. If modes with natural frequency beyond 33 Hz are to be considered, modal combination should be carried out only for modes up to 33 Hz. The effect of modes with natural frequency beyond 33 Hz should be included by considering the missing mass correction following well-established procedure.

5.8.2.1 Modal Analysis

Buildings with regular, or nominally irregular, plan configurations may be modelled as a system of masses lumped at the floor levels with each mass having one degree of freedom, that of lateral displacement in the direction under consideration. In the modal analysis, the variability in masses and stiffness is accounted for in the computation of lateral force coefficients. The following expressions are used for the computation of various quantities.

(a) *Modal mass* [clause 7.8.4.5 (a)] The modal mass M_k of mode k is given by

$$M_k = \frac{\left[\sum_{i=1}^{n} W_i \phi_{ik}\right]^2}{g \sum_{i=1}^{n} W_i (\phi_{ik})^2} \tag{5.10}$$

where g is the acceleration due to gravity, ϕ_{ik} is the mode shape coefficient at floor i in mode k, and W_i is the seismic weight of floor i.

(b) *Modal participation factor* [clause 7.8.4.5(b)] The modal participation factor P_k of mode k is given by

$$P_k = \frac{\sum_{i=1}^{n} W_i \phi_{ik}}{\sum_{i=1}^{n} W_i (\phi_{ik})^2} \tag{5.11}$$

(c) *Design lateral force at each floor in each mode* [clause 7.8.4.5(c)] The peak lateral force Q_{ik} at floor i in k^{th} mode is given by

$$Q_{ik} = A_k \phi_{ik} P_k W_i \tag{5.12}$$

where A_k is the design horizontal acceleration spectrum value (clause 6.4.2) using the natural period of vibration T_k of kth mode.

(d) *Storey shear forces in each mode* [clause 7.8.4.5(d)] The peak shear force V_{ik} acting in storey i in mode k is given by

$$V_{ik} = \sum_{j=i+1}^{n} \phi_{ik} \tag{5.13}$$

(e) *Storey shear forces due to all modes considered* [clause 7.8.4.5(e)] The peak storey shear force V_i in storey i due to all modes considered is obtained by combining those due to each mode as explained in modal combination.

1. Since the modal maximum values generally do not occur simultaneously, approximate methods such as Square Root of Sum of Squares (SRSS) or Maximum Absolute Response (ABS) methods are used.
2. The ABS method gives upper limit for maximum response and is therefore conservative and seldom used. The SRSS method provides reasonable estimate of total maximum response.

(f) *Lateral forces at each storey due to all modes considered* [clause 7.8.4.5(f)] The design lateral forces, F_{roof} and F_i, at roof and at floor i are given by

$$F_{\text{roof}} = V_{\text{roof}} \tag{5.14}$$

$$F_i = V_i - V_{i+1} \tag{5.15}$$

5.8.2.2 Modal Combination

The peak response quantities (e.g., member forces, displacements, storey forces, storey shears, and base reactions) should be combined as per the complete quadratic combination (CQC) method (clause 7.8.4.4).

$$\lambda = \sqrt{\sum_{i=1}^{r}\sum_{j=1}^{r} \lambda_i \rho_{ij} \lambda_j} \tag{5.16}$$

where r is the number of modes being considered, ρ_{ij} is the cross-modal coefficient given by Eqn (5.17), λ_i is the response quantity in mode i (including sign), and λ_j is the response quantity in mode j (including sign).

$$\rho_{ij} = \frac{8\delta^2 (1+\beta)\beta^{1.5}}{(1-\beta^2)^2 + 4\,\delta^2\,\beta\,(1+\beta)^2} \tag{5.17}$$

where ξ is the modal damping ratio (in fraction), 2% and 5% of critical for steel and RCC buildings, β is the frequency ratio and is equal to ω_j/ω_i, ω_i is the circular frequency in the ith mode, and ω_j is the circular frequency in the jth mode.

Alternatively, the peak response quantities may be combined by the SRSS method [clause 7.8.4.4(a)] as in case 1 and by ABS method [clause 7.8.4.4(b)] as in case 2 as follows.

Case 1 If the building does not have closely spaced modes, the peak response quantity λ due to all modes considered should be obtained as

$$\lambda = \sqrt{\sum_{k=i}^{r} (\lambda_k)^2} \tag{5.18}$$

where λ_k is the absolute value of the quantity in mode k, and r is the number of modes being considered.

Case 2 If the building has a few closely spaced modes, the peak response quantity λ^* due to these modes should be obtained as

$$\lambda^* = \sum_{c}^{r} \lambda_c \qquad (5.19)$$

where the summation is for the closely spaced modes only. This peak response quantity due to the closely spaced modes λ^* is then combined with those of the remaining well-separated modes by the method described above.

5.9 Time-history Method

Although the spectrum method, outlined in the previous section, is a useful technique for the elastic analysis of structures, it is not directly transferable to inelastic analysis because the principle of superposition is no longer applicable. Also, the analysis is subject to uncertainties inherent in the modal superimposition method. The actual process of combining the different modal contributions is a probabilistic technique and, in certain cases, it may lead to results not entirely representative of the actual behaviour of the structure. The THA technique represents the most sophisticated method of dynamic analysis for buildings. In this method, the mathematical model of the building is subjected to accelerations from earthquake records that represent the expected earthquake at the base of the structure. The method consists of a step-by-step direct integration over a time interval; the equations of motion are solved with the displacements, velocities, and accelerations of the previous step serving as initial functions.

The equation of motion can be represented as

$$\boldsymbol{k}x(t) + \boldsymbol{c}\,\dot{x}(t) + \boldsymbol{m}\,\ddot{x}(t) = \boldsymbol{p}(t)$$

where

$\boldsymbol{k}$ is the stiffness matrix, $\boldsymbol{c}$ is the damping matrix, and $\boldsymbol{m}$ is the diagonal mass matrix

$\boldsymbol{x}$, $\dot{x}$, and $\ddot{x}$ are the displacements, velocities, and accelerations of the structure and $\boldsymbol{p}(t)$ is the applied load.

In case of an earthquake, $\boldsymbol{p}(t)$ includes ground acceleration and the displacements, velocities, and accelerations are determined relative to ground motion.

The time-history method is applicable to both elastic and inelastic analyses. In elastic analysis the stiffness characteristics of the structure are assumed to be constant for the whole duration of the earthquake. In the inelastic analysis, however, the stiffness is assumed to be constant through the incremental time only. Modifications to structural stiffness caused by cracking, formation of plastic hinges, etc., are incorporated between the incremental solutions. Even with the availability of sophisticated computers, the use of this method is restricted to the design of special structures such as nuclear facilities, military installations, and base-isolated structures.

A brief outline of the method, which is applicable to both elastic and inelastic analysis, is given below.

The earthquake motions are applied directly to the base of the model of a given structure. Instantaneous stresses throughout the structure are calculated at small intervals of time for the duration of the earthquake or a significant portion of it. The maximum stresses that occur during the earthquake are found by scanning the computer output. The procedure usually includes the following steps.

1. An earthquake record representing the design earthquake is selected.
2. The record is digitized as a series of small time intervals of about 1/40 to 1/25 of a second.
3. A mathematical model of the building is set up, usually consisting of a lumped mass at each floor. Damping is considered proportional to the velocity in the computer formulation.
4. The digitized record is applied to the model as accelerations at the base of the structure.
5. The equations of motions are then integrated with the help of software program that gives a complete record of the acceleration, velocity, and displacement of each lumped mass at each interval.

The accelerations and relative displacements of the lumped masses are translated into member stresses. The maximum values are found by scanning the output record. This procedure automatically includes various modes of vibration by combining their effect as they occur, thus eliminating the uncertainties associated with modal combination methods. However, the time-history method has the following sources of uncertainty.

(a) The design earthquake must still be assumed.
(b) If the analysis uses unchanging values for stiffness and damping, it will not reflect the cumulative effects of stiffness variation and progressive damage.
(c) There are uncertainties related to the erratic nature of earthquakes. By pure coincidence, the maximum response of the calculated time-history could fall at either a peak or a valley of the digitized spectrum.
(d) Small inaccuracies in estimating properties of the structure will have a considerable effect on the maximum response.
(e) Errors latent in the magnitude of the time-step chosen are difficult to assess unless the solution is repeated with several smaller time-steps.

5.10 Torsion [IS 1893 (Part 1): 2002, Clause 7.9]

Torsional responses in a structure arise from two sources—(a) eccentricity in the mass and stiffness distribution, which causes a torsional response coupled with a translational response and (b) torsion arising from accidental causes, including the rotational component of ground motion about a vertical axis, the difference (eccentricity) between assumed and actual stiffness and mass, uncertain live load distribution, uncertainties in DL due to variations in workmanship and materials, asymmetrical patterns of non-linear force-deformation relations, and subsequent alternations that may be made in a building (e.g., addition of walls, which not only changes the DL but may change the position of the centre of rigidity).

For symmetrical buildings, the elementary analysis does not disclose the slightest torque, while actually, the probability that there will be such generalized forces during an earthquake is 1. Even non-linear behaviour can introduce torque that is not accounted for by conventional analysis. The current state of scientific advancement in this field precludes an accurate estimate of these accidental additional torsions.

To allow for effects such as the ones listed above, IS 1893 (Part 1): 2002 requires that buildings be designed to resist an additional torsional moment. Provision should be made in all buildings for increase in shear forces on the lateral force-resisting elements, which is a result of the horizontal torsional moment arising due to eccentricity between the centre of mass and the centre of rigidity. The design forces calculated are to be applied at the centre of mass, which is appropriately displaced so as to cause design eccentricity between the displaced centre of mass and the centre of rigidity. The design eccentricity, e_{di} to be used at floor i should be

$$e_{di} = \begin{cases} 1.5e_{si} + 0.05b_i \\ \quad \text{or} \\ e_{si} - 0.05b_i \end{cases} \tag{5.20}$$

whichever gives the more severe effect in the shear of any frame. Here e_{si} is the static eccentricity at floor i, defined as the distance between centre of mass and centre of rigidity, and b_i is the floor plan dimension of floor i, perpendicular to the direction of force. The factor 1.5 represents dynamic amplification factor, while the factor 0.05 represents the extent of accidental eccentricity. Figure 5.7 shows the plan of a building with the explanation of design eccentricity to be considered in the design. The dynamic amplification factor, also known as response amplification factor, is used to convert the static torsional responses to dynamic torsional responses. Highly irregular buildings are analysed according to modal analysis (Section 5.8.2).

The value of accidental eccentricity is assumed as 5% of the plan dimension of the building storey, particularly for the accidental torsional response during applied ground motion. Therefore, additive shears have been superimposed for a statically applied eccentricity of $\pm 0.05b_i$ with respect to the centre of rigidity. The factor 0.05 represents the extent of accidental eccentricity. This accidental eccentricity is considered for the following reasons:

1. The computation of eccentricity is approximate.
2. Due to change in usage of structure during its service life, the affected alterations may cause a change in the centre of mass.
3. Ground motion may produce some torsional component.

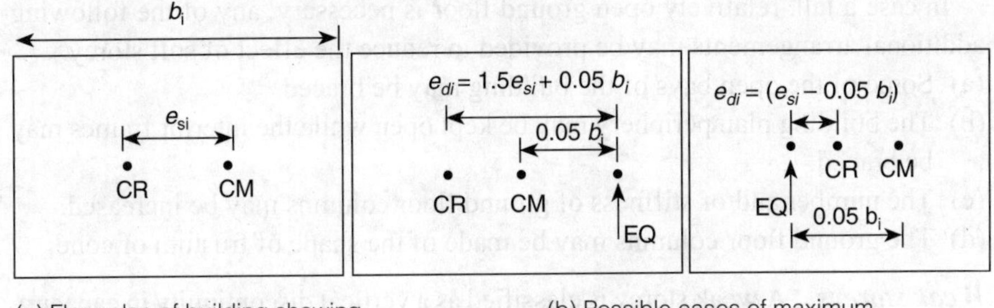

(a) Floor plan of building showing calculated positions of CR and CM

(b) Possible cases of maximum eccentricity, e_{di}

e_{si}: static eccentricity at i^{th} floor
b_i: i^{th} floor plan dimension perpendicular to the force direction
CR: centre of rigidity
CM: centre of mass

Fig. 5.7 Design eccentricities

5.11 Soft and Weak Storeys in Construction [IS 1893 (Part 1): 2002, Clause 7.10]

The essential distinction between a soft storey and a weak storey is that while a soft storey is classified based on stiffness or simply the relative resistance to lateral deformation or storey drift, the weak storey qualifies on the basis of strength in terms of force resistance (statics) or energy capacity (dynamics).

Soft storeys A soft storey is characterized by vertical discontinuity in stiffness. When an individual storey in a building (often the ground level storey) is made taller and/or more open in construction, it is called a *soft storey*. It is also sometimes called a *flexible storey*. The soft storey can—and sometimes does—occur at an upper level. However, it is more common at the ground level between a rigid foundation system and a relatively stiffer upper level system. This form of construction is quite common in residential and commercial buildings. Any storey for which the lateral stiffness is less than 60% of that of the storey immediately above, or less than 70% of the combined stiffness of the three storeys above, is classified as a soft storey. Buildings with a soft storey at the ground level having open spaces for parking are known as *stilt buildings*. Special arrangements should be made to increase the lateral strength and the stiffness of the soft storey. The beams and columns of soft storeys are designed to withstand two-and-a-half times the storey shears and moments calculated for specified seismic loads. The shear wall should be designed to withstand one-and-a-half times the lateral storey shear force, and should be placed symmetrically in both directions of the building, as far away from the centre of the building as feasible. Alternatively, dynamic analysis of a building is carried out, including the strength and stiffness effects of infill and inelastic deformations in the members, particularly those in the soft storey.

In case a tall, relatively open ground floor is necessary, any of the following additional arrangements may be provided to reduce the effect of soft storey.

(a) Some of the open bays of the building may be braced.
(b) The building plan periphery may be kept open while the interior frames may be braced.
(c) The number and/or stiffness of ground floor columns may be increased.
(d) The ground floor columns may be made of the shape of frustum of cone.

Weak storeys A weak storey is classified as a vertical discontinuity in capacity. Any storey of a structure in which the total lateral storey strength is less than 80% of that of the storey above is called a weak storey. The height of the building should be limited to 10 m if the weak storey has a strength less than 65% of the storey above.

A form of weak storey may be created by the overstiffening or overstrengthening of a lower floor. For example, to reduce the soft storey effect of an open ground floor, the correction factors adopted may result in a concentrated effect on the second floor, which may need more stiffness or strength than a simple static investigation may reveal.

5.12 Overturning Moment

Lateral forces in the direction of ground motion lead to overturning moments in each level of a building. This produces additional longitudinal stresses in columns and walls and additional upward or downward forces in foundation. Overturning moments may be calculated by considering the building to be a fixed-end cantilever beam loaded with the static lateral earthquake forces, accounting for shears, and acting simultaneously in the same direction. The overturning moment at one storey of a building is the product of the lateral forces and the distance from where the forces act on the storey under consideration, summed for all the storeys above that storey. Though the assumption is not entirely correct, it gives fairly reasonable results for low- or medium-height buildings. The application of the above concept to tall or slender structures is generally over-conservative since (a) the shear force due to an earthquake is actually an envelope of maximum shear forces induced during vibration, and in reality maximum shear forces will never occur simultaneously and (b) severe distress does not occur because of uplift, because moment is reduced with uplift as a result of the decrease in stiffness and the consequent decrease in induced force.

For the reasons described above, different codes prescribe different recommendations, a couple of which are given below.

(a) Only 75% of the DL be mobilized in resisting overturning to take account of vertical acceleration.
(b) Provision for overturning moment should be made for the specified earthquake forces in the top 10 storeys of the building or the top 40 m of other structures,

and the moments should be assumed to remain constant from these levels into the foundation.

Note: IS 1893 (Part 1) is silent about the consideration of overturning moment.

5.13 Other Structural Requirements

The basic design procedure for structure consists of the selection of lateral forces appropriate for the design purposes, and then providing a complete, appropriately detailed, lateral-force-resisting system to carry these forces from the top to the foundation. For good seismic performance, a building needs to have adequate lateral stiffness. Low lateral stiffness of a structure may lead to large deformations and strains, significant P-Δ effect and consequently cause damage to non-structural elements, discomfort to occupants, and pounding with adjacent structures. The cantilever projections, foundations, and compound walls, etc., need to be designed for forces as described in the following subsections.

5.13.1 Storey Drift [IS 1893 (Part 1): 2002, Clause 7.11.1]

Floor deflections are caused when buildings are subjected to seismic loads. These deflections are multiplied by the ductility factor, resulting in the total deflections which account for inelastic effects. The drift in a story is computed as a difference of deflections of the floors at the top and bottom of the story under consideration. The total drift in any storey is the sum of shear deformations of that storey, axial deformations of the floor systems, overall flexure of the building (axial deformation of columns), and foundation rotation (Fig. 5.8). It is normally specified at the elastic design level, although it will be greater for the maximum earthquake. Due to the minimum specified design lateral force, the storey drift in any storey with a partial load factor of 1.0 should not exceed 0.004 times the storey height. However, for a building that has been designed to accommodate storey drift, this limit of 0.004 is not placed.

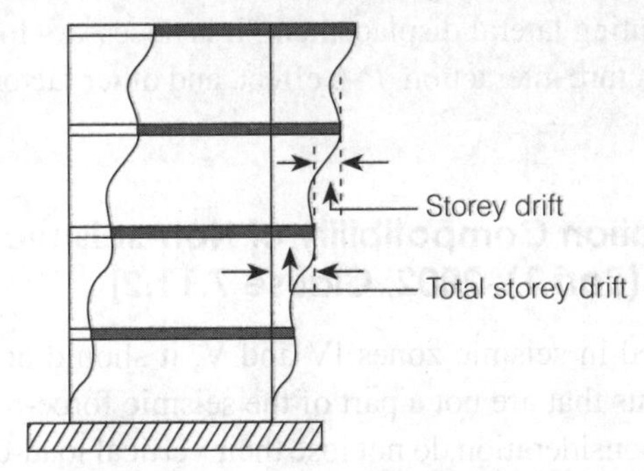

Fig. 5.8 Diagrammatic representation of storey drift

In the computation of lateral deflection, secondary members are considered to be unreliable, and their contribution is usually neglected. Expansion and contraction of columns in slender buildings, when bent, have a relatively large effect on storey drift as well as on total lateral deflection, and must be considered.

Storey drift or inter-storey displacement should be limited during earthquakes for the following reasons.

(a) To limit damage to non-structural elements and to provide human comfort.

(b) Limiting the drift, in turn, limits the effects of eccentric gravity loads, which otherwise magnify the effects of the lateral forces. Drift control is important in preserving the vertical stability of a structural system. If a structural system is excessively flexible, it may collapse due to P-Δ effects, especially if it is also massive. However, rather than limiting the drift, it is preferable to take the effects into consideration and to design the structure accordingly.

(c) Where buildings are constructed in close proximity to one another, damage due to pounding between the buildings is possible. Pounding may result in irregular response of buildings of different heights, local damage to columns as the floor of one building collides with columns of another, collapse of damaged floors, and in many cases, collapse of the entire structure. Damage due to pounding can be minimized by drift control, building separation, or as a last resort, aligning floors in adjacent buildings so that columns do not bear the blows of oncoming floor slabs. It is sometimes argued that drift limitations tend to limit pounding between adjoining structures. However, it is the total sway that constitutes the pertinent response and not the drift.

When two buildings are located very close to each other or when seismic expansion joints exist, it is necessary to leave enough space between neighbouring structures so that they do not pound each other. Extra distance is usually provided between two adjacent buildings in addition to the sway of the buildings. In computing lateral displacement, it is necessary to consider plastic deflection, soil-structure interaction, P-Δ effect, and other factors besides elastic deflection.

5.13.2 Deformation Compatibility of Non-seismic Members [IS 1893 (Part 1): 2002, Clause 7.11.2]

For buildings located in seismic zones IV and V, it should be ensured that the structural components that are not a part of the seismic force-resisting system in the direction under consideration do not lose their vertical load-carrying capacity. The induced moments resulting from storey deformations are equal to the response reduction factor, R, times the storey displacements.

For instance, consider a flat-slab building in which the lateral load resistance is provided by shear walls. Since the lateral load resistance of the slab–column system is small, it is often designed only for gravity loads, while all the seismic force is resisted by the shear walls. Even though the slabs and columns are not required to share the lateral forces, these deform with the rest of the structure under the seismic force. The concern is that under such deformation, the slab–column system should not lose its vertical load capacity.

5.13.3 Separation between Adjacent Units [IS 1893 (Part 1): 2002, Clause 7.11.3]

Two adjacent buildings, or two adjacent units of the same building with a separation joint in between them, should be separated by a distance equal to the response reduction factor, R, times the sum of each of their storey displacements, to avoid damaging contact when the two units deflect towards each other. When the two adjacent units hit each other due to lateral displacement, it is known as *pounding* or *hammering*. When floor levels of two similar adjacent units or buildings are at the same elevation levels, factor R in this requirement may be replaced by $R/2$.

5.13.4 Foundations [IS 1893 (Part 1): 2002, Clause 7.12.1]

The foundations vulnerable to significant differential settlement due to ground shaking should be avoided for structures in seismic zones III, IV, and V. In seismic zones IV and V, individual spread footings or pile caps should be interconnected with ties, except when individual spread footings are directly supported on rock. All ties should be capable of carrying, in tension and in compression, an axial force equal to $A_h/4$ times the larger of the column or pile cap load, in addition to the otherwise computed forces.

5.13.5 Cantilever Projections [IS 1893 (Part 1): 2002, Clause 7.12.2]

Towers, tanks, parapets, smoke stacks (chimneys), and other vertical cantilever projections attached to buildings and projecting above the roof, should be designed and checked for a stability of five times the design horizontal seismic coefficient A_h. In the analysis of the building, the weight of these projecting elements will be lumped with the roof weight.

All horizontal projections like cornices and balconies should be designed and checked for a stability of five times the design vertical coefficient specified as $(10/3)\ A_h$.

Note: The specifications for non-structural elements are under revision and are presented in Chapter 10.

5.13.6 Compound Walls [IS 1893 (Part 1): 2002, Clause 7.12.3]

Compound walls should be designed for the design horizontal seismic coefficient A_h with importance factor 1.0.

5.13.7 Connections between Parts [IS 1893 (Part 1): 2002, Clause 7.12.4]

All parts of the building, except between the separation sections, should be tied together to act as a single integrated unit. All connections between different parts, such as beams to columns and columns to their footings, should be made capable of transmitting a force, in all possible directions, of magnitude Q_i/W_i (or 0.05), whichever is greater, times the weight of the smaller part or the total of DL and IL reaction. Frictional resistance should not be relied upon for fulfilling these requirements.

5.14 Earthquake-resistant Design Methods

Conventional civil engineering structures are designed on the basis of two main criteria—*strength* and *stiffness*. The strength is related to damageability or ultimate limit state, whereas the stiffness is related to serviceability limit state for which the structural displacements must remain limited. In case of earthquake-resistant design, a new criterion, the *ductility* should also be added. The first two criteria, can be achieved by—(a) specifying severe (or moderate) design earthquake levels, (b) limiting the maximum stresses or internal forces in critical members, and (c) limiting the storey drift ratio. The third criterion, which is prevention of building collapse, is achieved by providing sufficient strength and ductility to ensure that the structures do not collapse in a service earthquake. Based on the three criteria, rigidity (serviceability), strength (damageability), and ductility (survivability), the methods of seismic design are classified as lateral strength design, displacement- or ductility-based design, capacity design method, and energy-based design. These design approaches are described below.

Lateral strength design Lateral strength design is the most common seismic design approach used today. IS 1893: 2002 is based on this approach. It is based on providing the structure with the minimum lateral strength to resist seismic loads, assuming that the structure will behave adequately in the non-linear range. Concept of response reduction factor is used with an intent to account for both ductility and damping inherent in a structural system. Further, some simple constructional detail rules are to be satisfied, such as material ductility, member slenderness, cross sections, etc., as laid down in IS 4326: 1993 and IS 13920: 1993.

Displacement or ductility-based design It is a well-known fact that due to economic reasons, structures are not designed to have sufficient strength to remain elastic in severe earthquakes. The structure is designed to possess adequate ductility so that it can dissipate energy by yielding and survive the shock. The ductility-based design operates directly with deformation quantities and, therefore, gives a better insight to the expected performance of structures, rather than simply providing strength, as the lateral strength design approach does.

Capacity design method The capacity design method is a design approach in which the structures are designed so that hinges can only form in predetermined positions and sequences. It is a procedure of the design process in which strengths and ductilities are allocated and the analyses are interdependent. The capacity design procedure stipulates the margin of strength that is necessary for elements to ensure that their behaviour remains elastic. The capacity design method is so called because in the yielding condition, the strength developed in the weaker member is related to the capacity of the stronger member.

Energy-based design One of the promising approaches for earthquake-resistant design in the future is the energy-based design approach. In this approach, it is recognized that the total energy input E_t can be resisted by the sum of the kinetic energy E_k, the elastic strain energy E_{es}, the energy dissipated through plastic deformations (hysteretic damping) E_p, and the equivalent viscous damping E_ξ. The energy equation for a single-mass vibrating system is the energy balance between the total input energy and the energies dissipated by viscous damping and inelastic deformations. The energy equation can be written as

$$E_t = E_k + E_p + E_{es} + E_\xi \tag{5.21}$$

5.15 Seismic Response Control

The conventional approach to seismic design of structures relies on the ductile behaviour of the structural system to dissipate the seismic energy through plastic deformation cycles (Fig. 5.9). This approach has the disadvantage that the structure, during high-intensity earthquakes suffers damage requiring costly repair. The damage can sometimes be so severe that the building may need to be demolished. This necessitated the search for alternative ways to achieve earthquake resistance. The dynamic interaction between the structure and earthquake ground motion can be modified in order to minimize structural damage and to control structural response. The control is based on two different approaches—modification of the dynamic characteristics or modification of the energy absorption capacity of the structure.

Fig. 5.9 Conventional structure without any artificial response control

Earthquakes cause high accelerations in stiff buildings and large inter-storey drifts in flexible structures. The possibility of artificially increasing both the period of vibration and the energy-dissipation capacity of a structure has to be regarded as an attractive way of improving its seismic resistance. Devices proposed and in use are either to prevent an earthquake force from acting on a structure (isolator), known as *input reduction approach*, or to absorb a portion of the earthquake energy (dampers) that is introduced in the structure, known as *damping augmentation approach*.

The basic principle of seismic base isolation is to increase the structure's natural time period, which leads to a decrease in its natural frequency of vibration as compared to that of its corresponding fixed-base structure and that of the predominant period of the soil at the site. Thus the idea of the seismic isolation is to shift the fundamental period of vibration of the building away from the predominant period of earthquake-induced ground motion, to reduce the forces transmitted into the structures. Decrease in fundamental frequency of vibration also decreases the pseudo-acceleration of structures, thereby also reducing the base shear. Isolation provides a means of limiting the earthquake entering the structure by decoupling the building's base from its superstructure. The concept of base isolation is most effective for low-rise, relatively stiff buildings (rigid structures) located on hard grounds and with large mass; it is not suited for use in a high-rise building because of large overturning moments. However, base isolation is technically complex and costly to implement.

The energy dissipation concept, on the contrary, allows seismic energy into the building and the capacity of the structure to absorb this energy is enhanced through appropriate devices. By incorporating suitable damping mechanisms into the lateral load-resisting system of the structure, the earthquake energy is dissipated as the structure sways back and forth due to seismic loading and helps in lowering the overall displacements of the structure. This technique is most effective in structures that are relatively flexible and have some inelastic deformation capacity. This approach too is technically not that simple but less costly than base isolation. The essential concepts of both techniques are appropriate and logical. The systems using the above concepts to control seismic responses are discussed in the following section.

5.16 Seismic Response Control Systems

From the foregoing discussion, it is evident that the seismic response of the structural systems can be controlled either by preventing or diverting a major portion of earthquake energy entering into the structure. The systems using these techniques are referred to as *seismic control systems*. These systems may be passive, active, hybrid, or semi-active control systems. These have been shown schematically in Fig. 5.10.

The techniques and the devices used to control the seismic response of the structures are described in the following sections.

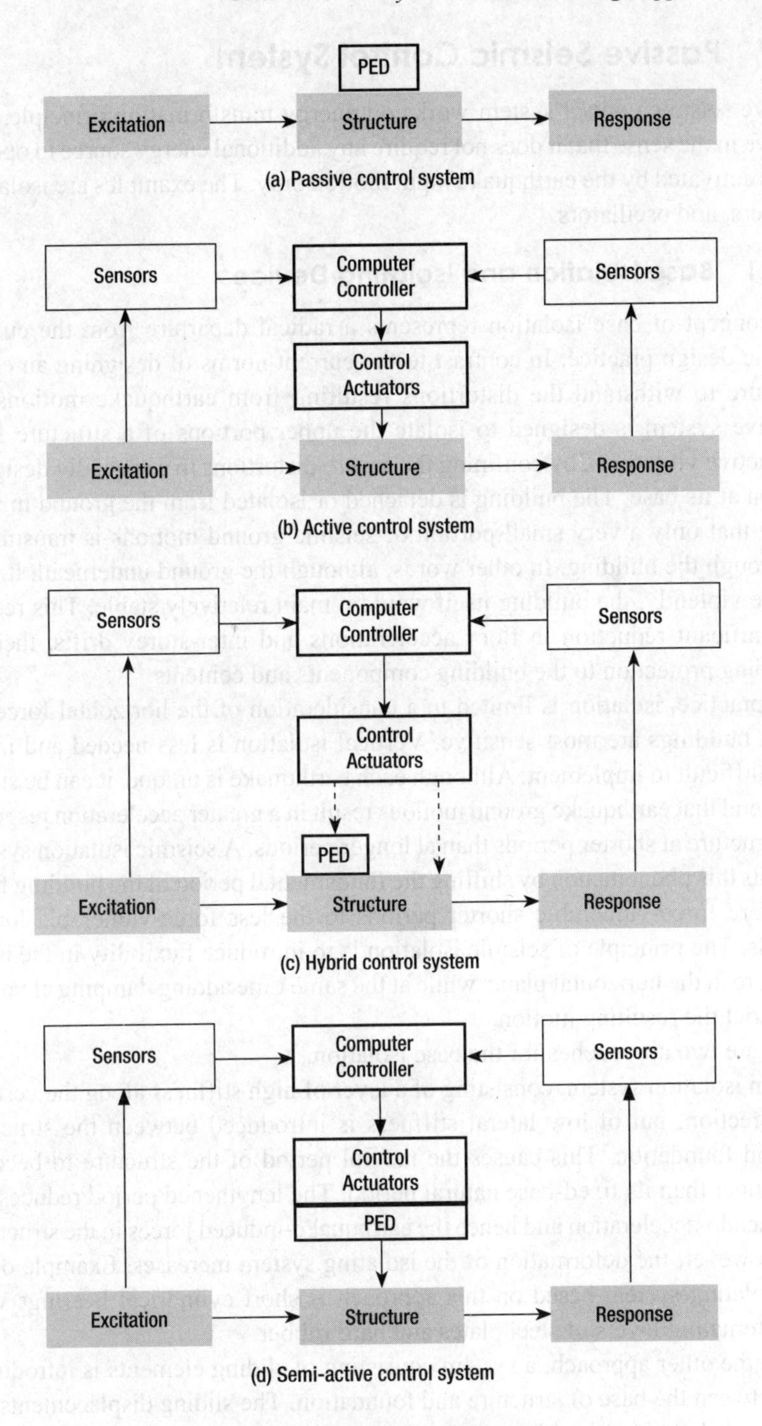

Fig. 5.10 Schematic representation of seismic control systems

5.17 Passive Seismic Control System

Passive seismic control system works on energy transformation principle. It is passive in the sense that it does not require any additional energy source to operate and is activated by the earthquake input motion only. The examples are isolators, dampers, and oscillators.

5.17.1 Base Isolation and Isolating Devices

The concept of base isolation represents a radical departure from the current seismic design practice. In contrast to the current norms of designing an entire structure to withstand the distortions resulting from earthquake motions, an adaptive system is designed to isolate the upper portions of a structure from destructive vibrations, by confining the severe distortions to a specially designed portion at its base. The building is detached or isolated from the ground in such a way that only a very small portion of seismic ground motions is transmitted up through the building. In other words, although the ground underneath it may vibrate violently, the building itself would remain relatively stable. This results in significant reduction in floor accelerations and inter-storey drifts, thereby providing protection to the building components and contents.

In practice, isolation is limited to a consideration of the horizontal forces to which buildings are most sensitive. Vertical isolation is less needed and much more difficult to implement. Although each earthquake is unique, it can be stated in general that earthquake ground motions result in a greater acceleration response in a structure at shorter periods than at longer periods. A seismic isolation system exploits this phenomenon by shifting the fundamental period of the building from the more force-vulnerable shorter periods to the less force-vulnerable longer periods. The principle of seismic isolation is to introduce flexibility in the basic structure in the horizontal plane, while at the same time adding damping elements to restrict the resulting motion.

There are two approaches for the base isolation.

1. An isolation system, consisting of a layer of high stiffness along the vertical direction, but of low lateral stiffness is introduced between the structure and foundation. This causes the natural period of the structure to become longer than its fixed-base natural period. The lengthened period reduces the pseudo-acceleration and hence the earthquake-induced forces in the structure. However, the deformation of the isolating system increases. Example of an isolating system based on this approach is short cylindrical bearings with alternating layers of steel plates and hard rubber.
2. In the other approach, a system consisting of sliding elements is introduced between the base of structure and foundation. The sliding displacements are controlled either by high-tension springs, laminated rubber bearings, or by making the sliding surface curved. Friction pendulum system is an example of this approach.

A practical base isolation system should consist of the following.

(a) A flexible mounting to increase the period of vibration of the building sufficiently to reduce forces in the structure above.

(b) A damper or energy dissipater to reduce the relative deflections between the building and the ground to a practical level.

(c) A method of providing rigidity to control the behaviour under minor earthquakes and wind loads.

Flexibility can be introduced at the base of the building by many devices such as elastomeric pads, rollers, sliding plates, cable suspension sleeved piles, rocking foundations, etc. Substantial reductions in acceleration with an increase in period and a consequent reduction in base shear are possible, the degree of reduction depends on the initial fixed-base period and shape of the response curve. However, the decrease in base shear comes at a price; the flexibility introduced at the base will give rise to large relative displacements across the flexible mount. Hence, the necessity of providing additional damping at the level of isolators arises. This can be provided through hysteretic energy dissipation of mechanical devices, which use the plastic deformation of either lead or mild steel to achieve high damping.

An isolation system should be able to support a structure, while providing additional horizontal flexibility and energy dissipation. The provision of this resisting element having adequate stiffness exhibits essentially linear elastic behaviour under the maximum wind loading, but yields when subjected to earthquake forces slightly greater than those corresponding to the maximum wind loading. By allowing the isolating mechanism at the base of a structure to yield at a predetermined lateral load, the structure above it is effectively isolated from forces which would otherwise cause inelastic deformations. The structure then need to be designed for only vertical and wind loads, with special attention for earthquake resistance focused only on the isolating mechanism at its base. An effective isolation system not only allows the structure above to remain elastic during a strong earthquake, but spares the non-structural elements from extensive distress. Since the non-structural components in a typical multi-storey structure account for about 80% of the building cost, significant savings in the repair and replacements cost can be achieved. Various types of isolation systems are described below.

5.17.1.1 Devices using Low Lateral Stiffness Layers

With devices that make use of interposition of a layer with low horizontal stiffness between the structure and the foundation, the structure is decoupled from the horizontal components of earthquake ground motion and the ground motion is prevented from being transmitted to the structure. The devices used are described as follows.

Rubber Pads

The traditional concept of structural base isolation, whereby laminated rubber pads prevent ground motion from being transmitted from the building foundation into the superstructure, is shown in Fig. 5.11. Figure 5.11(a) shows a steel building with a fixed base and its response to ground motion, whereas in Fig. 5.11(b) the building shown has been seismically isolated from the base. The expansion joint details at the ground level (GL) have been shown in Fig. 5.11(c). The main types of isolation pads in use represent a common means for introducing flexibility, with respect to the horizontal forces, into structure. They serve the following purposes.

(a) Increase the period of the fundamental mode.
(b) Make the higher-mode response insignificant by concentrating practically all of the masses in the mass of the fundamental mode, thereby drastically decreasing the input energy.

The more widely used type of base isolation is made of laminated rubber pads, similar to the pads used as bearing pads in non-seismic situations (bridge-

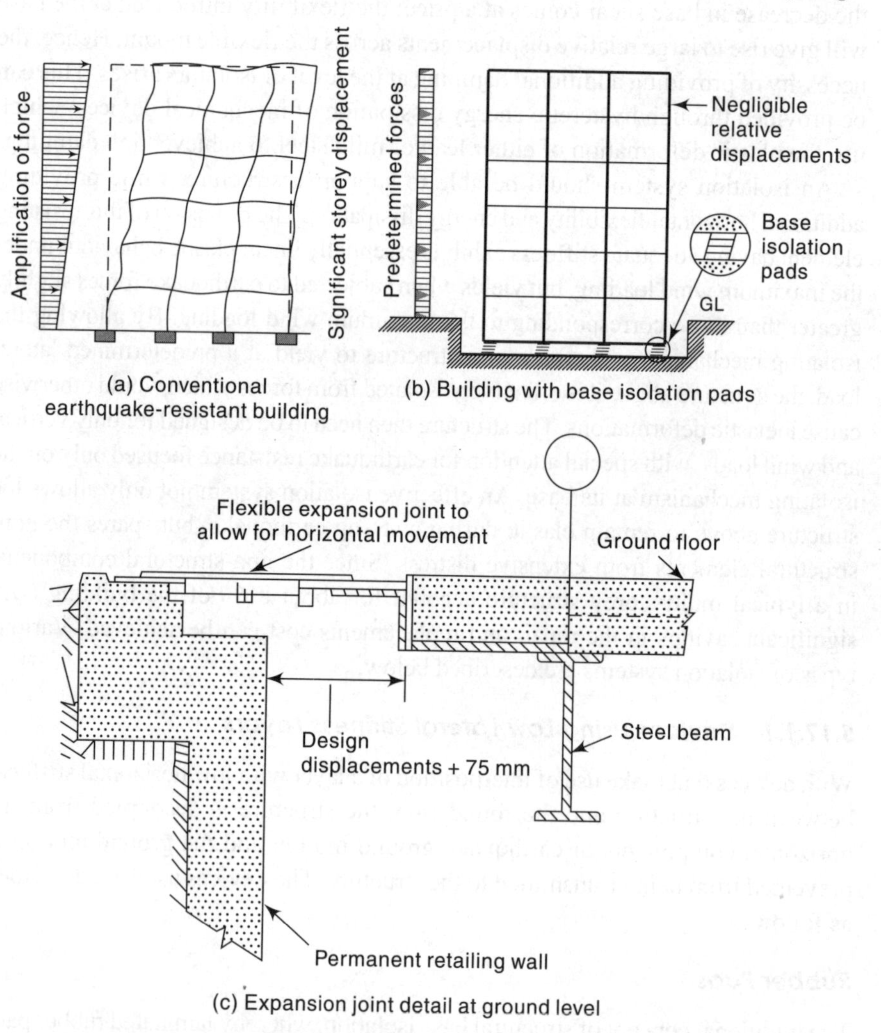

Fig. 5.11 Response of buildings with and without isolation

deck supports for instance). They consist of thin layers of natural rubber that are vulcanized and bonded to steel plates (Fig. 5.12). Due to the shear stress in rubber, such pads are very flexible in the horizontal direction. On the other hand, they are rather stiff vertically due to the presence of the steel plates, which result in a relatively high bearing capacity. This kind of pads show a substantial linear response, governed essentially by the properties of rubber. The rubber sheets may be made with natural rubber or artificial elastomers. The choice of rubber composition is extremely important as the pads should have the following properties.

(a) Good mechanical properties as concerns the dynamic behaviour—low shear strain modulus, high damping capacity—and resistance to large strain
(b) Large load-bearing capacity
(c) Low degradability with time

These properties may be obtained by the adjunction of fillers in the rubber.

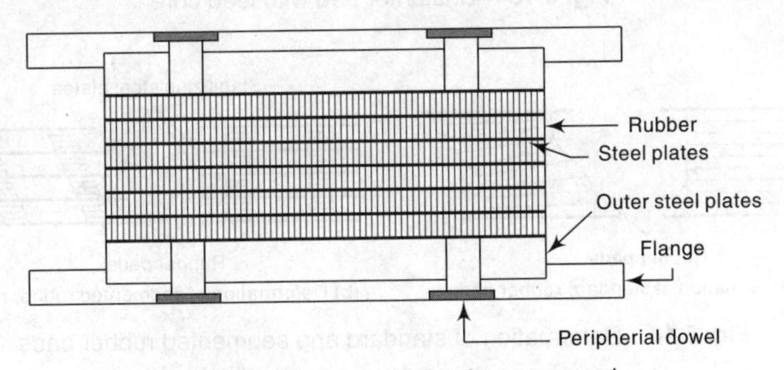

Fig. 5.12 Structure of an elastomer pad

The pad damping provided by the viscous behaviour is quite low, upto approximately 5%. For certain rubber compositions (using ferrite filler), damping values up to 20% may be obtained. High values (about 30%) may be obtained by arranging a core of lead in the centre of the pad, where the energy absorption is obtained by yielding of the lead, which remains hooped by the rubber (Fig. 5.13).

The seismic displacement of the rubber pads discussed above, is of the order of one-half of the pad dimension in plan. The allowable displacement can be increased by segmenting the pad and introducing stabilising plates, which may increase the height of the pad. The height of the pad is usually limited by its buckling. A way to increase the pad's height, while overcoming the stability problem, is to use a multistage rubber bearing (Fig. 5.14). Here, intermediate plates prevent the rubber bearings from rotation under horizontal displacements, thereby maintaining stability against buckling.

The most practical device proposed so far for isolation consists of a horizontal flexible mounting that supports the building, while allowing it to move freely in the horizontal direction. A load limiter resists movement due to small horizontal loads. The load limiter is a flexural or torsional member with various possible configurations and is made of steel, lead, or other materials. In Fig. 5.15, the

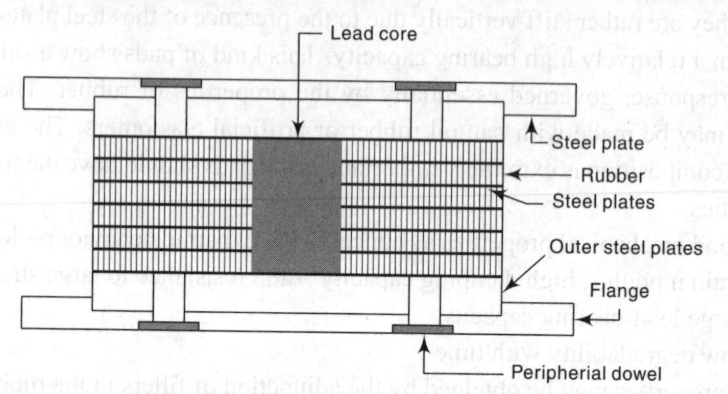

Fig. 5.13 Elastomer pad with lead core

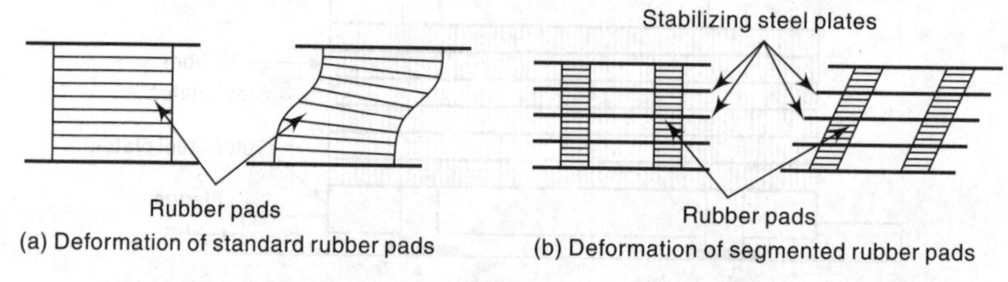

Fig. 5.14 Deformation of standard and segmented rubber pads

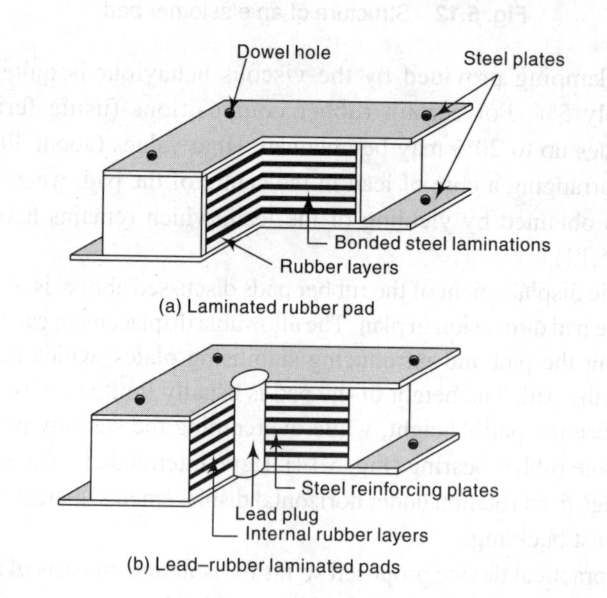

Fig. 5.15 Types of laminated rubber springs used in isolation

horizontal mounting and the load limiter are built as one piece, with the limiter being the central load cylinder that resists flexure. The horizontal flexible mounting is a sandwiched rubber-and-steel plate. These laminated plates have high vertical stiffness. Up to a certain yield load, the load limiter works elastically, and at higher loads it exhibits plastic deformation, which allows horizontal building movement. When deformed plastically, it works as a damper to minimize the response and limit the amount of horizontal movement.

5.17.1.2 Devices using Dry Friction

The simplest system consists of friction plates installed above a rubber pad (Fig. 5.16). The upper plate is made of stainless steel and is fixed to the superstructure. The lower plate is made of bronze and lead and is fixed above a classical rubber pad. The advantage of this system is that it avoids the self-sizing of the two plates and provides for a suitable friction factor, which induces reasonable accelerations while keeping the remaining displacements in reasonable limits. The friction plates may be placed above or under the rubber pad; in certain cases the second solution is favoured to avoid eccentricity of the weight on the rubber pad after sliding. Friction surfaces may also be associated to pendulum bearings.

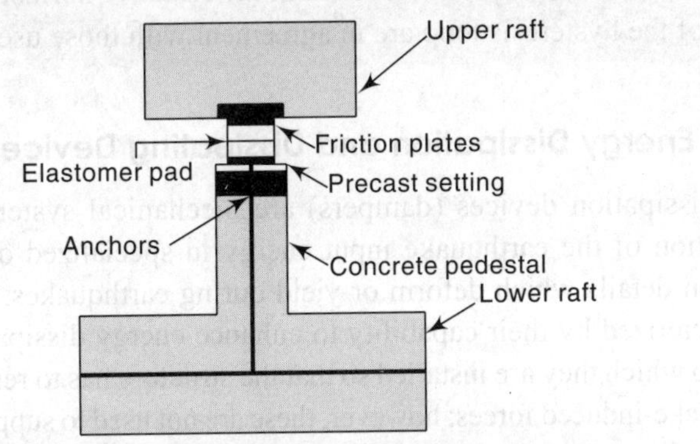

Fig. 5.16 Sliding plates associated with a delaminated elastomer pad

Friction pendulum systems also known as sliding isolators have been developed using the pendulum effect. In such a device, an upper rotating ball or an articulated slider moves on a concave surface (Fig. 5.17); any horizontal movement would,

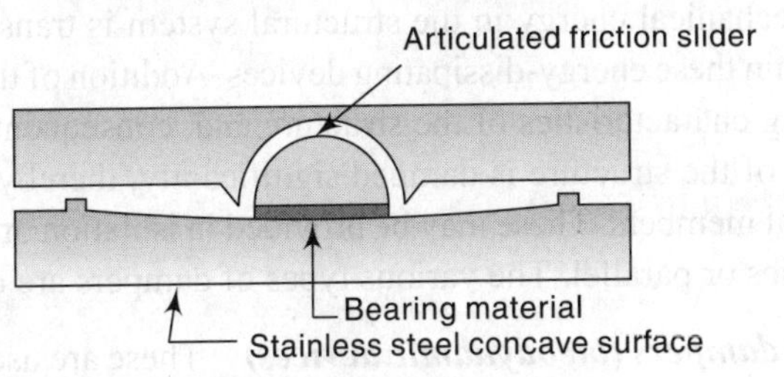

Fig. 5.17 Friction pendulum system

therefore, imply a vertical uplift of the superstructure when the movement occurs. The range of vertical load capacity, stiffness to lateral force, and period of vibration are of the same order of magnitude as those of lead–rubber pads of similar size.

- Access for inspection and replacement of bearings should bc provided at bearing locations.
- Stub walls or columns to function as backup systems should be provided to support the building in the event of isolator failure.
- A diaphragm capable of delivering lateral loads uniformly to each bearing is preferable. If the shear distribution is unequal, the bearing should be arranged such that larger bearings are under stiffer elements.
- Provisions must be made around the building to allow free-movement for the maximum predicted horizontal displacement.
- The isolator must be free to deform horizontally in shear and must be capable of transferring maximum seismic forces between the superstructure, the substructure, and the foundation.
- The bearings should be tested to ensure that they have lateral stiffness properties that are both predictable and repeatable. The tests should show that over a wide range of shear strains, the effective horizontal stiffness and area of the hysteresis loop are in agreement with those used in the design.

5.17.2 Energy Dissipation and Dissipating Devices—Dampers

Energy dissipation devices (dampers) are mechanical systems to dissipate a large portion of the earthquake input energy in specialized devices or special connection details which deform or yield during earthquakes. In general, these are characterized by their capability to enhance energy dissipation in structural systems to which they are installed so that the structure has to resist lesser amount of earthquake-induced forces; however, these are not used to support the structures. These involve a period shift of the structure from the predominant period of earthquake motion, when passive energy dissipation allows earthquake energy into the building. Through appropriate configuration of the lateral resisting system, the earthquake energy is directed towards energy-dissipation devices, located within the lateral resisting elements, to intercept this energy. Due to the earthquake the induced mechanical energy in the structural system is transformed into thermal energy within these energy-dissipating devices. Addition of these devices enhance the damping characteristics of the structure and, consequently, the amplitude of the motion of the structure is damped significantly, thereby reducing the forces on structural members. These may be provided in isolation or coupled with rubber pads in series or parallel. The various types of dampers are described below.

Hydraulic dampers (oleodynamic devices) These are used with the objective of permitting slowly developing displacements due to thermal movements, but limiting the response under dynamic actions. These systems dissipate energy by forcing a fluid through an orifice similar to the shock absorbers of an automobile. The fluid may be oil or very high molecular weight polymers. They may be constituted of a piston moving axially in the polymer, inside a cylinder. They may also be constituted of a piston moving in every direction in very viscous elastomers like silicon or bitumen. Oil dampers are not often recommended as these require frequent maintenance.

Electro-rheological fluid dampers (ERF-D) received considerable attention for vibration control of structures. These are passive fluid dampers inducing friction-type forces. These operate under shear flow. A typical ERF-D is shown in Fig. 5.18. Fluid viscous damping reduces stress and deflection because the force from the damping is completely *out of phase** with stresses due to seismic loading; this is because the damping force varies with stroking velocity. Other types of dampers, such as yielding elements, friction devices, plastic hinges, and visco–elastic elastomers, do not vary their output with velocity; hence they can increase column stress while reducing deflection.

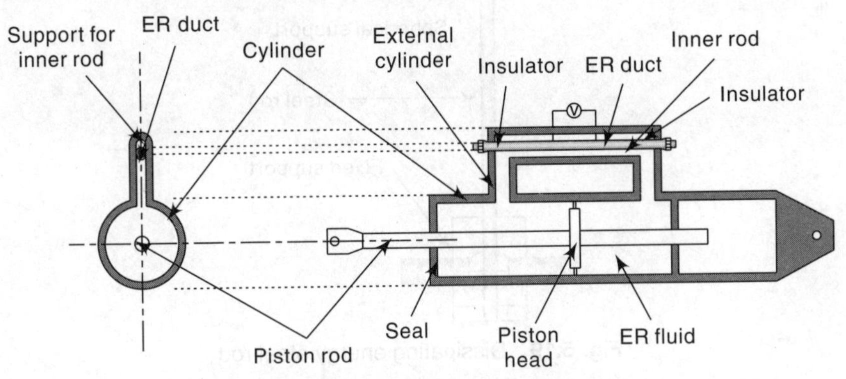

Fig. 5.18 Schematic diagram of ER damper

Metallic dampers Metallic dampers are based on the same concept as the lead cores in rubber pads. These can be fabricated from steel, lead, or special shape-memory alloys. These systems are referred to as amplitude-dependent systems

*To understand the out-of-phase response of fluid viscous dampers, consider a building shaking laterally back and forth during a seismic event. The stress in a lateral load-resisting element, such as a frame column, is at a maximum when the building has deflected the maximum amount from its normal position. This is also the point at which the building reverses direction to move back in the opposite direction. If a fluid viscous damper is added to the building, damping force will drop to zero at this point of maximum deflection. This is because the damper stroking velocity goes to zero as the building reverses direction. As the building moves in the opposite direction, maximum damping force occurs at maximum velocity, which occurs when the building goes through its normal, upright position. This is also the point where the stresses in the lateral load-resisting elements are at a minimum. This out-of-phase response is the most desirable feature of fluid viscous damping.

since the amount of energy dissipated, which is hysteretic in nature, is usually proportional to force and displacement. The devices are most often located within structural lateral load-resisting elements such as braced frames.

Steel dampers Steel dampers use the plastification of steel to dissipate energy. Steel hysteretic dampers display stable elastic–plastic behaviour and a long fatigue life. These dampers are made of a simple bar, a plate, or a profile with a specially studied shape. A typical steel damper is shown in Fig. 5.19. The most common device utilizes the plastic flexural or torsional deformation of steels. Such dampers are inserted in the bracings [Fig. 5.20(a)], at wall-to-wall joints, or at the borders of a wall and a surrounding frame. Isolation coupled with energy-absorbing damping is shown in Fig. 5.20(b). Sometimes steel plates with slits (in the form of shear walls) are placed between the walls in the surrounding frame (Fig. 5.21). Shear

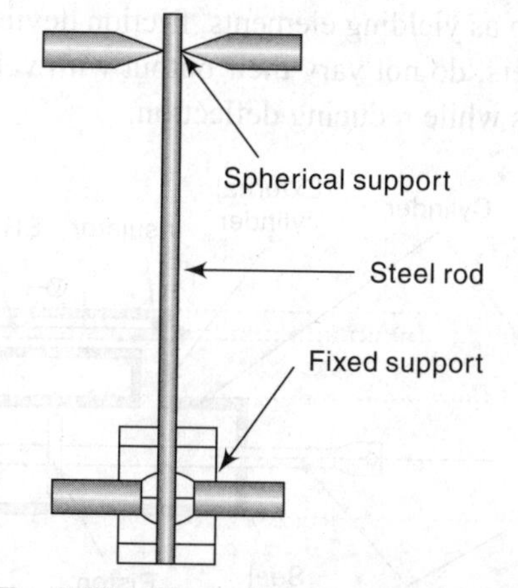

Fig. 5.19 Dissipating energy steel rod

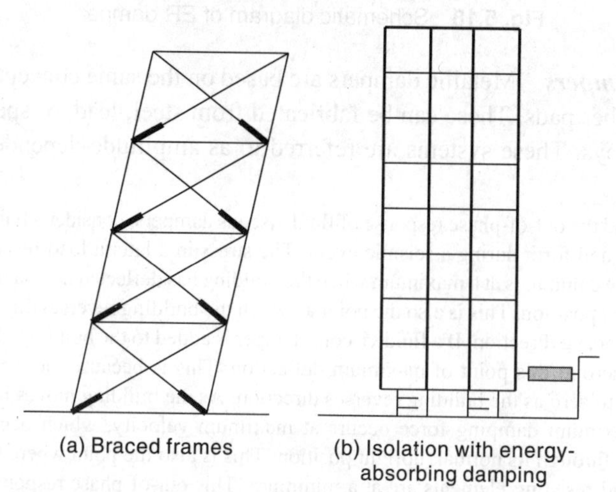

Fig. 5.20 Structural configurations with energy-absorbing dampers

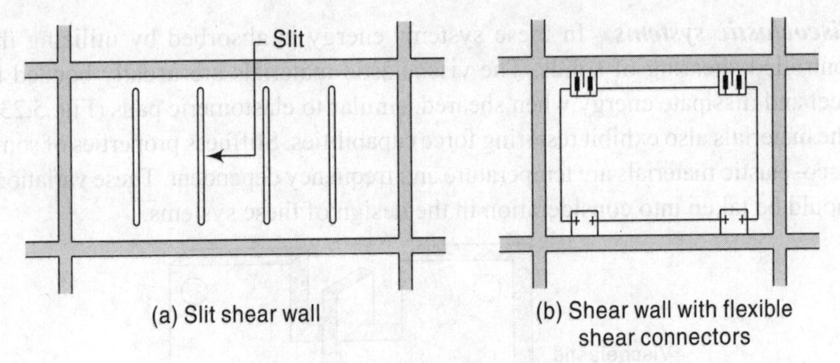

Fig. 5.21 Ductile shear walls

deformation of sandwiched materials of high viscosity is sometimes utilized as an energy-dissipation device.

Lead-extrusion dampers The energy dissipation in these dampers is provided by the processes that take place in the metal when it is forced through an orifice. The reduction of the cross-section of lead, forced to pass through the orifice, involves plastic deformation, with significant surface friction and the heat production. If the temperature increases, the extrusion force and the heat generated decreases. At the end of the dynamic process, lead goes through the physical process of recovery, re-crystallization, and grain growth, returning to its original form.

Shape-memory alloys Also known as smart alloys, are metals that, after being strained, revert back to their original shape. These enable large forces and large movement actuation, as they can recover from large strains. Shape-memory alloys have the potential capability of dissipating energy without incurring damage, as in the case of steel dampers, when they yield. The most effective and widely used alloys include NiTi, CuZnAl, and CuAlNi.

Friction dampers In such systems, the friction surfaces are clamped with prestressing bolts (Fig. 5.22). The characteristic feature of this system is that almost perfect rectangular hysteretic behaviour is exhibited. These systems are referred to as displacement-dependent systems, since the amount of energy dissipated is proportional to displacement. Contact surfaces used are lead–bronze against stainless steel, or teflon against stainless steel.

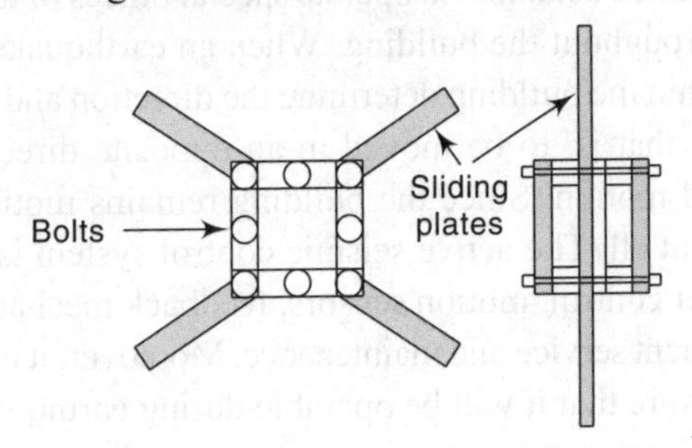

Fig. 5.22 Friction damper with sliding plates

Viscoelastic systems In these systems energy is absorbed by utilizing the controlled shearing of solids. The viscoelastic materials are usually bonded to steel and dissipate energy when sheared, similar to elastomeric pads (Fig. 5.23). The materials also exhibit restoring force capabilities. Stiffness properties of some visco–elastic materials are temperature and frequency dependent. These variations should be taken into consideration in the design of these systems.

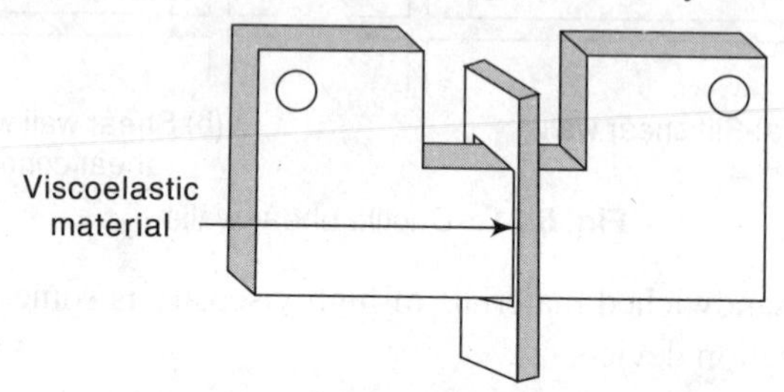

Fig. 5.23 Viscoelastic damper

5.17.3 Dynamic Oscillators

Dynamic oscillators are of recent origin and are used to transform energy among the vibrating modes of tall buildings during strong earthquakes; the supplemental oscillators act as dynamic absorbers. Tuned Mass Damper (TMD), for example, is an oscillator. A TMD, with its frequency of vibration tuned to the exciting frequency, is attached/connected to the main structural system. During excitation, a TMD simply moves in and out of phase to that of the main structural system, thereby imparting opposing inertial forces to that of the external vibrating forces acting on the structure. While doing so, the TMD is simply maintaining its inertial property.

5.18 Active Seismic Control System

The active seismic control system relies on counter-balancing the motion of the structural system by means of automated counter-weight system. This implies that additional forces are imposed on the structures to counter-balance the earthquake-imposed forces. These are active in the sense that they require an energy source and computer-controlled actuators to operate special braces of tuned mass dampers (TMDs) located throughout the building. When an earthquake hits the building, sensors located within the building determine the direction and weight of counter-balancing systems, that is, to be moved in an opposite direction to that of the earthquake-induced motion. Since the building remains motionless, it does not suffer any damage at all. The active seismic control system is a complex one as it relies on computer control, motion sensors, feedback mechanisms, and moving parts requiring frequent service and maintenance. Moreover, it needs an emergency power source to ensure that it will be operable during earthquake. The examples are active TMDs, tuned liquid dampers, active braces systems, and active tendon systems. Figure 5.24 shows various types of active seismic control devices on various storeys of a building. Their brief description follows.

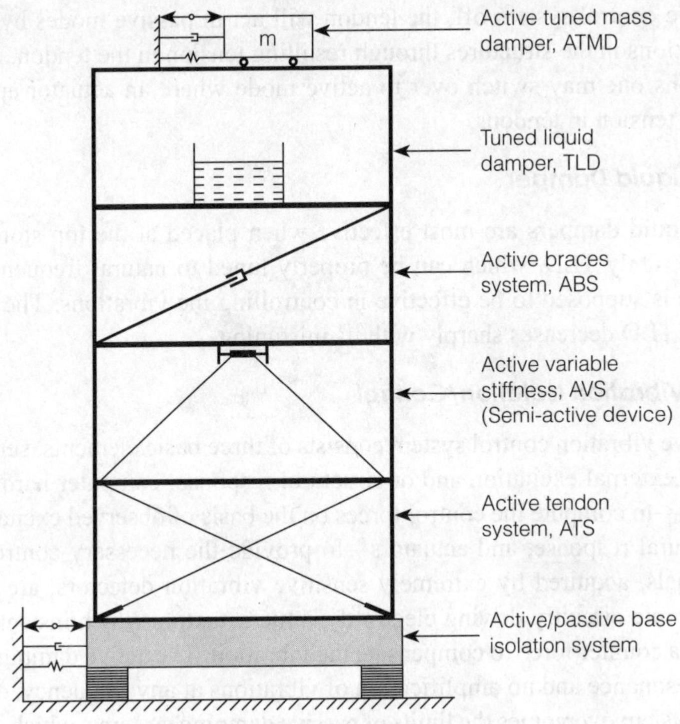

Fig. 5.24 Active control system

Active-tuned Mass Damper

Active TMD is a natural extension of passive TMD described in Section 5.17.3. In this system, when about 1% of the total building mass is directly excited by an actuator without any spring and dashpot, the building vibrations are reduced to about 75% by the use of ATMDs.

Active Braces System

Active braces system consists of a set of prestressed braces connected to a structure whose tensions are controlled by electro-hydraulic servo-mechanisms. The ABS can make use of existing structural members and thus minimize extensive additions or modifications of an as-built structure. The ABS steers the structural frequencies away from dominant (and damping) frequencies associated with the seismic power spectra at each time instant. This can be accomplished using active braces through length adjustment or position adjustment.

Active Tendon System

This system is based on active diagonals consisting in tendons having the role to provide the limitation of relative floor displacements. At low excitations, with the active control system off, the tendon will act in passive modes by resisting deformations in the structures through resulting tension in the tendon. At higher excitations one may switch over to active mode where an actuator applies the required tension in tendons.

Tuned Liquid Damper

Tuned liquid dampers are most effective when placed at the top storey of the structure. Only TLD, which can be properly tuned to natural frequency of the structure is supposed to be effective in controlling the vibrations. The damping effect of TLD decreases sharply with its mistuning.

Active Vibration Isolation/Control

The active vibration control system consists of three basic elements: sensors—to measure external excitation and/or structural response, computer hardware and software—to compute the control forces on the basis of observed excitation and/or structural response, and actuators—to provide the necessary control forces. The signals, acquired by extremely sensitive vibration detectors, are analysed by electronic circuitry driving electro-dynamic actuators which instantaneously produce a counter force to compensate the vibration. The active damping system has no resonance and no amplification of vibrations at any frequency.

The system overcomes the limits of passive damping systems, which are active only above 200 Hz. Vibration isolation starts from 1.0 Hz and is active until 200 Hz. Active isolation enables the structure to get rid of disturbing vibrations caused by machinery, air-conditioning systems, shaking of the building, and by the vibrations caused by nearby traffic, and so on.

5.19 Hybrid Seismic Control Systems

A hybrid seismic control system combines the features of both passive and active seismic control systems. In general, it has reduced power demands, improved reliability, and reduced cost when compared to fully active systems. Hybrid seismic control systems utilize the advantages of both the passive and active seismic control systems. An example of hybrid system is base isolation used in conjunction with hybrid mass damper (HMD)—computer-controlled actuators to operate TMDs—located throughout the building. Thus, HMD is a combination of a passive TMD and an active control actuator. The ability of this device to reduce structural responses relies mainly on the natural motion of the TMD. The forces from the control actuator are employed to increase the efficiency of the

HMD and to increase its robustness to changes in the dynamic characteristics of the structure. The energy and forces required to operate a typical HMD are far less than those associated with a fully active mass damper system of comparable performance. The HMD is the most common control device employed in structures. The principle of HMD is illustrated in Fig. 5.25.

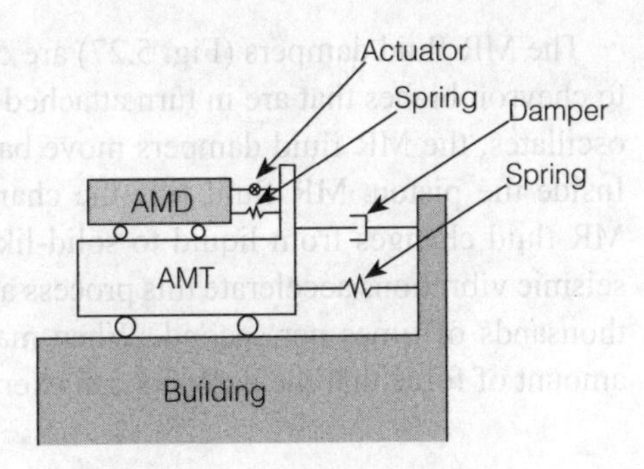

Fig. 5.25 Hybrid mass damper

5.20 Semi-active Control Systems

Semi-active seismic control system, a recent development, is also an example of hybrid system. In this system, external energy is used only for adjustment of the mechanical characteristics of the system and this requires less power than both the hybrid systems and active control systems. Appropriately implemented semi-active systems perform significantly better than passive devices and have the potential to achieve the majority of the performance of fully active systems. This allows for the possibility of effective response reduction during a wide array of dynamic loading conditions. The examples are the active variable stiffness (AVS) damper, active variable-orifice damper, and magneto-rheological (MR) fluid damper; AVS is the most commonly used damper. It is made from two very stiff inclined beams that can be moved to the left or to the right by the upper active piston and, thus, minimizing the relative floor displacement and changing the floor stiffness.

Active variable-orifice damper (Fig. 5.26) is a hydraulic damper system, whose damping coefficient can be changed by mechanically adjusting a valve. In both the systems, the stiffness and damping intensity is changed by small adjustments; thereby, significantly decreasing the amount of energy required in maintaining the systems.

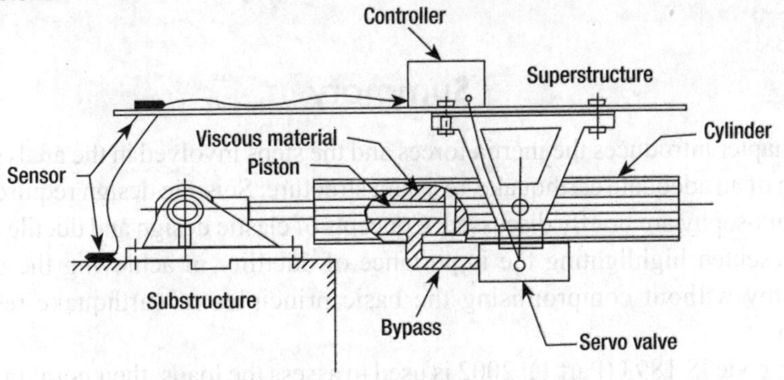

Fig. 5.26 Active variable-orifice damper

The MR fluid dampers (Fig. 5.27) are comparable to pistons and are attached to chevron braces that are in turn attached to a steel cross-beam. As the building oscillates, the MR fluid dampers move back and forth offsetting the vibrations. Inside the piston, MR fluid fills the chamber of an electromagnetic coil. The MR fluid changes from liquid to solid-like state when the coil is charged. The seismic vibrations accelerate this process and cause the MR fluid to change states thousands of times per second. When magnetized, the MR fluid increases the amount of force that the dampers can exert.

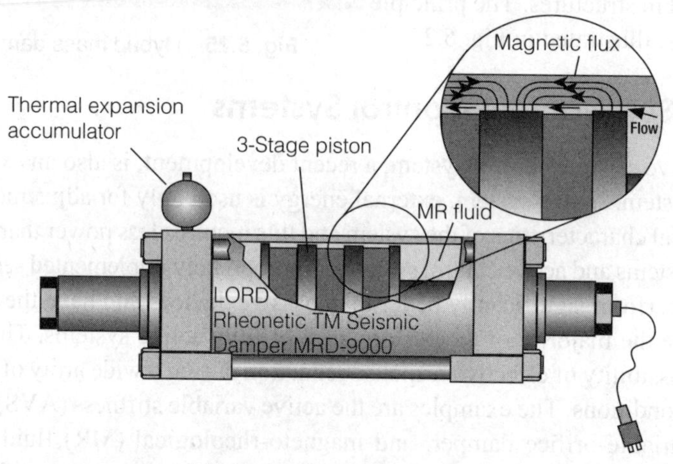

Fig. 5.27 MR fluid damper

Source: Adapted from Terry Stephens, 'Special dampers may shake up engineering field', *A&E Perspectives*, 20 November 2003, retrieved from http://www.djc.com/news/ae/11151055.html, accessed 19 September 2012.

Note: A side benefit of hybrid and semi-active control systems is that, in the case of a power failure, the passive components of the control still offer some degree of protection, unlike a fully active control system.

Summary

The chapter introduces the inertia forces and the steps involved in the analysis and design of an adequate earthquake-resistant structure. Seismic design requirements and philosophy are briefly discussed. Concepts of elastic design and ductile design are presented highlighting the importance of ductility in achieving the overall economy without compromising the basic principles of earthquake-resistant design.

The code IS 1893 (Part 1): 2002 is used to assess the loads, their combinations, and the formulations of analysis procedures. The chapter presents a brief account of principal methods of seismic structural analysis. Dynamic analysis can be performed by the response spectrum method or by time-history method. The methods of analysis, as per IS 1893 (Part 1): 2002, the seismic coefficient method and the response spectrum method, are described in detail. The applications of these methods to solve real structural problems are illustrated with the help of solved examples for beginners.

Principles involved in the conventional methods of earthquake-resistant design methods are discussed. The methodologies for earthquake-resistant design of structures are introduced. The code IS 1893 (Part 1): 2002 recommends the use of lateral strength design approach ensuring minimum lateral strength to resist seismic loads. Moreover, it is assumed that the structure will behave adequately in the non-linear range by following the specifications laid in IS 19320: 1993.

Methodologies of base isolation and energy dissipation are described to make the reader familiar with alternative design methods for making structures earthquake resistant. The aim of providing base isolation is to shift the natural period of vibration to the velocity, sensitive region of spectrum, where the pseudo-acceleration is much smaller resulting in small base shear. Devices used to prevent an earthquake force from acting on a structure, the isolators, and those used to absorb earthquake energy, the dampers, are introduced. Essential concepts of the techniques using these devices to control the response of the structures are discussed.

Solved Problems

5.1 The plan and elevation of a three-storey RCC school building is shown in Fig. 5.28. The building is located in seismic zone V. The type of soil encountered is medium stiff and it is proposed to design the building with a special moment-resisting frame. The intensity of DL is 10 kN/m^2 and the floors are to cater to an IL of 3 kN/m^2. Determine the design seismic loads on the structure by static analysis.

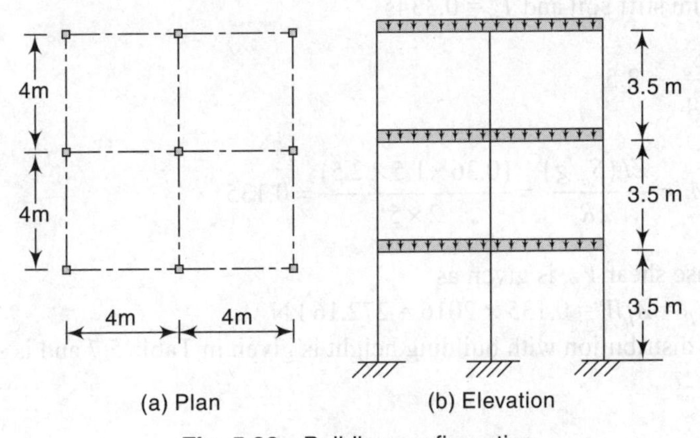

Fig. 5.28 Building configuration

Solution

Design parameters:

For seismic zone V, zone factor, $Z = 0.36$ (Table 5.2)

Importance factor, $I = 1.5$ (Table 5.3)

Response reduction factor $R = 5$ (Table 5.4)

Seismic weight:

Floor area = $8 \times 8 = 64$ m^2

For live load up to and including 3 kN/m^2,

percentage of live load to be considered = 25%.

The total seismic weight on the floors is

$$W = \Sigma W_i$$

where W_i is sum of loads from all the floors which includes DLs and appropriate percentage of live loads.

Seismic weight contribution from one floor = $64 \times (10 + 0.25 \times 3) = 688$ kN

Load from roof = $64 \times 10 = 640$ kN

Hence, the total seismic weight of the structure = $2 \times 688 + 640 = 2016$ kN

Fundamental natural period of vibration, is given as

$$T_a = \frac{0.09h}{\sqrt{d}}$$

where h is the height of the building in metres and d is the base dimension in metres at plinth level along the direction of the lateral load.

$$T_a = \frac{0.09 \times 10.5}{\sqrt{8}} = 0.334 \text{ s}$$

Since the building is symmetrical in plan, the fundamental natural period of vibration will be the same in both the directions.

For medium stiff soil and $T_a = 0.334$s

$$\frac{S_a}{g} = 2.5$$

$$A_h = \frac{ZI(S_a/g)}{2R} = \frac{(0.36 \times 1.5 \times 2.5)}{2 \times 5} = 0.135$$

Design base shear V_B, is given as

$$V_B = A_h W = 0.135 \times 2016 = 272.16 \text{ kN}$$

The force distribution with building height is given in Table 5.7 and is shown in Fig. 5.29.

Table 5.7 Lateral load distribution with height

Storey level	W_i (kN)	h_i (m)	$W_i h_i^2$		Lateral force V_i at i^{th} level for EL in x and y directions (kN)
3	640	10.5	70560	0.626	170.37
2	688	7	33712	0.299	81.38
1	688	3.5	8428	0.0748	20.41
			$\Sigma W_i h_i^2 = 112700$		$\Sigma V_i = 272.16$

Fig. 5.29 Design seismic forces by static analysis

5.2 For the building data given in solved problem 5.1 (Fig. 5.28), determine the dynamic properties (natural periods and mode shapes) for vibrations in both the directions.

Solution

Stiffness of each column:

Let us assume the size of column = 300 × 300 mm^2

Translation stiffness =12 EI_c/L^3 (when stiffness of beam is ∞, i.e., $I_b = \infty$)

Rotational stiffness = 3 EI_c/L^3 (when stiffness of beam is 0, i.e., $I_b = 0$)

$$\text{Moment of inertia of column } I_c = \frac{0.3 \times 0.3^3}{12} = 6.75 \times 10^{-4} \text{ m}^4$$

$$E = 5000\sqrt{f_{ck}} = 5000 \times \sqrt{20} = 22361 \times 10^3 \text{ kN/m}^2$$

Therefore,

$$\text{Translational stiffness of each column} = \frac{12 \times 22361 \times 10^3 \times 6.75 \times 10^{-4}}{3.5^3}$$

$$= 4224.4 \text{ kN/m}$$

$$\text{Rotational stiffness of each column} = \frac{3 \times 22361 \times 10^3 \times 6.75 \times 10^{-4}}{3.5^3}$$

$$= 1056.1 \text{ kN/m}$$

Hence, total stiffness of each column = 4224.4 + 1056.1 = 5280.5 kN/m
There are total nine columns in each floor.
Therefore, total stiffness of each floor = 9 × 5280.5 = 47524.5 kN/m
Natural frequencies and mode shapes can be calculated as follows.

$$m\ddot{x} + kx = 0 \quad \text{(for free vibration)}$$

Using the second law of motion in conjuction with free body diagram (Fig. 5.30) for finding mode shapes, the equations of motion are

$$m_1\ddot{x}_1 + (k_1 + k_2)\, x_1 - k_2\, x_2 = 0$$
$$m_2\ddot{x}_2 - k_2\, x_1 + (k_2 + k_3)\, x_2 - k_3\, x_3 = 0$$
$$m_3\ddot{x}_3 + k_3\, x_3 - k_3\, x_2 = 0$$

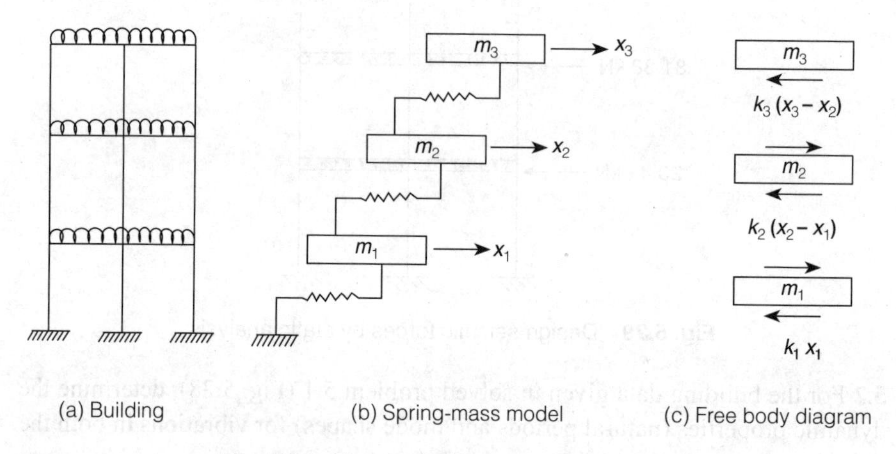

Fig. 5.30 Free body diagram of the structure

From Eqn (6) of Appendix VI, the natural frequencies and mode shapes can be determined

$$\left| k - \omega^2 m \right| = 0$$

The stiffness matrix,

$$k = \begin{bmatrix} k_1 + k_2 & -k_2 & 0 \\ -k_2 & k_2 + k_3 & -k_3 \\ 0 & -k_3 & k_3 \end{bmatrix}$$

Now, $k_1 = k_2 = k_3 = 47524.5$ kN/m $= 4.75245 \times 10^7$ N/m
Substituting the values,

$$k = \begin{bmatrix} 9.5049 & -4.75245 & 0 \\ -4.75245 & 9.5049 & -4.75245 \\ 0 & -4.75245 & 4.75245 \end{bmatrix} \times 10^7 \text{ N/m}$$

The mass matrix,

$$m = \begin{bmatrix} m_1 & 0 & 0 \\ 0 & m_2 & 0 \\ 0 & 0 & m_3 \end{bmatrix}$$

$$\text{Now, } m_1 = m_2 = \frac{688 \times 10^3}{9.81} = 7.0133 \times 10^4 \text{ kg}$$

$$m_3 = \frac{640 \times 10^3}{9.81} = 6.524 \times 10^4 \text{ kg}$$

Substituting the values,

$$m = \begin{bmatrix} 7.0133 & 0 & 0 \\ 0 & 7.0133 & 0 \\ 0 & 0 & 6.524 \end{bmatrix} \times 10^4 \text{ kg}$$

Now, $|k - \omega^2 m| = 0$

$$\begin{bmatrix} 9.5049 \times 10^7 - 7.0133 \times 10^4 \omega^2 & -4.75245 \times 10^7 & 0 \\ -4.75245 \times 10^7 & 9.5049 \times 10^7 - 7.0133 \times 10^4 \omega^2 & -4.75245 \times 10^7 \\ 0 & -4.75245 \times 10^7 & -4.75245 \times 10^7 - 6.524 \times 10^4 \omega^2 \end{bmatrix} = 0$$

By solving the above determinant,

$\omega_1 = 47.04$ Hz, $\omega_2 = 32.93$ Hz, $\omega_3 = 11.79$ Hz

Natural period $T = \dfrac{2\pi}{\omega}$

$$T_1 = \frac{2\pi}{47.04} = 0.134s,\ T_2 = \frac{2\pi}{32.93} = 0.191s,\ T_3 = \frac{2\pi}{11.79} = 0.533s$$

For determing the mode shapes, using Eqn. (5) of Appendix VI,

$$\left[k - \omega^2 m\right]\{\phi\} = 0$$

For $\omega_1 = 47.04$ Hz,

$$[\boldsymbol{k} - \omega^2 \boldsymbol{m}]\{\phi\} = \begin{bmatrix} -6013.9 & -4752.45 & 0 \\ -4752.45 & -6013.9 & -4752.45 \\ 0 & -4752.45 & -9683.6 \end{bmatrix} \begin{Bmatrix} \phi_{11} \\ \phi_{12} \\ \phi_{13} \end{Bmatrix} = 0$$

$$\begin{aligned} -6013.9\,\phi_{11} &- 4752.45\,\phi_{12} + 0 = 0 \\ -4752.45\,\phi_{11} &- 6013.9\,\phi_{12} - 4752.45\,\phi_{13} = 0 \\ 0 &- 4752.45\,\phi_{12} - 9683.6\,\phi_{13} = 0 \end{aligned}$$

Putting $\phi_{13} = 1$, we get

$\phi_{11} = 1.611, \phi_{12} = -2.038$

$$\phi_1 = \begin{Bmatrix} 1.611 \\ -2.038 \\ 1 \end{Bmatrix}$$

For $\omega_2 = 32.93$ Hz:

$$\left[k - \omega^2 m\right]\{\phi\} = \begin{bmatrix} 1899.78 & -4752.45 & 0 \\ -4752.45 & 1879.78 & -4752.45 \\ 0 & -4752.45 & -2322.08 \end{bmatrix} \begin{Bmatrix} \phi_{21} \\ \phi_{22} \\ \phi_{23} \end{Bmatrix} = 0$$

$$\begin{aligned} 1899.78\phi_{21} &- 4752.45\phi_{22} + 0 = 0 \\ -4752.45\phi_{21} &+ 1899.78\phi_{22} - 4752.45\phi_{23} = 0 \\ 0 &- 4752.45\,\phi_{22} - 2322.08\,\phi_{23} = 0 \end{aligned}$$

Putting $\phi_{23} = 1$, we get

$\phi_{21} = -1.223, \phi_{22} = -0.489$

$$\phi_2 = \begin{Bmatrix} -1.223 \\ -0.489 \\ 1 \end{Bmatrix}$$

For $\omega_3 = 11.79$ Hz

$$\left[k - \omega^2 m\right]\{\phi\} = \begin{bmatrix} 8530.02 & -4752.45 & 0 \\ -4752.45 & 8530.02 & -4752.45 \\ 0 & -4752.45 & -3845.59 \end{bmatrix} \begin{Bmatrix} \phi_{31} \\ \phi_{32} \\ \phi_{33} \end{Bmatrix} = 0$$

$$\begin{aligned} 8530.02\phi_{31} &- 4752.45\phi_{32} + 0 = 0 \\ -4752.45\phi_{31} &+ 8530.02\phi_{32} - 4752.45\phi_{33} = 0 \\ 0 &- 4752.45\phi_{32} + 3845.59\phi_{33} = 0 \end{aligned}$$

Putting $\phi_{33} = 1$, we get

$\phi_{32} = 0.81, \phi_{31} = 0.450$

$$\phi_3 = \begin{Bmatrix} 0.450 \\ 0.81 \\ 1 \end{Bmatrix}$$

The mode shapes are shown in Fig. 5.31.

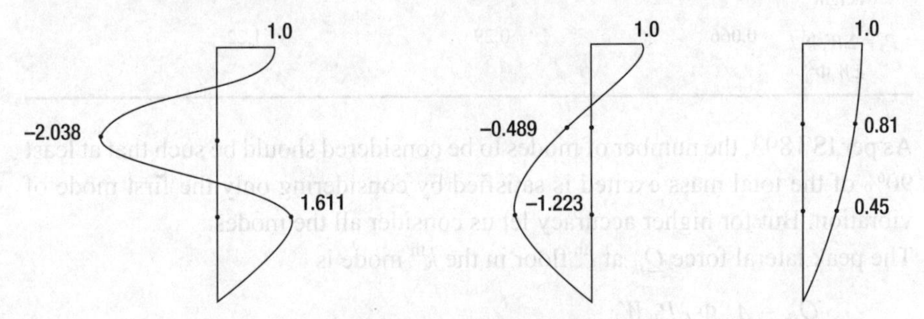

Fig. 5.31 Mode shapes

Table 5.8 Dynamic properties

Storey level	Natural period (s)	Mode 1	Mode 2	Mode 3
3	0.134	1.00	1.00	1.00
2	0.191	–2.038	–0.489	0.81
1	0.533	1.611	–1.223	0.45

5.3 Determine the design seismic forces for the building in Example 5.1 (Fig. 5.28) using dynamic analysis and show the distribution of lateral forces with building height. Consider the free vibration properties of the building as given in Table 5.8.

Also, determine the maximum moment and axial load in bottom-storey columns due to seismic forces. The building is symmetrical in the x and y directions and its properties in both the directions are the same.

Solution

The calculations for the three modes are given in Table 5.9.

Table 5.9 Calculation of mode shapes

Storey level	Weight W_i (kN)	Mode 1			Mode 2			Mode 3		
		Φ_{ik}	$W_i\Phi_{ik}$	$W_i\Phi_{ik}^2$	Φ_{ik}	$W_i\Phi_{ik}$	$W_i\Phi_{ik}^2$	Φ_{ik}	$W_i\Phi_{ik}$	$W_i\Phi_{ik}^2$
3	640	1.00	640	640	1.00	640	640	1.00	640	640
2	688	–2.038	–1402.14	2829.60	–0.489	–336.43	164.51	–0.81	557.30	451.40
1	688	1.611	1108.37	1785.58	1.223	841.424	1029.06	0.45	309.60	139.30
Σ	2016		346.23	5255.18		537.85	1833.57		1506.90	1230.70

(Contd)

Table 5.9 (*Contd*)

$M_k = (\Sigma W_i \Phi_{ik})^2 / g \Sigma W_i \Phi_{ik}^2$	$\dfrac{346.23^2}{9.81 \times 5255.18} = 2.33\,\text{kg}$	$\dfrac{537.85^2}{9.81 \times 1833.57} = 16.08\,\text{kg}$	$\dfrac{1506.90^2}{9.81 \times 1230.70} = 188.08\,\text{kg}$
% of total Weight	1.13%	7.79%	91.08%
$P_k = \Sigma W_i \Phi_{ik} / \Sigma W_i \Phi_{ik}^2$	0.066	0.29	1.22

As per IS 1893, the number of modes to be considered should be such that at least 90% of the total mass excited is satisfied by considering only the first mode of vibration. But for higher accuracy let us consider all the modes.
The peak lateral force Q_{ik} at i^{th} floor in the kth mode is

$$Q_{ik} = A_{hk} \Phi_{ik} P_k W_i$$

The values of A_{hk} for different modes are as given in Table 5.10.

Table 5.10 Design acceleration spectrum values

Mode 1	Mode 2	Mode 3
$T_1 = 0.0134$	$T_2 = 0.191$	$T_3 = 0.533$
$S_a/g = 1 + 15T = 3.01$	$S_a/g = 2.50$	$S_a/g = 2.50$
$A_1 = ZI(S_a/g)/2R$	$A_2 = ZI(S_a/g)/2R$	$A_3 = ZI(S_a/g)/2R$
$= \dfrac{(0.36 \times 1.5 \times 3.01)}{2 \times 3}$	$= \dfrac{(0.36 \times 1.5 \times 2.50)}{2 \times 3}$	$= \dfrac{(0.36 \times 1.5 \times 2.50)}{2 \times 3}$
$= 0.271$	$= 0.225$	$= 0.225$
$Q_{ik} = 0.271 \times 0.066 \times \Phi_{ik} \times W_i$	$Q_{ik} = 0.225 \times 0.29 \times \Phi_{ik} \times W_i$	$Q_{ik} = 0.225 \times 1.22 \times \Phi_{ik} \times W_i$

The storey shears in each mode are given in Table 5.11.

Table 5.11 Storey shear in each mode

Floor level (i)	Weight W_i (kN)	Mode 1 Φ_{i1}	Q_{i1}	V_{i1}	Mode 2 Φ_{i2}	Q_{i2}	V_{i2}	Mode 3 Φ_{i3}	Q_{i3}	V_{i3}
3	640	1.00	11.45	11.45	1.00	41.76	41.76	1.00	175.68	175.68
2	688	−2.038	−25.08	−13.63	−0.489	−21.95	19.81	0.81	152.97	328.65
1	688	1.611	−19.82	6.19	−1.223	−54.90	−35.09	0.45	84.98	413.63

Modal Combination

The contribution of different modes is obtained by the SRSS.

$$V_3 = (11.45^2 + 41.76^2 + 175.68^2)^{1/2} = 180.94 \text{ kN}$$

$$V_2 = (-13.63)^2 + 19.81^2 + 328.65^2)^{1/2} = 329.53 \text{ kN}$$

$$V_1 = (6.19^2 + (-35.09)^2 + 413.63^2)^{1/2} = 415.16 \text{ kN}$$

The design storey shears at different storeys are shown in Fig. 5.32.

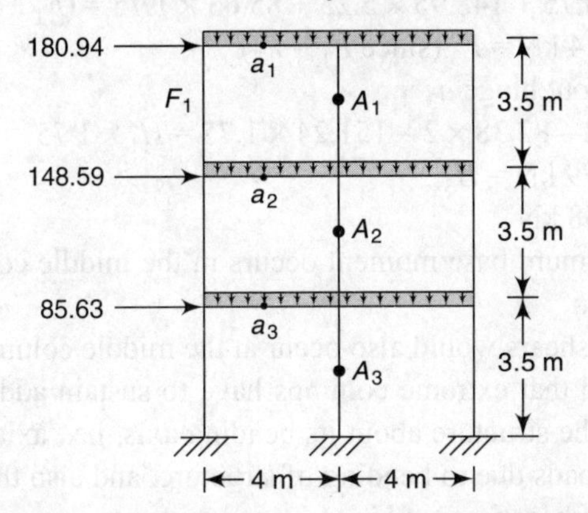

Fig. 5.32 Design seismic forces by dynamic analysis

$$Q_3 = V_3 = 180.94 \text{ kN}$$

$$Q_2 = V_2 - V_3 = 329.53 - 180.94 = 148.59 \text{ kN}$$

$$Q_1 = V_1 - V_2 = 415.16 - 329.53 = 85.63 \text{ kN}$$

Here, the base shear from the static analysis method (V_{BS}) is less than the base shear from the dynamic analysis method (V_{BD}).

$$V_{BS} < V_{BD}$$

There is no need to scale-up response quantities.

Base moments and shear forces (using cantilever method of analysis)

Bending axis of the structure in x direction is 4 m from exterior column.

For top floor, taking moments about the hinge A_1;

$$180.94 \times 3.5/2 = F_1 \times 4 + F_2 \times 0 + F_3 \times 4$$

and $F_1 = F_3$ (due to symmetry)

hence, $F_1 = 39.58$ kN

Taking moment about a_1

$$F_1 \times 2 = H_1 \times 3.5/2$$

$$H_1 = 45.23 \text{ kN} = H_3 \text{ and } H_2 = 2H_1 = 90.46 \text{ kN}$$

For 1st floor

Taking moment about A_2

$$180.94 \times 5.25 + 148.59 \times 1.75 = (F_1 + F_3) \times 4$$

$$F_1 = 151.24 \text{ kN (since } F_1 = F_3)$$

Taking moment about hinge a_2

$$151.24 \times 2 - 45.23 \times 3.5/2 - 39.58 \times 2 = 1.75\, H_1$$

Therefore, $H_1 = 82.38$ kN $= H_3$

and $H_2 = 2H_1 = 164.76$ kN

For ground floor

Taking moment about A_3

$$180.94 \times 8.75 + 148.95 \times 5.25 + 85.63 \times 1.75 = (F_1 + F_3) \times 4$$

$$F_1 = 314.14 \text{ kN} = F_3 \text{ (since } F_1 = F_3)$$

Taking moment about hinge a_3

$$314.14 \times 2 - 82.38 \times 2 - 151.24 \times 1.75 = H_1 \times 1.75$$

$$H_1 = 103.79 \text{ kN} = H_3$$

$$H_2 = 207.58 \text{ kN}$$

Therefore, maximum base moment occurs in the middle column = 207.58 × 1.75 = 363.26 kN/m.

Maximum base shears would also occur at the middle column = 207.58 kN.

It must be noted that extreme columns have to sustain additional axial load due to rotation of the structure about its bending axis, i.e., axial load from dead weight plus axial loads due to bending of structure, and also the additional base moment due to the shear force at hinge.

5.4 A 10-storey OMRF building has plan dimensions as shown in Fig. 5.33. The storey height is 3.0 m. The DL per unit area of the floor, consisting of the floor slab, finishes, etc., is 4 kN/m^2. Weight of the partitions on the floor can be assumed to be 2 kN/m^2. The intensity of live load on each floor is 3 kN/m^2 and on the roof is 1.5 kN/m^2. The soil below the foundation is hard and the building is located in Delhi. Determine the seismic forces and shears at different floor levels.

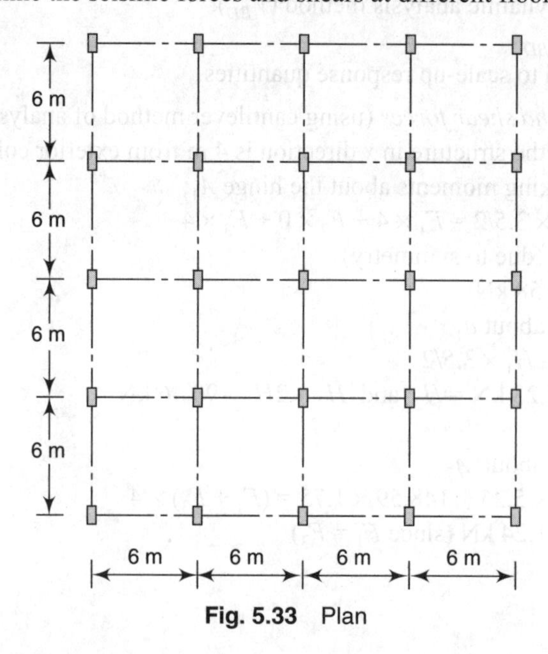

Fig. 5.33 Plan

Solution

Design parameters:

For Delhi (Zone IV), zone factor $Z = 0.24$

Importance factor $I = 1.0$

Response reduction factor $R = 3.0$ (OMRF)

Seismic weight:

Floor area = $24 \times 24 = 576$ m^2

Dead load = 4 kN/m^2

Weight of partitions = 2 kN/m^2

For live load upto and including 3 kN/m^2,

percentage of live load to be considered = 25%

Total seismic weight on the floors,

$$W = \Sigma W_i$$

where ΣW_i is the sum of loads from all the floors, which includes dead loads and appropriate percentage of live loads.

Effective weight at each floor except the roof = $4.0 + 2.0 + 0.25 \times 3 = 6.75$ kN/m^2, and at the roof = 4.0 kN/m^2.

Weight of the beams at each floor and the roof = $0.3 \times 0.6 \times 240 \times 25 = 1080$ kN

Weight of the columns at each floor = $0.3 \times 0.6 \times 2.4 \times 25 \times 25 = 270$ kN

Weight of the column at the roof = $\frac{1}{2} \times 270 = 135$ kN

Total plan area of the building is 24 m × 24 m = 576 m^2

Equivalent load at roof level = $4 \times 576 + 1080 + 135 = 3519$ kN

Equivalent load at each floor = $6.75 \times 576 + 1080 + 270 = 5238$ kN

Seismic weight of the building $W = 3519 + 5238 \times 9 = 50661$ kN

Base shear:

Fundamental natural period of vibration of a moment-resisting frame without infill

$T_a = 0.075h^{0.75} = 0.075\,(30)^{0.75} = 0.96$ s

Average response acceleration coefficient S_a/g for 5% damping and type I soil is 1.04.

Design horizontal seismic coefficient,

$$A_h = \frac{ZI(S_a/g)}{2R} = \frac{0.24 \times 1.0 \times 1.04}{2 \times 3} = 0.0416$$

Base shear $V_B = A_h W = 0.0416 \times 50661 = 2107.5$ kN

Lateral load and shear force at various floor levels

Design lateral force at floor i, $Q_i = V_B \dfrac{W_i h_i^2}{\sum_{j=1}^{n} W_i h_i^2}$

The calculation of design lateral forces at each floor level is shown in Table 5.12.

Table 5.12 Lateral loads and shear forces at different floors levels

Mass No.	W_i (kN)	h_i (m)	$W_ih_i^2$ (kN-m^2)	$\frac{W_ih_i^2}{\sum_{i=1}^{n} W_ih_i^2}$	Q_i (kN)	V_i (kN)
1	3519	30.0	3167100	0.1907	402.0	402.0
2	5238	27.0	3818502	0.2299	484.6	886.6
3	5238	24.0	3017088	0.1817	382.9	1269.5
4	5238	21.0	2309958	0.1391	293.3	1562.8
5	5238	18.0	1697112	0.1022	215.5	1778.3
6	5238	15.0	1178550	0.0709	149.5	1927.8
7	5238	12.0	754272	0.0454	95.7	2023.5
8	5238	9.0	424278	0.0255	54.0	2077.5
9	5238	6.0	188568	0.0114	24.0	2101.5
10	5238	3.0	47142	0.0028	6.0	2107.5
$\Sigma W_ih_i^2$ = 16602570						

The design seismic forces at different floor levels are shown in Fig. 5.34.

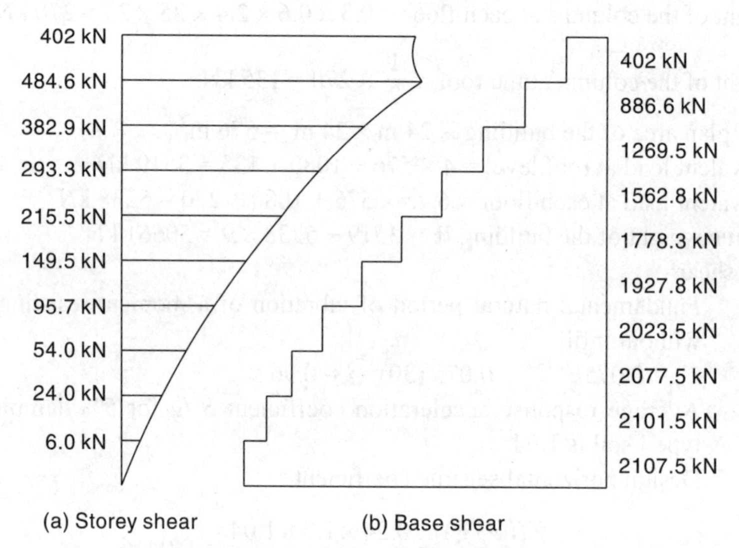

Fig. 5.34 Design seismic forces

5.5 A simple one-storey building has two shear walls in each direction as shown in Fig. 5.35. It has some gravity columns that are not shown. All four walls are in M-25 grade concrete, 200-mm thick, and 5-m long. The storey height is 4 m. The RC floor is cast in situ. Design shear force on the building is 200 kN in either direction. Compute design lateral forces on different shear walls.

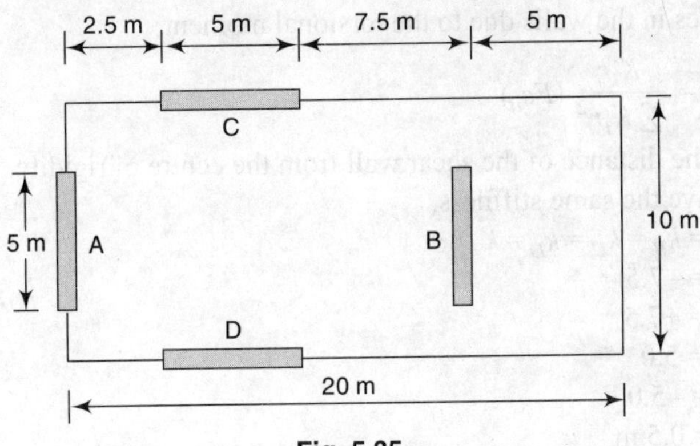

Fig. 5.35

Solution

Design shear force = 200 kN

Grade of concrete: M-25

$$E = 5000\sqrt{25} = 25000\ \text{N/mm}^2$$

Storey height $h = 4500$ mm

Thickness of wall $t = 200$ mm

Length of walls $L = 5000$ mm

All the four shear walls are of the same size and section, and hence have the same lateral stiffness k.

Centre of mass will be the geometric centre of the floor slab, i.e., (10.0, 5.0).

Centre of rigidity will be at (7.5, 5.0)

Earthquake force in the x-direction:

Because of symmetry in this direction, the eccentricity = 0.0 m (e_{si})

Lateral forces in the walls due to translation:

$$F_{CT} = \frac{k_C}{k_C + k_D} F = \frac{1}{1+1} \times 200 = 100.0\ \text{kN}$$

$$F_{DT} = \frac{k_D}{k_C + k_D} F = \frac{1}{1+1} \times 200 = 100.0\ \text{kN}$$

Design eccentricity

$$e_{di} = \begin{Bmatrix} 1.5e_{si} + 0.05b_i \\ \text{or} \\ e_{si} - 0.05b_i \end{Bmatrix}$$

whichever of these conditions gives the more severe effect in the shear of a frame.

$$e_d = \begin{Bmatrix} 1.5 \times 0.00 + 0.05 \times 10 = 0.5 \\ \text{or} \\ 0.00 - 0.05 \times 10 = -0.5 \end{Bmatrix}$$

Lateral forces in the walls due to the torsional moment,

$$F_{iR} = \frac{K_i r_i}{\sum K_i r_i^2}(Fe_d)$$

where r_i is the distance of the shear wall from the centre of rigidity.
All walls have the same stiffness,

$k_A = k_B = k_C = k_D = k$

$r_A = -7.5$

$r_B = +7.5$

$r_C = 5.0$

$r_D = -5.0$

and $e_d = 0.5$ m

Therefore,

$$F_{AR} = \frac{r_A}{(r_A^2 + r_B^2 + r_C^2 + r_D^2)} Fe_d$$

$$= \frac{-7.5}{[(-7.5)^2 + 7.5^2 + 5^2 + (-5)^2]} \times 200 \times (\pm 0.5) = \pm 4.62 \text{ kN}$$

$$F_{BR} = \frac{7.5}{[(-7.5)^2 + 7.5^2 + 5^2 + (-5)^2]} \times 200 \times (\pm 0.5) = \pm 4.62 \text{ kN}$$

$$F_{CR} = \frac{5}{[(-7.5)^2 + 7.5^2 + 5^2 + (-5)^2]} \times 200 \times (\pm 0.5) = \pm 3.08 \text{ kN}$$

$$F_{DR} = \frac{-5}{[(-7.5)^2 + 7.5^2 + 5^2 + (-5)^2]} \times 200 \times (\pm 0.5) = \pm 3.08 \text{ kN}$$

Total lateral forces in the walls due to seismic load in the x direction

$$\left.\begin{aligned} F_A &= 4.62 \text{ kN} \\ F_B &= 4.62 \text{ kN} \\ F_C &= \max(100 \pm 3.08) = 103.08 \text{ kN} \\ F_D &= \max(100 \pm 3.08) = 103.08 \text{ kN} \end{aligned}\right\} \tag{5.5.1}$$

Earthquake force in the y direction

Lateral forces in the walls due to translation:

$$F_{AT} = \frac{k_A}{k_A + k_B} F = \frac{1}{1+1} \times 200 = 100 \text{ kN}$$

$$F_{BT} = \frac{k_B}{k_A + k_B} F = \frac{1}{1+1} \times 200 = 100 \text{ kN}$$

Calculated eccentricity = (10 – 7.5) = 2.5 m

Design eccentricity

$$e_d = \begin{cases} 1.5 \times 2.5 + 0.05 \times 20 = 4.75 \\ \text{or} \\ 2.5 - 0.05 \times 20 = 1.5 \end{cases}$$

$e_d = 4.75$ m

Lateral force in the walls due to torsional moment,

$$F_{AR} = \frac{r_A}{(r_A^2 + r_B^2 + r_C^2 + r_D^2)} Fe_d = -43.84 \text{ kN}$$

$$= \left[\frac{-7.5}{(-7.5)^2 + (7.5)^2 + (5)^2 + (-5)^2}\right] \times 200 \times 4.75 = -43.84 \text{ kN}$$

$$F_{BR} = \left[\frac{7.5}{(-7.5)^2 + (7.5)^2 + (5)^2 + (-5)^2}\right] \times 200 \times 4.75 = +43.84 \text{ kN}$$

$$F_{CR} = \left[\frac{5.0}{(-7.5)^2 + (7.5)^2 + (5)^2 + (-5)^2}\right] \times 200 \times 4.75 = 29.23 \text{ kN}$$

$$F_{DR} = \left[\frac{-5.0}{(-7.5)^2 + (7.5)^2 + (5)^2 + (-5)^2}\right] \times 200 \times 4.75 = -29.23 \text{ kN}$$

Total lateral forces in the walls

$F_A = 100 - 43.84 = 56.16$ kN
$F_B = 100 + 43.84 = 143.84$ kN
$F_C = 0 + 29.23 = 29.23$ kN
$F_D = 0 - 29.23 = -29.23$ kN

Similarly, when $e_d = 1.5$ m, then the total lateral forces in the walls

$F_A = 100 - 43.84 \times (1.5/4.75) = 86.16$ kN
$F_B = 100 + 43.84 \times (1.5/4.75) = 113.84$ kN
$F_C = 0 + 29.23 \times (1.5/4.75) = 9.23$ kN
$F_D = 0 - 29.23 \times (1.5/4.75) = 9.23$ kN

Maximum forces in walls due to seismic load in the y direction

$$\left.\begin{aligned} F_A &= \max(56.16, 86.16) = 86.16 \text{ kN} \\ F_B &= \max(143.84, 113.84) = 143.84 \text{ kN} \\ F_C &= \max(29.23, 9.23) = 29.23 \text{ kN} \\ F_D &= \max(29.23, 9.23) = 29.23 \text{ kN} \end{aligned}\right\} \quad (5.5.2)$$

The forces obtained from seismic loading in the x and y directions.

$F_A = 86.16$ kN
$F_B = 143.84$ kN
$F_C = 103.08$ kN
$F_D = 103.08$ kN

The design lateral forces in the shear walls will be the maximum of the values obtained from Eqns (5.5.1) and (5.5.2).

Note: As per IS 1893 (Part1): 2002, negative torsional shear should be neglected. Hence, wall *A* should be designed for not less than 100 kN lateral load.

Exercises

5.1 Explain the following:
- (a) Inertial force
- (b) Response spectrum factor
- (c) Provisions for torsion
- (d) Storey drift
- (e) Soft storey

5.2 State the assumptions made in the analysis of earthquake-resistant design of buildings.

5.3 What are the two seismic design requirements an engineer has to account for in the analysis and design of earthquake-resistant buildings? Discuss briefly how these are incorporated to achieve the objective.

5.4 Discuss the factors required for assessing
- (a) the lateral design forces
- (b) the design response spectrum

5.5 Discuss the ways and means to prevent an earthquake force from acting on the superstructure of a building.

5.6 Write short notes on the following:
- (a) Isolating devices
- (b) Energy-dissipation devices
- (c) Properties of construction materials for earthquake resistance.

5.7 Determine the natural frequencies and mode shapes of the three-storey structure shown in Fig. 5.36.

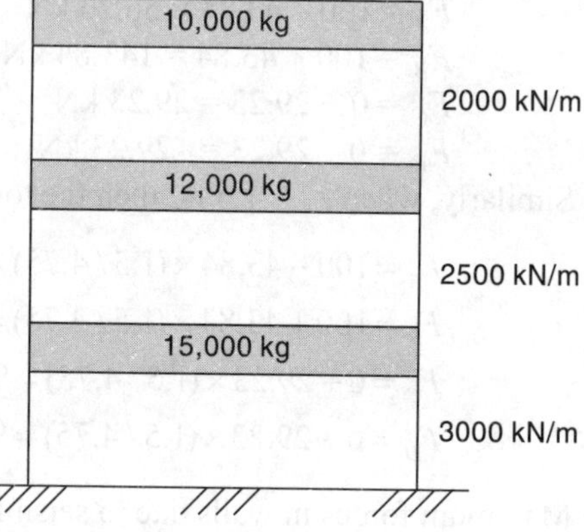

Fig. 5.36 Plan

Ans:

Mode / Storey \ Frequency	1	2	3
Frequency	0.5316	0.5373	0.771
3	1	1	1
2	0.7540	–0.4650	–2.0215
1	–0.3880	0.9959	1.4243

5.8 Plan of a five-storey building is shown in Fig. 5.37. Dead load including self weight of slab, finishes, etc. can be assumed as 3 kN/m^2 and live load as 4 kN/m^2 on each floor and as 1.5 kN/m^2 on the roof. Weight of partitions is 2 kN/m^2. Determine the lateral forces and shears at different storey levels.

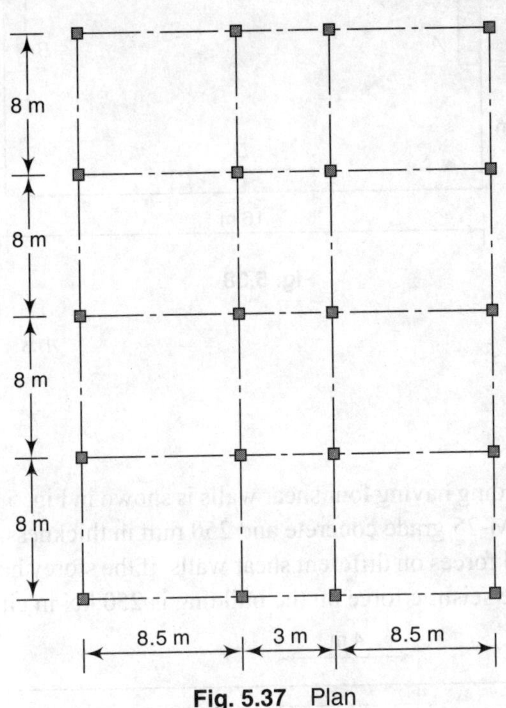

Fig. 5.37 Plan

Ans: Assuming $z = 0.24$, $I = 1$, $R = 5$, soil type = 2, storey height = 3.5 m

Lateral forces (Q)	Storey shear (V)
250.215	250.215
266.956	517.171
150.144	667.315
66.739	734.054
16.666	750.72

5.9 Plan of a single-storey building having two shear walls in each direction is shown in Fig. 5.38. All the four walls are of M-20 grade concrete, 200 mm in thickness and 6 m in length. Height of the building is 3.6 m. Design shear force on the building is 120 kN in either direction. Determine the design lateral force for different shear walls using the torsion provisions of the code.

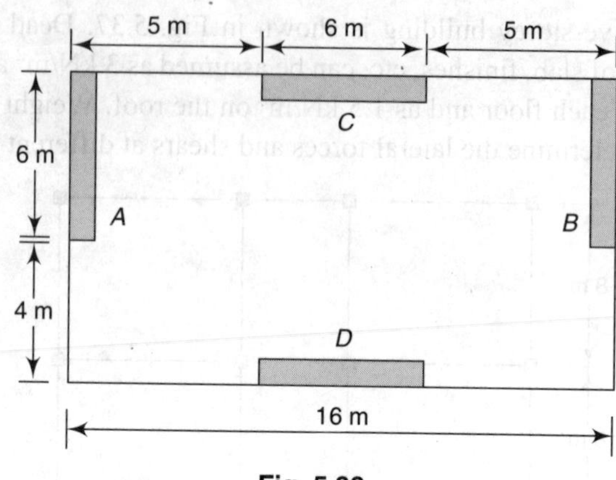

Fig. 5.38

Ans: $F_A = 64.31$ kN
$F_B = 64.31$ kN
$F_C = 61.68$ kN
$F_D = 61.68$ kN

5.10 Plan of a building having four shear walls is shown in Fig. 5.39. All the four walls are of M-25 grade concrete and 250 mm in thickness. Determine the design lateral forces on different shear walls, if the storey height is given as 4.5 m and the seismic force on the building is 250 kN in either direction.

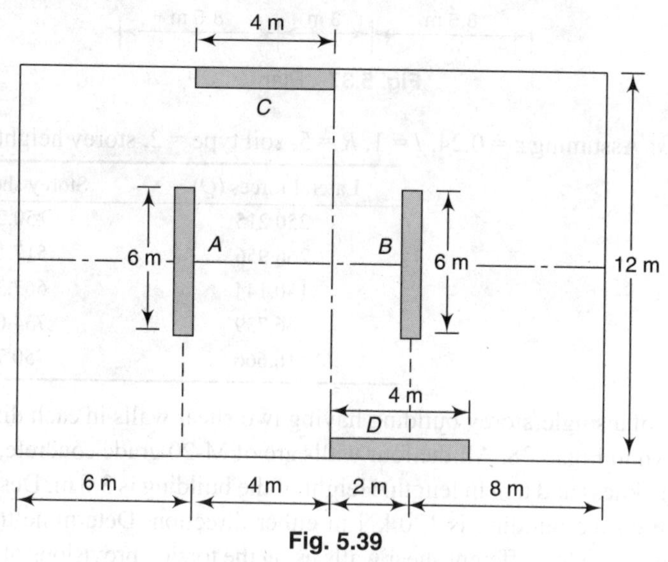

Fig. 5.39

Ans: $F_A = 104.263$ kN
$F_B = 145.737$ kN
$F_C = 137.481$ kN
$F_D = 137.481$ kN

CHAPTER

6

Masonry Buildings

Till the early twentieth century, most buildings were masonry constructions. Gradually, RC and steel constructions became popular because of their inherent advantage. However, since it is economical, is good for insulation, has a good finishing, and is easy to procure, masonry is still in use in most countries. Primarily, it is used in walls for buildings. It is also used as infill panels, partitions, etc., in framed buildings, where it is subjected to forces from the displacement of the frame and inertia forces. In such situations, the designer must be concerned for the interaction between the masonry and the frame, which may modify the frame's response and the forces acting on it.

Masonry covers a very wide range of materials, such as bricks, stones, blocks, etc., joined with different types of mortars such as lime mortar, cement mortar, etc., that exhibit different mechanical properties. Masonry may be used with or without reinforcement; the latter is not suitable for use in seismic areas. Properly detailed reinforced masonry may be used as a primary structural system and can be designed to resist earthquake forces.

In many countries, most masonry as well as wooden buildings are constructed in the traditional manner with little or no intervention by qualified engineers and architects. Such buildings are constructed spontaneously and informally without any due regard to the stability of the system under horizontal seismic forces and are called *non-engineered buildings*. Since more than 90 per cent of the population still lives and works in non-engineered buildings, a structural engineer is, therefore, concerned with laying down the guidelines for their design and construction, and restoration and strengthening. A masonry building designed and detailed for all the stipulated forces is termed as an *engineered building*. The design procedure for engineered buildings is explained in Section 6.10.

Due to poor performance of some forms of masonry in earthquakes, the official attitude towards masonry is generally cautious in most moderate to strong motion seismic areas. The poor performance of masonry buildings in earthquakes is because of the following reasons.

(a) The material itself is brittle and its strength degradation due to load repetition is severe.
(b) Masonry has great weight because of thick walls. Consequently, the inertia forces are large.
(c) Large stiffness of the material, which leads to large response to earthquake waves of short natural period.
(d) Quality of construction is not consistent because of quality of the locally manufactured masonry units (bricks), unskilled labour, etc., that leads to large variability in strength.

Masonry buildings are characterized by high stiffness and great weight. As a result, energy dissipation into the ground acceleration is large. Maximum acceleration response can easily rise to as much as three times the ground acceleration. Masonry buildings can be designed economically if energy dissipation can be assumed to occur as a result of ductile behaviour. This can be achieved by combining steel effectively with masonry and proper designing details. The general principles to be followed for the analysis and design of masonry buildings are the same as those outlined in Chapters 4 and 5.

In the sections to follow, the behaviour of masonry walls is discussed, followed by the ways in which their seismic behaviour can be improved. Thereafter, the step-by-step procedure for the seismic design of masonry buildings is outlined. At the end of the chapter, methods of restoring and strengthening of dilapidated and existing masonry walls are described.

6.1 Categories of Masonry Buildings

As per IS 4326: 1993, masonry buildings have been categorized into five classes from A to E from the viewpoint of earthquake-resistant features. The classification is based on the value of the design seismic coefficient (α_h) for the building and is given in Table 6.1.

$$\alpha_h = \alpha_o I \beta \qquad (6.1)$$

where α_o is the basic seismic coefficient for the seismic zone in which the building is located (Appendix VII), I is the importance factor (Appendix VIII), and β is the soil-foundation factor (Appendix IX).

IS 4326: 1993 has still not been revised. All the recommendations made in it are basically for non-engineered buildings (masonry and wooden buildings). Hence, for design of engineered buildings the recommendations of IS 1893: 2002 may be followed.

6.2 Behaviour of Unreinforced Masonry Walls

Masonry buildings are vulnerable to strong earthquake shaking. A masonry building has three components—the roof, the wall, and the foundation (Fig. 6.1). The inertia forces travel through the roofs and walls to the foundation. These inertia forces are developed both in x and y directions. A wall topples down easily if pushed horizontally at the top in a direction perpendicular to its plane (weak direction). This is called *out-of-plane failure* [wall B shown in Fig. 6.2(a)]. However, a wall [wall A shown in Fig. 6.2(b)] offers much greater resistance if pushed along its length (strong direction). This is called *in-plane resistance*. This is because of the wall's large dimension in the plane of bending. Such a wall, carrying horizontal loads in its own plane, is known as a *shear wall*.

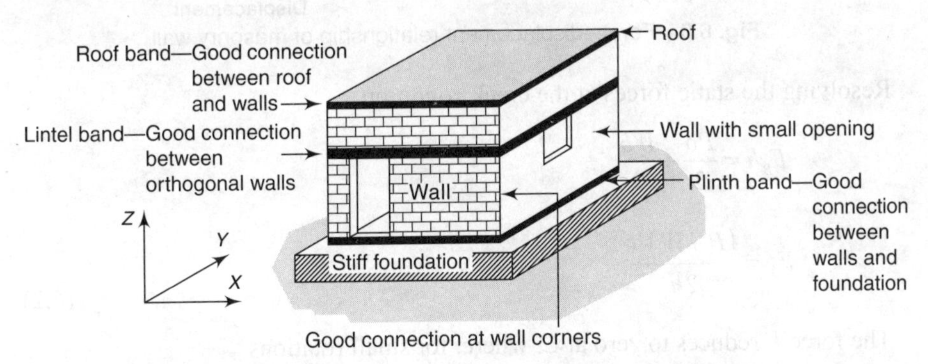

Fig. 6.1 Principal directions of a building and essential requirements to ensure box action in a masonry building (adapted from Murty, 2005)

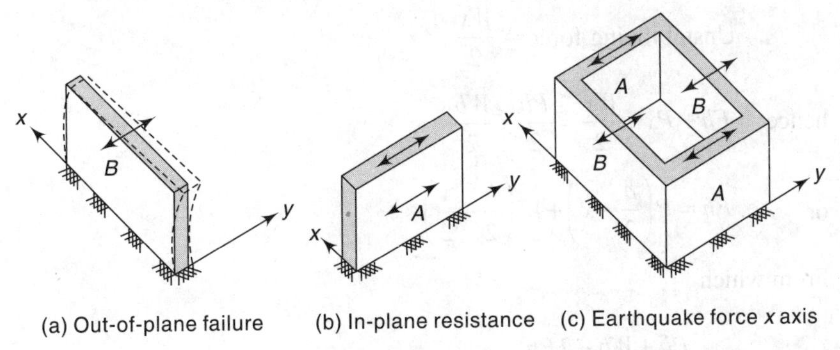

Fig. 6.2 Box action in masonry building (adapted from Murty, 2005)

The seismic capacity for unreinforced masonry is most commonly based on stability and energy considerations rather than stress levels. Neither elastic, nor ultimate strength analysis adequately predicts the seismic capacity—both methods produce over-conservative results.

Figure 6.3 shows the force–displacement relationship for a masonry wall subjected to static lateral loading. The wall behaves elastically up to a point A, where the base cracks and the force immediately drops from F_A to F_B.

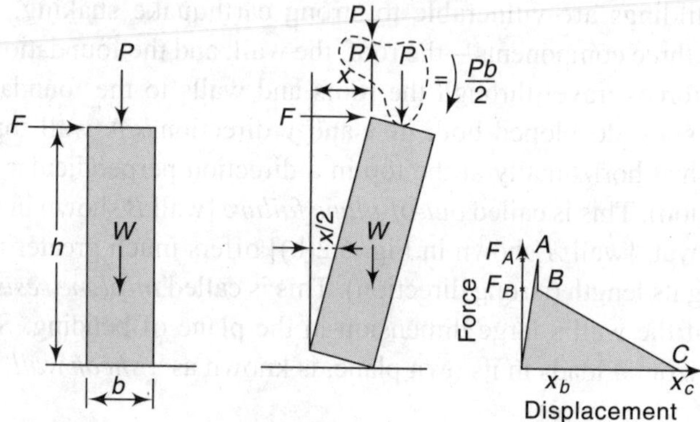

Fig. 6.3 Force–displacement relationship of masonry wall

Resolving the static forces at the crack condition

$$F_B h = \frac{Pb}{2} + \frac{Wb}{2}$$

$$F_B = \frac{(P+W)b}{2h} \tag{6.2}$$

The force F reduces to zero at C, where, for small rotations

$$\text{Stabilizing force} = \frac{Pb}{2} + \frac{Wb}{2}$$

$$\text{Unstabilizing force} = \frac{Wx}{2}$$

hence $$Fh + Px + \frac{Wx}{2} = \frac{Pb}{2} + \frac{Wb}{2}$$

or $$Fh = P\left(\frac{b}{2} - x\right) + W\left(\frac{b}{2} - \frac{x}{2}\right)$$

From which

$$x = \frac{Pb + Wb - 2Fh}{2P + W} \tag{6.3}$$

When $F = 0$

$$x_c = \frac{Pb + Wb}{2P + W} \tag{6.4}$$

From Fig. 6.3, at point A, the incremental stiffness of the wall becomes negative so that for a steadily applied force F_A, collapse will occur unless the force F_A is transferred by an alternative load path to other stiffer structural elements. For a ground acceleration pulse, this is not necessarily the case because the pulse which initiated rocking will have to be continued for a sufficient time to reach failure. If the ground acceleration reverses soon after rocking has started, the wall will stabilize again. Under earthquake loading, the displacement may even exceed x_c and return to a stable state if a sufficiently strong reverse pulse occurs.

It has been shown by a number of analytical and practical studies that failure in masonry is closely related to energy. Furthermore, it has been demonstrated that the energy requirement to cause in-plane stability failure of masonry walls is so high that failure is normally by shear rather than by instability. Out-of-plane failure, however, is customarily by instability and Fig. 6.4 shows the simplified response of the masonry wall in Fig. 6.3 to cyclic loading. For simple buildings, ultimate loading can be assessed approximately by replacing the wall by an equivalent elastic structure, the response of which is shown by the broken line. The energy required to cause failure is approximately equal for the actual and equivalent structures, x_b being much smaller than x_c for practical conditions. The equivalent stiffness can be found from the expression

$$K_c = \frac{F_B}{x_c} \tag{6.5}$$

and the equivalent natural frequency

$$f_c = \frac{1}{2\pi}\left(\frac{k_c}{W + P}\right)^{\frac{1}{2}} \tag{6.6}$$

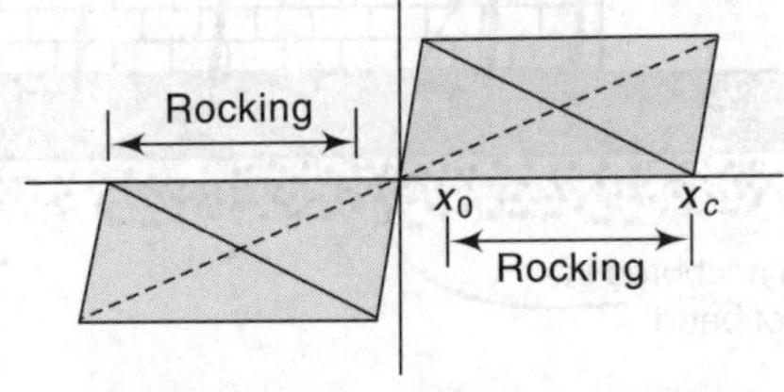

Fig. 6.4 Equivalent structure response

The maximum displacement can be derived from the appropriate response spectrum or other methods that are applicable to elastic structures, the criteria for stability being $x \leq x_c$. For multi-storey unreinforced masonry buildings, the problem of the assessment of stability becomes more complex and reference should be made to the energy approach.

6.3 Behaviour of Reinforced Masonry Walls

The reinforced masonry walls are used and designed for lateral out-of-plane loads and axial loads. Most reinforced masonry walls are designed to span vertically and transfer the lateral loads to the roof, floor, or foundation. Normally these walls are designed as simple beams spanning between structural supports. So far as axial loads are concerned these are transferred directly to the foundation except for the case of eccentric loading that may cause tension in the wall. In the vertically reinforced masonry construction, the vertical reinforcing bars start from the foundation concrete. These must pass through all seismic bands and tied to the horizontal band reinforcement with binding wires and finally embedded in the roof band/roof slab. The vertical reinforcement should be bent using a 300-mm 90° bend. Embedding vertical reinforcement bars in the edges of the wall piers and their anchorage in the foundation and in the roof band (Fig. 6.5), forces the slender masonry piers to undergo bending instead of rocking. In wider wall piers, the vertical bars enhance their capability to resist horizontal earthquake forces and delay the cross-cracking. Further, the vertical bars also help to protect the wall from sliding as well as from collapsing in the weak direction. A reinforced masonry shear wall may fail in flexure or shear as discussed further.

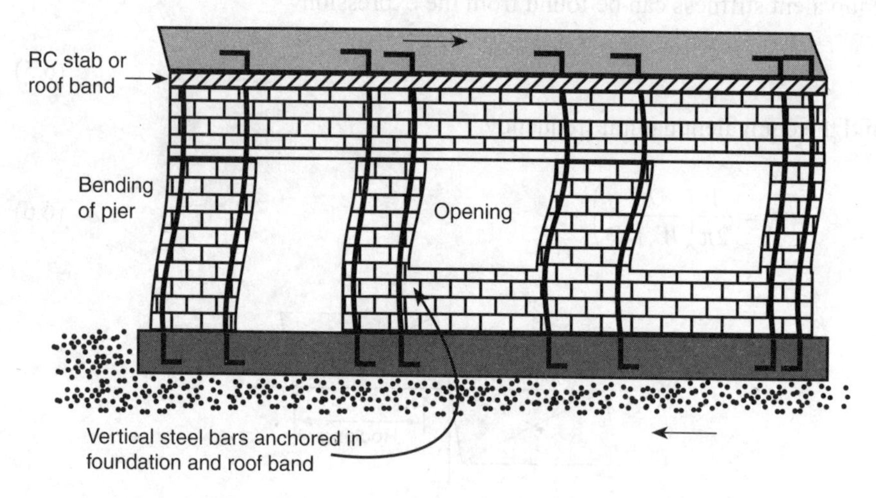

Fig. 6.5 Vertical reinforcement causes bending of masonry piers and checks their rocking

Flexural failure When the ratio of height to length of masonry wall is large and the vertical reinforcement is small, flexural failure takes place. The hysteresis behaviour of such a wall under repeated in-plane bending with low axial force is approximately of the elastoplastic type and shows high ductility and little strength degradation. A masonry wall failing in flexure and subjected to high axial force is not necessarily ductile, and degradation is severe. The behaviour of a reinforced masonry wall subjected to out-of-plane bending is similar to an RC wall, and ductility is very large.

Shear failure In masonry walls without openings, shear failure often takes place as shown in Fig. 6.6(a), while in wall piers and spandrels, in walls with openings, failure is as shown in Fig. 6.6(b). Shear failure is likely to occur in wall elements with small height-to-length ratio. Shear failure tends to be brittle, with low energy dissipation capacity and severe strength degradation due to load repetition.

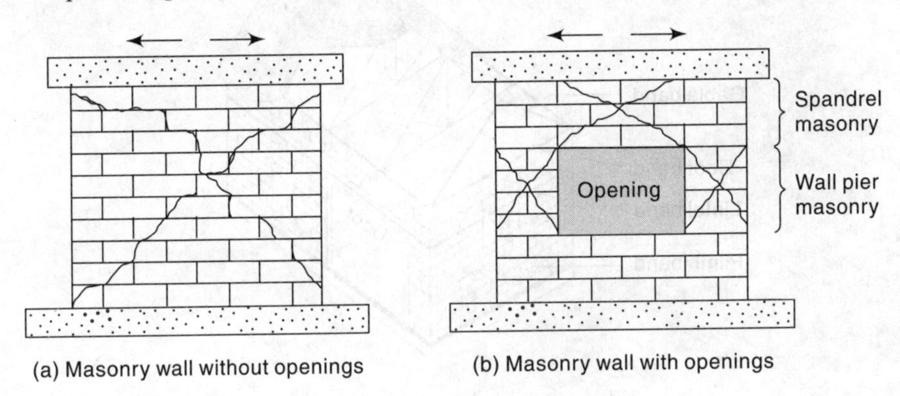

Fig. 6.6 Patterns of cracks in masonry walls

6.4 Behaviour of Walls—Box Action and Bands

The box-type construction consists of walls along both the axes of the building as shown in Fig. 6.2(c). All traditional masonry construction falls under this category. The walls support vertical loads and also act as shear walls for lateral loads acting in any direction. Figure 6.2 illustrates the box action of the walls of a building. For the loading case shown in Fig. 6.2(c), walls *A* will act as shear walls and walls *B* will topple over. Besides offering resistance themselves, however, walls *A* offer resistance against collapse of walls *B*, if both the walls *A* and walls *B* are properly tied up like a box [Fig. 6.2(c)].

The walls *B* of Fig. 6.2(c) may be considered to act as vertical slabs supported on two vertical sides and at the bottom, and subjected to inertia force on their own mass for ground motion along *y*-axis. Near the vertical edges the walls will carry bending moments in the horizontal plane for which the masonry strength may not be adequate. This may result in cracking and separation of walls. If, however, a flexural member is introduced at a suitable level (say lintel level) in walls *B*, and continued in walls *A*, it will take care of the bending tensions in the horizontal plane. The situation will be the same for walls A for ground motion along the x axis. Thus a flexural member is also required in walls A. This implies that the flexural member is required all around the walls. This flexural member is known as a *band*. Such bands may be incorporated at the roof, lintel, and plinth levels, as shown in Figs 6.1 and 6.7. These bands, also known as *bond beams,* integrate the components of the masonry building into a structural unit. In addition to their horizontal reinforcing action, the bands also tend to distribute the vertical concentrated loads that might be placed on a wall.

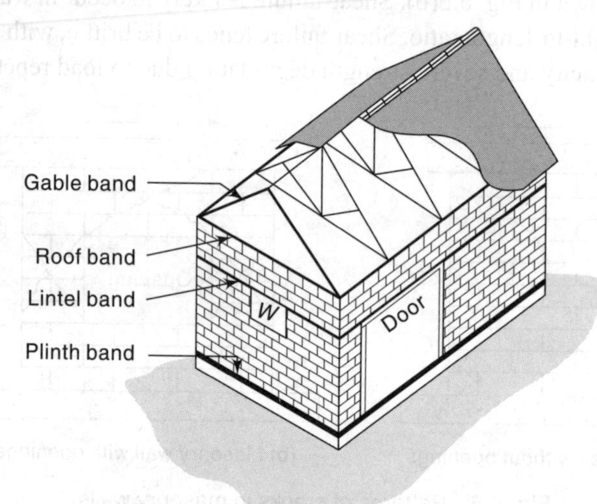

Fig. 6.7 Bands in pitched roof masonry buildings required for box action (adapted from Murty, 2005)

During earthquake shaking, a masonry wall gets grouped into three sub-units; namely, spandrel masonry, wall pier masonry, and sill masonry, as shown in Fig. 6.8(a). When the ground shakes, the inertia force causes the small-sized masonry wall piers to disconnect from the masonry above and below. These units rock back and forth, developing contact only at opposite diagonals, as shown in Fig. 6.8(b), and may crush the masonry at corners. Rocking is possible when masonry piers are slender and when the weight of the structure above them is small; otherwise the piers are more likely to develop diagonal cracks as shown in Fig. 6.8(c). These diagonal cracks can be checked by providing vertical reinforcement anchored to the foundations (between roof band and lintel band, lintel band and plinth band, and plinth band and foundation).

Openings in a wall cause reduction in the cross-sectional area of the wall. This may sometimes lead to sliding of the masonry just under the roof, below the lintel band, at the sill level, or at the plinth level as shown in Fig. 6.8(d). However, this effect is rarely observed.

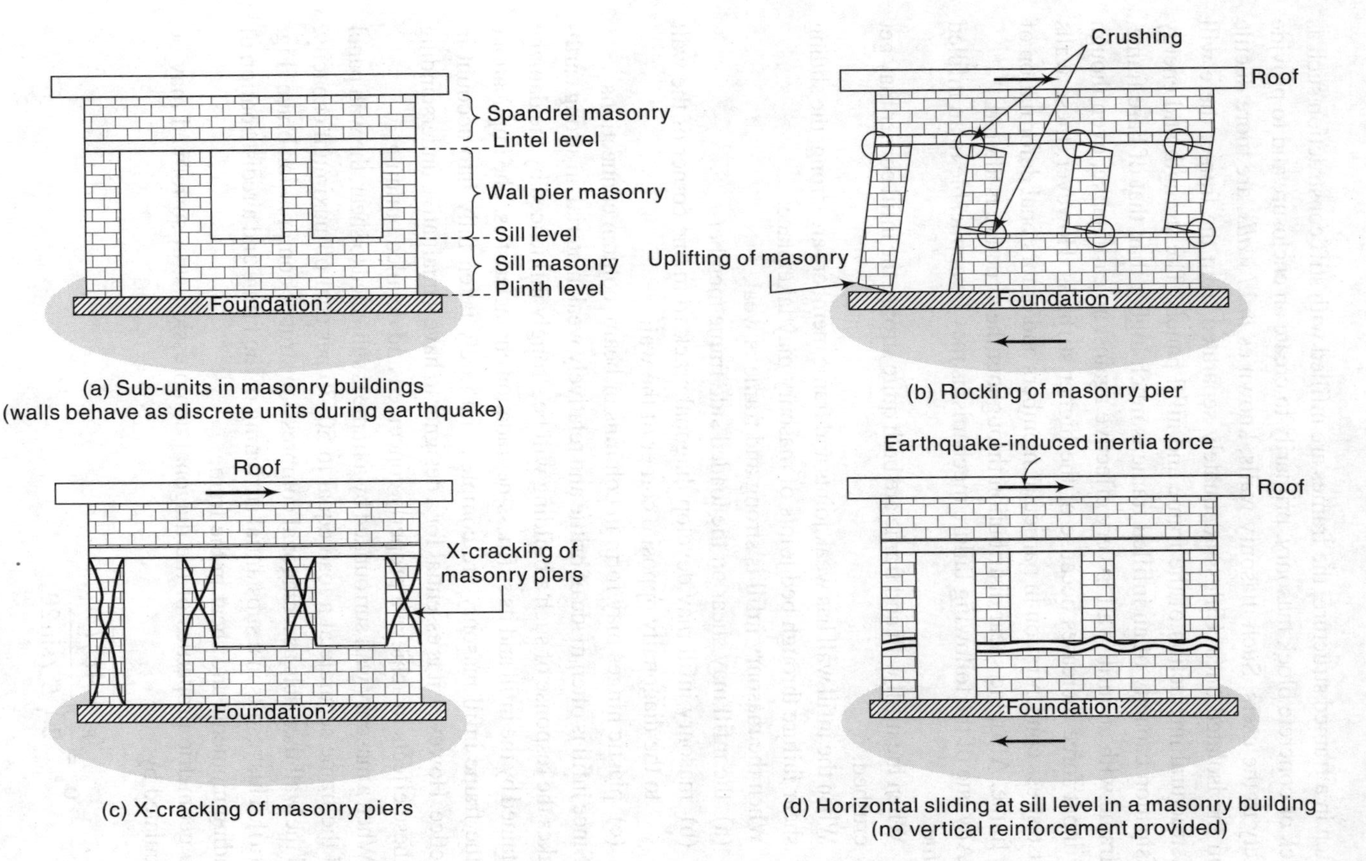

(a) Sub-units in masonry buildings (walls behave as discrete units during earthquake)

(b) Rocking of masonry pier

(c) X-cracking of masonry piers

(d) Horizontal sliding at sill level in a masonry building (no vertical reinforcement provided)

Fig. 6.8 Damages in masonry buildings (adapted from Murty, 2005)

6.5 Behaviour of Infill Walls

Often in a framed structure, the frames are infilled with stiff construction such as brick or concrete block masonry, primarily to create an enclosure and to provide safety to the users. Such masonry walls, known as *infill walls*, are more ductile than the isolated ones. Unless adequately separated from the frame, there will be structural interaction of the frame and infill panels. The strength and energy dissipation capacity of an infilled frame is much higher than that of bare frame. A frame with an infill wall is very effective against an earthquake, even though the input force increases because of the higher stiffness. However, these walls cause stress concentration in particular members and/or torsional deformation of the frame. Also, the shear distribution throughout the structure is altered.

Any one of the following failure mechanisms may be applicable to an infilled frame:

1. When the infill wall and frame are both quite strong, the infill corner may get crushed.
2. When the infill wall has weak joints and frame members are strong, the sliding shear failure through bed joints of masonry may take place.
3. When the masonry infill is strong and frame is weak
   (a) the infill may shear on the loaded side frame member
   (b) masonry infill may develop diagonal crack from one corner of the wall to the diagonally opposite corner of the wall
   (c) plastic hinges may form in columns or beam–column connections

Since infill is often made of brittle and relatively weak material, in strong earthquakes the response of such a building will be strongly influenced by the damage sustained by the infill and its stiffness-degradation characteristics. The implications of the frame infill masonry are complex and rarely taken fully into account in practice. However, it is essential for a designer to have a qualitative understanding of these effects to achieve a properly conceived and detailed building.

When a masonry wall surrounded by a frame is subjected to shear, the wall panel and the frame separate at a load equal to 50–70 per cent of maximum capacity, and the wall then acts as a diagonal compression strut or compression brace [Fig. 6.9(a)]. This results in a substantial stiffening of the frame and a redistribution of bending moments and shear in the frame.

The geometric properties of the diagonal compression strut so formed may be calculated by:

$$\alpha_h = \frac{\pi}{2}\sqrt[4]{\frac{4E_f I_c h}{E_m t \sin 2\theta}}$$

$$\alpha_L = \frac{\pi}{2}\sqrt[4]{\frac{4E_f I_b L}{E_m t \sin 2\theta}}$$

where E_m and E_f are the elastic moduli of masonry wall and frame material, respectively;

t, h, L are the thickness, height, and length of the infill wall, respectively; and

I_c and I_b are the moments of inertia of column and beam of the frame, respectively.

$$\theta = \tan^{-1}(h/L)$$

The effective strut width as suggested by Hendry is

$$w = \frac{1}{2}\sqrt{\alpha_h^2 + \alpha_L^2}$$

Some researchers have suggested the width w to be equal to one-third and one-fourth of the diagonal length L_1 of the infill wall panel. The failure in this case may occur and be noticed by crushed corner of the infill wall panel.

When the resistance to sliding is smaller than the strength of the diagonal strut, the wall panel may fail by sliding as shown in Fig. 6.9(b). Once the panel has sheared, the effect of the diagonal compression is lost. The resistance to external shear will then be provided only by columns since the friction at the sliding surfaces becomes very small. The interaction between a frame and partial infill masonry is shown in Fig. 6.9(c).

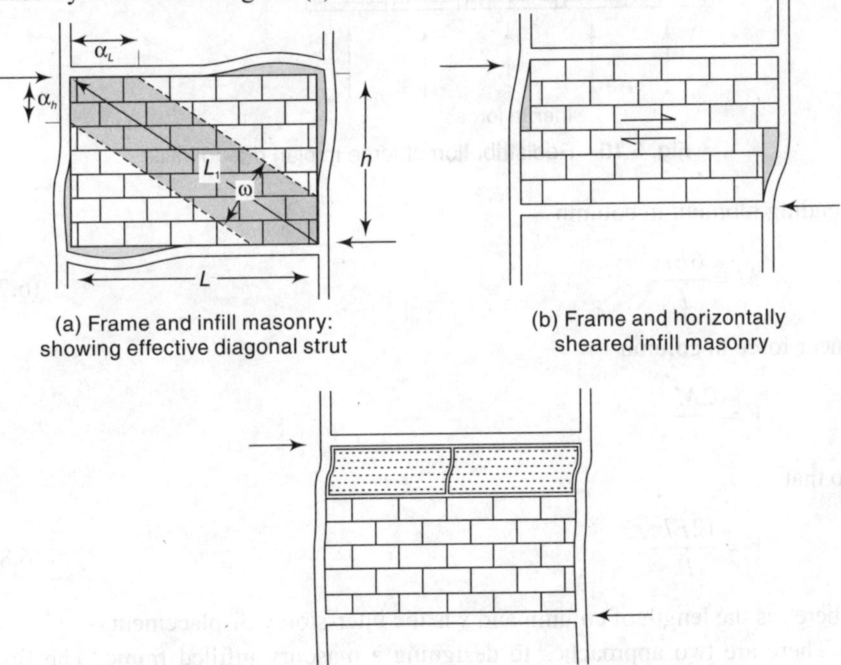

(a) Frame and infill masonry: showing effective diagonal strut

(b) Frame and horizontally sheared infill masonry

(c) Frame and partial infill masonry

Fig. 6.9 Interaction between a frame and infill masonry

Figure 6.10 illustrates the redistribution of forces in plan, due to the stiffening effect of infill masonry. The whole of the lateral force is resisted by the two end frames. This situation continues as long as masonry panels retain their strength. If the masonry at one end is damaged, high torsional effects will result.

By stiffening frames with infill masonry, the natural period of vibration is reduced, which increases the effective lateral force. The local effect of stiffening is to redistribute the forces on to stiffened frames, possibly producing undesirable eccentricity. Consequently the forces on a frame increase over many times. Contact at the frame masonry interface modifies the distribution of frame forces. The effective length of a beam/column is reduced so that the ratio of shear to bending force is increased.

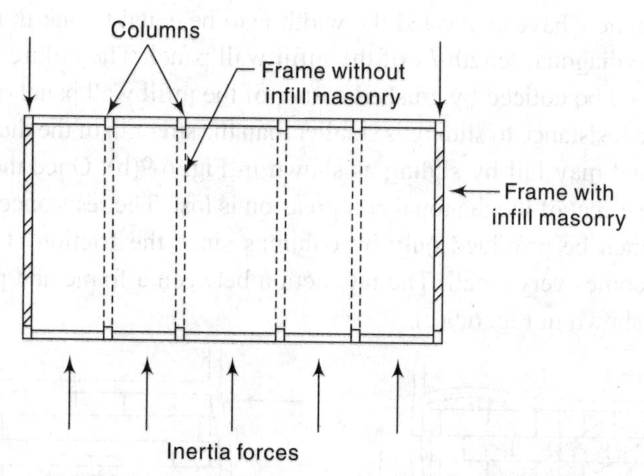

Fig. 6.10 Redistribution of force in plan due to infill

Bending moment in column

$$M = \frac{6EIx}{l^2} \tag{6.7}$$

Shear force in column

$$V = \frac{2M}{l}$$

So that

$$V = \frac{12EIx}{l^3} \tag{6.8}$$

where l is the length of column and x is the inter-storey displacement.

There are two approaches to designing a masonry infilled frame. The first approach, the qualitative design approach, leads to heavy reinforcement in both the frame and the masonry. However, it provides an advantage in case of a major earthquake, as it makes full use of additional stiffness provided and of energy

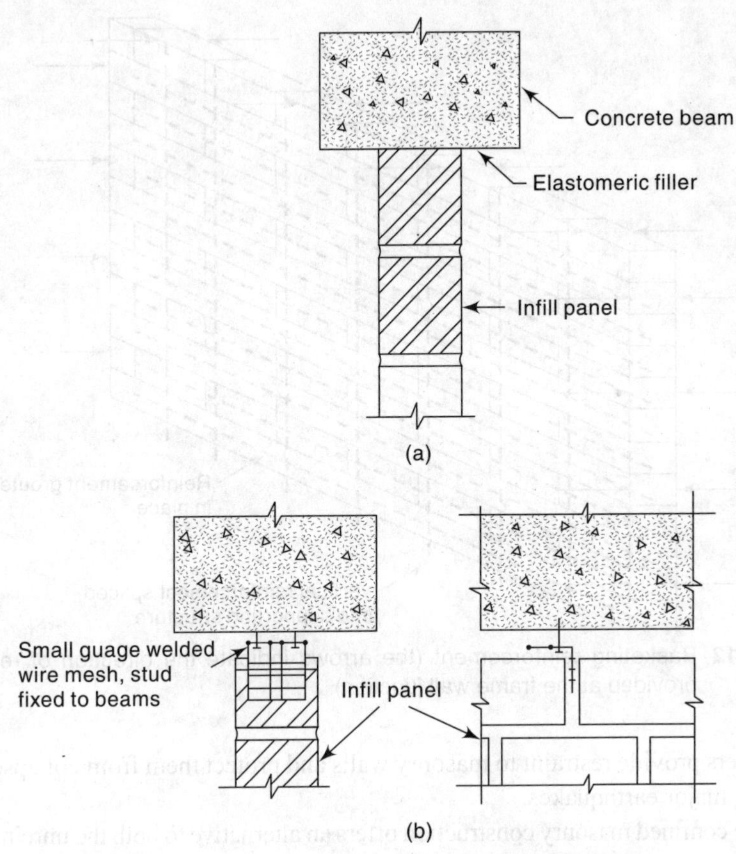

Fig. 6.11 Lateral restraint details to a free infill panel

absorbed by the reinforced masonry. The other approach involves a full separation joint between the masonry and the frame at the ends and the top (Fig. 6.11). The out-of-plane failure may then be dealt with either by reinforced masonry to act as a vertical cantilever or by providing basketting reinforcement as shown in Fig. 6.12.

6.6 Confined Masonry Construction

So far the behaviour of plain masonry wall, reinforced masonry wall, and infilled masonry walls in RC frame construction has been discussed. Another type of masonry construction commonly used is the *confined masonry construction*. In this type of construction, the plain masonry walls are confined on all the four sides by RC members that improve the ductility. In addition to enhancement of ductility, the strength and the stability of the masonry wall also increases. Further, an improvement in the integrity and containment of earthquake-damaged wall is achieved. The masonry walls transmit the gravity loads, and the confining

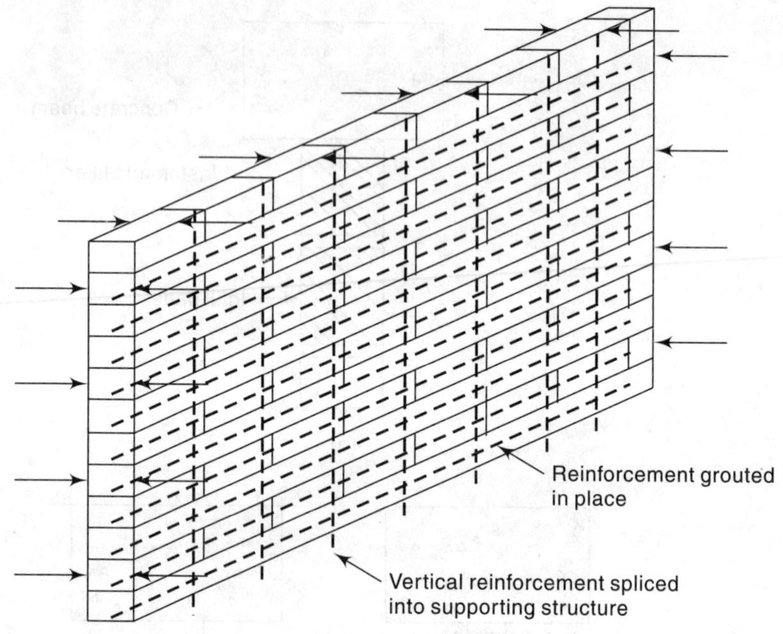

Fig. 6.12 Basketing reinforcement (the arrows indicate the direction of restraint provided at the frame wall junction)

members provide restraint to masonry walls and protect them from collapse even during major earthquakes.

The confined masonry construction offers an alternative to both the unreinforced masonry and RC frame construction with infill masonry walls, by using the best features of both the technologies. It consists of masonry walls and horizontal and vertical RC members, called *confining members*. They confine the masonry wall panel from all the four sides. The horizontal RC members are called *bond beams* or *tie-beams* (resembling beams in RC frames), and the vertical members are called *tie-columns* (resembling columns in RC frames. It may be noted that the tie-beams and the tie-columns do not constitute a framing system, i.e., these are neither intended nor designed as a frame. However, adequate splicing and anchoring of the reinforcement is required (Fig. 6.13). Unlike RC frame members, they are quite small in dimension and do not provide any lateral resistance. Since tie columns and tie-beams do not contribute to the lateral resistance of the structure, the specific design calculations for confining elements is not required. In this type of construction, the masonry walls are expected to resist both the gravity and lateral loads. The minimum size of tie-column is kept 150 × 150 mm with a maximum spacing of 4 m.

It is important to highlight the distinction between the confined masonry construction and the RC frame infill masonry wall construction, as they look alike (Fig. 6.14) once the structure is finished. Although, the two types of construction appear to be the same, they are nevertheless different in the following two ways.

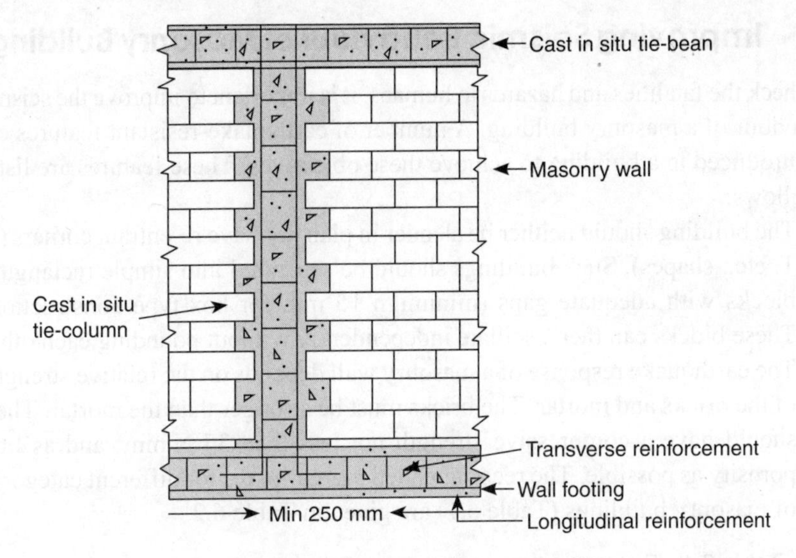

Fig. 6.13 Confined masonry wall

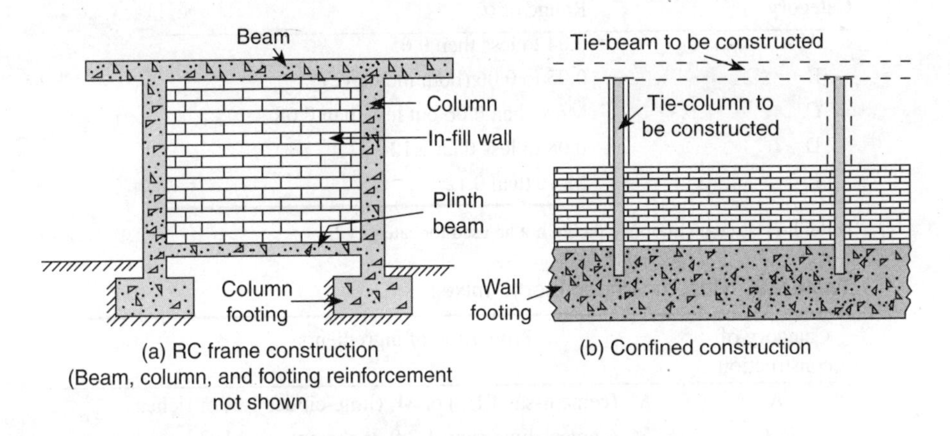

(a) RC frame construction (Beam, column, and footing reinforcement not shown)

(b) Confined construction

Fig. 6.14 Masonry constructions

1. In confined masonry construction, masonry walls are constructed first. Subsequently, tie-columns are cast in place [Fig. 6.14(b)]. Thereafter, horizontal confining elements are constructed on top of the walls along with the construction of floors/roofs. In RC frame construction, the frame is constructed first. At a later stage, the masonry walls are constructed and therefore no bond is developed between the frame members and the infill wall [Fig. 6.14(a)]. The masonry walls are used for space partitioning only.
2. In confined masonry construction, the masonry walls are the main load-carrying members and are expected to resist both gravity and lateral loads. Since no bond is developed between the frame members and the infill masonry walls in the RC frame construction, the walls are not the load-carrying members.

6.7 Improving Seismic Behaviour of Masonry Buildings

To check the fatalities and hazards to humans, it is important to improve the seismic behaviour of a masonry building. A number of earthquake-resistant features can be introduced in a building to achieve these objectives. These features are listed as follows:

(a) The building should neither be slender in plan nor have re-entrant corners (H, T, etc., shapes). Such buildings should be separated into simple rectangular blocks with adequate gaps (minimum 15 mm for box-type construction). These blocks can then oscillate independently without pounding each other.

(b) The earthquake response of a masonry wall depends on the relative strengths of the bricks and mortar. The bricks must be stronger than the mortar. These should have a compressive strength not less than 35 N/mm^2 and as little porosity as possible. The recommended mortar mixes for different categories of masonry buildings (Table 6.1) are given in Table 6.2.

Table 6.1 Building categories for earthquake-resisting features

Category	Range of α_h
A	0.04 to less than 0.05
B	0.05 to 0.06 (both inclusive)
C	More than 0.06 but less than 0.08
D	0.08 to less than 0.12
E	More than 0.12

Note: Low strength masonry should not be used for category E.

Table 6.2 Recommended mortar mixes

Category of construction	Proportion of ingredients
A	M_2 (cement-sand 1:6) or M_3 (lime-cinder 1:3) or richer
B, C	M_2 (cement-lime-sand 1:2:9 or cement-sand 1:6) or richer
D, E	H_2 (cement-sand 1:4) or M_1 (cement-lime-sand 1:1:6) or richer

(c) Good interlocking of masonry courses at the junctions should be ensured as the walls transfer loads to each other at their junctions. To obtain full bond between perpendicular walls, it is necessary to make a stepped joint by making the corners first to a height of 600 mm and then building the wall in between them. Otherwise, a toothed joint can be made in both the walls, in lifts of about 450 mm (Fig. 6.15).

(d) For single-storey construction the wall thickness should not be less than one brick. In buildings of up to three storeys the wall thickness should not be less than one and a half bricks for the bottom storeys, and one brick in the top storey. It should also not be less than one-sixteenth of the length of wall between two consecutive perpendicular walls.

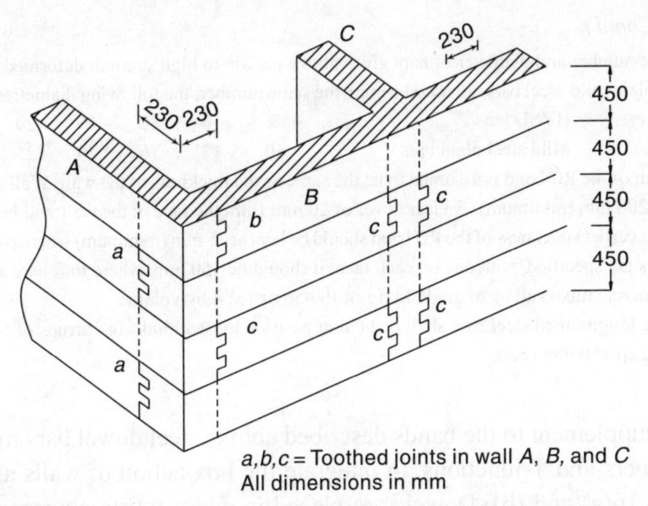

Fig. 6.15 Alternating toothed joints in walls at corner and T-joints

(e) Horizontal reinforcement should be provided in walls to strengthen them against horizontal in-plane bending. This also helps to tie together the perpendicular walls. Provisions of horizontal bands should be made at various levels (Figs 6.1 and 6.7), in particular at the lintel level. The lintel band ties the walls together and creates a support to the walls loaded in the weak direction. This band also reduces the unsupported height of the walls and improves their stability in the weaker direction. A band at the roof level prevents out-of-plane failure of walls. The longitudinal steel in bands should be provided as given in Table 6.3.

Table 6.3 Recommended longitudinal steel in RC bands

Span (m)	Building category B		Building category C		Building category D		Building category E	
	No. of bars	Diameter (mm)	No. of bars	Diameter (mm)	No. of bars	Diameter (mm)	No. of bars	Diameter (mm)
≤ 5	2	8	2	8	2	8	2	10
6	2	8	2	8	2	10	2	12
7	2	8	2	10	2	12	4	10
8	2	10	2	12	4	10	4	12

Notes:

1. Span of wall will be the distance between the centre lines of its cross walls or buttresses. For spans greater than 8 m it is desirable to insert pilasters or buttresses to reduce the span or to make special calculations to determine the strength of the wall and the section of the band.

(Contd)

Table 6.3 (*Contd*)

2. The number and diameter of bars given above pertain to high strength deformed (HYSD) bars. If plain mild-steel bars are used keeping the same number, the following diametres may be used:

HYSD bars	8	10	12	16	20
Mild steel plain bars	10	12	16	20	25

3. Width of the RC band is assumed to be the same as the thickness of the wall. Wall thickness shall be 200 mm (minimum). A clear cover of 20 mm from the face of the wall will be maintained.
4. The vertical thickness of the RC band should be kept at 75 mm (minimum) where two longitudinal bars are specified, with one on each face; it should be 150 mm, where four bars are specified.
5. Concrete mix shall be of grade M-15 of IS 456 or 1:2:4 by volume.
6. The longitudinal steel bars shall be held in position by steel links or stirrups of 6 mm diameter spaced 150 mm apart.

(f) As a supplement to the bands described above, steel dowel bars may be used at corners and T-junctions, to integrate the box action of walls as shown in Fig. 6.16(a) and (b). Dowels are placed in every fourth course, or at about 50-cm intervals, and taken into the walls to sufficient length so as to provide full bond strength. As an alternative, strengthening of T-junction and corner can be done by introducing wire mesh as shown in Fig. 6.16(c) and (d), respectively.

(g) Tension occurs in the jambs of openings, at corners, and junction of walls. Therefore, at corners and junctions of walls, vertical reinforcing bars should be provided. The amount of vertical steel will depend upon the number of storeys, storey heights, the effective seismic coefficient, importance of the building, and soil type. The vertical reinforcement should be properly embedded in the plinth masonry of the foundation and the roof slab or roof band, so as to develop its tensile strength in bond. The reinforcement should pass through the lintel bands and floor slabs or floor level bands in all storeys. For walls up to 350 mm thick, the vertical reinforcement as specified in Table 6.4 should be provided. For thicker walls, the area of bars should be increased proportionately.

Table 6.4 Vertical steel reinforcement in rectangular masonry units

No. of storeys	Storey	Diameter of HYSD single bar in mm at each critical section			
		Category B	Category C	Category D	Category E
One	—	Nil	Nil	10	12
Two	Top	Nil	Nil	10	12
	Bottom	Nil	Nil	12	16
Three	Top	Nil	10	10	12
	Middle	Nil	10	12	16
	Bottom	Nil	12	12	16
Four	Top	10	10	10	Four-storey building not permitted
	Third	10	10	12	
	Second	10	12	16	
	Bottom	12	12	20	

Notes: 1. The diameters given earlier are for HYSD bars. For mild steel plain bars, the equivalent diameter as given under Table 6.3, Note 2 is used.

2. The vertical bars will be covered with concrete M-15 or mortar 1:3 grade in suitably created pockets around the bars. This will ensure their safety from corrosion and good bond masonry.

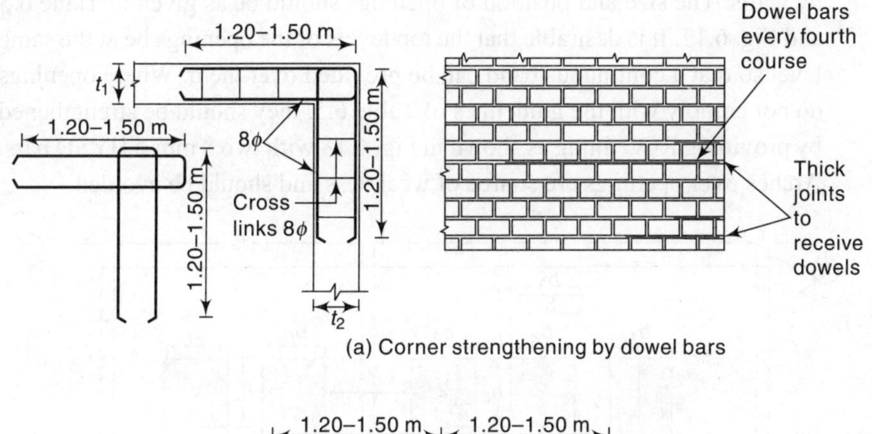

(a) Corner strengthening by dowel bars

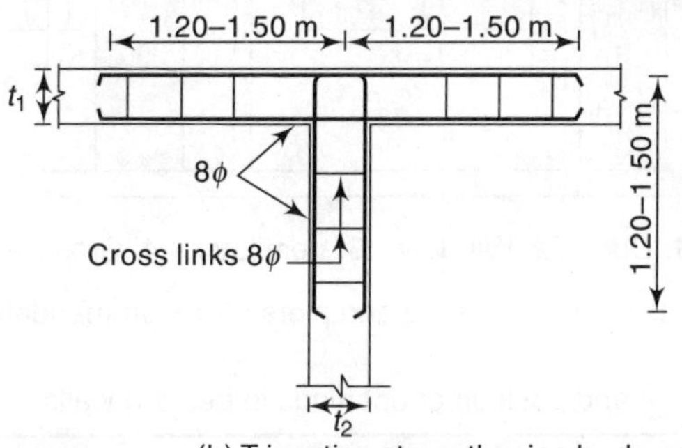

(b) T-junction strengthening by dowel bars

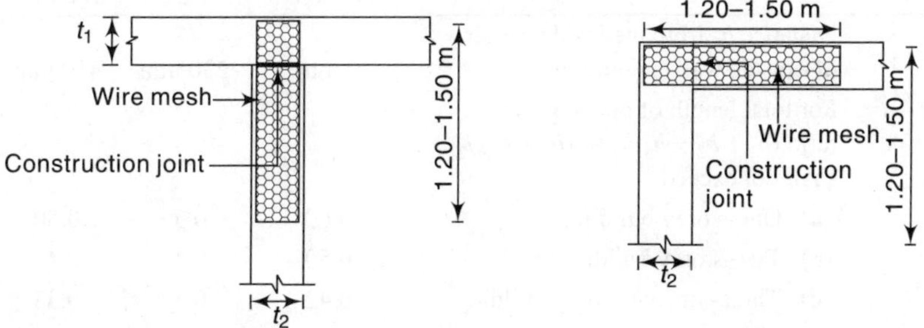

(c) T-junction strengthening by wire fabric (d) Corner strengthening by wire fabric

t_1, t_2 Wall thicknesses

Fig. 6.16 Strengthening of corners and T-junctions

(h) Location and size of openings is of great significance in deciding the performance of masonry buildings subjected to earthquake forces. The flow of force from one wall to the other is hampered, if the openings are too close to the junction of walls. Openings should not be eccentrically located on the structural plan, since eccentricity of the centre of stiffness relative to the centre of gravity causes torsional moment. The size of door and window openings needs to be kept smaller; the smaller the opening, the larger is the resistance offered by walls. The size and position of openings should be as given in Table 6.5 and Fig. 6.17. It is desirable that the top levels of the openings be at the same level so that a continuous band can be provided over them. Where openings do not comply with the guidelines of Table 6.5, they should be strengthened by providing RCC lining as shown in Fig. 6.18 with two 8 mm ϕ HYSD bars. Arches over openings are source of weakness and should be avoided.

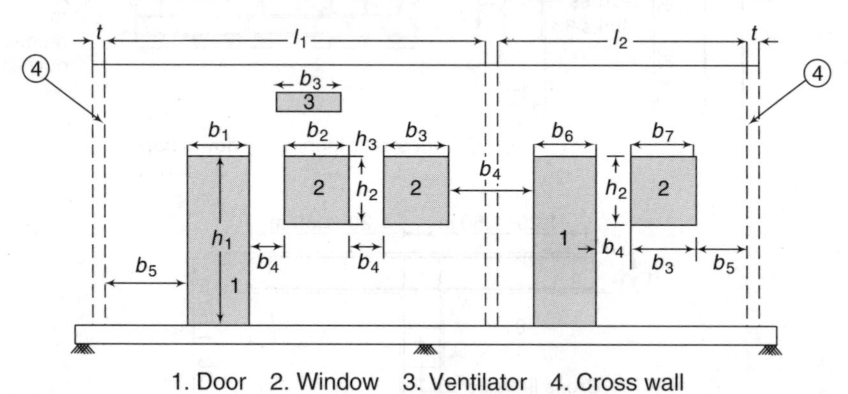

Fig. 6.17 Dimensions of opening and piers for recommendations in Table 6.5

Table 6.5 Size and position of openings in bearing walls

S.No.	Position of opening	Details of opening for building category		
		A and B	C	D and E
1	Distance b_5 from the inside corner of outside wall (minimum)	0 mm	230 mm	450 mm
2	For total length of openings, the ratio $(b_1 + b_2 + b_3)/l_1$ or $(b_6 + b_7)/l_2$ shall not exceed:			
	(a) One-storey building	0.60	0.55	0.50
	(b) Two-storey building	0.50	0.46	0.42
	(c) Three- or four-storey building	0.42	0.37	0.33
3	Pier width between consecutive openings, b_4 (minimum)	340 mm	450 mm	560 mm
4	Vertical distance between two openings one above the other, h_3 (minimum)	600 mm	600 mm	600 mm

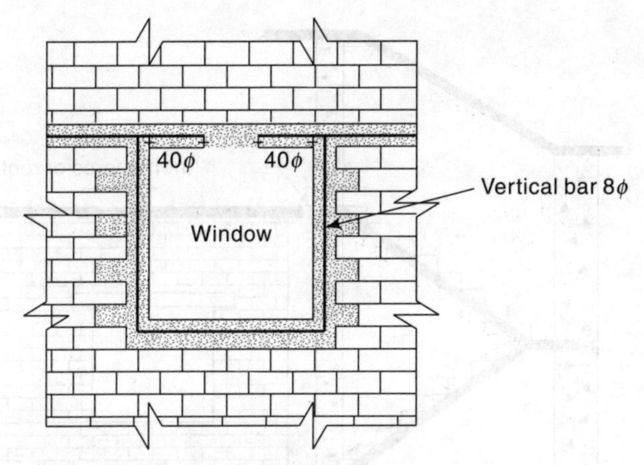

Fig. 6.18 Strengthening masonry with reinforced concrete, around opening

(i) Shear reinforcement should be provided in walls to ensure their ductile behaviour.

(j) Inclined flights of stairs joining different floor levels will act like a cross brace between various floors. They transfer large horizontal forces at the roof and lower levels and cause damage during earthquakes [Fig. 6.19(a)]. To check this, the staircase should be completely separated as shown in Fig. 6.19(b).

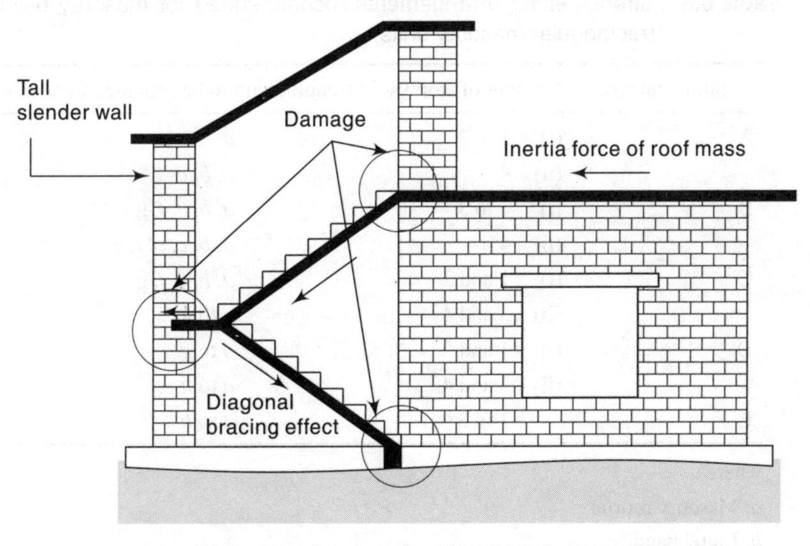

(a) Damage in a building with rigidly built-in staircase

Fig. 6.19 (*Contd*)

(*Contd*)

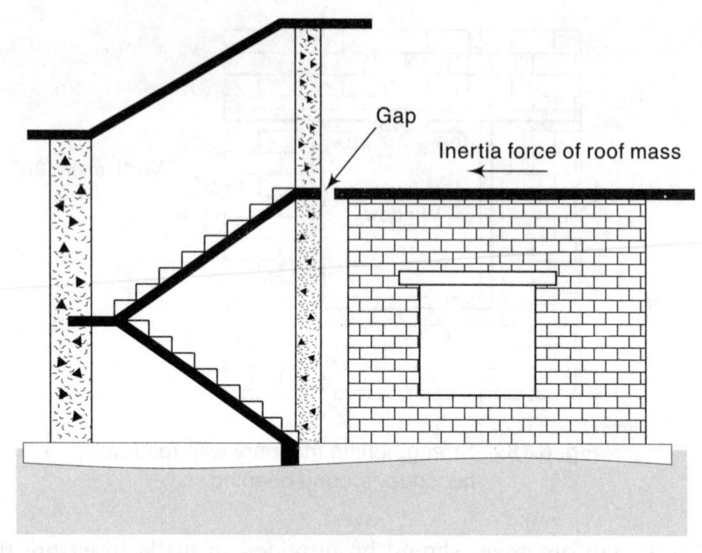

(b) A building with separated staircase

Fig. 6.19 Staircases in masonry buildings (adapted from Murty, 2005)

(k) Stiff, strong, and continuous footings should be used for the foundations.

(l) All the masonry buildings should be strengthened by the methods as specified for the various categories of buildings listed in Table 6.6.

Table 6.6 Strengthening arrangements recommended for masonry buildings (rectangular masonry units)

Building category	Number of storeys	Strengthening to be provided in all storeys
A	(i) 1 to 3	*a*
	(ii) 4	*a*, *b*, *c*
B	(i) 1 to 3	*a*, *b*, *c*, *f*, *g*
	(ii) 4	*a*, *b*, *c*, *d*, *f*, *g*
C	(i) 1 and 2	*a*, *b*, *c*, *f*, *g*
	(ii) 3 and 4	*a* to *g*
D	(i) 1 and 2	*a* to *g*
	(ii) 3 and 4	*a* to *h*
E	1 to 3*	*a* to *h*

where

a: Masonry mortar

b: Lintel band

c: Roof band and gable band where necessary

d: Vertical steel at corners and junctions of walls

e: Vertical steel at jambs of openings

(*Contd*)

Table 6.6 (*Contd*)

f: Bracing in plan at tie level of roof

g: Plinth band where necessary

h: Dowel bars

*4th storey not allowed in category E.

Note: In case of four-storey buildings of category B, the requirements of vertical steel may be checked through a seismic analysis using a design seismic coefficient equal to four times the one given in IS 1893. (This is because the brittle behaviour of masonry in the absence of a vertical steel results in much higher effective seismic force than that envisaged in the seismic coefficient provided in the code.) If this analysis shows that vertical steel is not required, the designer may take a decision accordingly.

6.8 Load Combinations and Permissible Stresses

Dead loads of walls, columns, floors, and roofs; ILs from floors and roofs; and wind loads (WLs) on walls and sloping roofs should be calculated as specified in IS codes 1911 and 875. Seismic loads (ELs) should be determined in accordance with IS 1893 (Part I): 2004. When a building is subjected to other loads, such as vibration from railway or machinery, these should be taken into consideration by judgement. Allowable stress design has been recommended by the code for the design of masonry structures. The adequacy of the masonry structure and its members is investigated for the following load combinations.

1. DL + IL
2. DL + IL + WL (or EL)
3. DL + WL
4. 0.9 DL + EL

The permissible stresses may be increased by one-third, when wind or earthquake forces are considered along with normal loads (cases 2, 3, and 4).

6.9 Seismic Design Requirements

Seismic design provisions contained in IS 4326: 1993 are empirical in nature; they are based on successful applications in the past and do not require any rational analysis. Small-sized buildings of up to three storeys may be designed as per requirements of IS 4326: 1993. However, other important buildings and those located in seismic zones IV and V should be designed for forces listed in IS 1893 (Part I) and provisions of IS 1905.

Masonry bearing walls should be straight and symmetrical in plan. Unreinforced masonry walls should not be more than 15 m in height, subject to a maximum of four storeys, unless rationally designed. For reinforced masonry walls, the provision of reinforcement should be made as given in the Tables 6.3 and 6.4.

The performance of masonry buildings relies on the performance of shear walls for the lateral load resistance. As per IS 1905 the shear walls are divided into four classes, on the basis of their capacity to resist earthquakes by inelastic behaviour and energy dissipation. These classes are described below.

Ordinary unreinforced masonry shear walls These are unreinforced walls and have poor post-elastic response. These are used only in low seismic regions (zone II) and for buildings of minor importance. The response reduction factor R to be used is 1.5.

Detailed unreinforced masonry shear walls These walls are designed as unreinforced masonry but contain minimum reinforcement in horizontal and vertical directions. Because of the reinforcement, walls display improved inelastic response and energy dissipation potential. These walls can be used for low to moderate seismic risk zones (zones II and III). The response reduction factor R to be used is 2.25.

A minimum of 100 mm^2 of vertical reinforcement should be provided at a maximum spacing of 3 m at the centre of the wall at critical sections and at corners, within 400 mm of each side of openings, and within 200 mm of the end of the walls. Horizontal reinforcement should be as follows.

(a) At least two 6 mm ϕ bars spaced not more than 400 mm apart or bond beam reinforcement of at least 100 mm^2 in cross-sectional area spaced not more than 3 m apart.

(b) At the bottom and top of openings and extending 500 mm, or 40 ϕ past the openings, continuously at structurally connected roof and floor levels, and within 400 mm of the top of walls.

Ordinary reinforced masonry shear walls These walls follow the same steel requirements as that of detailed unreinforced masonry shear walls. They can be subjected to large inelastic deformation and loss of strength and stability of the system, thus dissipating less energy. These walls are recommended in zones IV and V with an R-value of 3.0.

Special reinforced masonry shear walls These walls are meant to meet the seismic demands of zones IV and V with an R-value of 4.0. The masonry should be uniformly reinforced in both the horizontal and vertical directions such that the sum of the reinforcement area in both the directions should be at least 0.2 per cent of the gross cross-sectional area of the wall, with a minimum of 0.07 per cent of the gross cross-sectional area of the wall. The maximum spacing of the horizontal and vertical reinforcement should be the lesser of one-third of the length of shear walls or one-third of the height of shear walls or 1.20 m, whichever is least.

- Minimum area of steel in the vertical direction should be one-third of shear reinforcement required.
- Shear reinforcement should be anchored around vertical reinforcing bars with a 135° or 180° standard hook.

6.10 Seismic Design of Masonry Buildings

The specifications and limits specified in IS: 1905 apply to the design and construction of masonry to improve its performance when subjected to earthquake loads. The provisions are in addition to general requirments of IS: 1893 (Part 1). For the seismic design of a masonry building, the procedure given below may be followed.

1. The lateral loads are determined by the procedure outlined in Section 5.8. Then the base shear is calculated and distributed vertically to different floor levels. This is called storey shear. The procedure for the calculation and distribution of the base shear is illustrated through Solved Problem 6.4.
2. In case of rigid diaphragms, the storey shear is distributed to the vertical-resisting elements in direct proportion to their relative rigidities, whereas for a flexible diaphragm the exterior vertical-resisting elements share half the shear of that shared by the interior ones. As already discussed in the previous section since these vertical lateral load-resisting elements are masonry shear walls, it becomes imperative to calculate the relative wall rigidities. For this purpose, the masonry shear wall may be assumed to behave like a cantilever. The segments of wall between the adjacent openings (doors, windows, ventilators, etc.), called piers, may be assumed to be fixed at their top and bottom. However, both of these may be considered as cantilever or fixed, depending on the relative rigidities of walls and floor diaphragms.

   The deflection Δ_c of the wall or pier, fixed at the bottom and free at the top, is given by

$$\Delta_c = \frac{P}{E_m t}\left[4\left(\frac{h}{d}\right)^3 + 3\left(\frac{h}{d}\right)\right] \tag{6.9}$$

   and the rigidity R_c of this cantilever pier by

$$R_c = \frac{1}{\Delta_c} \tag{6.10}$$

   where P is the lateral force on the pier or wall, E_m is the modulus of elasticity of masonry in compression, h is the height of the pier, d is the width of the pier panels, and t is the thickness of the pier or wall.

   For a wall or pier fixed at the top and bottom, the deflecting Δ_f and rigidity R_f, are given by

$$\Delta_f = \frac{P}{E_m t}\left[\left(\frac{h}{d}\right)^3 + 3\left(\frac{h}{d}\right)\right] \tag{6.11}$$

   and

$$R_f = \frac{1}{\Delta_f} \tag{6.12}$$

3. If the masonry shear wall segments are combined horizontally, the combined rigidity is given by

$$R_c = R_{c1} + R_{c2} + R_{c3} + \cdots \tag{6.13}$$

For combining the rigidities of segments vertically, the expression for the combined rigidity is

$$\frac{1}{R_c} = \frac{1}{R_{c1}} + \frac{1}{R_{c2}} + \frac{1}{R_{c3}} + \ldots \tag{6.14}$$

4. Usually, the walls have openings as shown in Fig. 6.20. For calculating the rigidity of such walls, the following steps may be followed.
   (i) The deflection of the solid wall (of dimensions L, h) as a cantilever is calculated as, say, Δ_{so}.
   (ii) An opening having a height equal to that of the largest opening is selected. The deflection of this height of strip (strip A of size $L \times h_1$) of the wall is calculated as, say, Δ_{st}.
   (iii) Deflection (say, Δ_p) of all the piers numbered 2, 3, 4, 5, 6, and 7 is worked out.

   The total deflection of the shear wall is

$$\Delta = \Delta_{so} - \Delta_{st} + \Delta_p \tag{6.15}$$

   and rigidity

$$R = \frac{1}{\Delta} \tag{6.16}$$

This step is illustrated with the help of solved problem 6.5.

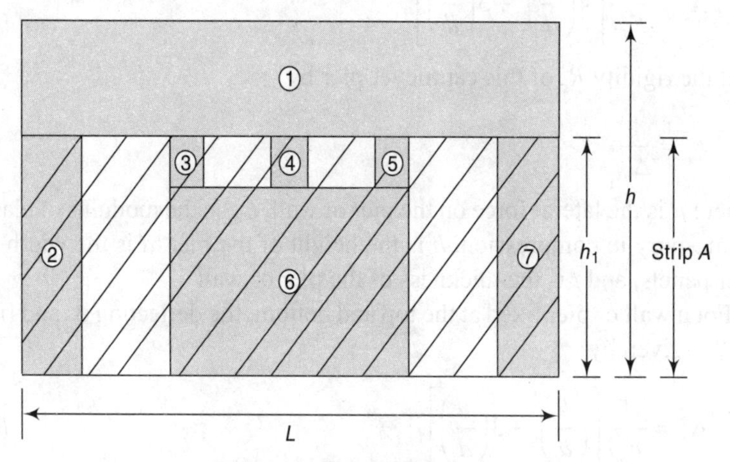

Fig. 6.20

5. The direct shear force in the wall, say i, is given as $R_i P_{x/y}$, where $P_{x/y}$ is the lateral force applied at the top of the pier and R_i is the relative stiffness of wall given by

$$R_i = \frac{K_i}{\sum_{i=1}^{n} K_i} \tag{6.17}$$

6. For a building, the centres of mass and rigidity may not coincide. Eventually torsional shear forces are generated and consequently torsional moments are induced. The concept of centres of mass and rigidity has already been explained through solved problems in Chapter 4.
   The torsional shears are given by

$$(P_y)_i = \frac{R_y \bar{x}}{J} P_y e_x \tag{6.18}$$

$$(P_x)_i = \frac{R_x \bar{y}}{J} P_x e_y \tag{6.19}$$

   where $(P_y)_i$ and $(P_x)_i$ are the torsional shears due to the seismic forces P_y and P_x along the y axis and x axis of the building, respectively; R_y and R_x are the relative rigidity of each wall along y axis and x axis of the building, respectively; e_y, e_x are the respective eccentricities between the centre of mass and centre of rigidity; and J is the relative rotational stiffness of all the walls in the storey under consideration.

   Even in symmetrical structures, a provision of 5 per cent eccentricity, called the accidental eccentricity, has been made in the specifications and should be considered for the analysis and design.

7. Horizontal bands of reinforcement are provided at critical levels to strengthen the masonry buildings. Bands can be made up of reinforced brick work in cement mortar not leaner than 1:3, or of RCC, the latter being preferred. The band should be of full width of the wall, not less than 75 mm in depth, and reinforced with steel as indicated in Table 6.3. The area of steel for reinforced brick work bands is kept the same as that for RC bands. The minimum size of the band and amount of reinforcement will depend upon the unsupported length of wall between cross walls, on the effective seismic coefficient based on the seismic zone, the importance of the building, and type of soil. Bands are designed for the calculated base shear (Solved Problem 6.2). For full integrity of walls at corners and junctions of walls, and for effective horizontal bending resistance of bands, the reinforcement should be made continuous as shown in Fig. 6.21.
8. Sometimes, the lateral forces from winds or earthquakes cause severe overturning moments on buildings, causing tension at the ends of piers of shear walls. These may also induce high compressive forces in the piers of the wall, thereby increasing the axial load in the wall. This effect should be examined carefully.
9. The shear walls are also checked for in-plane bending and the transverse or flexural walls are checked for out-of-plane forces along with gravity loads.
10. Vertical reinforcement at jambs of windows and door openings as shown in Fig. 6.18, and in RCC bands (Fig. 6.21) is provided. Typical details of

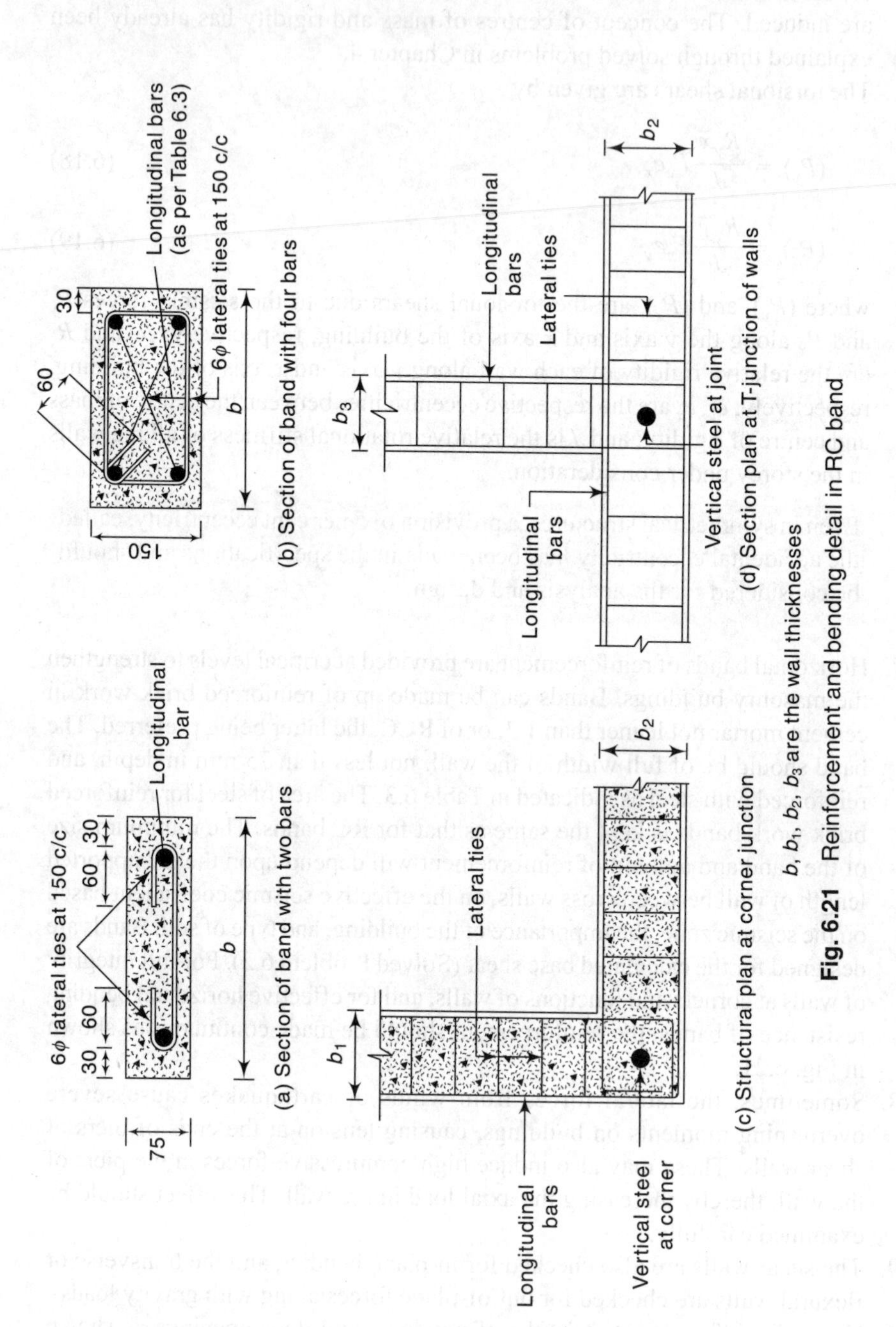

Fig. 6.21 Reinforcement and bending detail in RC band

providing vertical steel in brickwork masonry with rectangular solid units at corners and T-junctions are shown in Fig. 6.22.

6.11 Restoration and Strengthening of Masonry Walls

The need to restore the strength of a damaged masonry building after an earthquake or strengthening of an existing masonry building to sustain future earthquakes is a challenge for engineers. The restoration is defined as the restitution of the strength the building had before the occurrence of damage. This is achieved by structural repairs to load-bearing elements, either by adding more materials, e.g., reinforcing meshes and mortar, or by injecting epoxy-like materials. On the other hand, the old masonry buildings that have almost completed their useful service life or buildings to which additions and alterations have been affected during the passage of time may need strengthening. The lateral strength of such buildings can be improved by increasing the strength and stiffness of existing individual walls, whether they are cracked or uncracked. Some of the methods to achieve this goal are discussed further.

6.11.1 Grouting

Cracks in load-carrying masonry members must be given due cognizance since they reduce the strength drastically. For cracks with a small crack width of less than 6 mm, the original tensile strength of the cracked element may be restored by pressure injection of epoxy or cement mortar, known as *grouting*. However, grouting cannot be relied upon for improving the connection between orthogonal walls. The procedure for grouting is described below.

The external surfaces are cleaned of non-structural materials. Then, plastic injection ports are placed along the surface of the crack on both sides of the member, as shown in Fig. 6.23, and are secured in place with epoxy sealants. The centre-to-centre spacing of the ports may be approximately equal to the thickness of the element. After the sealant has cured, a low-viscosity epoxy resin is injected into one port at a time, beginning at the lowest part of the crack if it is vertical, or inclined, or at any one end of the crack if it is horizontal.

The resin is injected till it is seen flowing from the opposite sides of the member at the corresponding port or from the next higher port on the same side of the member. The injection port should be closed at this stage and the injection equipment moved to the next port and so on.

The smaller the crack, the higher the pressure or more closely spaced the ports should be, so as to obtain complete penetration of the epoxy material through the depth and width of the member. This technique is appropriate for all masonry elements.

6.11.2 Guniting

The gunite is placed pneumatically on the surface of masonry in the form of a slab and may be an expansive cement mortar, quick-setting cement mortar, or

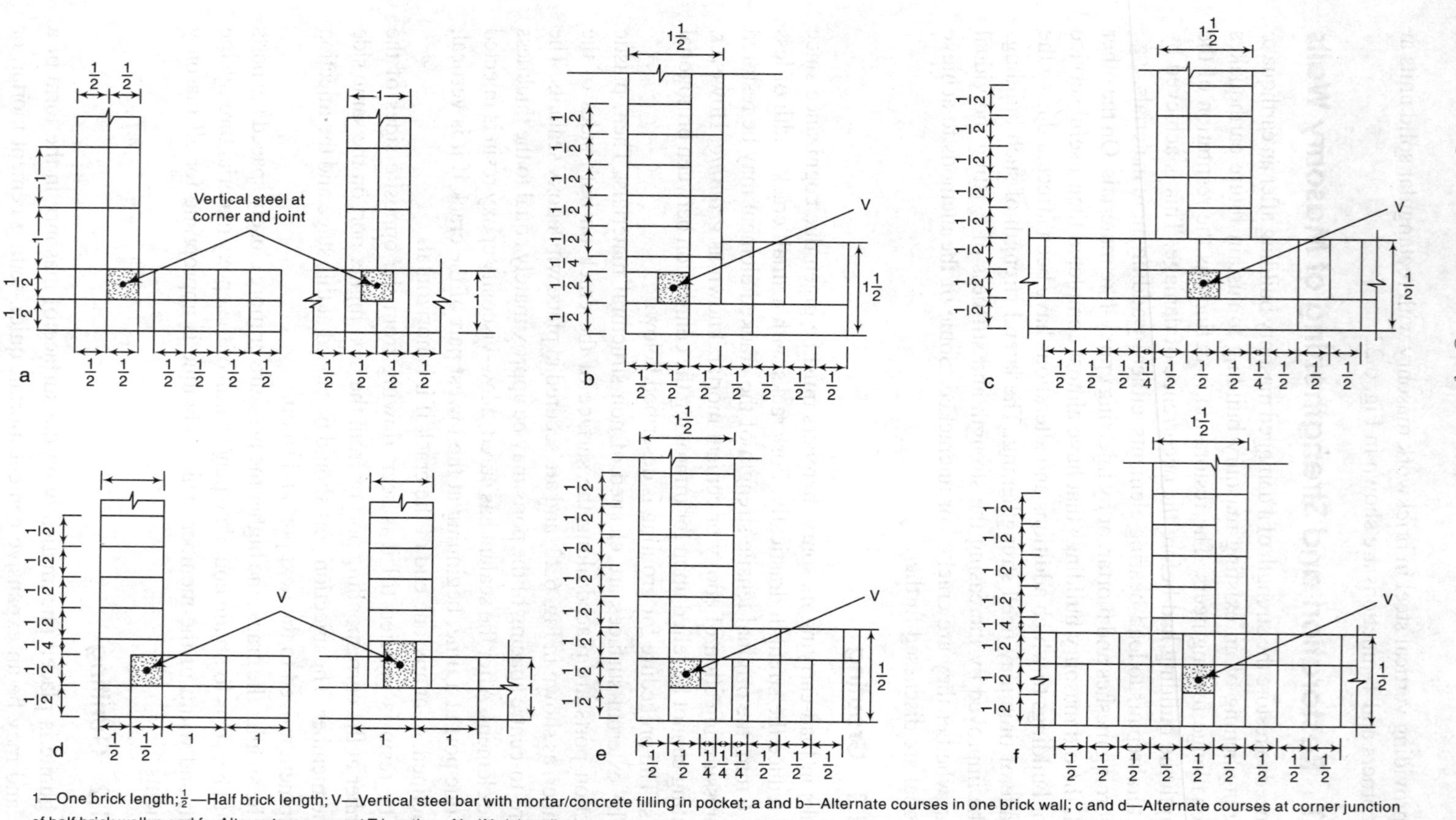

1—One brick length; $\frac{1}{2}$—Half brick length; V—Vertical steel bar with mortar/concrete filling in pocket; a and b—Alternate courses in one brick wall; c and d—Alternate courses at corner junction of half brick wall; e and f—Alternate courses at T-junction of half brick wall

Fig. 6.22 Typical details of providing vertical steel bars in brick masonry

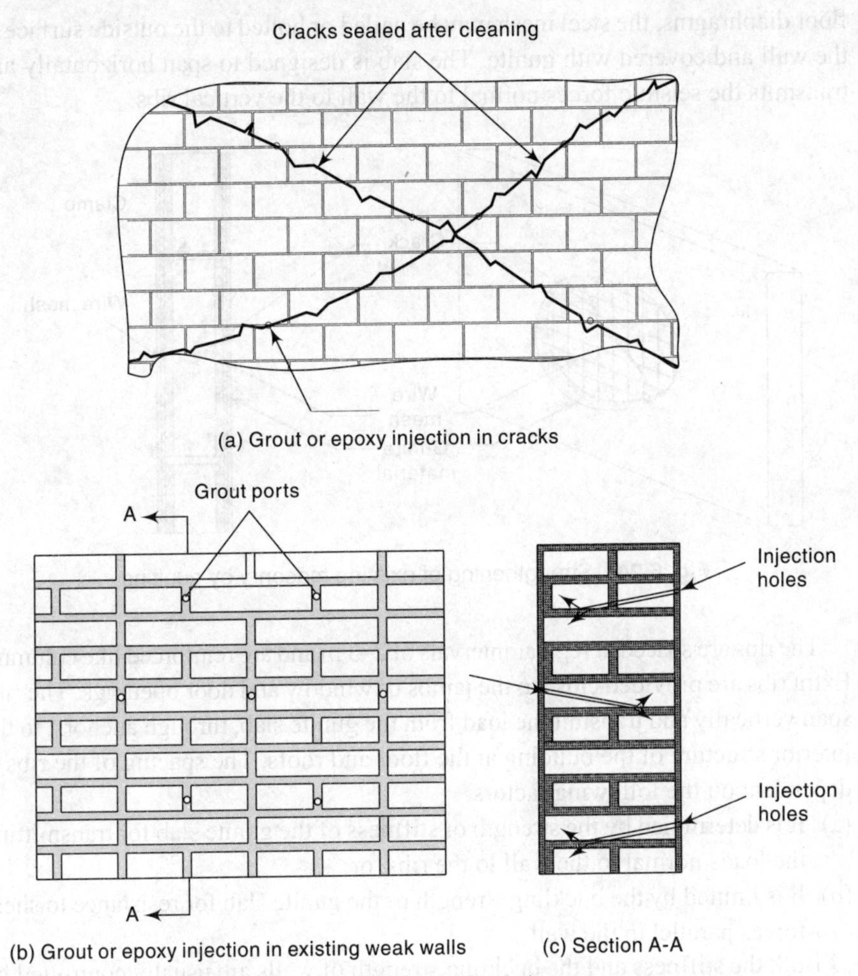

Fig. 6.23 Strengthening of existing masonry

gypsum cement mortar. It is used widely for repairing and strengthening masonry buildings. Guniting may be done on the exterior or interior surface of masonry as the conditions permit. For cracks wider than 6 mm or for regions in which the masonry has crushed, the following procedure is adopted for strengthening of the wall.

Exterior guniting consists of cutting groves in the exterior surface of wall and then laying gunite in thickness of about 8–10 cm on the wall surface (Fig. 6.24). This makes a slab with ribs of gunite. For walls not less than one-and-half brick in thickness, the outer course of brick is removed and gunite of about half-brick thickness is applied so as to bring its outer surface to the original face of the wall. Where necessary, additional shear reinforcement may be provided in the gunite slab and covered with mortar (Fig. 6.24). In case of damage of walls and floor diaphragms, the steel mesh may be nailed or bolted to the outside surface of the wall and covered with gunite. The slab is designed to span horizontally and transmits the seismic forces normal to the wall to the vertical ribs.

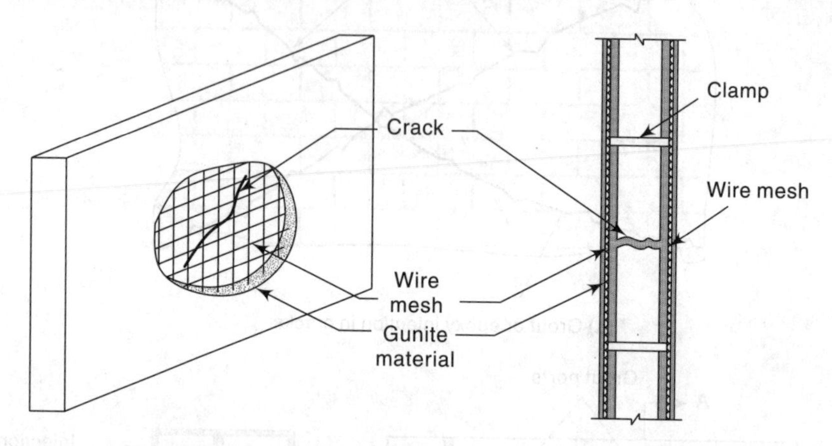

Fig. 6.24 Strengthening of existing masonry by guniting

The ribs are spaced at regular intervals of 2–3 m and are reinforced like columns. Extra ribs are provided close to the jambs of window and door openings. The ribs span vertically and transmit the load from the gunite slab, through anchors, to the interior structure of the building at the floor and roofs. The spacing of the ribs is dependant on the following factors.

(a) It is determined by the strength or stiffness of the gunite slab for transmitting the loads normal to the wall to the ribs, or

(b) It is limited by the buckling strength of the gunite slab for resistance to shear forces parallel to the wall.

Both the stiffness and the buckling strength of walls are usually controlled by the maximum span-to-thickness ratio only. The composite action of gunite and masonry need not be considered. For masonry walls of quality construction, without openings, the gunite slabs may be omitted and the wall may be strengthened with gunite ribs only. However, such walls should have sufficient tensile strength in flexure in the direction of the running bond. Also, it should have sufficient shear and diagonal tension resistances to act as a shear wall.

When there is no access to the outer face of the masonry wall, guniting is done on the inner surface of the wall. The vertical gunite ribs are cut into the walls between the beams. If the beams are parallel to the wall, ribs may be placed anywhere in the wall since there will be no beam interference. In such a case, the slab is gunited on the face of the wall. Removal of a course of bricks for thick walls, as described above, is not done here, as it may reduce or remove the end bearing of floor beams.

When the wall is gunited from inside, there is no gunite slab between the top and bottom of the floor beams. This has no significant effect on the resistance of the wall to the forces normal to it. The shear resistance of the strip of unsupported masonry plus the shear resistance of the concrete portion of the gunite ribs determine the resistance of the wall to forces parallel to it. Therefore, the concrete portion of the ribs should have closely spaced ties around the vertical bars. This total shear resistance should generally be adequate, but this is uncertain. That is why it is preferred to gunite the wall from outside wherever possible.

For thick walls, plugs of gunite are extended from the gunite slab to securely hold the inner units of masonry. These plugs are placed at regular intervals.

6.11.3 Prestressing

Prestressing is a technique by which internal stresses of suitable magnitude and distribution are introduced so that the stresses resulting from external loads are counteracted to a desired degree. This concept is used to its advantage for retrofilling and strengthening of damaged and old walls. A horizontal compression state induced by horizontal tendons can be used to increase the shear strength of walls. Moreover, this will also improve considerably the connections of orthogonal walls (Fig. 6.25). The easiest way of affecting the precompression is to place two steel rods on two sides of the wall and strengthening them by turnbuckles. Fairly good effects can be obtained by slight horizontal prestressing (about 0.1 MPa) on the vertical section of the wall. Prestressing is also useful to strengthen the spandrel beam between two rows of openings, in case no rigid slab exists.

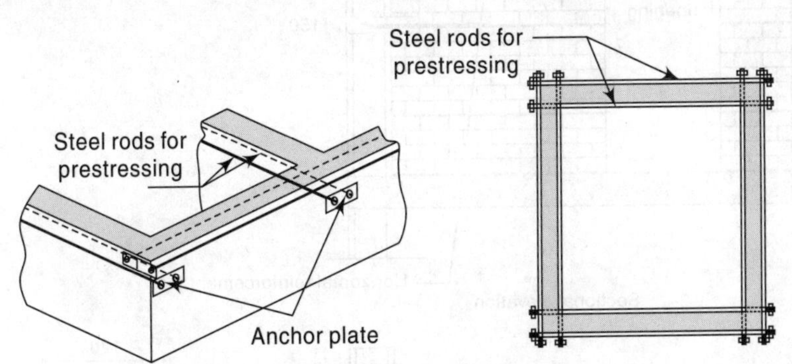

Fig. 6.25 Strengthening of walls by prestressing

6.11.4 External Binding

Opposite parallel walls can be held to internal cross walls by prestressing bars as shown in Fig. 6.25. Anchoring is done against horizontal steel channels instead of steel plates. The steel channels running from one cross wall to the other will hold the walls together and improve the integral box-like action of the walls.

6.11.5 Inserting New Wall

In case an existing building shows any type of dissymmetry, which may produce dangerous torsional effects during earthquakes, the centre of mass should be made coincident with the centre of stiffness. This can be achieved by separating parts of the building, making each individual unit symmetric, or by inserting new vertical-resisting elements (masonry or RC walls, such as internal shear walls or external buttresses). Figures 6.26 and 6.27 show the connections of the new and the old walls. Figure 6.26 refers to a T-junction and Fig. 6.27 to a corner junction. For bracing the longitudinal walls of long barrack-type buildings, a cross wall may be inserted to provide transfer supports to longitudinal walls, else a portal type of framework can be inserted transverse to the walls and connected to them. Alternatively, masonry buttresses or pilasters may be added externally as shown in Fig. 6.28. In framed buildings, inserting knee braces or full diagonal braces, or inserting infill walls, can improve the lateral resistance.

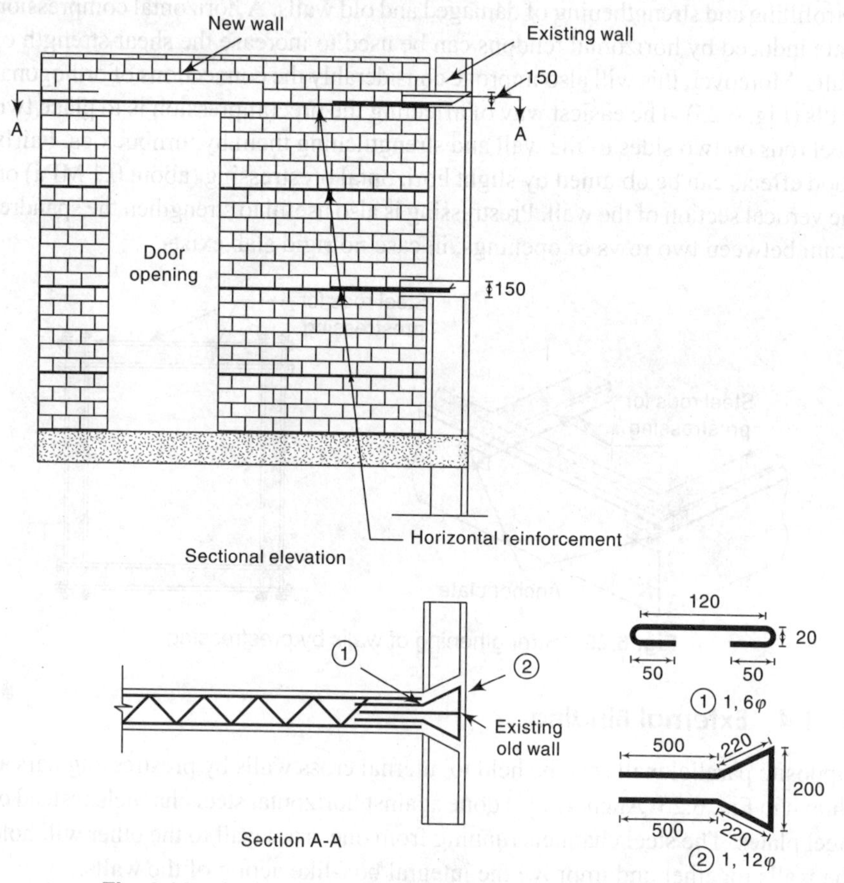

Fig. 6.26 Connection of new and old brick walls (T-junction)

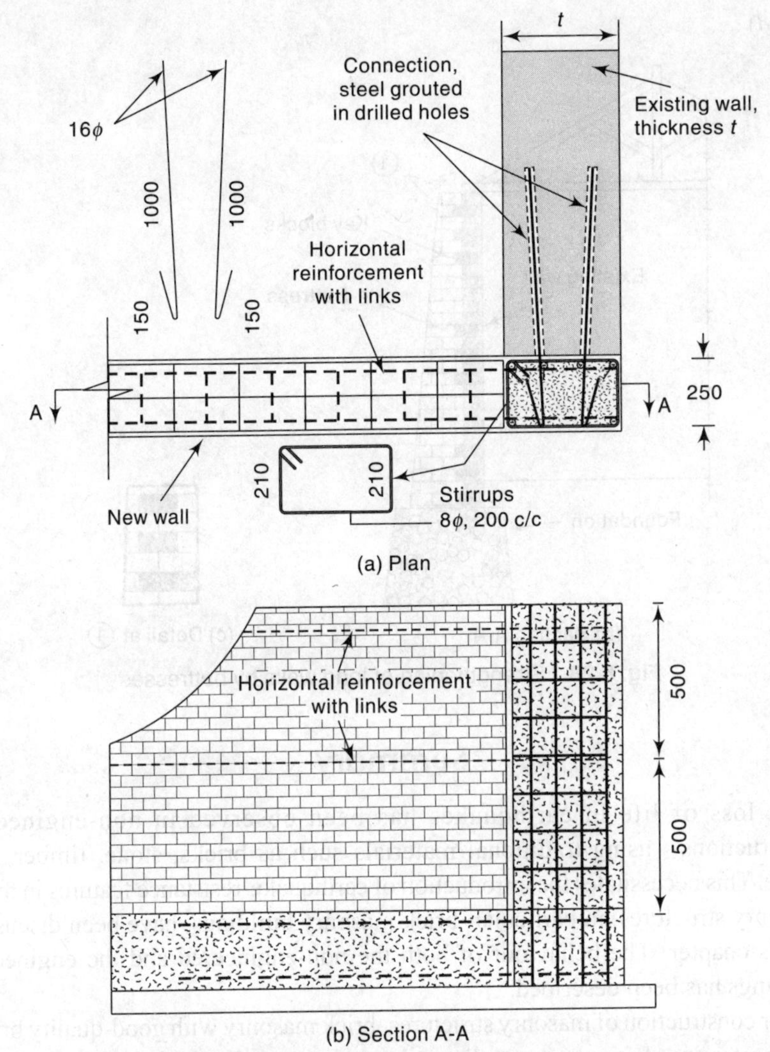

Fig. 6.27 Connection of new and old walls (corner junction)

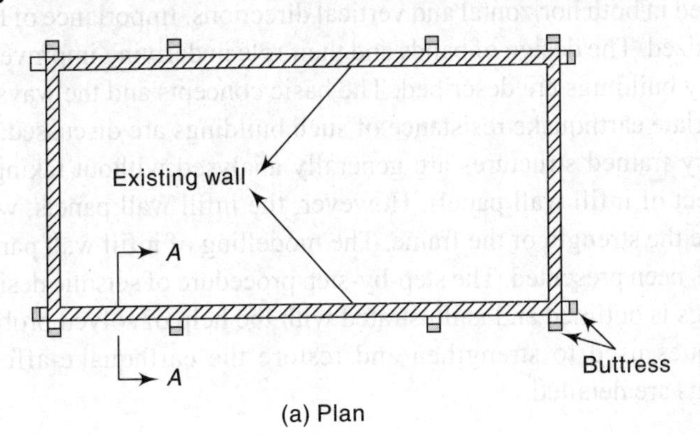

Fig. 6.28 (*Contd*)

(*Contd*)

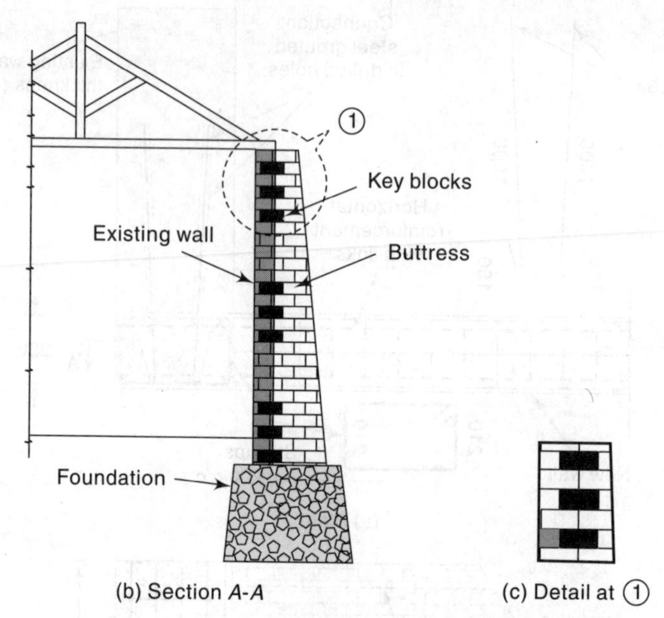

(b) Section *A-A*

(c) Detail at 1

Fig. 6.28 Strengthening of long walls by buttresses

Summary

Most loss of life in earthquakes has been observed in non-engineered constructions—using traditional materials such as bricks, stone, timber, and adobe. This necessitates the introduction of earthquake-resistance features in them. Masonry structures, in particular, brick masonry buildings, have been discussed in this chapter. The behaviour of both the non-engineered and the engineered buildings has been described.

For construction of masonry structures, brick masonry with good-quality bricks and mortar, involving cement or lime, should be used. Steel reinforcement should be placed in both horizontal and vertical directions. Importance of bands has been emphasized. The design of bands and their role in deriving improved behaviour of masonry buildings are described. The basic concepts and the ways for achieving appropriate earthquake resistance of such buildings are discussed.

Masonry framed structures are generally analysed without taking into account the effect of infill wall panels. However, the infill wall panels, when provided, increase the strength of the frame. The modelling of infill wall panel as diagonal strut has been presented. The step-by-step procedure of seismic design of masonry buildings is outlined and is illustrated with the help of solved problems. Various techniques used to strengthen and restore the earthquake-affected masonry buildings are detailed.

Solved Problems

6.1 Determine the frequency and design seismic coefficient for an ordinary masonry shear wall in a school building at Allahabad, for the following data.

Roof load $P = 15$ kN/m
Height of wall $h = 3.0$ m
Width of wall $b = 0.2$ m
Unit weight of wall $w = 19.2$ kN/m^3
Soil is medium

Solution

Given data as follows

height of wall $h = 3.0$ m
width of wall $b = 0.2$ m
roof load $P = 15$ kN/m

Self weight of wall, $W = 0.2 \times 3.0 \times 19.2 = 11.52$ kN/m

From Eqn (6.2),

$$F_B = \frac{(15 + 11.52) \times 0.2}{2 \times 3} = 0.884 \text{ kN}$$

When $x = x_c$, $F = 0$ and from Eqn (6.4)

$$x_c = \frac{11.52 \times 0.2 + 15 \times 0.2}{2 \times 15 + 11.52} = 0.128 \text{ m}$$

From Eqn (6.5)

$$k_c = \frac{0.884}{0.128} = 6.91 \text{ kN/m}$$

From Eqn (6.6), frequency $f = \dfrac{1}{2\pi}\left(\dfrac{6.91 \times 9.81}{15 + 11.52}\right)^{\frac{1}{2}} = 0.254$ Hz

Time period $T = \dfrac{1}{f} = 3.92$ s

For Allahabad (Zone II), zone factor $Z = 0.10$
Importance factor $I = 1.5$ (for a school building)
Response reduction factor $R = 1.5$
Assuming the damping coefficient for masonry to be 5%,
S_a/g value [from Eqn (5.7) of Section 5.6] for time period $T = 3.92$ s and medium soil

$$\frac{S_a}{g} = \frac{1.36}{T} = \frac{1.36}{3.92} = 0.347$$

Design seismic coefficient, $A_h = \dfrac{ZI(S_a/g)}{2R}$

$$A_h = \frac{0.1 \times 1.5 \times 0.347}{2 \times 1.5} = 0.01735; 0.02$$

6.2 In a single-room building, as shown in Fig. 6.29, the walls are built with 200-mm modular bricks in 1:6 cement–sand mortar. The self weight of the roof is 6500 N/m^2. Check the wall for vertical bending and design an RC lintel band. The design seismic coefficient is 0.18. Assume the unit weight of masonry to be 19200 N/m^3.

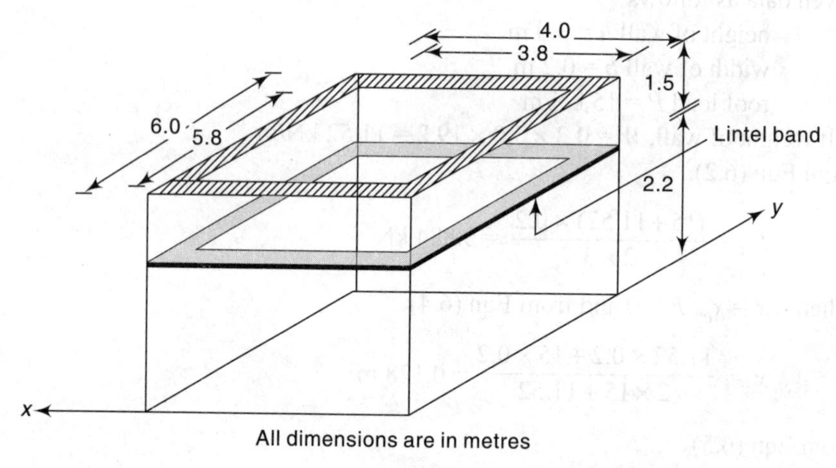

Fig. 6.29

Solution

Let the earthquake force act along the x axis. The lintel band divides the long wall in two parts. The lower part, 2.2 m in height, spans the plinth band and the lintel band. The upper part, 1.5 m in height, spans the lintel band and the roof band.

Vertical bending of wall

Vertical span = 2.2 m, thickness of wall = 200 mm

Self weight of wall = 19200 × 0.20 = 3840 N/m^2

Design force $W_1 = \alpha_h$, $W = 0.18 \times 3840 = 691.2$ N/m^2

Considering the design force as uniformly distributed load

$$\text{Bending moment } M = \frac{W_1 l^2}{8} = \frac{691.2 \times (2.2)^2}{8}$$

$$= 418.17 \text{ Nm/m}$$

$$\text{Bending stress } \sigma_{bct} = \pm\frac{M}{Z} = \frac{418.17 \times 10^3}{\dfrac{1000 \times (200)^2}{6}} = \pm 0.062 \text{ N/mm}^2$$

Self weight of the wall and roof per metre (above the lintel band)
Self weight of wall = 3840 × 1.5 = 5760.0 N/m
Assuming the weight of slab is transferred uniformly on wall

$$\text{Self weight of roof} = \frac{6500 \times (5.8 + 2 \times 0.2) \times (3.8 + 2 \times 0.2)}{2 \times [(5.8 + 0.2) + (3.8 + 0.2)]}$$

= 8463 N/m

Total self weight W_2 = 5760 + 8463 = 14223 N/m

$$\text{Axial compressive stress, } \sigma_{ac} = \frac{14223}{1000 \times 200} = 0.071 \text{ N/mm}^2$$

Combined stress = 0.0711 ± 0.062
= 0.133 N/mm² (compressive) and
0.009 N/mm² (compressive)

Hence, safe.

Lintel band

Horizontal load on lintel band (neglecting openings),

$$q_h = 691.2 \times \left(\frac{2.2 + 1.5}{2}\right) = 1278.72 \text{ N/m}$$

Assuming continuity of band at corners

$$\text{Bending moment } M = \frac{1278.72 \times (6.0)^2}{10} = 4603.4 \text{ Nm}$$

$$\text{Shear force } F = \frac{1278.72 \times 5.8}{2} = 3708.3 \text{ N}$$

Width of band = 200 mm

Using M-20 concrete and Fe 415 steel

$$A_{st} = \frac{4603.4 \times 1000}{230 \times 0.9 \times (200 - 25 - 5)} = 131 \text{ mm}^2$$

Therefore, provide 150-mm thick band and two 10 ϕ bars on each face, i.e., provide total 4 bars of 10 ϕ, tied with stirrups 6 mm ϕ at 150 c/c as shown in Fig. 6.30.

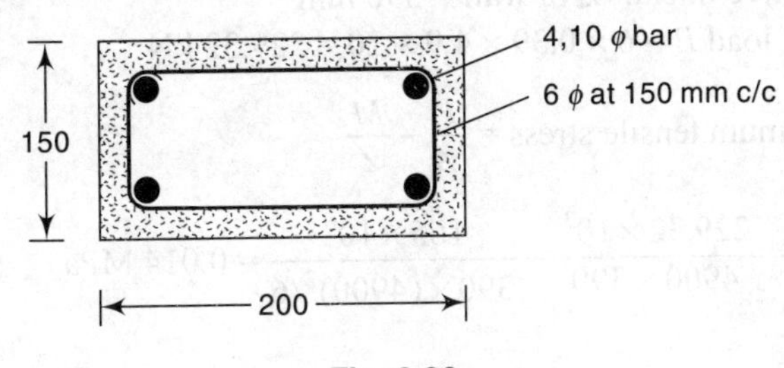

Fig. 6.30

6.3 Design an unreinforced 6-m high and 4.9-m wide masonry shear wall (centre lines of walls), as shown in Fig. 6.31, based on the following data.

Unit weight of wall = 20,000 N/m^3

Prism strength of masonry f_m = 10 MPa

Seismic force at roof level H = 30 kN

Height above roof level = 0.5 m

No superimposed load is applied on the wall.

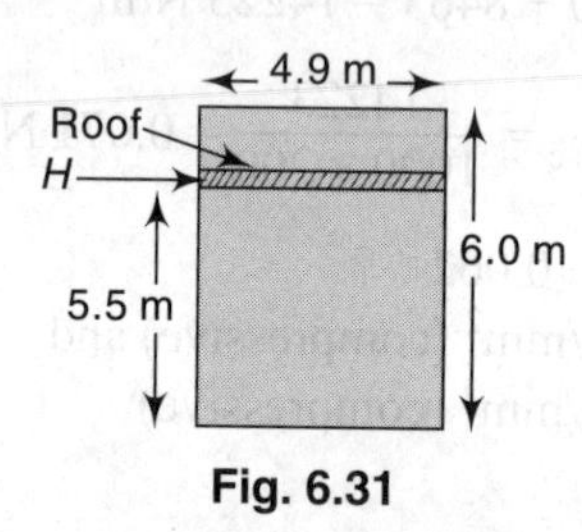

Fig. 6.31

Solution

Let us assume one-brick thick wall of 0.20 m with joints raked to a depth of 10 mm.

Thickness of wall = 200 mm

Effective thickness = 190 mm

Axial load at the base of wall $P = 0.19 \times 6.0 \times 4.9 \times 20 = 111.72$ kN

Bending moment $M = 3 \times 5.5 = 165$ kNm

Check for tension

$$\text{Maximum tensile stress} = \frac{P}{A} - \frac{M}{Z}$$

$$= \frac{111.72 \times 10^3}{4900 \times 190} - \frac{165 \times 10^6}{190 \times (4900)^2/6} = -0.09 \text{ MPa}$$

As no tension is allowed in an unreinforced masonry wall, the design should be modified.

Assume that a two-brick thick wall is provided with joints raked to a depth of 10 mm.

Nominal thickness of wall = 400 mm

Effective thickness of wall = 390 mm

Axial load $P = 6 \times 0.39 \times 4.9 \times 20 = 229.32$ kN

$$\text{Maximum tensile stress} = \frac{P}{A} - \frac{M}{Z}$$

$$= \frac{229.32 \times 10^3}{4900 \times 390} - \frac{165 \times 10^6}{390 \times (4900)^2/6} = 0.014 \text{ MPa} \quad \text{(compressive)}$$

Since no tension occurs in this section, the section is all right.

Check for shear

Shear force due to earthquake = 30 kN

$$\text{Maximum shear stress} = \frac{3}{2}\left(\frac{V}{A}\right) = \frac{3}{2} \times \frac{30}{0.39 \times 4.9}$$

$$= 0.024 \text{ MPa}$$

Allowable shear stress is given by the least of the following.

(a) 0.5 MPa

(b) $0.1 + 0.2f_d = 0.1 + 0.2\,\dfrac{P}{A} = 0.1 + 0.2 \times \dfrac{229.32 \times 10^3}{4900 \times 390} = 0.124$ MPa

(c) $0.125\sqrt{f_m} = 0.125\sqrt{10} = 0.395$ MPa

Allowable shear stress = 1.333 × 0.124 = 0.165 > 0.024 MPa

Hence the designed section is safe.

6.4 Determine the lateral forces on a two-storey unreinforced brick masonry building, as shown in Fig. 6.32, situated near Allahabad (zone III) for the following data:

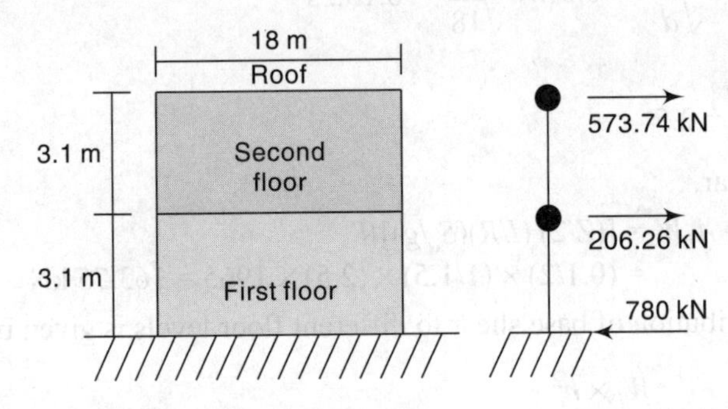

Fig. 6.32

Plan size = 18 m × 8 m
Total height of the building = 6.2 m
Storey height = 3.1 m
Weight of roof = 2.5 kN/m²
Weight of wall = 5.0 kN/m²
Live load on roof = 0
Live load at floor = 1.0 kN/m²
Zone factor = 0.10
Importance factor = 1.0
Response reduction factor = 1.5
Soil: (Type II) medium soil

Solution

Seismic DL at roof level W_r:

Weight of roof = 2.5 × 18 × 8 = 360 kN

Weight of wall = [5 × (18 + 8) × 2 × 3.1]/2 = 403 kN

W_r = weight of roof + weight of wall

= 360 + 403 = 763 kN

Seismic DL at second floor level W_2 :

Weight of second floor = 2.5 × 18 × 8 = 360 kN

Weight of wall = 5 × (18 + 8) × 2 × 3.1 = 806 kN

Live load = 1 × 18 × 8 × 0.25 = 36 kN

W_2 = 360 + 806 + 36 = 1202 kN

As per Table 8 of IS 1893 (Part1): 2002, 25% of the imposed load (live load) is considered only if the imposed load is less than 3.0 kN/m².

Total building weight = 763 + 1202 = 1965 kN

The fundamental period of building.

$$T = \frac{0.09\,h}{\sqrt{d}} = 0.09 \times \frac{6.2}{\sqrt{18}} = 0.132 \text{ s}$$

$$\frac{S_a}{g} = 2.5$$

The base shear,

$$V_B = A_h W = [(Z/2)\,(I/R)(S_a/g)]W$$

$$= (0.1/2) \times (1/1.5) \times (2.5) \times 1965 = 163.75 \text{ kN}$$

Vertical distribution of base shear to different floor levels is given by

$$Q_i = V_B \frac{W_i \times h_i^2}{\sum_{i=1}^{n} W_i \times h_i^2}$$

where Q_i is design lateral forces at floor i, W_i is seismic weight of floor i, h_i is height of floor i measured from the base, and n is the number of storeys in the building.

Shear at roof level

$$Q_r = \frac{163.75 \times 763 \times (6.2)^2}{[763 \times (6.2)^2 + 1202 \times (3.1)^2]} = 117.48 \text{ kN}$$

Shear at second floor level

$$Q_2 = \frac{163.75 \times 1202 \times (3.1)^2}{[763 \times (6.2)^2 + 1202 \times (3.1)^2]} = 46.26 \text{ kN}$$

6.5 Determine the rigidity of the shear wall shown in Fig. 6.33 in terms of Et.

(E is the modulus of elasticity and t is the thickness of wall)

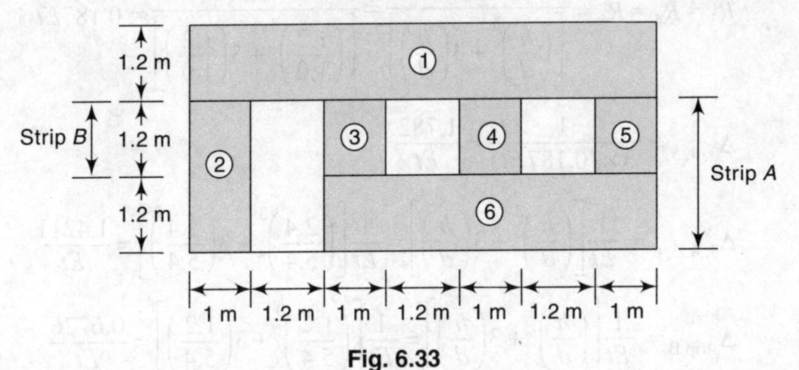

Fig. 6.33

Solution

Δ_{wall} = deflection of the given wall

$\Delta_{solid\ wall}$ = deflection of the solid wall as cantilever

$\Delta_{strip\ A}$ = deflection of the strip A (strip of wall of maximum high opening)

$\Delta_{2,\,3,\,4,\,5,\,6}$ = deflection of the fixed solid wall portions, i.e., piers 2, 3, 4, 5, and 6 as shown in Fig. 6.33.

$$\Delta_{wall} = \Delta_{solid\ wall} - \Delta_{strip\ A} + \Delta_{2,\,3,\,4,\,5,\,6} \tag{6.5.1}$$

Now $$\Delta_{2,\,3,\,4,\,5,\,6} = \frac{1}{R_{2,3,4,5,6}} \text{ (where } R \text{ is the relative rigidity)} \tag{6.5.2}$$

But $$R_{2,\,3,\,4,\,5,\,6} = R_2 + R_{3,\,4,\,5,\,6} \tag{6.5.3}$$

$$R_{3,\,4,\,5,\,6} = \frac{1}{\Delta_{3,4,5,6}} \tag{6.5.4}$$

Now, deflection of portions 3, 4, 5, 6 as a fixed wall

$$\Delta_{3,\,4,\,5,\,6} = \Delta_{solid\ 3,\,4,\,5,\,6} - \Delta_{strip\ B} + \Delta_{3,\,4,\,5} \tag{6.5.5}$$

Here, $\Delta_{3,\,4,\,5} = \dfrac{1}{R_3 + R_4 + R_5}$

Now, $$\Delta_{solid} = \frac{1}{Et}\left[4\left(\frac{h}{d}\right)^3 + 3\left(\frac{h}{d}\right)\right] = \frac{1}{Et}\left[4\left(\frac{3.6}{7.6}\right)^3 + 3\left(\frac{3.6}{7.6}\right)\right] = \frac{1.846}{Et} \tag{6.5.6}$$

and $$\Delta_{strip\ A} = \frac{1}{Et}\left[4\left(\frac{h}{d}\right)^3 + 3\left(\frac{h}{d}\right)\right] = \frac{1}{Et}\left[4\left(\frac{2.4}{7.6}\right)^3 + 3\left(\frac{2.4}{7.6}\right)\right] = \frac{1.0733}{Et} \tag{6.5.7}$$

$$\Delta_{2,\,3,\,4,\,5,\,6} = \frac{1}{R_{2,3,4,5,6}}$$

The rigidities of piers 3, 4, 5 are same.

$$R_3 = R_4 = R_5 = \frac{Et}{\left[\left(\frac{h}{d}\right)^3 + 3\left(\frac{h}{d}\right)\right]} = \frac{Et}{\left[\left(\frac{1.2}{1.0}\right)^3 + 3\left(\frac{1.2}{1.0}\right)\right]} = 0.187Et$$

$$\Delta_{3,\,4,\,5} = \frac{1}{3\times(0.187Et)} = \frac{1.782}{Et}$$

$$\Delta_{3,\,4,\,5,\,6} = \frac{1}{Et}\left[\left(\frac{h}{d}\right)^3 + 3\left(\frac{h}{d}\right)\right] = \frac{1}{Et}\left[\left(\frac{2.4}{5.4}\right)^3 + 3\left(\frac{2.4}{5.4}\right)\right] = \frac{1.4211}{Et}$$

$$\Delta_{\text{strip B}} = \frac{1}{Et}\left[\left(\frac{h}{d}\right)^3 + 3\left(\frac{h}{d}\right)\right] = \frac{1}{Et}\left[\left(\frac{1.2}{5.4}\right)^3 + 3\left(\frac{1.2}{5.4}\right)\right] = \frac{0.6776}{Et}$$

Substituting the above obtained values in Eqn (6.5.5)

$$\Delta_{3,\,4,\,5,\,6} = \frac{1.4211}{Et} - \frac{0.6776}{Et} + \frac{1.7825}{Et} = \frac{2.52598}{Et}$$

The rigidity of 3, 4, 5, 6 is obtained as

$$R_{3,\,4,\,5,\,6} = 0.39588\ Et$$

$$R_2 = \frac{Et}{\left[\left(\frac{h}{d}\right)^3 + 3\left(\frac{h}{d}\right)\right]} = \frac{Et}{\left[\left(\frac{2.4}{1.0}\right)^3 + 3\left(\frac{2.4}{1.0}\right)\right]} = 0.04756\ Et$$

Substituting the above two values in Eqn (6.5.3)

$$R_{2,\,3,\,4,\,5,\,6} = 0.04756\ Et + 0.39588\ Et$$

$$= 0.44344\ Et$$

Hence, $$\Delta_{2,\,3,\,4,\,5,\,6} = \frac{2.25507}{Et} \tag{6.5.8}$$

Substituting the values of Eqns (6.5.6), (6.5.7), (6.5.8) in Eqn (6.5.1)

$$\Delta_{\text{wall}} = \frac{1.846}{Et} - \frac{1.0733}{Et} + \frac{2.25507}{Et}$$

$$\Delta_{\text{wall}} = \frac{3.02777}{Et}$$

Rigidity of the shear wall

$$R_{\text{wall}} = \frac{1}{\Delta_{\text{wall}}}$$

$$= 0.330276Et$$

The rigidity of the wall is 0.330276*Et*.

6.6 Calculate the torsional shear forces in a one-storey shear wall masonry structure with a rigid diaphragm roof for the following data. There are four shear walls, with relative rigidity of each wall as shown in Fig. 6.34.

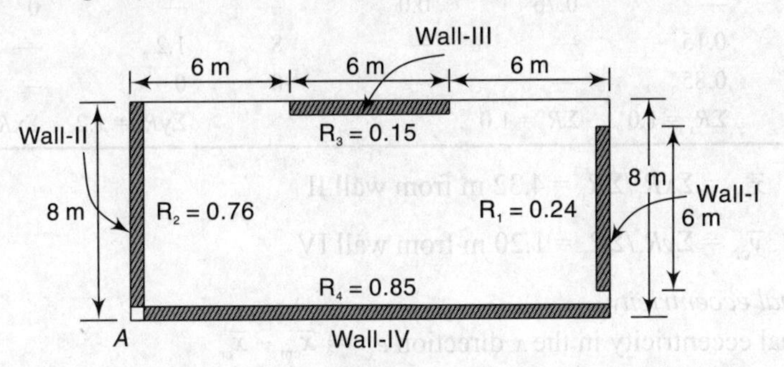

Fig. 6.34

Height of parapet wall = 1 m
Height of walls up to roof levels = 3 m
Seismic zone V
$Z = 0.36, I = 1.0, R = 1.5, S_a/g = 2.5$
Self weight of the roof = 3.0 kN/m^2
Self weight of the wall = 5.0 kN/m^2
Base shear $V = 300$ kN

Solution

Location of the centre of mass

Let the centre of masses in the x and y directions be x_{cm} and y_{cm}, respectively. Take static moments about a point, say A, the left corner of the building (Fig. 6.34).

Slab	Weight W (kN)	x (m)	y (m)	Wx (kNm)	Wy (kNm)
Roof slab	$18 \times 8 \times 3$	9	4	3888	1728
Wall I	$6 \times 4 \times 5$	18	4	2160	480
Wall II	$8 \times 4 \times 5$	0	4	0	640
Wall III	$6 \times 4 \times 5$	9	8	1080	960
Wall IV	$18 \times 4 \times 5$	9	0	3240	0
	$\Sigma W = 1192$			$\Sigma Wx = 10368$	$\Sigma Wy = 3808$

$$\bar{x}_{cm} = \Sigma Wx/\Sigma W = 8.69 \text{ m from wall II}$$
$$\bar{y}_{cm} = \Sigma Wy/\Sigma W = 3.19 \text{ m from wall IV}$$

Location of centre of rigidity

Let the centres of rigidity in the x and y directions be x_{cr} and y_{cr}, respectively.

Take static moments about a point, say A, the left corner of the building. The stiffness of the slab and parapet heights are not considered in the calculation of the centre of rigidity.

Item No.	R_x	R_y	x(m)	y(m)	yR_x	xR_y
Wall I	—	0.24	18	—	—	4.32
Wall II	—	0.76	0.0	—	—	0
Wall III	0.15	—	—	8	1.2	—
Wall IV	0.85	—	—	0	0	—
	$\Sigma R_x = 1.0$	$\Sigma R_y = 1.0$			$\Sigma yR_x = 1.2$	$\Sigma xR_y = 4.32$

$\overline{x}_{cr} = \Sigma xR_y/\Sigma R_y = 4.32$ m from wall II

$\overline{y}_{cr} = \Sigma yR_x/\Sigma R_x = 1.20$ m from wall IV

Torsional eccentricity

Torsional eccentricity in the x direction $e_x = \overline{x}_{cm} - \overline{x}_{cr}$

$= 8.69 - 4.32 = 4.37$ m

Accidental eccentricity (5%) $= 0.05 \times 18 = 0.9$ m

Total eccentricity $= 4.37 + 0.9 = 5.27$ m

Torsional eccentricity in y direction $e_y = \overline{y}_{cm} - \overline{y}_{cr}$

$e_y = 3.19 - 1.2 = 1.99$ m

Accidental eccentricity (5%) $= 0.05 \times 8 = 0.4$ m

Total eccentricity $= 1.99 + 0.4 = 2.39$ m

Torsional moment

The torsional moment due to the seismic force in the I–II direction will rotate the building in the y direction, hence

$M_{Tx} = V_x e_y = 300 \times 2.39 = 717$ kNm

Similarly, the seismic force in the III–IV direction will rotate the building in the y direction, hence

$M_{Ty} = V_y e_x = 300 \times 5.27 = 1581$ kNm

Distribution of direct shear forces and torsional shear forces

If we consider the seismic force only in the I–II direction, then the walls in the III–IV direction will resist the forces and the walls in the I–II direction may be ignored. Similarly for the seismic force in the III–IV direction, the walls in the I–II direction will resist the forces and the walls in the III–IV direction may be ignored. The distribution of forces in the walls will be as below.

$$\text{Direct shear in wall III} = \frac{Rx}{\sqrt{\Sigma Rx}} \times V_x$$

$$= 0.15 \times 300 = 45 \text{ kN}$$

$$\text{Direct shear in wall IV} = \frac{Rx}{\sqrt{\Sigma Rx}} \times V_x$$

$$= 0.85 \times 300 = 255 \text{ kN}$$

$$\text{Torsional shear forces in wall III} = \frac{R_x d_y}{\sum R_x d_y^2} \times V_x e_y = \frac{1.02}{8.16} \times 717$$

$$= 89.62 \text{ kN}$$

$$\text{Torsional shear forces in wall IV} = \frac{R_x d_y}{\sum R_x d_y^2} \times V_x e_y = -\frac{1.02}{8.16} \times 717$$

$$= -89.62 \text{ kN}$$

Distribution of force in shear walls III and IV

Distance of the wall III from centre of gravity = 8 – 1.2 = 6.8 m

Distance of the wall IV from centre of gravity = 0 – 1.2 = –1.2 m

Wall	R_x	d_y (m)	$R_x d_y$	$R_x d^2_y$	Direct shear force (kN)	Torsional shear force (kN)	Total shear (kN)
Wall III	0.15	6.8	1.02	6.936	45	89.62	134.62
Wall IV	0.85	–1.2	–1.02	1.224	255	–89.62*	255
				$\Sigma R_x d_y^2 = 8.16$			

*Negative torsional shear force is neglected.

$$\text{Direct shear in wall I} = \frac{R_y}{\Sigma R_y} \times V_y$$

$$= 0.24 \times 300 = 72 \text{ kN}$$

$$\text{Direct shear in wall II} = \frac{R_y}{\Sigma R_y} \times V_y$$

$$= 0.76 \times 300 = 228 \text{ kN}$$

$$\text{Torsional forces in wall I} = \frac{R_y d_x}{\Sigma R_y d_x^2} \times V_y e_x$$

$$= \frac{3.28}{59.09} \times 1581 = 87.75 \text{ kN}$$

$$\text{Torsional forces in wall II} = \frac{R_y d_x}{\Sigma R_y d_x^2} \times V_y e_x$$

$$= \frac{3.28}{59.09} \times 1581 = 87.75 \text{ kN}$$

Distribution of force in shear walls I and II

Distance of the wall I from centre of gravity = 18 – 4.32 = 13.68 m

Distance of the wall II from centre of gravity = 0 – 4.32 = –4.32 m

Wall	R_y	d_x(m)	$R_y d_x$	$R_y d^2_x$	Direct shear force (kN)	Torsional shear force (kN)	Total shear (kN)
Wall I	0.24	13.68	3.28	44.91	72	+87.75	159.75
Wall II	0.76	–4.32	–3.28	14.18	228	–87.75	228
				$\Sigma R_y d_x^2 = 59.09$			

6.7 Determine the increase in axial load due to overturning effects of lateral forces in the wall, as shown in Fig. 6.35. The thickness of wall may be taken as 0.25 m.

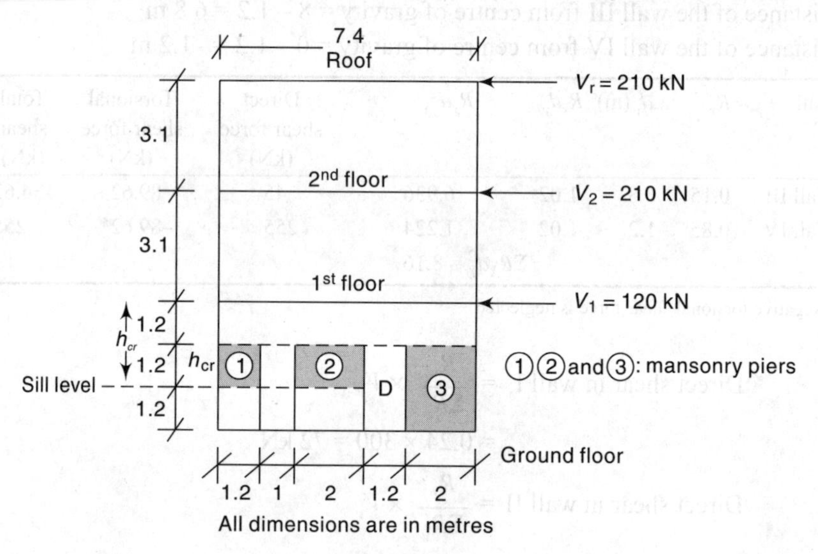

Fig. 6.35

Solution

Thickness of wall = 0.25 m

Taking overturning moment M at first floor level

$$M = V_r \times (3.1 + 3.1) + V_2 \times 3.1$$
$$= 210 \times 6.2 + 210 \times 3.1$$
$$= 1953 \text{ Nm}$$

Total shear force above sill level

$$V = V_r + V_2 + V_1$$
$$= 210 + 210 + 120 = 540 \text{ kN}$$

Total overturning moment M_0 on masonry piers in the ground floor

$$M_0 = M + V \times h_{cr}$$
$$= 1953 + 540 \times (1.2 + 0.6)$$
$$= 2925 \text{ kNm}$$

(h_{cr} is distance of the first floor level from critical level of the pier in the ground floor; this may be assumed at the sill level)

Let A_i be the area of ith pier, l_i be the distance of centroid of ith pier from the left edge of the wall, L_i be the distance of centroid of pier from the centre of gravity of the net wall section, I_i be the moment of inertia of the ith pier about its own axis, I_n be the moment of inertia of the ith pier about the axis under consideration and P_o be the increase in the axial load due to overturning.

Centroid of net section of wall

Pier No.	A_i (m²)	Distance l_i (m)	$A_i \times l_i$ (m³)
1	$1.2 \times 0.25 = 0.3$	0.6	0.18
2	$2 \times 0.25 = 0.5$	$1.2 + 1 + 1 = 3.2$	1.6
3	$2 \times 0.25 = 0.5$	$1.2 + 1 + 2 + 1.2 + 1 = 6.4$	3.2
	$\Sigma A_i = .3$		$\Sigma A_i l_i = 4.98$

$$\text{Distance of centroidal from left edge} = \frac{\Sigma A_i l_i}{\Sigma A_i}$$

$$= \frac{4.98}{1.3}$$

$$= 3.83 \text{ m}$$

Moment of inertia of net section of wall

Pier	A_i(m²)	L_i(m)	$A_i L_i^2$(m⁴)	$I_i = td^3/12$	$I_n = A_i L_i^2 + I_i$	$A_i L_i$	P_o(kN)
1	0.3	$3.83 - 0.6 = 3.23$	3.130	0.036	3.167	0.969	404.90
2	0.5	$3.83 - 3.2 = 0.63$	0.198	0.167	0.365	0.315	131.62
3	0.5	$6.4 - 3.83 = 2.57$	3.302	0.167	3.469	1.285	536.94
					$\Sigma I_n = 7.0$		

Increase in axial load on the individual pier

$$P_o = \frac{M_o \times A_i L_i}{I_n}$$

$$= \frac{2925\, A_i L_i}{7.0}$$

$$= 417.86\, A_i L_i$$

Exercises

6.1 (a) State the reasons for the poor performance of masonry buildings in seismic areas.

(b) Strong bricks and weak mortar are recommended for masonry buildings. Why?

6.2 Discuss the behaviour of the following masonry walls in seismic regions.

(a) Unreinforced masonry walls

(b) Reinforced masonry walls

(c) Infill walls

6.3 Describe the various earthquake-resistant features that can be introduced in a masonry building to make it earthquake resistant.

6.4 Write notes on the following:
(a) Categories of masonry buildings
(c) Strengthening of masonry walls
(b) Types of masonry walls
(d) Box action of walls

6.5 (a) What are the various methods of restoring an earthquake damaged masonry building?
(b) How can an old wall be strengthened by
(i) inserting a new wall
(ii) prestressing
Draw neat sketches to support your answer.

6.6 Define bands. At what levels in a masonry building would you provide them? Give justifications for each of them.

6.7 (a) How can the rocking of masonry piers in a masonry wall be prevented?
(b) What special precautions should be exercised during planning and construction of openings in a masonry wall?

6.8 Determine the frequency and design seismic coefficient for an ordinary masonry shear wall in a primary health centre at Dehradun, given the following data.
Roof load = 20 kN/m
Height of wall = 3.0 m
Width of wall = 0.3 m
Unit weight of wall = 20 kN/m^3
The building is situated on rocky soil.
Ans: Frequency = 0.25 Hz, sesismic coefficient = 0.03018

6.9 For a room of 8 m × 4 m internal dimensions, the walls are constructed with 200 mm thick modular bricks, having wall thickness of 300 mm, in cement mortar (1:6). The load on the roof is 8 kN/m^2. Check the long wall for vertical bending and design the lintel band (RC) for the following data.
Design seismic coefficient = 0.10
Height of wall = 4.2 m
Lintel height from plinth = 2.4 m
Unit weight of masonry = 19.2 kN/m^3

6.10 Design an unreinforced masonry wall for the following data.
Unit weight of wall = 20 kN/m^3
Prism strength of masonry = 7.5 N/mm^2
Seismic force at roof level = 20 kN at a height of 4.0 m from the base
Length of wall = 4.5 m
Height of wall = 4.6 m

6.11 Design a shear wall of 6.0 m width using 230-mm modular bricks with the following building geometry.
Total height of building = 5.5 m
Roof height = 5.0 m

Prism strength of masonry = 10 N/mm^2
Grade of mortar = M-2
Compressive strength of masonry units = 15 N/mm^2
Unit weight of masonry = 19.2 kN/m^3
Axial load from beam on wall = 50 kN
Lateral seismic load causing in-plane flexure in the wall =100 kN

6.12 A simple one-storey building having two shear walls in each direction is shown in Fig. 6.36. All the four walls are in M-25 grade concrete and 200 mm thick. Two walls are 5 m long and the remaining two are 4 m long. Storey height is 3.5 m. The floor consists of cast in situ reinforced concrete. Design shear force on the building is 100 kN in either direction. Compute design lateral forces on different shear walls using the torsion provisions of IS 1983 (Part 1) for the following data.

Grade of concrete: M-25
E = 25000 N/mm^2
Thickness of wall t = 200 mm
Length of walls L = 4000 mm
Self weight of the roof = 3.0 kN/m^2
Self weight of the wall = 5 kN/m^2
All the walls have same lateral stiffness k.

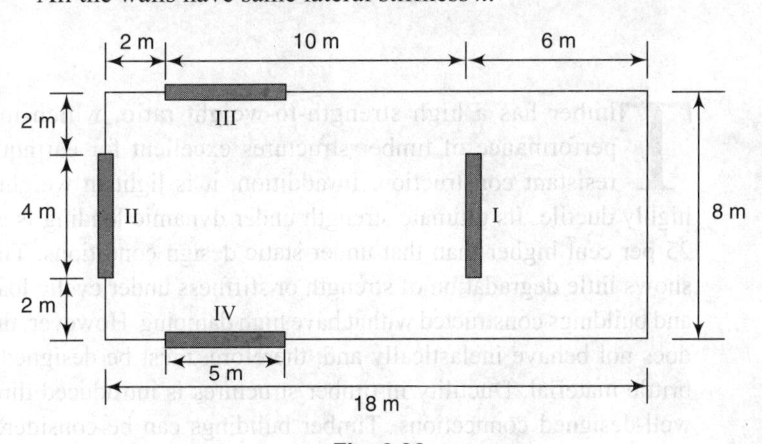

Fig. 6.36

Ans: F_{I} = 47.24 kN
F_{II} = 67.13 kN
F_{III} = 11.42 kN
F_{IV} = 11.42 kN

CHAPTER 7

Timber Buildings

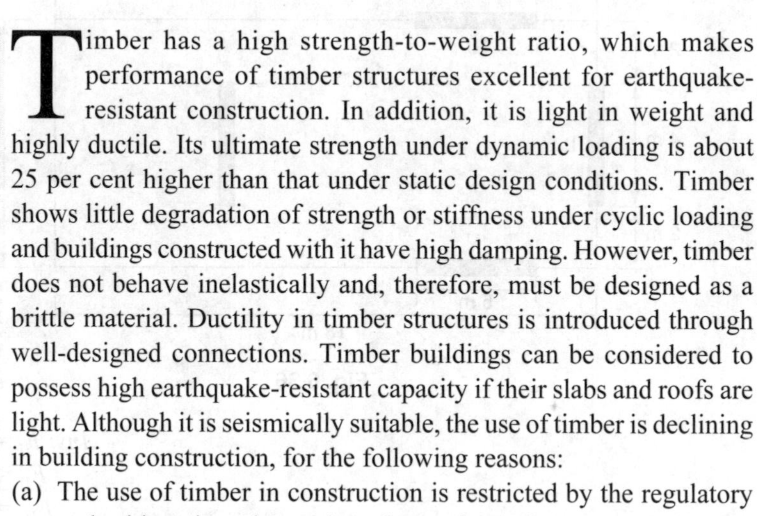

Timber has a high strength-to-weight ratio, which makes performance of timber structures excellent for earthquake-resistant construction. In addition, it is light in weight and highly ductile. Its ultimate strength under dynamic loading is about 25 per cent higher than that under static design conditions. Timber shows little degradation of strength or stiffness under cyclic loading and buildings constructed with it have high damping. However, timber does not behave inelastically and, therefore, must be designed as a brittle material. Ductility in timber structures is introduced through well-designed connections. Timber buildings can be considered to possess high earthquake-resistant capacity if their slabs and roofs are light. Although it is seismically suitable, the use of timber is declining in building construction, for the following reasons:

(a) The use of timber in construction is restricted by the regulatory authorities, since the cutting of trees for timber at a very fast rate is leading to an ecological imbalance. Timber buildings thus may only be used in those areas where it is available in abundance and only if the situation demands it.

(b) Great lateral loads are imposed over timber frames from the heavy cladding walls.
(c) Since timber is highly combustible, post-earthquake fires may be hazardous.

Timber buildings are classified as light constructions. Timber houses, which require non-structural calculations, are so-called non-engineered constructions. These buildings are usually composed of diaphragms and shear walls (e.g., wood-stud walls, brick-nogged walls). The lateral stability of a timber building is of prime concern and depends upon the design of diaphragms, shear walls, and their anchorage to the foundations. Appropriate connections of the different building elements play a vital role in the stability of a building subjected to seismic loads. Despite the high earthquake-resistant capacity of timber structures, many non-engineered timber structures have displayed inadequate performance for the following reasons:

(a) Asymmetry of the structural form
(b) Inadequate structural connections
(c) Use of heavy roofs
(d) Lack of integrity of substructure
(e) Site response
(f) Inadequate resistance to post-earthquake fires
(g) Timber decay

All the above aspects of earthquake resistance of timber structures are discussed in the sections that follow.

7.1 Structural Form

Asymmetry in the structural form leads to instability and should be avoided as far as possible. The building layout should be regular in plan and elevation with regards to mass and rigidity considerations. The building should not be large in plan. Roofing should be as light as possible. Walls should be arranged as symmetrically as possible to minimize any torsional moments to which the structure may be subjected. The plan of the building should be surrounded and divided by bearing wall lines, as shown in Fig. 7.1. The bearing walls may have stud-wall-type construction (Section 7.6) or brick-nogged-type frame construction (Section 7.7). These walls are braced diagonally in the vertical plane to resist wind and seismic forces. The height of timber buildings is generally limited to two storeys plus attic.

All bearing walls of the upper storey should be supported by the bearing walls of the lower storey. The maximum spacing of the bearing walls is restricted to 8 m, but preferably should be less than 6 m. Large openings in walls are not desirable. The maximum width of openings in the bearing walls is 4 m. The openings should be located at least 50 cm away from the corners. Adjacent openings should be at least 50 cm apart.

Care should be taken during the installation of window glass to ensure safety against structural deformation. Shear walls and columns should be supported on

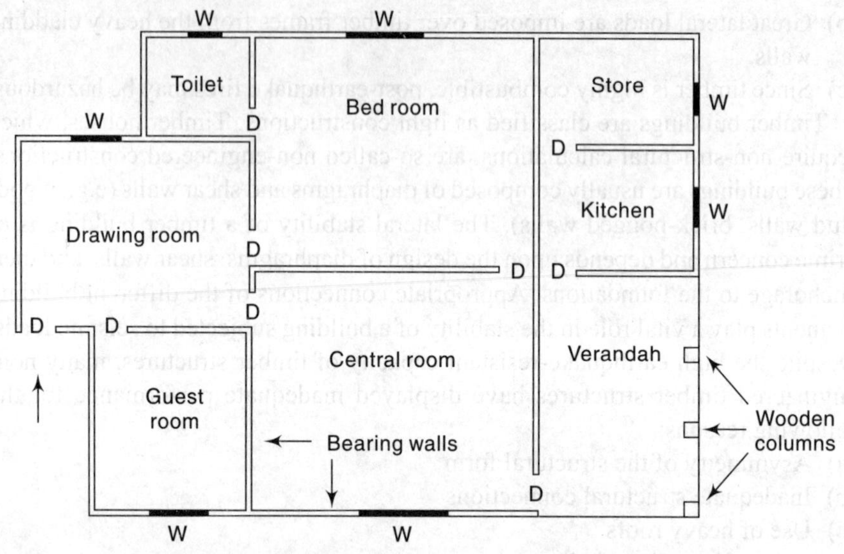

Fig. 7.1 Plan of a typical single-storey residential building

RCC footings and reinforcing bars in the walls and columns must be securely anchored in the footing. Steel plates should be used to connect walls or columns of the first and second storeys.

Horizontal diaphragms should be arranged to prevent relative horizontal deflection between vertical walls and columns. A diagonal brace should be formed in the adjoining vertical members and nailed. A hole drilled for nailing should be slightly smaller than the nail diameter, so that the brace is not split at the nail hole.

7.2 Connections

During an earthquake loading, joints in timber structures are inferior to most other types of joints. The failure of joints connecting columns and girders frequently occurs, causing the finishings to fall. Joint failure results in a change of angle between the columns and beams and the building starts titling progressively till it collapses. Generally, glue, nails, screws, bolts, metal straps, metal plates, or toothed metal connectors are used to make connections between timber members. Metal corner plates and toothed steel connectors are preferred for light timber construction. Framed connections require lot of notching and cutting. This causes reduced effective area of the member and hence framed connections are not recommended for timber buildings in seismic areas. For the connections to be able to dissipate energy, the parts connected must have thickness at least eight times the diameter of the connector, and the connector diameter should be 12 mm or less. For a well-designed connection between plywood and timber frame in shear walls, the hysteresis equivalent viscous damping ratio is about 8 to10 per cent.

7.2.1 Nailed Joints

Nailed joints are the most common and suitable type of joints for light and medium timber framings up to 15 m spans. However, nailed joints require careful attention. A nail driven parallel to the timber grain should be designed for not more than two-thirds of the lateral load, which would be allowed for the same size of nail driven normal to the grain. Nails driven parallel to the grain should not be expected to resist withdrawal forces. A minimum of two nails for node joints and four nails for lengthening joints are desirable. Also, two nails in a horizontal row are better than using the same number of nails in a vertical row. The arrangement of nails, their end-distance, edge-distance, and spacing of nails is given in Table 7.1. The details are shown in Fig. 7.2 for lengthening joints and in Figs 7.3 and 7.4 for node joints. The edge- or end-distance of the nails should not be less than half the required nail penetration.

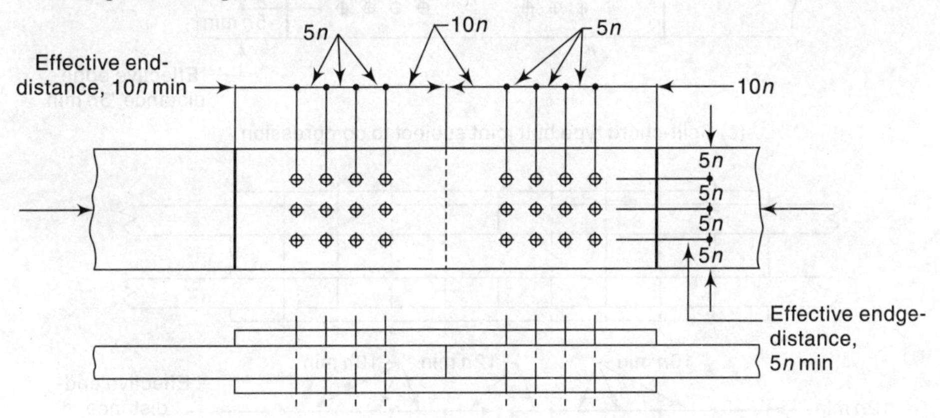

(a) Monochord-type butt joint subject to compression

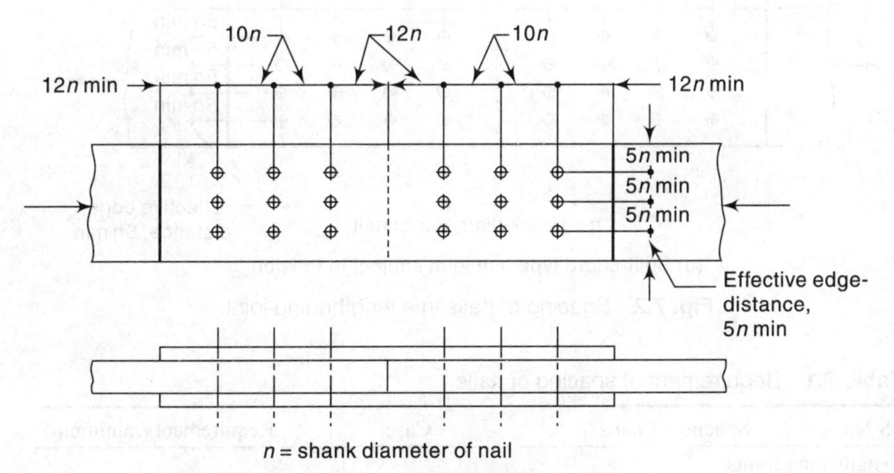

(b) Monochord-type butt joint subject to tension

Fig. 7.2 (*Contd*)

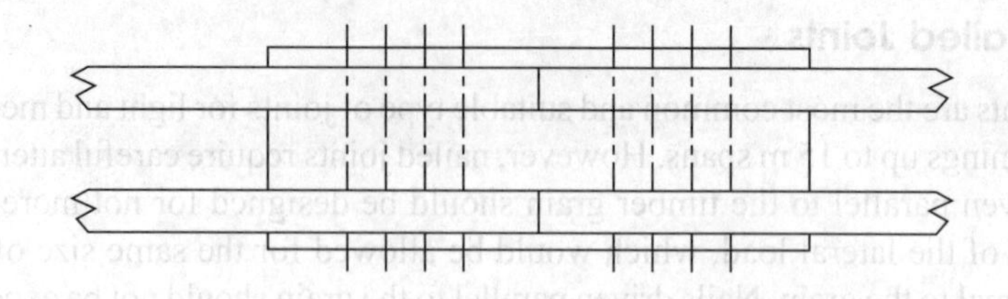

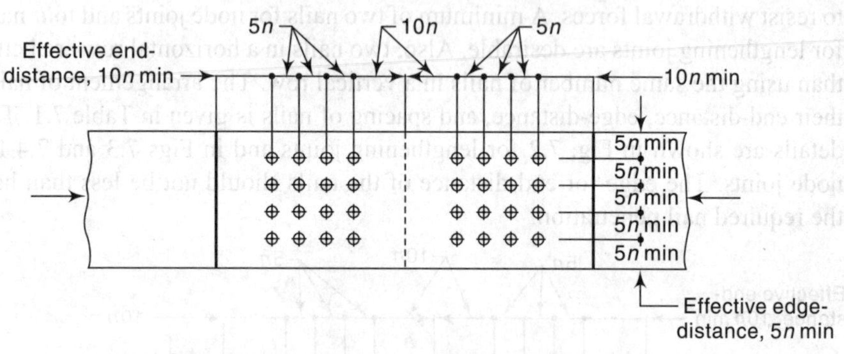

(c) Split-chord type butt joint subject to compression

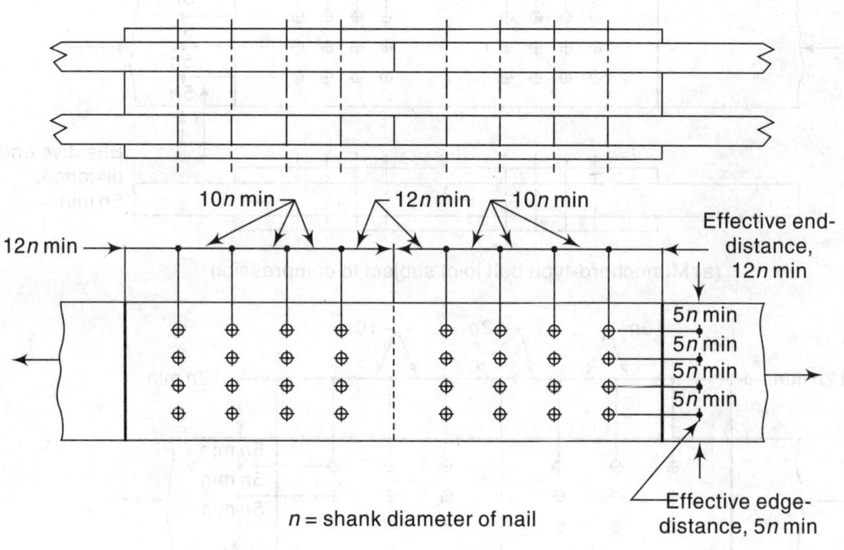

(d) Split-chord type butt joint subject to tension

Fig. 7.2 Spacing of nails in a lengthening joint

Table 7.1 Requirement of spacing of nails

S.No.	Spacing of nails	Case	Requirement (minimum)
Lengthening Joints			
1.	End-distance	Tension	$12n$*
		Compression	$10n$

(*Contd*)

Table 7.1 (*Contd*)

2.	Edge-distance		$5n$
3.	Spacing in direction of grain	Tension	$10n$
		Compression	$5n$
4.	Spacing between rows of nails perpendicular to grain		$5n$

*n is the diameter of nail

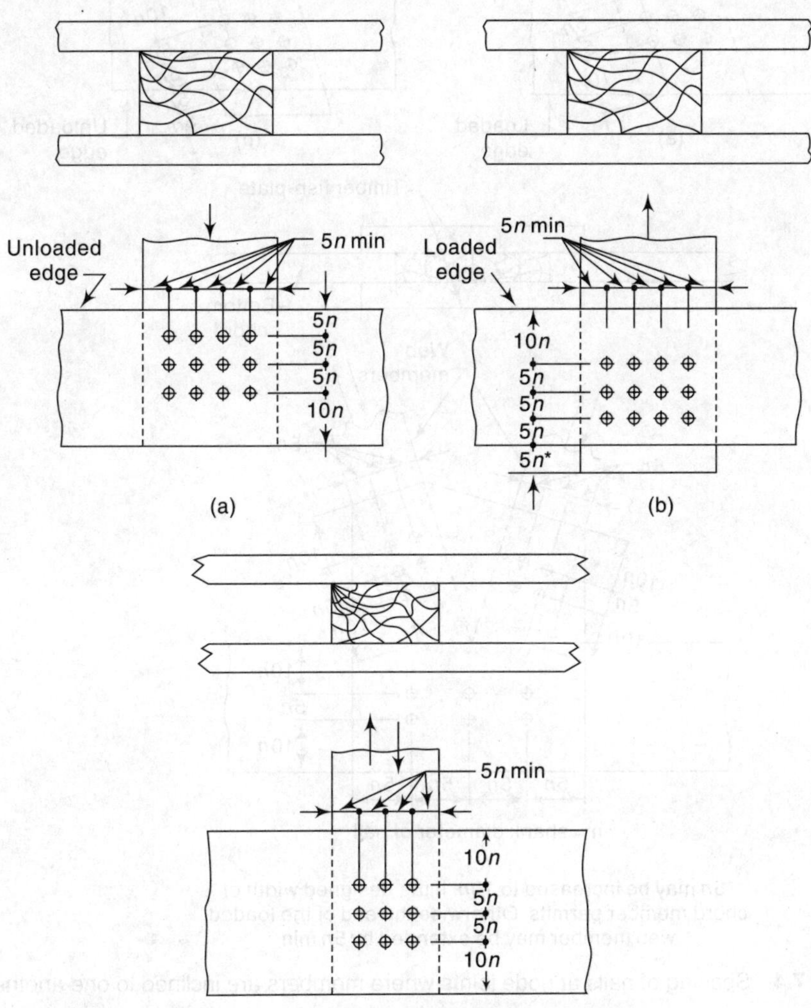

Fig. 7.3 Spacing of nails where members are at right angles to one another

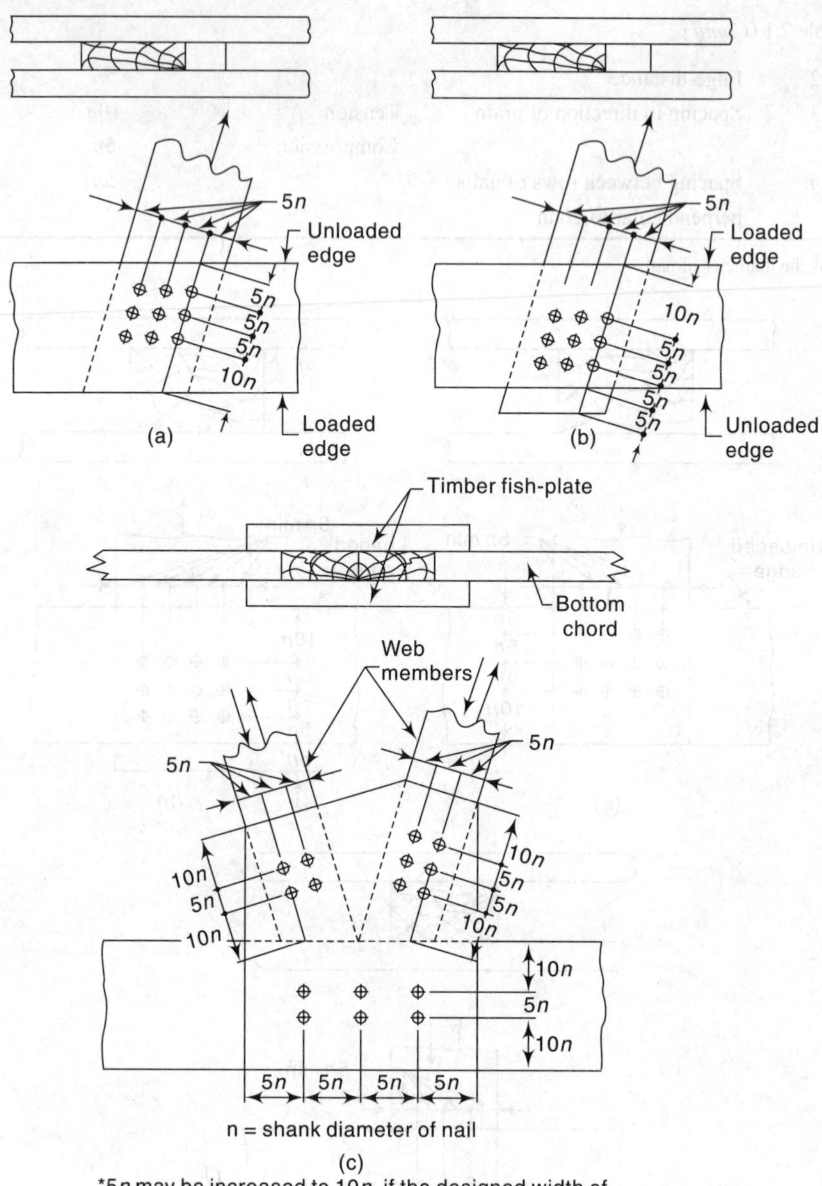

Fig. 7.4 Spacing of nails at node joints where members are inclined to one another

7.2.2 Bolted Joints

Bolted joints suit the requirements of prefabrication in small- and medium-span timber structures for speed and economy in construction. Bolted joint constructions offer better facilities as regards to workshop ease, mass production of components, transport convenience, and reassembly at site of work. The design of bolted joints follows the pattern of riveted joints in steel structures. A minimum of two bolts for nodal joints and four bolts for lengthening joints are provided. More rows of bolts are preferred than more bolts in a row. Further, more small-diameter bolts are desirable rather than a small number of large-diameter bolts in a joint. The arrangement of bolts, spacing between rows of bolts, end-distance, and edge-distance are given in Table 7.2. The bolts in the joint are preferably staggered if the load is acting perpendicular to the grain of wood. However, staggering of bolts is avoided for members loaded parallel to the grain of wood. Spacing of bolts in structural joints is shown in Fig. 7.5. Some of the timber structure bolted connections are shown in Figs 7.6, 7.7, and 7.8.

Table 7.2 Requirements of spacing of bolts

S.No.	Spacing of bolts	Case	Requirement (Minimum)	Type of wood
1.	End-distance	Tension	$7\,d$*	Soft wood
		Tension	$5\,d$	hard wood
		Compression	$4\,d$	
2.	Edge-distance	Parallel to grain	$1.5\,d_3$ or half the distance between rows of bolts, whichever is greater	
		Perpendicular to grain	$4\,d$	
3.	Spacing of bolts in a row	Parallel/perpendicular to grain	$4\,d$	
4.	Spacing between rows of bolts	Parallel to grain loading	Minimum of $(N-4)d$ or $2.5d$	
		Perpendicular to grain loading	$2.5d$ for $\frac{t^{**}}{d} = 2$ $5d$ for $\frac{t}{d} = 6$ or more (For the ratios between 2 to 6 linear interpolation may be done).	

**d* is the diameter of the bolt in mm
***t* is the thickness of main member in mm

Figure 7.6 shows the jointing of chord members of timber diaphragms. The perimeter framing may need jointing that is capable of carrying longitudinal forces arising from seismic loading. Pole frame buildings are usually jointed using bolts, steel straps, and clouts (Fig. 7.7). An effective means of obtaining resistance to lateral shear forces is to create moment-resisting triangles at the knees of portals, using steel rods as diagonal members. Connections between shear walls and the foundation, or between successive storeys of shear walls, must be capable of transmitting the horizontal shear forces and the overturning moments applied to them. Details for these connections are illustrated in Fig. 7.8.

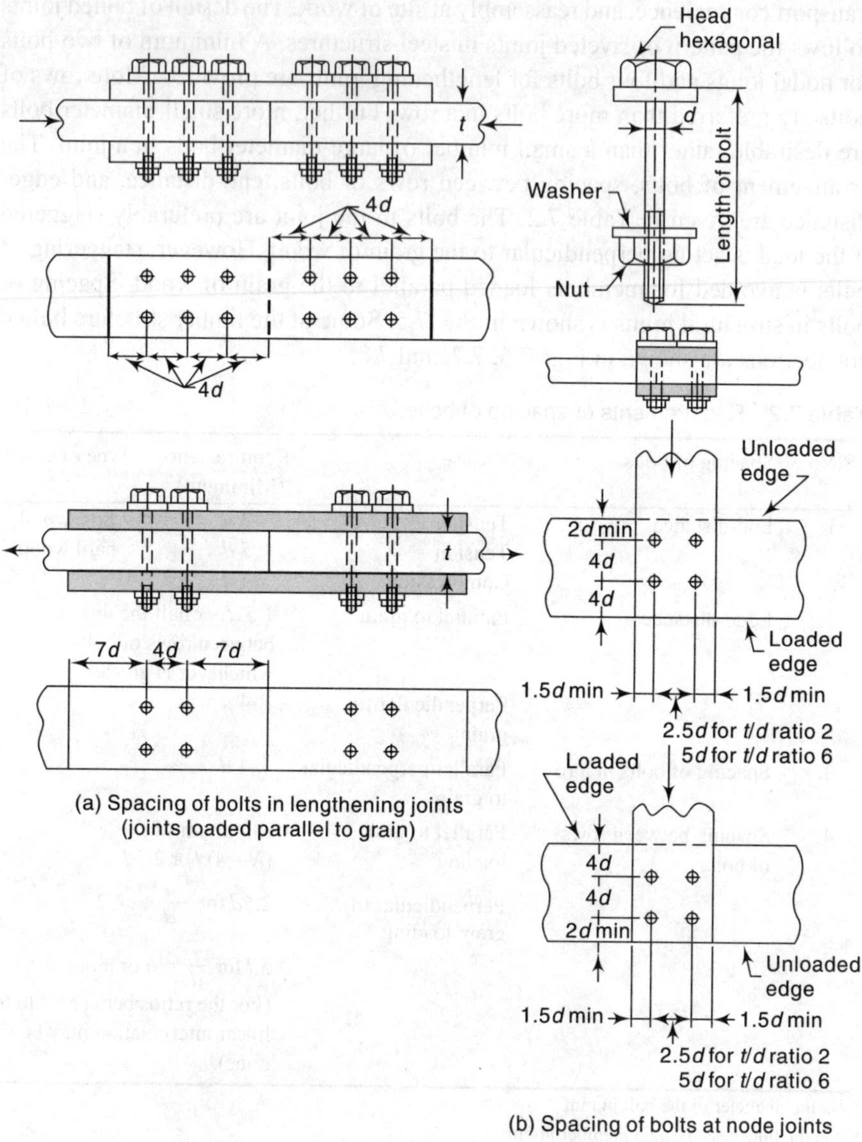

(a) Spacing of bolts in lengthening joints (joints loaded parallel to grain)

(b) Spacing of bolts at node joints

Fig. 7.5 Typical spacing of bolts in structural joints

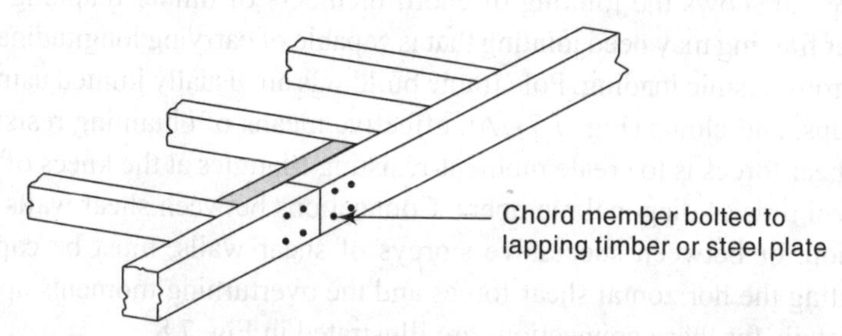

Fig. 7.6 Method of jointing chord members of timber diaphragms

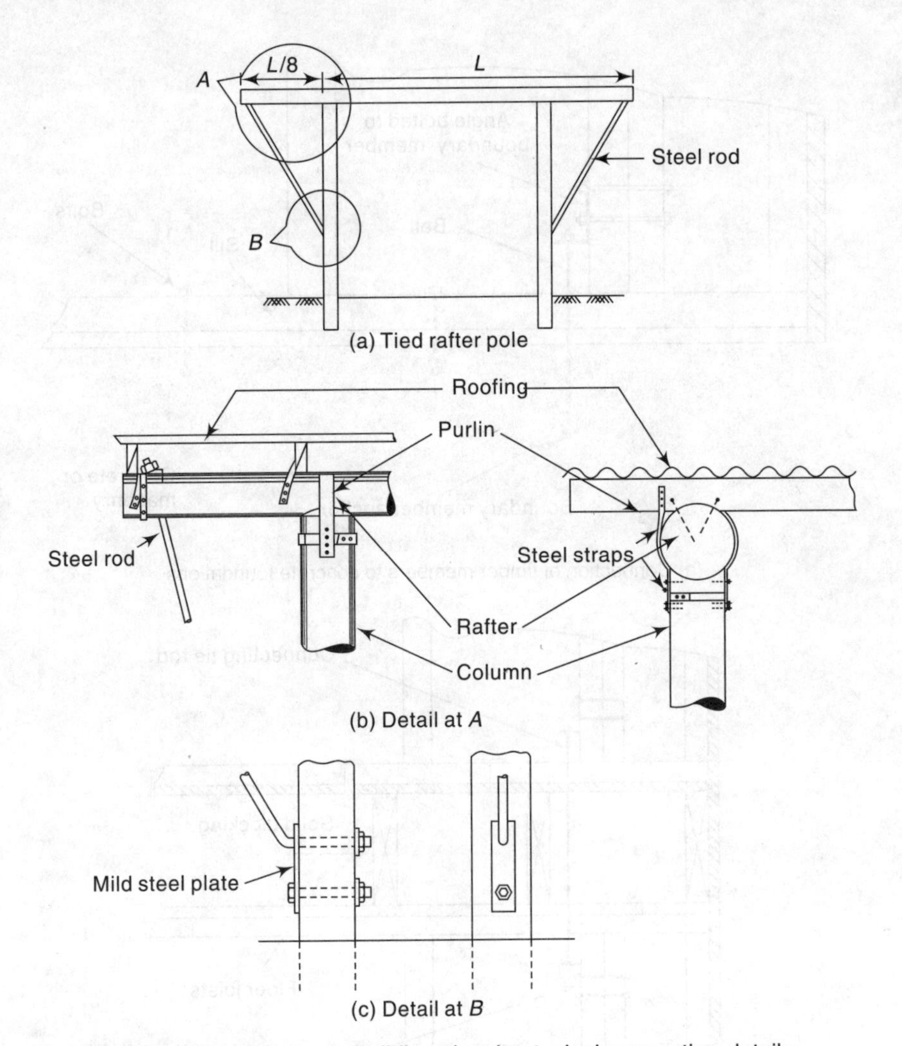

Fig. 7.7 Tied rafter pole building showing typical connection details

7.2.3 Connector Joints

In large-span structures, the members have to transmit very heavy stresses requiring stronger jointing techniques. Metallic rings or wooden disc dowels may be used to achieve this.

Metallic ring connector It is a split circular band of steel made from mild steel pipes. This is placed in the grooves cut into the contact faces of the timber members to be jointed, the assembly being held together by means of a connecting bolt as shown in Fig. 7.9.

Wooden disc-dowel joints It is a circular hardwood disc generally tapered each way from the middle so as to form a double conical frustum. Such a disc is made

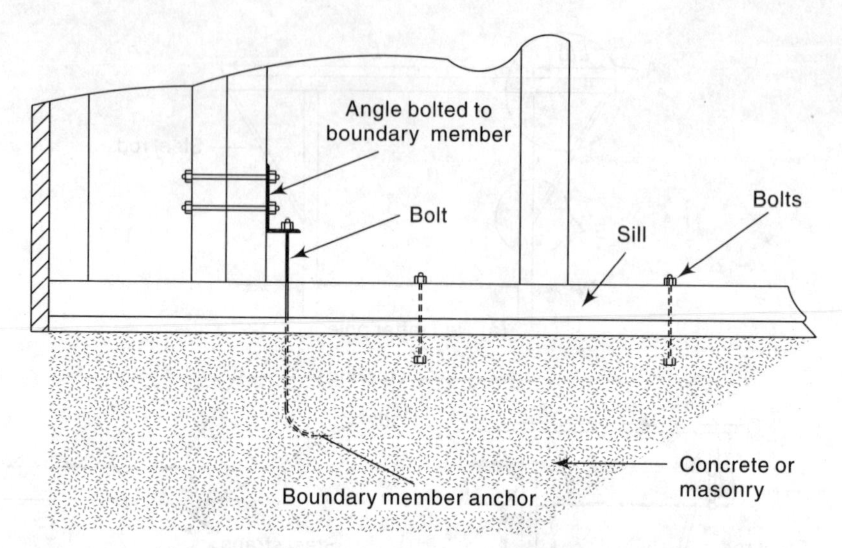

(a) Connection of timber members to concrete foundations

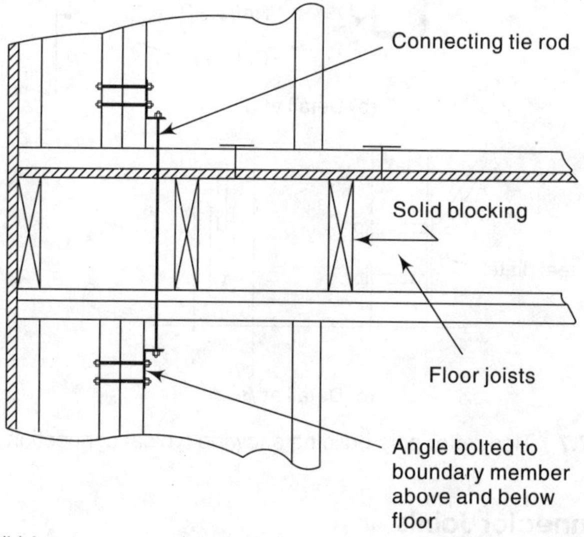

(b) Inter-storey connections of shear walls in timber buildings

Fig. 7.8 Connection details for plywood shear walls

to fit into recesses, half in one member and the other half in another, the assembly being held by one mild steel bolt through the centre of the disc to act as a coupling for keeping the jointed wooden members from spreading apart. Wooden disc-dowel joints and stress distributions for lap and butt joints are shown in Fig. 7.10.

7.2.4 Finger Joints

These are glued joints connecting timber members end to end (Fig. 7.11). Finger joints provide long lengths of timber, ideal for upgrading timber by permitting

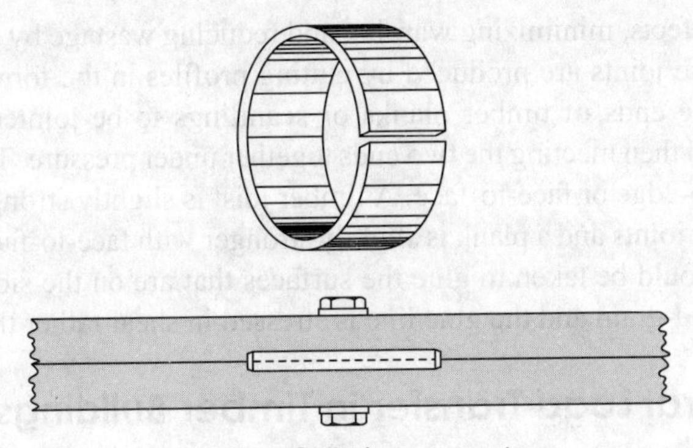

Fig. 7.9 Split ring connector

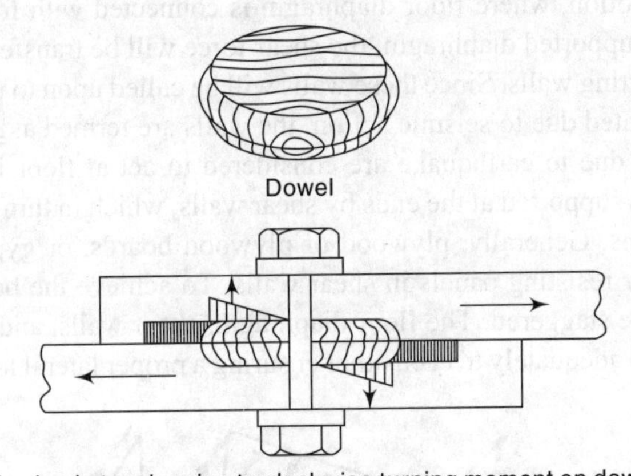

Bolt in simple tension due to clockwise turning moment on dowel

(a) Lap joint

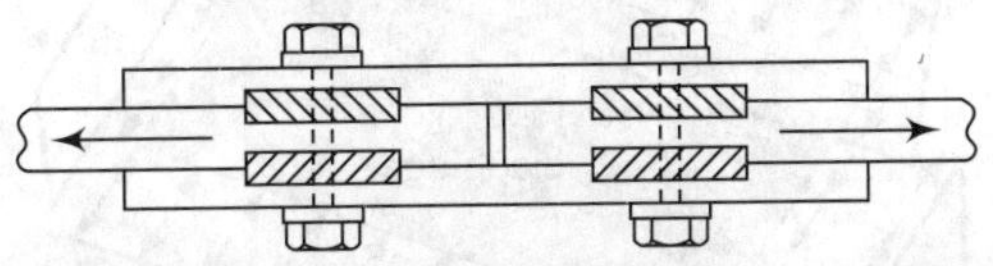

No tilting moment in dowel due to balancing effect [dowels are in shear (no bending, shearing, or tensile stress on bolts)]

(b) Butt joint

Fig. 7.10 Dowel joint and stress distributions

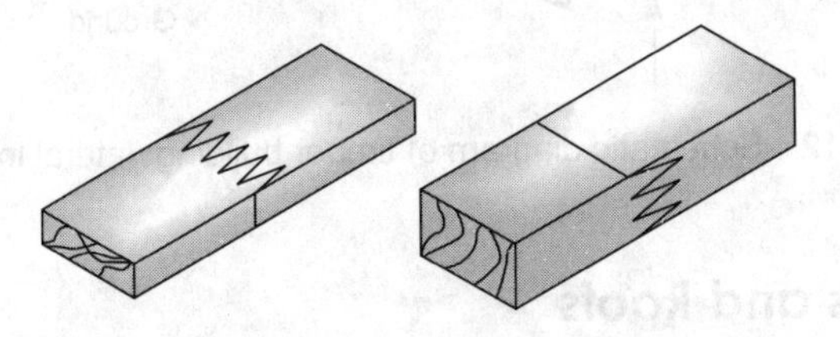

Orientation of finger joints

Fig. 7.11 Typical finger joint geometry

removal of defects, minimizing warping and reducing wastage by avoiding short off-cuts. These joints are produced by cutting profiles in the form of V-shaped grooves to the ends of timber planks or scantlings to be jointed, glueing the interfaces, and then meeting the two ends together under pressure. The figures can be cut edge-to-edge or face-to-face. A timber joist is slightly stronger with edge-to-edge finger joints and a plank is slightly stronger with face-to-face finger joint. Precaution should be taken to glue the surfaces that are on the side-grain rather than on the end-grain and the glue line is stressed in shear rather than in tension.

7.3 Lateral Load Transfer in Timber Buildings

Figure 7.12 shows the free body diagram of a box-like timber building subjected to ground motion, where floor diaphragm is connected with four timber walls. For simply supported diaphragm, the shear force will be transferred to the edges of the supporting walls. Since these walls will be called upon to transfer the shear forces generated due to seismic action, the walls are termed as *shear walls*. The lateral loads due to earthquake are considered to act at floor levels. The floor diaphragm is supported at the ends by shear walls, which in turn transfer the load to foundations. Generally, plywood or plywood boards, or gypsum boards are used as shear resisting panels in shear walls. To achieve the best performance, the panels are staggered. The floor diaphragm, shear walls, and foundations are all connected adequately to each other ensuring a proper lateral load transfer path.

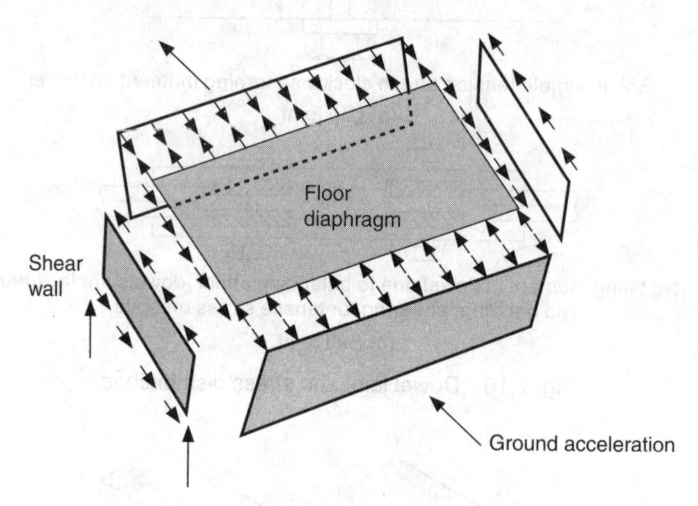

Fig. 7.12 Schematic diagram of timber building: lateral load path

7.4 Floors and Roofs

As discussed in the previous section, floors are used to transfer lateral loads to shear walls. Floors are usually constructed with timber joists connected, by nails or screws, to structural timber panels made of plywood, or plywood board, or particle board as floor-panel material. The floor diaphragm may be assumed to act as a deep I-beam, which is supported by walls that are parallel to the direction of the lateral force in conformation to the condition mentioned in Fig. 7.13. The panels of the diaphragm floor act as web of the I-beam resisting shear and the chords act like flanges resisting compressive and tensile forces caused by the bending moment. The *chord* is either the bottom or top plate of the wall frame or the end joist of the floor running in the direction of the wall and perpendicular to the main floor joists. The bottom plate usually consists of two timber sections nailed together as well as to the floor.

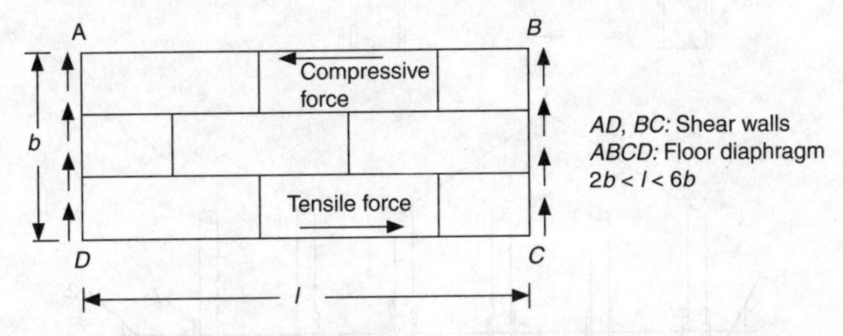

Fig. 7.13 Floor diaphragm acting as deep I-beam

Roofs are the source of a great deal of earthquake damage in timber buildings. Timber frames have often proven to be inadequate for forces caused by heavy roof construction. Roof tiles easily slide down during earthquakes and may injure people. Some of the modifications that can be made to improve seismic resistance of roofs are as follows:

(a) Slates and roofing tiles are brittle and easily dislodge. Wherever possible, they should be replaced with corrugated iron or asbestos sheeting.
(b) False ceilings of brittle materials are dangerous. Non-brittle materials, such as hesian cloth, bamboo matting, or foam may be used.
(c) Roof truss frames should be braced by welding or clamping suitable diagonal bracing members in the vertical as well as horizontal planes.
(d) Anchors of roof trusses to supporting walls should be improved, and the roof thrust on the walls should be minimized by the modification as shown in Figs 7.14 and 7.15.
(e) Where the roof or floor consists of prefabricated units such as rectangular, T, or channel units made of RCC, or wooden poles and joists carrying brick tiles, their integration is necessary. Timber elements could be connected to diagonal planks by being nailed to them and spiked to an all-round wooden frame at the ends. Reinforced concrete elements may either have 40-mm cast in situ concrete topping with 6-mm ϕ bars that are 150 mm c/c both ways, or

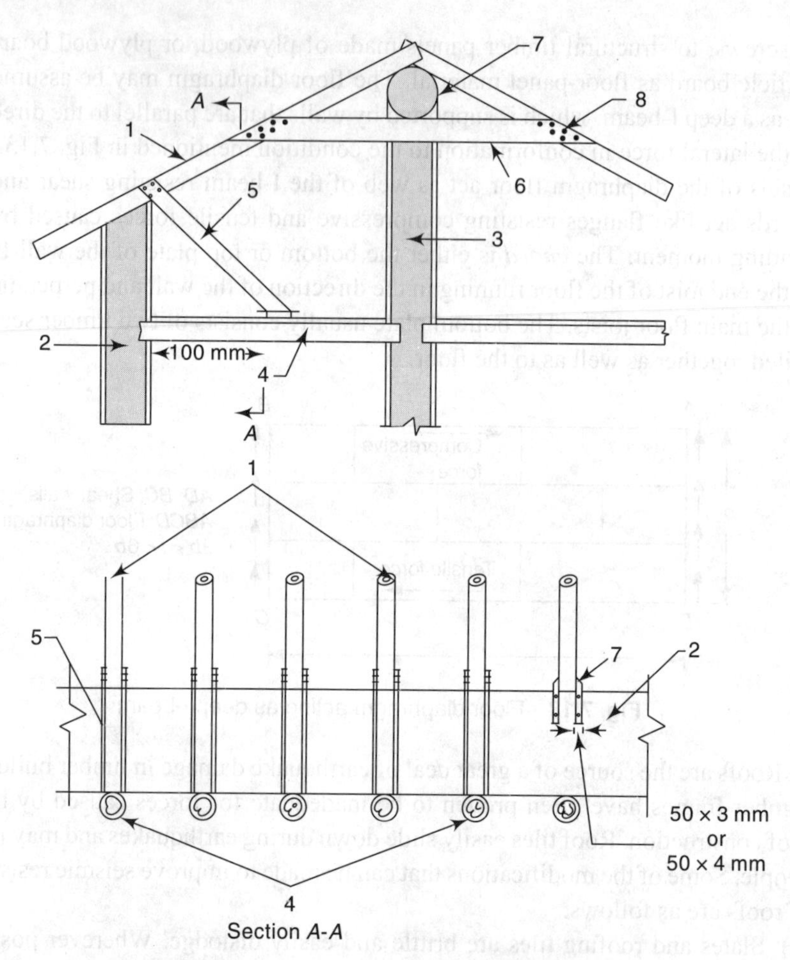

Fig. 7.14 Roof modification to reduce thrust on walls

a horizontal cast in situ RCC ring beam all around, into which the ends of the RCC elements are embedded. Figure 7.16 shows one such detail.

(f) Roofs or floors consisting of steel joists and flat or segmental arches must have horizontal ties that hold the joists horizontally in each arch span, so as to prevent the spreading of joists. If such ties do not exist, these should be installed by welding or clamping.

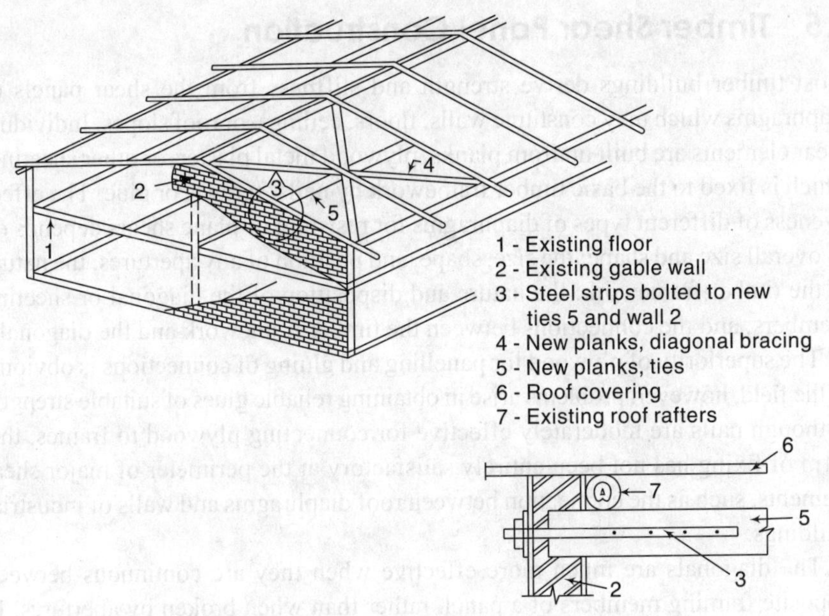

Fig. 7.15 Details of new roof bracing

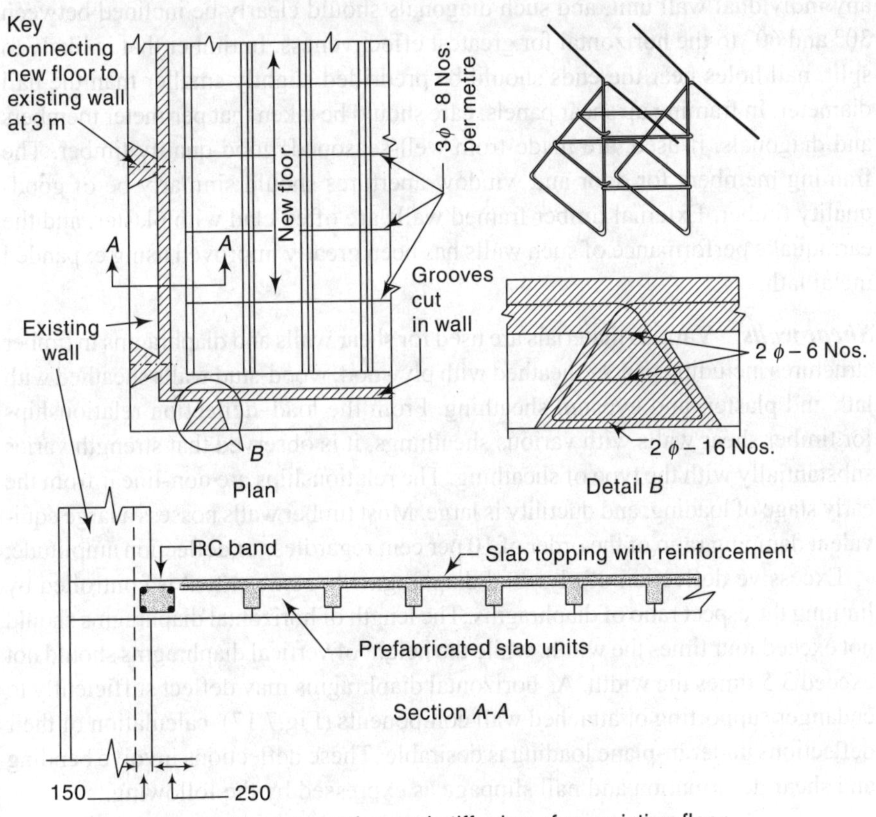

Fig. 7.16 Integration and stiffening of an existing floor

7.5 Timber Shear Panel Construction

Most timber buildings derive strength and stiffness from the shear panels or diaphragms which may constitute walls, floors, ceilings, or roof slopes. Individual shear elements are built up from planks, plywood metal plaster, or other sheeting, which is fixed to the basic timber framework by nails, screws, or glue. The effectiveness of different types of diaphragms for resisting in-plane shears depends on its overall size and shape; the size, shape, and position of any apertures; the nature of the timber framework; the nature and disposition of the diagonal or sheeting members; and the connections between the timber framework and the diagonals.

The superiority of plywood for panelling and gluing of connections is obvious. In the field, however, problems arise in obtaining reliable glues of suitable strength. Although nails are moderately effective for connecting plywood to frames, this form of fixing has not been entirely satisfactory at the perimeter of major shear elements, such as the connection between roof diaphragms and walls of industrial buildings.

The diagonals are much more effective when they are continuous between opposite framing members of a panel, rather than when broken by apertures. In residential buildings it is common for only one or two diagonals to be used within any individual wall unit, and such diagonals should clearly be inclined between 30° and 60° to the horizontal for greatest effectiveness. In timber that is likely to split, nail holes near the ends should be predrilled slightly smaller than the nail diameter. In framing up shear panels, care should be taken that perimeter members and diagonals, if used, are made from well-seasoned, good-quality timber. The framing members for door and window apertures should similarly be of good-quality timber. External timber framed walls are often clad with plaster, and the earthquake performance of such walls has been greatly improved using expanded metal lath.

Shear walls Various materials are used for shear walls and diaphragms in timber structures including panels sheathed with plywood, wood-stud walls sheathed with lath and plaster, and gypsum sheathing. From the load–deflection relationships for timber shear walls with various sheathings, it is observed that strength varies substantially with the type of sheathing. The relationships are non-linear from the early stage of loading, and ductility is large. Most timber walls possess a large equivalent damping ratio of the order of 10 per cent regardless of deflection amplitude.

Excessive deflection of plywood diaphragms to some extent is controlled by limiting the aspect ratio of diaphragms. The length of horizontal diaphragms should not exceed four times the width, while the height of vertical diaphragms should not exceed 3.5 times the width. As horizontal diaphragms may deflect sufficiently to endanger supporting or attached wall components (Fig.7.17), calculation of their deflections under in-plane loading is desirable. These deflections involve bending and shear deformation and nail slippage as expressed by the following.

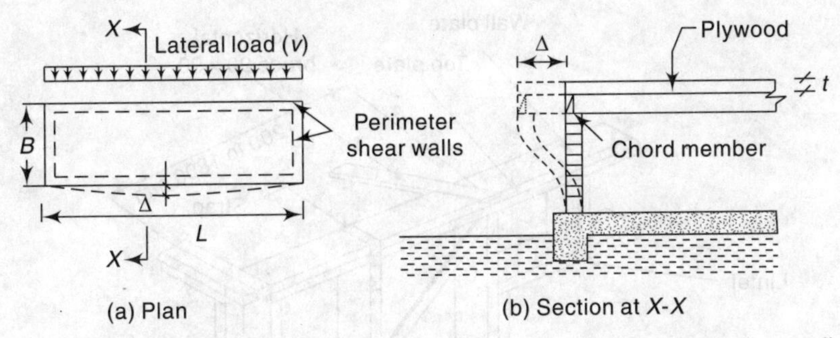

Fig. 7.17 A typical horizontal timber diaphragm—the effect on supporting walls on deflection under lateral loading

$$\Delta = \frac{52vL^3}{EAB} + \frac{vL}{4Gt} + 0.308Le_n \tag{7.1}$$

where Δ is the deflection (mm), v the applied shear loading (N/m), L the length of diaphragm (m), B the width of diaphragm (m), A the cross-sectional area of the chord (mm^2), E the modulus of elasticity of chords (N/mm^2), G the shear modulus of plywood (N/mm^2), t the thickness of plywood (mm), and e_n the nail deformation (mm).

7.6 Stud-wall Construction

A stud-wall construction consists of timber studs and corner posts framed into sills, top plates, and wall plates. Horizontal struts and diagonal braces are used to stiffen the frame against lateral loads. The wall covering may consist of matting made from bamboo reeds and timber boarding, or the like. Typical details of stud walls, with and without openings, are shown in Figs 7.18 and 7.19, respectively. If wooden sheathing boards used are properly nailed to the timber frame, the diagonal bracing may be omitted. Minimum sizes and spacing of various members should be as follows:

(a) The timber studs for use in load-bearing walls should have a minimum finished size of 40 mm × 90 mm and their spacing should not exceed those given in Table 7.3. For non-load-bearing walls the timber studs should not be less than 40 mm × 70 mm in finished cross-section. Their spacing should not exceed 1 m.

(b) There should be at least one diagonal brace for every 1.6 m × 1 m area of load-bearing walls. Their minimum finished sizes should be in accordance with Table 7.4.

(c) The horizontal struts should be spaced not more than 1 m apart. They must have a minimum size of 30 mm × 40 mm for all locations.

(d) The finished sizes of the sill, the wall plate, and the top plate should not be less than the size of the studs used in the wall.

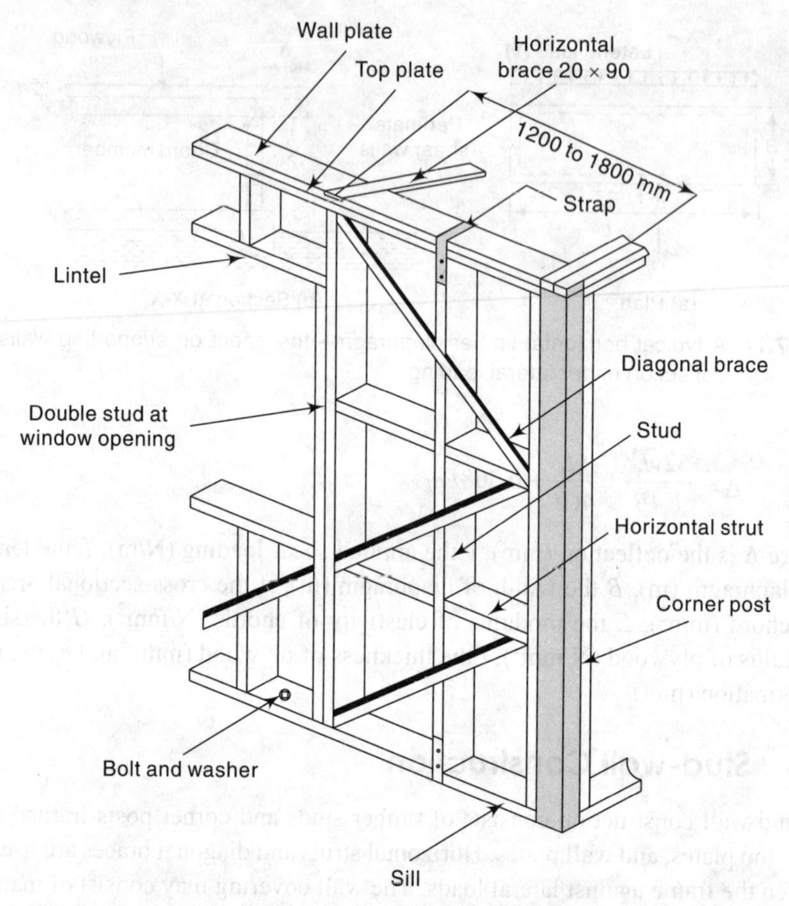

Fig. 7.18 Timber framing in stud-wall construction with opening in wall

Group of timber (Grade I*)	Single-storey or first floor of double-storey buildings		Ground floor of double-storey buildings	
	Exterior wall (cm)	Interior wall (cm)	Exterior wall (cm)	Interior wall (cm)
A and B	100	80	50	40
C	100	100	50	50

*Grade I timber as defined in Table 5 of IS 883: 1992

(e) The corner posts should consist of three timber pieces, two being equal in size to the studs used in the walls meeting at the corner and the third timber being of a size to fit so as to make a rectangular section (Fig. 7.19).

(f) The diagonal braces should be connected at their ends with the stud-wall members by means of wire nails having 6 gauge (4.88 mm diameter) and 10 cm length. Their minimum number should be four nails for 20 mm × 40 mm braces and six nails for 30 mm × 40 mm braces. The far end of the nails may be clutched as far as possible.

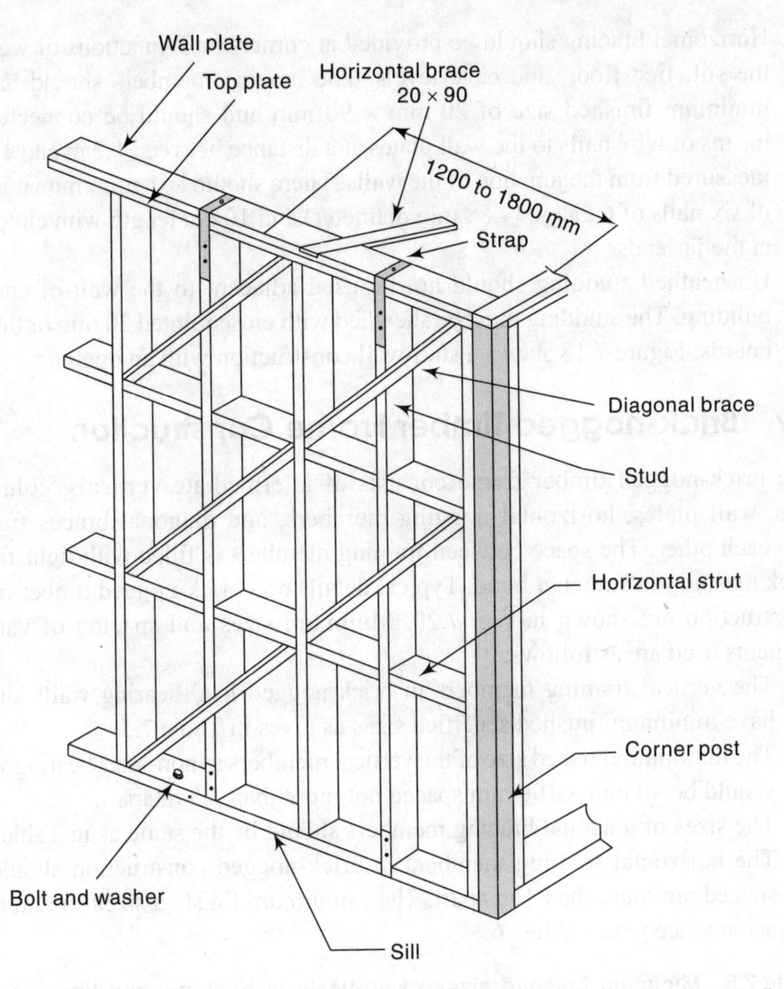

Fig. 7.19 Timber framing in stud-wall construction without opening in wall

Table 7.4 Minimum finished sizes of diagonal braces

Building category	Group of timber (Grade I*)	Single-storey or first floor of double-storey buildings		Ground floor of double-storey buildings	
		Exterior wall (mm × mm)	Interior wall (mm × mm)	Exterior wall (mm × mm)	Interior wall (mm × mm)
A, B, C	All	20 × 40	20 × 40	20 × 40	20 × 40
D and E	A and B	20 × 40	20 × 40	20 × 40	30 × 40
C	C	20 × 40	30 × 40	30 × 40	30 × 40

*Grade I timber as defined in Table 5 of IS 883: 1992

(g) Horizontal bracing should be provided at corners or T-junctions of walls at the sill, first floor, and eave levels. The bracing members should have a minimum finished size of 20 mm × 90 mm and should be connected by means of wire nails to the wall plates at a distance between 1.2 m and 1.8 m, measured from the junction of the walls. There should be a minimum number of six nails of 6 gauge (4.88 mm diameter) and 10 cm length with clutching at the far ends.

(h) Unsheathed studding should not be used adjacent to the wall of another building. The studding must be sheathed with close jointed 20 mm or thicker boards. Figure 7.18 shows a stud-wall construction with an opening.

7.7 Brick-nogged Timber Frame Construction

The brick-nogged timber frame consists of intermediate verticals, columns, sills, wall plates, horizontal nogging members, and diagonal braces framed into each other. The space between framing members is filled with tight-fitting brick masonry in stretcher bond. Typical details of a brick-nogged timber frame construction are shown in Fig. 7.20. Minimum sizes and spacing of various elements used are as follows:

(a) The vertical framing members in brick-nogged load-bearing walls should have minimum finished specified sizes as gives in Table 7.5.

(b) The minimum finished size of the vertical members in non-load-bearing walls should be 40 mm × 100 mm spaced not more than 1.5 m apart.

(c) The sizes of diagonal bracing members should be the same as in Table 7.4. The horizontal framing members in brick-nogged construction should be spaced not more than 1 m apart. Their minimum finished sizes should be in accordance with Table 7.6.

Table 7.5 Minimum finished sizes of verticals in brick-nogged timber frame construction

Spacing (m)	Group of timber (Grade I*)	Single-storey or first floor of double-storey buildings		Ground floor of double-storey buildings	
		Exterior wall (mm × mm)	Interior wall (mm × mm)	Exterior wall (mm × mm)	Interior wall (mm × mm)
1	A, B	50 × 100	50 × 100	50 × 100	50 × 100
	C	50 × 100	70 × 100	70 × 100	90 × 100
1.5	A, B	50 × 100	70 × 100	70 × 100	80 × 100
	C	70 × 100	80 × 100	80 × 100	100 × 100

*Grade I timber as defined in Table 5 of IS 883: 1992

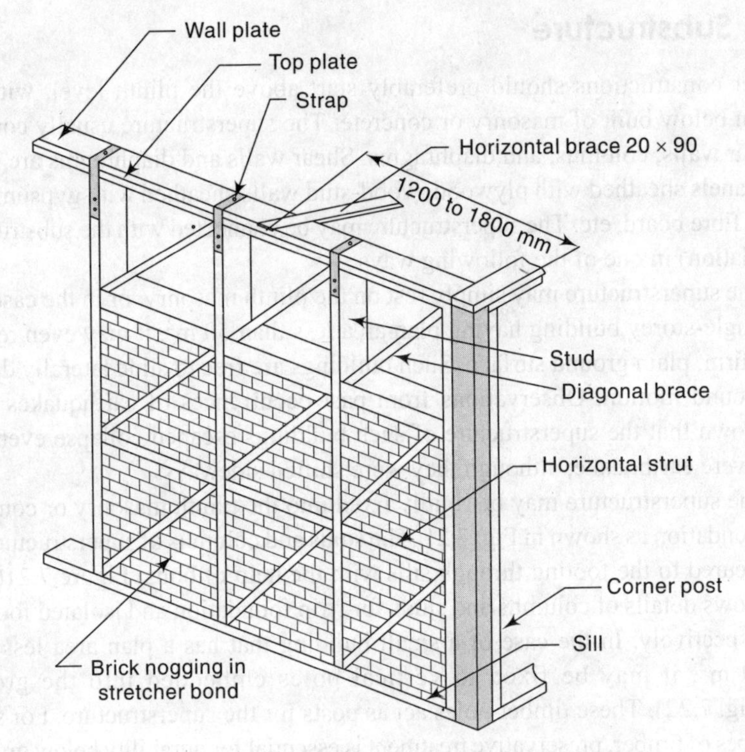

Fig. 7.20 Brick-nogged timber frame construction

Table 7.6 Minimum finished sizes of horizontal nogging members

Spacing of verticals (m)	Size (mm × mm)
1.5	70 × 100
1	50 × 100
0.5	25 × 100

(d) The finished sizes of the sill, wall plate, and top plate should not be less than the size of the vertical members used in the wall.

(e) Corner posts should consist of three vertical timbers.

(f) The diagonal braces should be connected at their ends with the other members of the wall by means of wire nails.

(g) Horizontal bracing members at corners or T-junctions of the wall should be as specified for stud-wall construction.

7.8 Substructure

Timber constructions should preferably start above the plinth level, with the portion below built of masonry or concrete. The superstructure usually consists of shear walls, columns, and diaphragms. Shear walls and diaphragms are made with panels sheathed with plywood, wood-stud walls sheathed with gypsum wall board, fibre board, etc. The superstructure may be connected with the substructure (foundation) in one of the following ways:

(a) The superstructure may simply rest on the plinth masonry, or in the case of a single-storey building having plan area less than 50 m^2, it may even rest on a firm, plain ground surface. Such buildings are free to slide laterally during ground motion. Observations from past occurrences of earthquakes have shown that the superstructure of such buildings escaped collapse even in a severe earthquake, although they were shifted sideways.

(b) The superstructure may be rigidly fixed into the plinth masonry or concrete foundation as shown in Fig. 7.21. The studs and columns of superstructure are secured to the footing through sills with the help of bolts. Figure 7.21(a, b) shows details of columns and studs for strip foundation and isolated footing, respectively. In the case of a small building that has a plan area less than 50 m^2, it may be fixed to vertical poles embedded into the ground (Fig. 7.22). These timber poles act as posts for the superstructure. For some types of timber, preservative treatment is essential for durability below ground. The embedded portion of the poles is painted with tar to protect the timber from white ants. In each case, the building is likely to move along with its

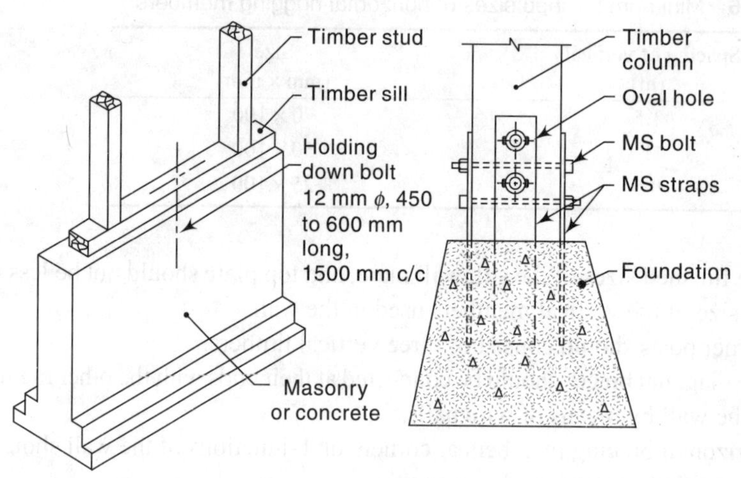

(a) Suitable for strip foundations (b) Suitable for isolating column footings

All dimensions in mm

Fig. 7.21 Typical column to foundation connections

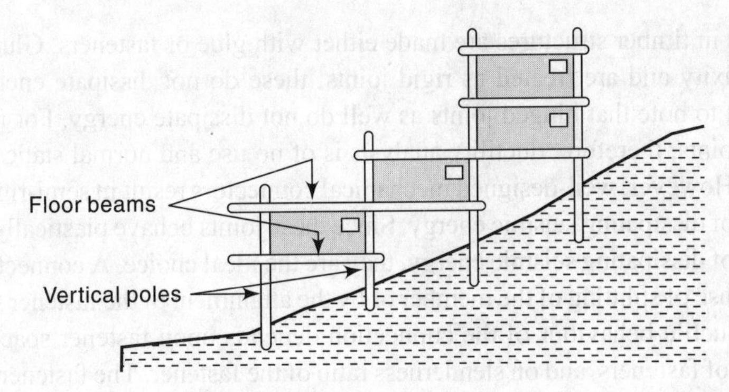

Fig. 7.22 Pole frame apartments

foundation. Therefore, the superstructure should be designed to carry the resulting earthquake shears.

7.9 Site Response

Damage to timber building structures is considerably influenced by the ground condition on which they stand. Timber buildings suffer more earthquake damage when located on soft ground rather than on hard ground. The reasons for this occurrence are uncertain. For instance, the possible role of resonance is obscure. As most one- and two-storey timber buildings have fundamental periods of vibration in the range 0.1 s–0.6 s, resonance with the ground seems more likely on thin layers rather than on thick layers of soft ground. After heavy shaking a timber building loosens at the joints and its natural periods are likely to lengthen, but the manner of its vibration is uncertain and it is unlikely to have well-defined modes in which resonance can occur.

It is possible that timber houses on soft ground are weakened by seasonal ground movements, making them more vulnerable to earthquakes. Also, the differential earthquake-ground movements in softer soils are larger than in firm soils, and this is likely to affect timber buildings with light foundations more than buildings with stiffer construction. If timber buildings are to be built on soft ground in a seismic area, extra measures should be taken to ensure structural integrity, particularly at the foundation level.

7.10 Ductile Behaviour of Joints

Under cyclic loading, wood, in general behaves linearly and elastically. It has low energy-dissipation capacity since it does not deform plastically, and therefore possesses no plastic strain. Because of natural defects in timber such as knots, it fails in a brittle manner. Therefore, performance of timber structures during earthquakes is dependent upon behaviour of joints.

Joints in timber structures are made either with glue or fasteners. Glued joints impart fixity and are treated as rigid joints; these do not dissipate energy. It is pertinent to note that hinged joints as well do not dissipate energy. For rigid and hinged joints, therefore, ductility analysis is of no use and normal static analysis is done. However, well-designed mechanical connectors result in semi-rigid joints capable of dissipating seismic energy. Since these joints behave plastically and are capable of dissipating seismic energy, they are the ideal choice. A connection may fail because of splitting of the member or by the attainment of the fastener strength.

The ductile behaviour of the connection depends upon fastener spacing, end distance of fasteners, and on slenderness ratio of the fastener. The fastener spacing and end distance check the splitting of wood and its brittle failure, and therefore allow the structural members to display ductile behaviour. Splitting may also be checked by reinforcing the connection areas with plywood panels, consequently enhancing the tensile strength in the direction perpendicular to the wood grains. Besides this, the reinforcement ensures yielding of the connectors and increases connection ductility.

Fasteners with high slenderness ratios are preferred since they dissipate more seismic energy than those with low slenderness ratios. It is so because for fasteners with large slenderness ratios, the plastic yield points are formed in the fasteners, while those with low slenderness ratios behave more elastically and do not dissipate much seismic energy. Further, to achieve ductility, joints in timber buildings are so detailed that pull-out failure of fastener and brittle failure of member do not take place.

7.11 Fire Resistance

The capacity of the members of a building to carry the design loads and enclosing the premises under the conditions of fire is known as *fire resistance*. The danger from post-earthquake fires, resulting from electrical short-circuiting and kitchen fires, is immense and timber constructions are particularly vulnerable in this respect. Timber construction is fire resistant with the following limits:

1. Up to the failure of the load-carrying structural components.
2. Up to the formation of openings in partitions, walls, doors, etc.
3. Until a temperature of 150°C is reached.

The fire resistance of timber can be enhanced either by impregnating it with chemicals or by designing it to provide slow-burning construction. Timber of substantial dimension offers high resistance to fire. Heavy timber on burning forms a protective coating of charcoal, which retards the penetration of heat to the interior and delays the ignition.

Pole frame structures have a relatively low fire risk for timber construction. Poles are difficult to ignite because of two reasons—firstly, poles have a large volume-to-surface area ratio and a smooth exterior and secondly, poles have wide spacing and fire cannot spread easily from one structural member to another. The loss of strength due to surface charring will not generally be critical.

Covering wooden members with heat-insulating layers (stucco or plaster) raises the limit of fire resistance. Most fire retarding chemicals are considered to reduce the strength of timber and, therefore, design stresses must be reduced by 10 per cent.

The following structural measures should be taken to protect timber structures from fire.

- Fire-precaution breaks should be left between adjacent timber buildings at the planning and layout stage itself. Large-size buildings should be divided into sections by fire-proof walls and noncombustible zones.
- Load-carrying open timber structures should be designed with massive, solid members as these will take time to ignite.
- Empty spaces should be isolated so that the forced draughts are not developed.
- Roofs are the most dangerous from the point of view of life safety. Accordingly, the most important preventive measure is to change over to roofing of corrugated asbestos cement sheets or tiles.

7.12 Decay

The destruction of wood owing to the activities of fungi is known as *decay*. The development of the process of decay starts in wood with moisture content not less than 18–20 per cent in the presence of air and a temperature between 5°C and 45°C. In wood with a very high moisture content, the fungi develop slowly. Under water, the decay does not occur at all, owing to the absence of free air. Of the known species of fungi groups—moulds and forest, cellar, and house fungi—only moulds actually do not reduce the mechanical strength of wood. It is house fungi and some cellar fungi that destroy the basic skeleton of the timber—the cellulose—and initiate destructive decay. This is characterized by the appearance of transverse as well as longitudinal cracks on the infected surfaces. Timber decay can be controlled by high-temperature seasoning and by the protection of the seasoned material against dampness during storage, transport, erection of the building, and during its working life. Further, the timber to be used in structures should be subjected to an effective process of preservation to prevent decay. The building structure may collapse even in the smallest-size earthquake, if proper measures to protect timber from decay are not adopted.

7.13 Permissible Stresses

There are large varieties of timber in use. Therefore, it is not practicable to present the strength properties of all of them. These properties will depend on—wood species; direction of loading relative to wood grain; defects like knots, checks, cracks; the moisture content; condition of the timber structure; and location of use (inside protected, outside alternate wetting and drying). The permissible stresses of timbers, placed in three groups A, B, and C for different locations applicable to Grade I structural timber, are given in Table 7.7 provided that the following conditions are satisfied.

(a) The timber should be of high or moderate durability and be given suitable treatment where necessary.
(b) Timber of low durability should be used after proper preservative treatment.
(c) The load should be continuous and permanent and not of impact type.

Modification Factors for Permissible Stresses

The permissible stresses obtained from Table 7.7 may have to be modified due to change of slope of grain or due to change in duration of loading.

When the timber has not been graded and has major defects such as slope of grain, shakes, knots, etc., the permissible stresses of Table 7.7 are reduced by using the modification factors of Table 7.8.

Table 7.7 Basic permissible stresses for timber group*

Types of stress	Location	Permissible stress (MPa)		
		Group A	Group B	Group C
(i) Bending and tension along grain	inside	18	12	8
	outside	15	10	7
	wet	12	8	6
(ii) Shear in beams	all	1.2	0.9	0.6
Shear along grains	all	1.7	1.3	0.9
(iii) Compression parallel to grain	inside	12	7	6
	outside	11	6	6
	wet	9	6	5
(iv) Compression perpendicular to grain	inside	6	2.2	2.2
	outside	5	1.8	1.7
	wet	4	1.5	1.4

*Based on IS 883.

Notes: 1. Groups A, B, and C are classified according to Young's modulus of elasticity as follows:
Group A: more than 12,600 MPa
Group B: 9,800 to 12,600 MPa
Group C: 5,600 to 9,800 MPa

2. Permissible stresses given in the table should be multiplied by the following factors to obtain the permissible stresses for other grades:
(a) For Selected Grade Timber 1.16
(b) For Grade II Timber 0.84

3. For low durability timbers used on outside locations, the permissible stresses for all grades of timber should be multiplied by 0.8.

Table 7.8 Modification factor to allow for slope of grain

Slope	Modification factor k_1	
	Strength of beams, ties	Strength of posts, columns
1 in 10	0.80	0.74
1 in 12	0.90	0.82
1 in 14	0.98	0.87
1 in 15 and flatter	1.00	1.00

For different durations of design load, the permissible stresses given in Table 7.7 are enhanced by multiplying the basic permissible stress with the modification factors of Table 7.9.

Table 7.9 Modification factor to allow for duration of loading

Duration of loading	Modification factor
Continuous (normal)	1.00
Two months	1.15
Seven days	1.25
Wind or earthquake	1.33
Instantaneous or impact	2.00

7.14 Restoration and Strengthening

Since wood is highly workable, it will be easy to restore the strength of wooden members such as beams, columns, struts, and ties by splicing additional material. The weathered or rotten wood should be removed before restoration and strengthening. Nails, screws, or steel bolts will be most convenient as connectors. It will be advisable to use steel straps to cover all such splices and joints so as to keep them tight and stiff.

7.14.1 Strengthening of Slabs

Strengthening is the most common seismic improvement strategy for buildings with inadequate lateral-force-resisting systems. For timber buildings, typical systems employed for strengthening diaphragm are—increasing the existing nailing in the sheathing, replacing the sheathing with stronger material, or overlaying the existing sheathing with plywood. Some of the methods to enhance the strength of timber diaphragms are given below.

Insertion of a new slab A rigid slab inserted into existing walls plays an important role in the resisting mechanism of the building, in keeping the walls together, and in distributing seismic forces among the walls. The slab has to be properly connected to the walls through appropriate keys. Figure 7.23 shows a typical arrangement for integration and stiffening of an existing floor and some details of an inserted slab. A small RC band is inserted into the existing wall. The band is keyed at least every 3 m.

Existing wooden slabs In case existing slab is not removed, the following procedure is undertaken—the slab is stiffened either by planks nailed perpendicularly to the existing ones (Fig. 7.24) or by placing a reinforced concrete thin slab over the old one (Fig. 7.25). In this case, a steel network is nailed to the wooden slab and connected to the walls by a number of distributed steel anchors. These can be hammered into the interstices of the wall and a local hand-cement grouting is applied for sealing.

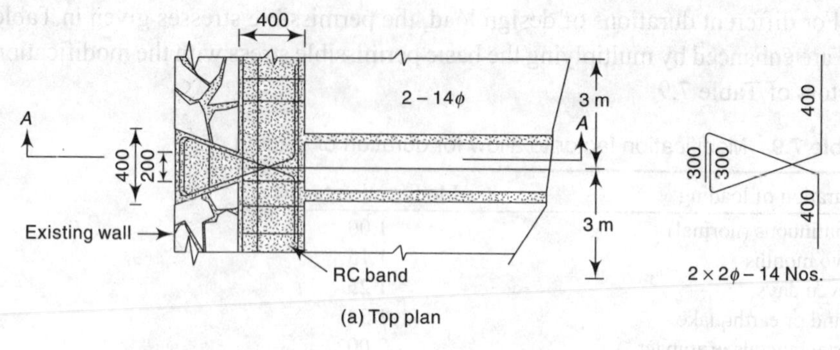

(a) Top plan

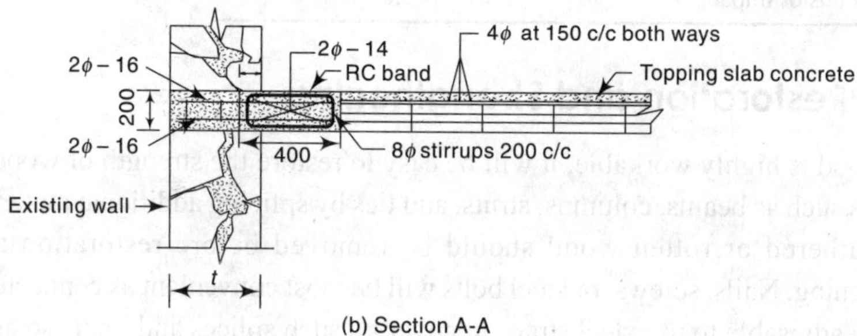

(b) Section A-A

All dimensions are in mm unless noted otherwise

Fig. 7.23 Detail of inserted slab

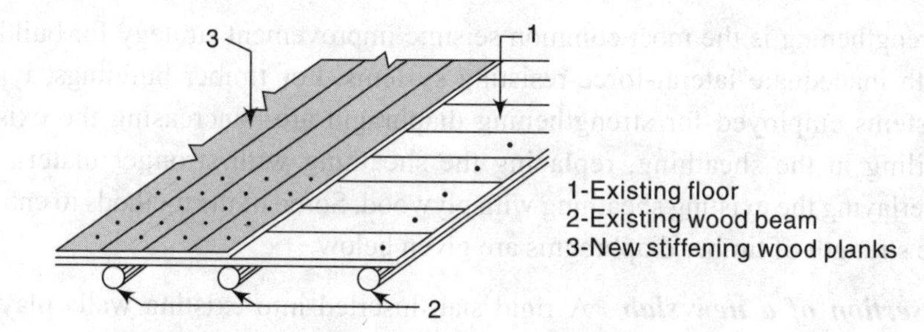

Fig. 7.24 Stiffening of wooden floor by wooden planks

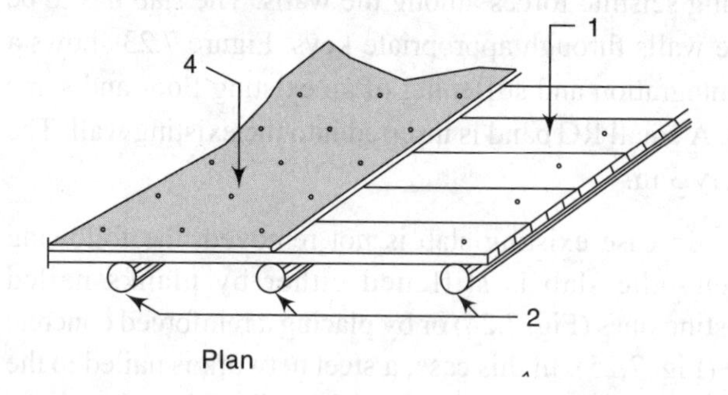

Fig. 7.25 Stiffening of wooden floor using RC slab and a connection to the wall

(*Contd*)

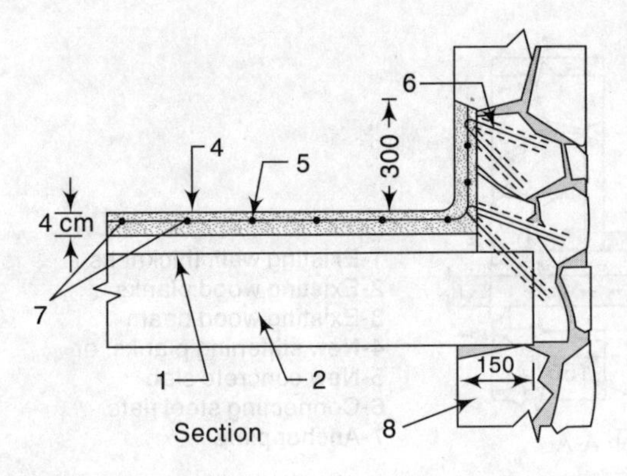

Fig. 7.25 Stiffening of wooden floor using RC slab and a connection to the wall

Connection of slab to walls A proper link can be obtained by means of the devices shown in Figs 7.26 and 7.27. They consist of flat steel bars nailed to the wooden supporting beams and to the wooden slab. Holes drilled in the walls to anchor them are infilled with cement.

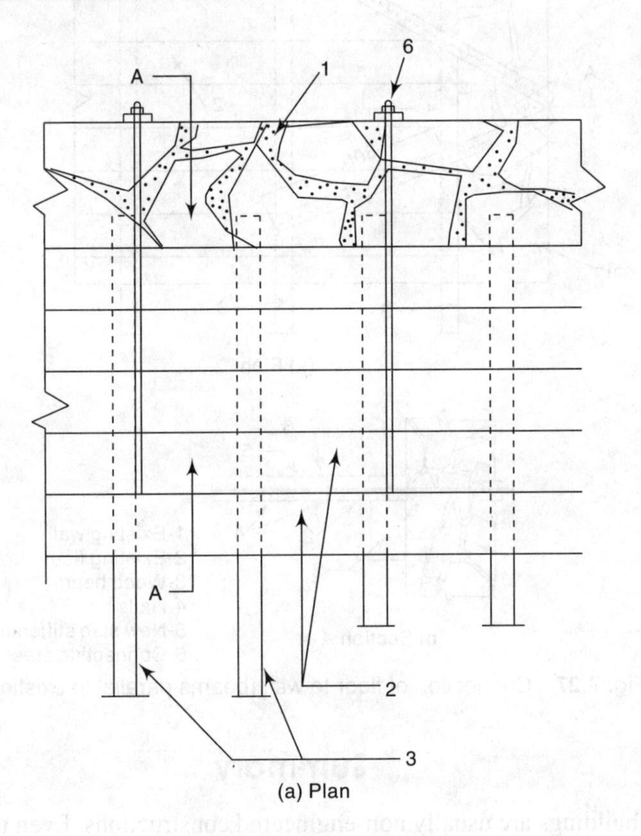

Fig. 7.26 (*Contd*)

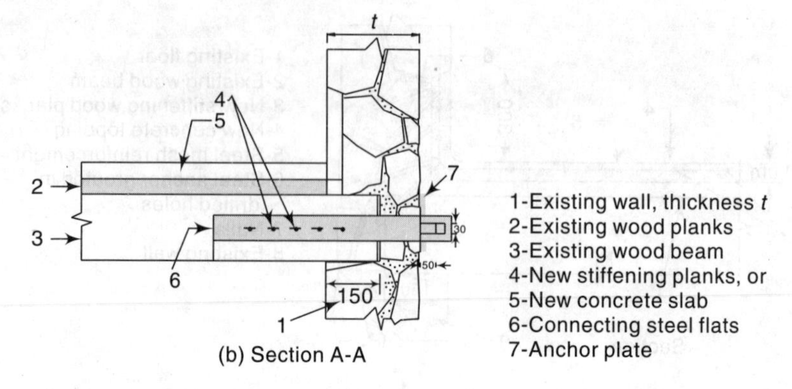

Fig. 7.26 Connection of floor to wall (beams orthogonal to walls)

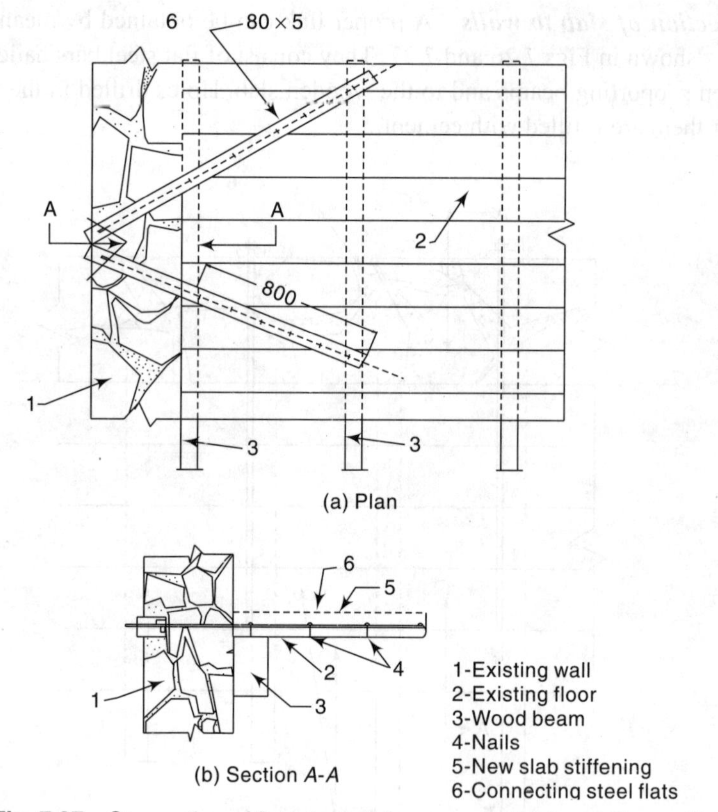

Fig. 7.27 Connection of floor to wall (beams parallel to existing wall)

Summary

Timber buildings are usually non-engineered constructions. Even though timber has a high strength-to-weight ratio, making it the most suitable material for earthquake-resistant construction, timber buildings display poor performance because of brittle failure, low capability of energy dissipation, inadequate connections, lack of integrity of substructure, and affinity to fire and decay.

Ways to check brittle failures of timber structural elements and enhancing energy-dissipation capacity are described. Semi-rigid connections with slender fasteners make an ideal choice since these impart ductility. All the structural elements should be adequately tied together to provide a lateral load path for transferring loads to the foundations. In seismic design of timber buildings, fire safety should be given due consideration, as fires may occur during an earthquake due to short-circuiting of electric wires, cutting of gas pipes, etc. Since timber is susceptible to decay under conducive environment and may lose strength, proper treatment should be affected to the timber and structural elements from time to time as and when needed.

The chapter describes the types of timber building systems briefly. The restoration and strengthening of earthquake-affected buildings is also discussed.

Exercises

7.1 Why is timber supposed to be one of the best materials for construction of earthquake-resistant buildings? What are its limitations?

7.2 With regards to the inadequate performance of timber buildings in earthquake-prone areas, discuss the following.
(a) Structural connections
(b) Post-earthquake fires
(c) Roofs
(d) Site response

7.3 Describe the following briefly with neat sketches.
(a) Timber shear wall construction
(b) Stud-wall construction

7.4 Describe the construction procedure and precautions to be exercised for brick-nogged timber frame constructions. Draw neat diagrams to support your answer.

7.5 Write notes on
(a) Timber shear walls
(b) Lessons learned from failures of timber buildings

7.6 Structural elements of timber are brittle in nature. Comment. How ductility can be introduced in timber structures?

7.7 Which structural element of timber structures is most affected by earthquakes? Explain methods to restore and strengthen it.

CHAPTER 8

Reinforced Concrete Buildings

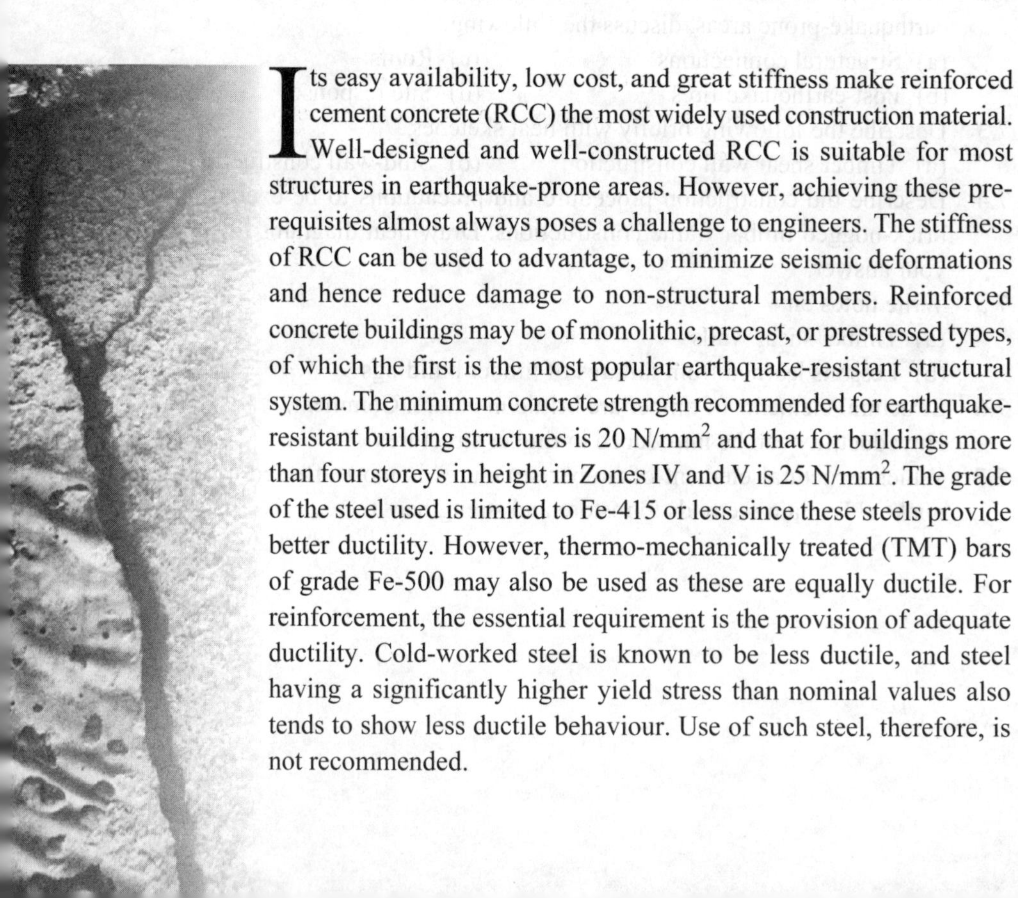

Its easy availability, low cost, and great stiffness make reinforced cement concrete (RCC) the most widely used construction material. Well-designed and well-constructed RCC is suitable for most structures in earthquake-prone areas. However, achieving these pre-requisites almost always poses a challenge to engineers. The stiffness of RCC can be used to advantage, to minimize seismic deformations and hence reduce damage to non-structural members. Reinforced concrete buildings may be of monolithic, precast, or prestressed types, of which the first is the most popular earthquake-resistant structural system. The minimum concrete strength recommended for earthquake-resistant building structures is 20 N/mm^2 and that for buildings more than four storeys in height in Zones IV and V is 25 N/mm^2. The grade of the steel used is limited to Fe-415 or less since these steels provide better ductility. However, thermo-mechanically treated (TMT) bars of grade Fe-500 may also be used as these are equally ductile. For reinforcement, the essential requirement is the provision of adequate ductility. Cold-worked steel is known to be less ductile, and steel having a significantly higher yield stress than nominal values also tends to show less ductile behaviour. Use of such steel, therefore, is not recommended.

Orthodox criteria for design of reinforced concrete (RC) members are almost exclusively concerned with strength, while ductility and energy absorption receive little consideration. However, the most severe likely earthquake can only be survived if the members are sufficiently ductile to absorb and dissipate seismic energy by inelastic deformations. For RC members to have adequate strength and ductility to withstand earthquakes, their design and detailing must conform to IS 456: 2000 and IS 13920: 2002. Good detailing enables even an ill-conceived structural form to survive a strong earthquake. The detailing provisions in IS 13920: 2002 provide RC members with adequate toughness and ductility and make them capable of undergoing extensive inelastic deformations and dissipating seismic energy in a stable manner. The purpose of this chapter is to describe in detail how the frame members and shear walls can be designed to maintain both strength and ductility, so as to achieve the desired objectives in the earthquake-resistant design of RC buildings. Characteristics and behaviour of precast concrete and prestressed concrete constructions are introduced at the end of the chapter.

8.1 Damage to RC Buildings

Although damage caused by earthquakes has decreased owing to improvements of earthquake design codes, the poor form of the buildings and the failure of beams, columns, shear walls, and joints are still potential causes of damage. Typical damage to reinforced elements includes cracking in the tension zone, diagonal cracking in the core and loss of concrete cover, concrete core crushing by reversal, diagonal cracking, stirrups bursting outwards, and buckling of main reinforcement.

It is evident that the root cause of failure, of even well-reinforced concrete members, is the cracking of concrete, which leads to degradation in the cracked zone. The reinforcement elongates permanently in the crack and the tensile stress drops. Consequently, the cracks do not close and the interlocking of aggregate is destroyed. In hinge and joint zones, the reversal cracking breaks down the concrete between the cracks completely, and sliding shear failure (slip of reinforcement) occurs (Fig. 8.1). The stiffness of the concrete members is reduced, and, consequently, their energy absorption capacity may reduce drastically. In addition, the following forms of failure occur in RC buildings.

(a) Bond failure
(b) Direct shear failure of short elements
(c) Shear cracking in the beam–column intersection zone.

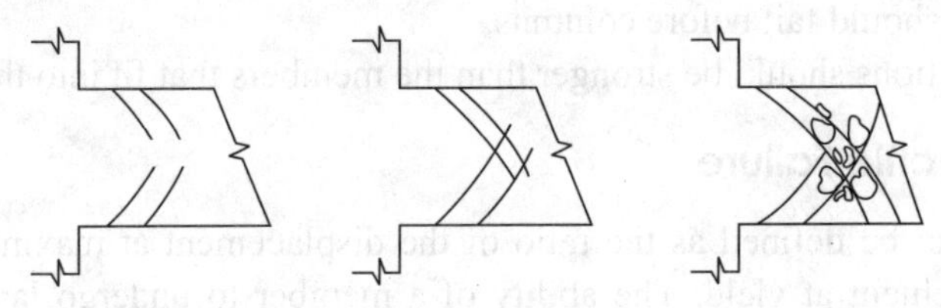

Fig. 8.1 Progressive failure of RC hinge zone under seismic loading

(d) Diagonal cracking of shear walls
(e) Tearing of slabs at discontinuities, and junctions with stiff vertical elements.

Most of the harmful effects of seismic loading can be prevented by the use of closely spaced ties or stirrups in zones of plastic hinging. However, there is a need to learn from earthquake-damage studies and to apply good engineering sense. An RC structure must adhere to the following provisions:

(a) All frame elements must be detailed so that they can respond to strong earthquakes in a ductile manner. Elements which are necessarily incapable of ductile behaviour must be designed to remain elastic at extreme load conditions.
(b) Non-ductile modes such as shear and bond failures must be avoided. This implies that the anchorage and splicing of bars should not be done in areas of high stress, and a high resistance to shear should be provided.
(c) Rigid elements should be attached to the structure with ductile or flexible fixings.
(d) A high degree of structural redundancy should be provided so that as many zones of energy-absorbing ductility as possible are developed before a failure mechanism is created. For framed structures this means that the yielding should occur first in the beams. Failure in columns should be avoided; they should remain elastic at the maximum design earthquake level.
(e) Joints should be provided at discontinuities, with adequate provision for movement, so that pounding of the two faces against each other is avoided.

8.2 Principles of Earthquake-resistant Design of RC Members

The lateral loads used in seismic design are highly unpredictable. In case of strong earthquakes, the magnitudes of lateral loads experienced by buildings are so large that an elastic design under these loads yields very large size of members. Thus, the structural members cost so much more that the risk of buildings going beyond their elastic limit is accepted with the stipulation that they do not fall down or collapse. The collapse of RC buildings is generally preventable if the following principles of earthquake-resistant design are observed.

(a) Failure should be ductile rather than brittle—ductility with large energy-dissipation capacity (with less deterioration in stiffness) must be ensured.
(b) Flexure failure should precede shear failure.
(c) Beams should fail before columns.
(d) Connections should be stronger than the members that fit into them.

8.2.1 Ductile Failure

Ductility can be defined as the ratio of the displacement at maximum load to the displacement at yield. The ability of a member to undergo large inelastic deformations with little decrease in strength is called *ductile behaviour*. The available ductility of a member increases with an increase in compression steel content, concrete compressive strength, and ultimate concrete strain. However, it decreases with an increase in tension steel content, steel yield strength, and axial load.

Reinforced concrete buildings that are properly designed in accordance with IS 456: 2000 and detailed as per IS 13920: 2002, have the desired strength and ductility to resist major earthquakes. The important points to which attention must be paid to achieve ductility are as follows.

(a) The ultimate concrete strain increases by confining concrete with stirrups or spiral reinforcement. The confining reinforcement further increases the shear resistance and provides additional lateral support to the main reinforcement. It also makes the strength in shear greater than the ultimate strength in flexure. Figure 8.2 illustrates the effects of axial load and confinement on rotational ductile capacity.

(b) Limitations on the amount of tensile reinforcement or the use of compression reinforcement increase energy-absorbing capacity.

(c) Use of confinement by hoops or spirals at critical sections of stress concentration, such as column–beam connections, increase the ductility of columns under combined axial load and bending.

(d) Special attention must be given to details, such as splices in reinforcement and the avoidance of planes of weakness that might be caused by bending or terminating all bars at the same section.

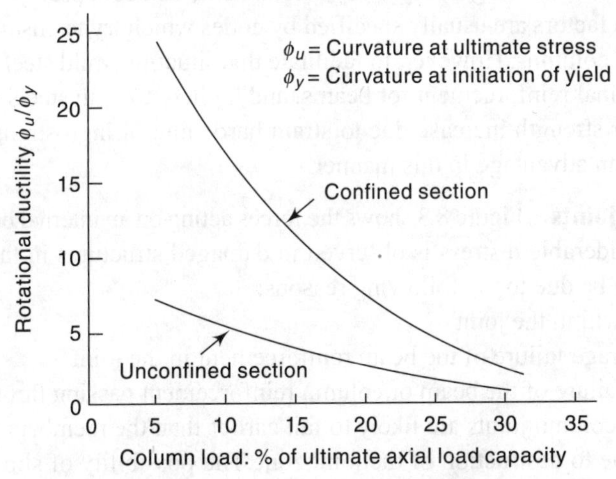

Fig. 8.2 Variation in rotational ductility for tied columns

If the designer observes the above points, it is expected that the building can withstand even strong earthquakes with little or no structural damage.

Flexural failure The load–deflection characteristics of earthquake-resistant structures are mainly dependent on the moment–curvature relationships of the sections. When the tension steel content is low and/or the compression steel content is high, the tension steel reaches the yield strength and then a large increase in curvature can occur at near constant bending moment. This type of failure is known as *tension failure*. Conversely, with high content of tension steel and low content of compression steel, the tension steel does not yield and section fails in a brittle manner if the concrete is unconfined. This is known as *compression failure*. This implies that the beams should be proportioned so as to exhibit the ductile characteristics of a tension failure. Further, to prevent shear failure occurring before bending failure, the design should be such that the flexural reinforcement in a member yields, while the shear reinforcement is at a stress less than yield. In beams, a conservative approach to ensure safety in shear is to make the shear strength equal to the maximum shear demand.

Weak-beam–strong-column design Structures should be proportioned to yield in locations most capable of sustaining inelastic deformations. Observations of failure due to yielding in columns have led to the formulation of the weak-beam–trong-column design, in which column strengths are made at least equal to beam strengths. The intended result is columns that form a stiff, unyielding spine over the height of the building, with inelastic action limited largely to beams. In RC frame buildings, attempts should be made especially to minimize yielding in columns, because of the difficulty of detailing for ductile response in the presence of high axial loads, and the possibility that column yielding may result in the formation of demanding storey-sway mechanisms and collapse.

Strength factors are usually specified by codes which try to ensure that beams fail prior to columns. However, to facilitate this situation, mild steel may be used as longitudinal reinforcement for beams, and higher-strength steels for columns. The greater strength increase due to strain hardening of high-strength steel can be used to an advantage in this manner.

Failure of joints Figure 8.3 shows the forces acting on an internal beam–column joint. Considerable distress is observed in damaged structures in this zone. The failure may be due to the following reasons:

(a) Shear within the joint
(b) Anchorage failure of the beam reinforcement in the joint
(c) Bond failure of the beam or column reinforcement passing through the joint

The beam–column joints are likely to fail earlier than the members framing into the joint due to destruction of the joint zone.The possibility of slip of anchored bars is shown in Fig. 8.1. This is particularly true for corner columns. The shear can be carried through the broken concrete zone by inclining the main reinforcement through the hinge zone towards a point of contraflexure at the centre of the beam.

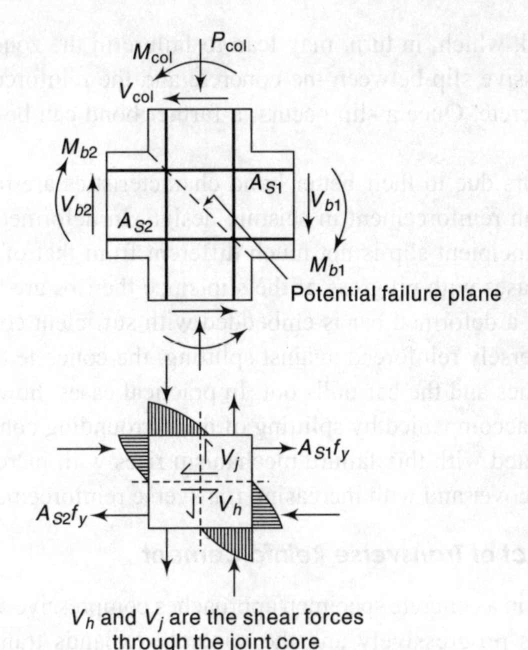

Fig. 8.3 Forces on an interior beam–column joint

Joints at discontinuities Proper joints should be made at discontinuities to avoid pounding of adjoining parts of the building. During seismic shaking two adjacent buildings or two adjacent units of the same building may hit (pound or hammer) each other. The pounding buildings/units can alter the dynamic response of both the buildings/units. Such pounding buildings/units should be separated by a distance equal to R times the sum of the storey displacements of each of them, to avoid damaging contact when the two buildings/units deflect towards each other. When floor levels of two similar adjacent buildings/units are at the same level, the response reduction factor R may be replaced by $R/2$.

8.3 Interaction between Concrete and Steel

Reinforced cement concrete is the most widely used construction material for structures. Concrete is markedly strong in compression but remarkably weak in tension. The steel bars, in the tension zone of the concrete, provide necessary resistance to the tension. Moreover, the reinforcing bars make the concrete a ductile material which otherwise is brittle. The bond between steel and surrounding concrete ensures strain compatibility. Some of the considerations to be made so that the concrete and reinforcing bars behave in unision and the desired ductility is achieved are discussed as follows.

Bond between Reinforcing Bars and Concrete

The bond strength between reinforcing bars and concrete is provided by chemical adhesion and friction. Repeated yielding of the longitudinal reinforcement and diagonal cracking often lead to concrete spalling. This phenomenon causes a partial loss of the bond, which, in turn, may lead to failure in the zones of anchorage through progressive slip between the concrete and the reinforcement or due to split of the concrete. Once a slip occurs, a further bond can be developed only by friction.

Deformed bars due to their better bond characteristics are recommended to be used for main reinforcement in seismic design. In deformed bars, the bond strength at the incipient slip is not much different from that of round bars, but resistance increases with progress of the slip since the ribs are wedged into the concrete. When a deformed bar is embedded with sufficient cover in concrete, which is transversely reinforced against splitting, the concrete between the ribs eventually crushes and the bar pulls out. In practical cases, however, pullout of the bar is often accompanied by splitting of the surrounding concrete. The bond strength associated with this failure mechanism rises with increasing thickness of the concrete cover and with increasing transverse reinforcement.

Confining Effect of Transverse Reinforcement

When the stress in a concrete specimen approaches compressive strength, internal cracking occurs progressively and the concrete expands transversely. If the compression zone is confined by transverse reinforcement such as spirals and hoop ties, the ductility of the concrete is greatly improved. If square hoop ties are used in the member, the concrete along the diagonals of the tie is confined, as shown in Fig. 8.4. A square hoop tie can generally apply confining pressure only near the corners because the pressure of concrete tends to bow the sides of the tie outwards. However, spirals confine the concrete more effectively because the circular shape enables it to provide a continuous confining pressure around

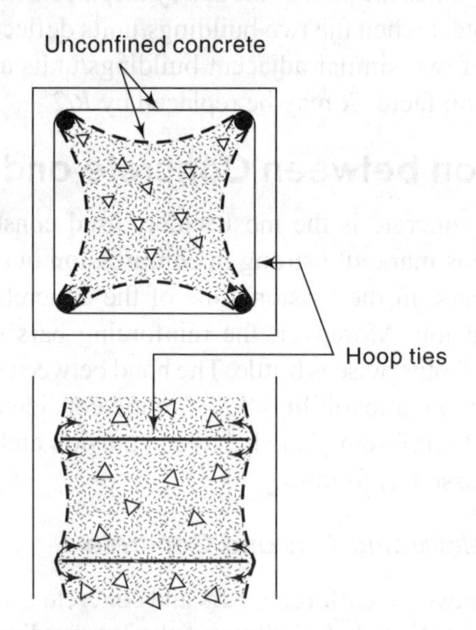

Fig. 8.4 Confinement of concrete by square hoop ties

the entire circumference. The resulting stress–strain relationship for the confined concrete is shown in Fig. 8.5. It is observed that increasing amounts of the confining reinforcement reduces the slope of the descending branch of the curve.

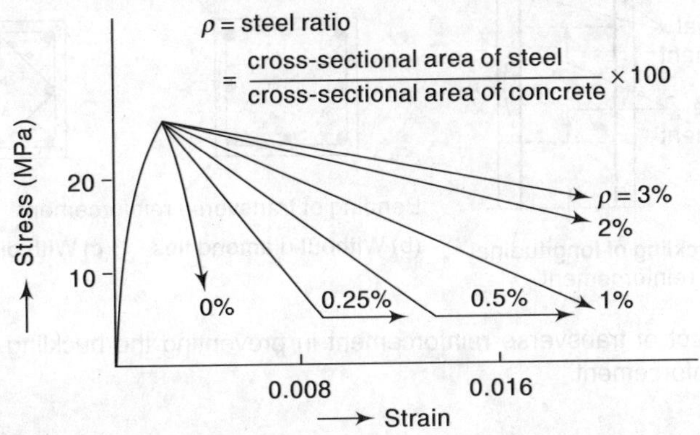

Fig. 8.5 Influence of the quantity of hoops on the stress–strain curves of concrete in members (Kent 1971)

Buckling of Reinforcing Bars

Longitudinal reinforcing bars under compression in beams and columns are prevented from buckling by the lateral restraint provided by concrete. Under cyclic loading that does not involve alternating flexure, the compression steel in straight members does not ordinarily buckle out of concrete, even at high strains or in the absence of restraining stirrups and ties. In fact, the concrete cover is normally sufficient for the purpose; moreover, the curvatures induced where there is important bending make the longitudinal steel bend inwards. The latter phenomenon does not apply to corner reinforcement, but a moderate number of stirrups or ties suffice to keep this steel in place. Another exception is found in the longitudinal steel that undergoes the smaller compressive stresses in columns under combined longitudinal force and bending, especially if the loads are sustained over long periods of time, so that the concrete creeps and gives place to large compressive strains. However, when covering concrete subjected to high compressive stresses becomes unstable, the restraining effect is reduced and the bar buckles as shown in Fig. 8.6(a), and the axial force carried by the compression bar is reduced. This reduces the load-carrying capacity of the member. Also, where the sign of the bending moment alternates and reaches values such that the steel yields in tension during part of a cycle and acts in compression during another part of the cycle, the bars tend to buckle out of the member and need transverse reinforcement for confinement.

In order to minimize reduction of carrying capacity and to ensure sufficient ductility, it is necessary to set a limit to the effective length of the longitudinal reinforcing bar, i.e., the distance between the lateral supports provided by transverse reinforcement. Thus, code limits are placed on the ratio of the distance between transverse reinforcement to the diameter of the longitudinal reinforcing

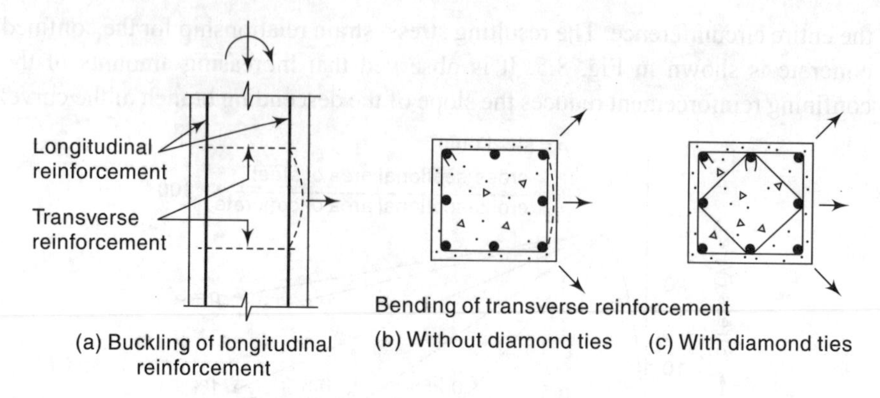

Fig. 8.6 Effect of transverse reinforcement in preventing the buckling of the main reinforcement

bar. The transverse reinforcement does not effectively support longitudinal reinforcing bars located at intermediate points between the corners, since the transverse reinforcement bends outward, as shown in Fig. 8.6(b). Diamond-shaped reinforcement, as shown in Fig. 8.6(c), is needed to make the effective length of such longitudinal bars equal to the distance between the transverse reinforcement. This type of reinforcement is effective for confining concrete and increasing maximum strength and failure strain of concrete.

8.4 Concrete Detailing—General Requirements

The design, construction, and detailing of RC buildings should be governed by the provisions of IS 456: 2000, except for the modifications suggested by the provisions below and the sections to follow.

To enable the elements of RC structures to be detailed in a consistent and satisfactory manner for earthquake resistance, the following rules must be observed strictly. These details should be satisfactory in regions of medium to high seismic risk. In low risk regions, relaxations may be made to the following requirements, but the principles of lapping, containment, and continuity must be retained if adequate ductility is to be obtained.

Cover Necessary minimum cover to reinforcement is provided to develop the required bond strength and to protect the reinforcement against corrosion. When high-strength deformed bars are used, especially greater than 36 mm, the development length (or bond strength) may be governed by the cover. Increased cover may have to be provided in case of members subjected to post-earthquake forces. Minimum cover for reinforcement should comply with Tables 16 and 16A of IS 456: 2000.

Concrete quality The minimum recommended characteristic strength for structural concrete is 20.0 N/mm^2. However, for all buildings that are more than four storeys in height in Zones IV and V, the minimum grade of concrete should be

M-25. Quality control, workmanship, and supervision are of the utmost importance in obtaining earthquake-resistant concrete.

The use of lightweight aggregates for structural purposes in seismic zones should be used very cautiously, as these may prove very brittle in earthquakes. Appropriate advice should be sought in selecting the type of aggregate, the mix proportions, and strengths in order to obtain a suitably ductile concrete.

Reinforcement quality Suitable quality of reinforcement must be ensured for achieving adequate earthquake resistance. As the properties of reinforcement vary greatly between manufacturers, much depends on knowing the source of the bars, and on applying the appropriate tests. The following points should be observed.

(a) An adequate minimum yield stress of steel should be ensured. Grades of steel with characteristic strength in excess of 415 N/mm^2 are not permitted. However, high-strength deformed bars, produced by the thermo-mechanical treatment process, of grades Fe-500 and Fe-550 having elongation more than 14.5 per cent, may also be used for the reinforcement. Cold worked steel is not recommended.

(b) The actual yield strength, based on a tensile test of steel, must not exceed the specified yield strength by more than 120 N/mm^2. If the difference is more, the shear or bond failure may precede the flexural hinge formation and the capacity design concept may not work.

(c) The ratio of the actual ultimate strength to the actual yield strength should be at least 1.25. To develop inelastic rotation capacity, a structural member needs an adequate length of yield region along the axis of the member; the larger the ratio of ultimate to yield moment, the longer the yield region.

(d) The elongation test is particularly important for ensuring adequate steel ductility.

(e) The bend and rebend test is most important for ensuring sufficient ductility of reinforcement.

(f) Welding of reinforcing bars may cause embrittlement and hence should only be allowed for steel of suitable chemical properties, using an approved welding process.

(g) Galvanizing may cause embrittlement and needs special consideration.

(h) Welding steel fabric (mesh) is unsuitable for earthquake resistance because of its potential brittleness.

Splices Laps of reinforcement in earthquake-resistant frames must continue to function while the members or joints undergo large deformations. As the stress transfer is accomplished through the concrete surrounding the bars, it is essential that there be adequate space in a member to place and compact good-quality concrete. Therefore, splices should ideally be staggered and located away from sections of maximum tension. Lapped splices should never be located in potential plastic hinge zones. In columns of buildings the splices should be positioned in the mid-height region between floors.

Laps should preferably not be made in regions of high stress, such as near beam–column connections, as the concrete may crack under larger deformations and thus adversely affect the transfer of stress by bond. In regions of high stress, laps should be considered as an anchorage problem rather than a lap problem, i.e., the transfer of stress from one bar to another is not considered; instead the bars required to resist tension should be extended beyond the zone of expected large deformation, in order to develop their strength by anchorage.

Both the contact laps and the spaced laps perform equally well, because the stress transfer is primarily through the surrounding concrete. Contact laps usually reduce the congestion and give better opportunity to obtain well-compacted concrete over and around the bars. Laps should preferably be staggered, but where this is impracticable and large number of bars are lapped at one location (e.g., in columns), adequate links or ties must be provided to minimize the possibility of splitting of the concrete. In columns and beams, even when laps are made in regions of low stress, at least two links should be provided.

Anchorage It is assumed that the longitudinal reinforcement will consist of deformed bars. Satisfactory anchorage may be achieved by extending bars as straight lengths or by using 90° and 180° bends, but anchorage efficiency will be governed largely by the state of stress of the concrete in the anchorage length. Tensile reinforcement should not be anchored in zones of high tension. If this cannot be achieved, additional reinforcement in the form of links should be added, especially where high shears exist, to help to confine the concrete in the anchorage length. It is especially desirable to avoid anchoring bars in the panel zone of beam–column connections.

Large amounts of the reinforcement should not be curtailed at any one section. Bars should not be cut off at points in the span where anchorage would be required in a region of tension under earthquake effects. If cut-offs of this kind cannot be avoided, additional transverse reinforcement should be provided because of discontinuity.

Confinement The ductility and the strength of the concrete is greatly enhanced by confining the compression zone with closely spaced steel. The rectangular, all-enclosing links, are moderately effective in small columns, but are of little use in large columns. Spirals are greatly superior to rectangular ones.

8.5 Flexural Members in Frames

To ensure sufficient ductility in beams, good design details are necessary; the critical design details are as follows. These requirements apply to the frame members which have factored axial stress in the member, under earthquake loading, not exceeding $0.1f_{ck}$. If the factored axial stress exceeds this value the member will be considered to be in significant compression, and should be detailed as explained in Section 8.6.

8.5.1 Dimensions

The following three limitations are imposed on the dimensions of beams:

(a) b/D not less than 0.3

(b) b not less than 200 mm

(c) D not more than one-fourth of the clear span

The first two limits are to check that difficulties do not arise in confining concrete through stirrups in narrow beams, which exhibit poor performance in comparison to well confined concrete. The third one is related with the structural behaviour of the member; when the ratio of the total depth of member to the clear span is appreciable, the member behaves like a deep beam. The behaviour of a deep beam is significantly different from that of a relatively slender member under cyclic inelastic deformations. Therefore design rules for relatively slender members do not apply to members with l/D less than four, especially with respect to shear strength.

8.5.2 Longitudinal Reinforcement

In order to ensure adequate ductility in RC beams, the amount of longitudinal reinforcement must be limited in relation to the dimensions of the beam, the quality of concrete, and the yield stress of reinforcement. In so far as earthquake-resistant design is concerned, the critical sections for the longitudinal reinforcement in frames occur at the face of the beam–column and girder–column connections, and at the beam–girder connection immediately adjacent to the columns. Since the distribution of bending moment along the beams/girders framing into columns may be quite different in a severe earthquake from that under gravity loads, the cut-off points of the bars require special consideration. It is desirable that only straight bars are used; however, bent bars may be used in beams that do not frame into columns. The following are the specifications for longitudinal reinforcement.

(a) The minimum bar diameter permissible is 12 mm. There must be at least two bars in both the top and the bottom face.

(b) The upper limit for the reinforcement ratio is 0.025 of the gross area, above which the ductility is inadequate. Also, beyond this limit, there will be considerable congestion of reinforcement. This may cause insufficient compaction or a poor bond between concrete and reinforcement.

(c) A lower limit on the tension steel ratio on any face at any section is placed as $0.24\sqrt{f_{ck}}/f_y$. This provision is derived on following considerations.

For small loads on RC members, the entire concrete section participates. As the load increases, tension cracks develop in the concrete. The concrete must then transfer the tensile force to the reinforcement present in the tension region. If the tensile reinforcement available is not adequate to carry the tensile force thus transferred, the section will fail suddenly, causing a brittle failure. Hence, adequate tensile reinforcement should be there to take the tensile force that was carried by the concrete prior to cracking.

This provision governs for the members having large cross-section from architectural requirements. It prevents the possibilities of sudden failure of members, by ensuring that the moment of resistance of the section is greater than the cracking moment of the section.

(d) The positive steel at a joint face must be equal to at least half the negative steel at the face (Fig. 8.8). This provision covers the following two aspects:

(i) The seismic moments are reversible. Further, design seismic loads may be exceeded by a considerable margin during strong earthquake shaking. Therefore, substantial sagging moments may develop at beam ends during strong shaking, which may not be reflected in an analysis (Fig. 8.7).

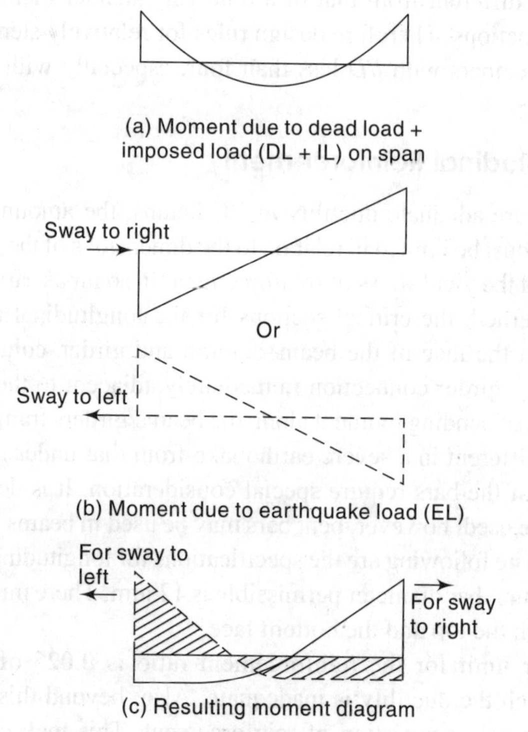

Fig. 8.7 Reversal of moments due to earthquake loading

(ii) Compression reinforcement increases ductility and, therefore, adequate compression reinforcement at the location of potential yielding is ensured. The application of this provision is illustrated in Fig. 8.8.

(e) The steel provided at each of the top and bottom faces of the member at any section along its length should be equal to at least one-fourth of the maximum negative moment steel provided at the face of either joint (Fig. 8.8) for two reasons. Firstly, sufficient reinforcement should be available at any section

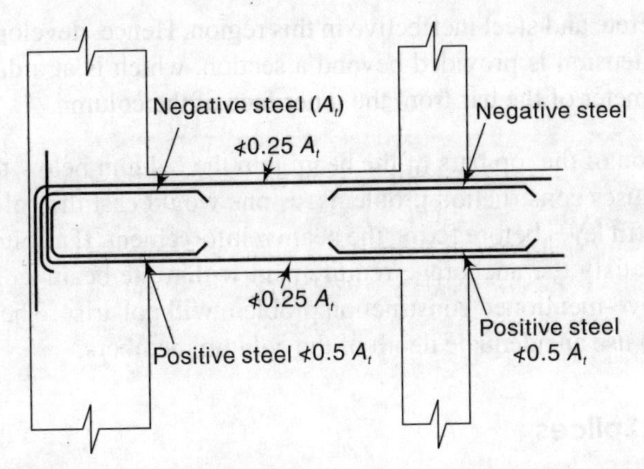

Fig. 8.8 Longitudinal reinforcement at the joints in a beam

along the length of the member to take care of reversal of loads or unexpected bending moment distribution. Secondly, the amount of steel should not be reduced abruptly, away from the joint. Hence, the code specifies that the steel to be provided at each of the top and bottom face of the member, at any section along its length, should be some fraction of the maximum negative moment steel provided at the face of either joint.

In an external joint, both the top and the bottom bars of the beam should be provided with anchorage length, beyond the inner face of the column, equal to the development length in tension plus 10 times the bar diameter minus the allowance for 90° bend(s) (as shown in Fig. 8.9). In an internal joint, both face bars of the beam should be taken continuously through the column.

During an earthquake, the zone of inelastic deformation that exists at the ends of a beam may extend for some distance into the column. This makes the bond

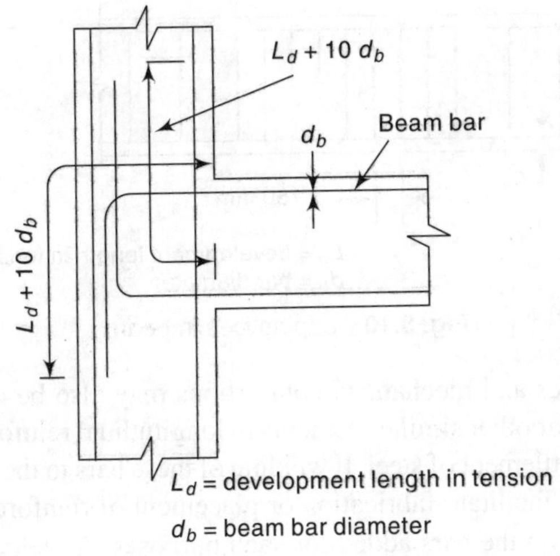

Fig. 8.9 Anchorage of beam bars in an external joint

between concrete and steel ineffective in this region. Hence, development length of the bar in tension is provided beyond a section, which is at a distance of 10 times the diameter of the bar from the inner face of the column.

The extension of the top bars of the beam into the column below the soffit of the beam causes construction problems, as one would cast the columns up to the beam soffit level before fixing the beam reinforcement. If a column is wide enough to satisfy the anchorage requirement within the beam–column joint, then the above-mentioned construction problem will not arise. Therefore, it is important to use an adequate depth of the column members.

8.5.3 Lap Splices

Lap splices are not reliable under cyclic inelastic deformations, and, hence, should not to be provided in critical regions. Lap splices of main bars should be located as far as possible in the zones of low stress. These are neither acceptable within the column zone nor within zones of a potential plastic hinge. Closely spaced hoops help improve performance of the splice when the cover concrete spalls off.

(a) The longitudinal bars should be spliced only if the hoops are provided over the entire splice length at a spacing not exceeding 150 mm (Fig. 8.10). The lap length should not be less than the bar development length in tension. Lap splices should not be provided in any of the following cases:
   - within a joint
   - within a distance of 2*d* from joint face
   - within a quarter length of the member, where flexural yielding may generally occur under the effect of earthquake forces

Note: Not more than 50 per cent of the bars should be spliced at one section.

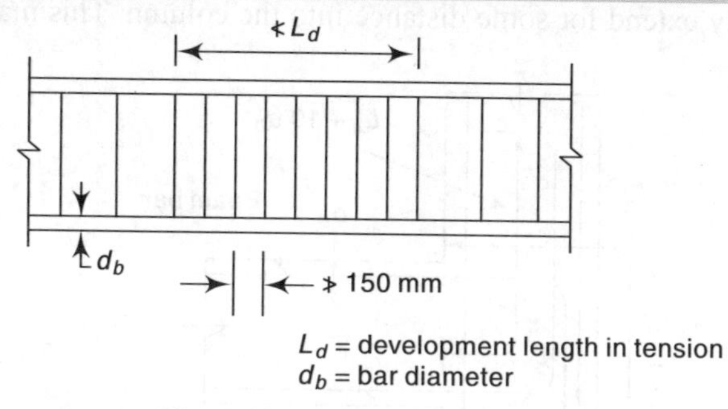

Fig. 8.10 Lap splices in beam

(b) Welded splices and mechanical connections may also be used. Welding of stirrups, ties, or other similar elements to longitudinal reinforcement can lead to local embrittlement of steel. If welding of these bars to the longitudinal bars is required to facilitate fabrication or placement of reinforcement, it should be done only on the bars added for such purposes. A welded splice reduces the need to depend on the concrete for stress transfer but may introduce discontinuity in the chemical and physical properties of reinforcement in the weld area, and impair its ductility. Fillet-welded splices will usually require adequate transfer reinforcement. Butt-welded splices, however, may be treated as continuous bars.

8.5.4 Web Reinforcement

Sufficient transverse web reinforcement must be provided in beams of an earthquake-resistant frame to ensure that its capacity will be governed by flexure and not by shear. Whenever reinforcing bars are called upon to act as compression reinforcement, ties should be provided to restrain these from buckling after spalling of the concrete cover. Stirrups in RC beams help in the following three ways.

(a) They carry the vertical shear force and thereby prevent the diagonal shear cracks.
(b) They confine the concrete.
(c) They prevent the buckling of compression bars by providing sufficient anchorage.

The following are the specifications for web reinforcement.

(a) Web reinforcement should consist of vertical hoops. A vertical hoop is a closed stirrup having a 135° hook and a 6 diameter extension (but not < 65 mm) at each end that is embedded in a confined core [Fig. 8.11(a)]. In compelling circumstances, it may also be made up of two pieces of reinforcement: a U-stirrup with a 135° hook and a 10 diameter extension (but not < 65 mm) at each end, embedded in the confined core, and a cross tie [Fig. 8.11(b)]. The hook shall engage peripheral longitudinal bars. Consecutive cross ties engaging the same longitudinal bars should have their 90° hooks at opposite sides of the flexural member. If the longitudinal reinforcement bars secured by the cross ties are confined by a slab on only

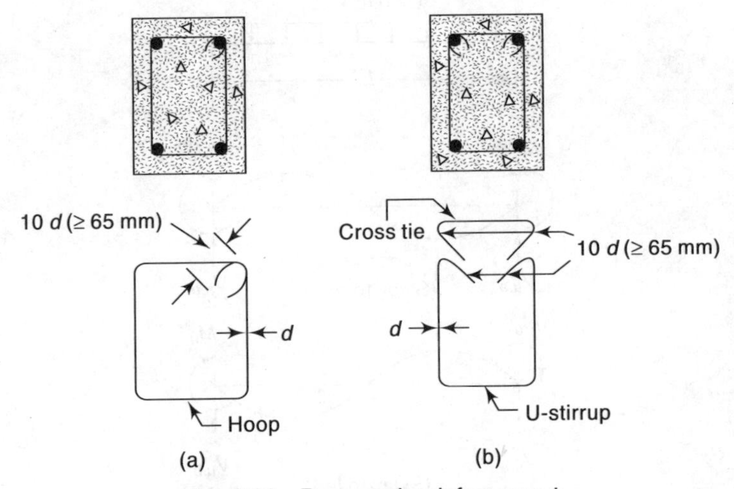

Fig. 8.11 Beam web reinforcement

one side of the flexural frame member, the 90° hook of the cross ties should be placed on that side.

(b) The minimum diameter of the bar forming a hoop should be 6 mm. However, in beams with a clear span exceeding 5 m, the minimum bar diameter should be 8 mm.

(c) The shear force to be resisted by the vertical hoops should be the maximum of:
- the calculated factored shear force as per analysis, and
- the shear force due to formation of plastic hinges at both ends of the beam, plus the factored gravity load on the span. This is given by the following:

For sway to right

$$V_{u,a} = V_a^{D+L} - 1.4\left[\frac{M_u^{As} + M_u^{Bh}}{L_{AB}}\right] \tag{8.1}$$

$$V_{u,b} = V_b^{D+L} + 1.4\left[\frac{M_u^{As} + M_u^{Bh}}{L_{AB}}\right] \tag{8.2}$$

For sway to left

$$V_{u,a} = V_a^{D+L} + 1.4\left[\frac{M_u^{Ah} + M_u^{Bs}}{L_{AB}}\right] \tag{8.3}$$

$$V_{u,b} = V_b^{D+L} - 1.4\left[\frac{M_u^{Ah} + M_u^{Bs}}{L_{AB}}\right] \tag{8.4}$$

where M_u^{As}, M_u^{Ah} and M_u^{Bs}, M_u^{Bh} are the sagging and hogging moments of resistance of the beam section at ends A and B, respectively (Fig. 8.12). These

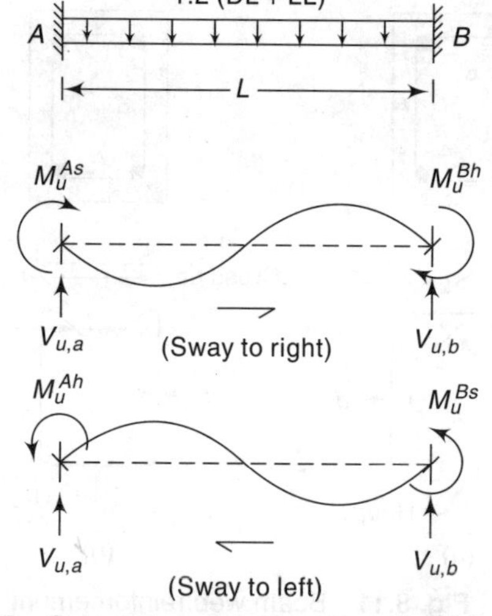

Fig. 8.12 Calculation of design shear force for beam

are to be calculated as per IS 456: 2000. L_{AB} is the clear span of the beam. V_a^{D+L} and V_b^{D+L} are the shears at ends A and B, respectively, due to vertical loads with a partial safety factor of 1.2 on the loads. The design shear at end A should be the larger of the two values of $V_{u,a}$ computed above. Similarly, the design shear at end B should the larger of the two values of $V_{u,b}$ computed earlier.

This provision ensures that brittle shear failure does not precede the actual yielding of the beam in flexure. This provision simplifies the process of calculating plastic moment capacity of a section by taking it to be 1.4 times the calculated moment capacity with the usual partial safety factors. This factor of 1.4 is based on the consideration that plastic moment capacity is usually calculated by assuming that the stress in flexural reinforcement is $1.25f_y$, as against $0.87f_y$ in the moment capacity calculation.

(d) Due to the cyclic nature of seismic loads, the shear force can change direction. The inclined hoops and bent up bars, effective in resisting shear force in one direction, will not be effective in the opposite direction. So the contribution of bent up bars and inclined hoops to shear resistance of the section should not be considered.

(e) Closely spaced hoops at the two ends of the beam are recommended to obtain large energy-dissipation capacity and better confinement. The spacing of hoops over a length of $2d$ at either end of the beam should not exceed (i) $d/4$, and (ii) 8 times the diameter of the smallest longitudinal bar. However, it should not be less than 100 mm (Fig. 8.13) to ensure space for the needle vibrator.

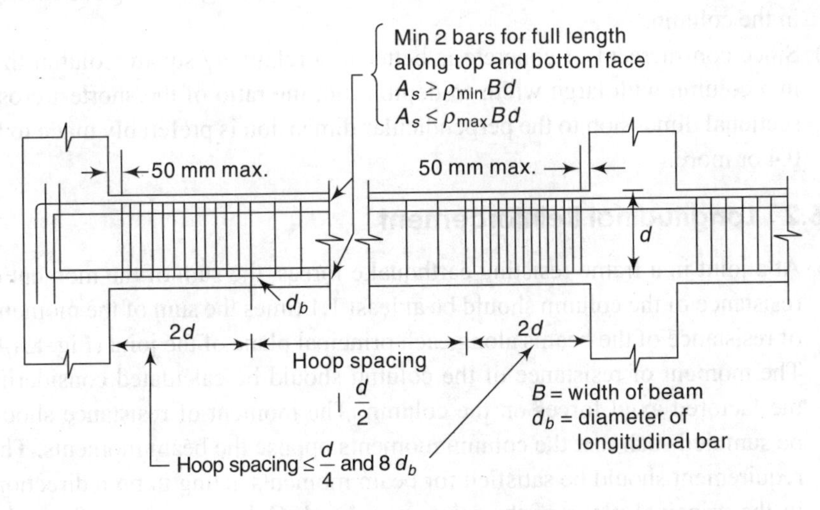

Fig. 8.13 Beam reinforcement

The first hoop should be at a distance not exceeding 50 mm from the joint face. Vertical hoops at the same spacing as above should also be provided over a length equal to $2d$ on either side of a section where flexural yielding may occur under the effect of earthquake forces. Elsewhere, the beam should have vertical hoops at a spacing not exceeding $d/2$. The hoop spacing is specified as $d/2$ over the remaining length of the beam to prevent the occurrence of an unexpected shear failure in this region. IS 456 permits $3d/4$ as against the requirement of $d/2$ in this provision. It must be remembered that the provisions of IS 13920 are over and above those in IS 456.

8.6 Columns and Frame Members Subjected to Bending and Axial Load

The proportioning of columns and their reinforcement in earthquake-resistant frames should receive very careful considerations. These requirements apply to the frame members which have a factored axial stress in excess of $0.1f_{ck}$ under the effect of earthquake forces. If the factored axial load is less than the specified limit, the frame member will be considered as a flexural member and should be detailed as explained in Section 8.5.

8.6.1 Dimensions

(a) The minimum dimension of the column should not be less than 300 mm or 15 times the largest beam bar diameter of the longitudinal reinforcement passing through or anchoring into the column joint. A small column width may lead to two problems. First, the moment capacity of the column section may be very low since the lever arm between the compression steel and tension steel would be very small. Second, the beam bars may not get enough anchorage in the column.

(b) Since confinement of concrete is better in a relatively square column than in a column with large width-to-depth ratio, the ratio of the shortest cross-sectional dimension to the perpendicular dimension is preferably made to be 0.4 or more.

8.6.2 Longitudinal Reinforcement

(a) At a joint in a frame resisting earthquake forces, the sum of the moment of resistance of the column should be at least 1.1 times the sum of the moments of resistance of the beams along each principal plane of the joint (Fig. 8.14). The moment of resistance of the column should be calculated considering the factored axial forces on the column. The moment of resistance should be summed such that the column moments oppose the beam moments. This requirement should be satisfied for beam moments acting in both directions in the principal plane of the joint considered. Columns not satisfying this requirement should have special confining reinforcements over their full height, and not just in the critical end regions.

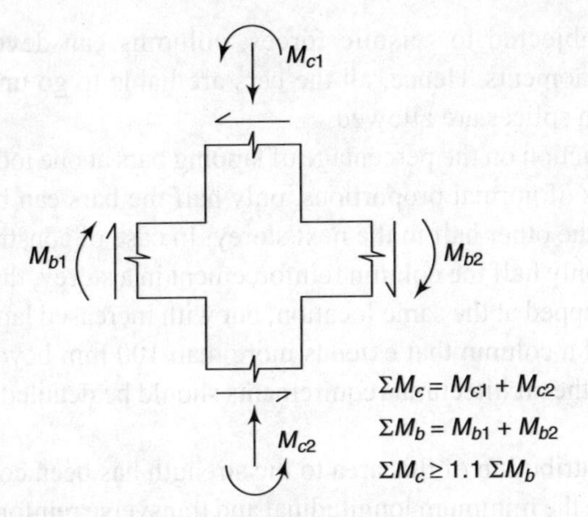

Fig. 8.14 Weak-beam–strong-column concept

This provision is based on the strong-column–weak-beam theory. It is meant to make the building fail in the beam-hinge mechanism (beams yield before columns) and not in the storey mechanism (columns yield before the beams). Storey mechanism must be avoided as it causes greater damage to the building. Therefore, columns should be stronger than the beams meeting at a joint.

(b) At least one intermediate bar should be provided between corner bars along each column face. This implies that rectangular columns in lateral-resisting frames should have a minimum of eight bars. Intermediate bars are required to ensure the integrity of the beam–column joint and to increase confinement to the column core.

(c) Lap splices should be provided only in the central half of the member length. They should be proportioned as tension splices. Hoops should be provided over the entire splice length at spacings not exceeding 150 mm, centre to centre. Preferably, not more than 50 per cent of the bars should be spliced at one section. If more than 50 per cent of the bars are spliced at one section, the lap length should be $1.3L_d$, where L_d is the development length of the bar in tension as per IS 456: 2000.

Since seismic moments are maximum in columns just above and just below the beam, the reinforcement must not be changed at those locations. Also, since the seismic moments are minimum away from the ends, lap splices are allowed only in the central half of the columns. This also implies that the structural drawings must specify the column reinforcement from one mid-storey height to the next mid-storey height.

This provision has a very important implication for dowels that are to be left for future extension. Inadequate projected length of column reinforcement for future vertical extension is a very serious seismic threat. This creates a very weak section in all the columns at a single location, and all upper storeys are prone to collapse at that level.

When subjected to seismic forces, columns can develop substantial reversible moments. Hence, all the bars are liable to go under tension, and only tension splices are allowed.

The restriction on the percentage of lapping bars at one location means that in buildings of normal proportions, only half the bars can be spliced in one storey and the other half in the next storey. In case of construction difficulty in lapping only half the column reinforcement in a storey, the code allows all bars to be lapped at the same location, but with increased lap length of $1.3L_d$.

(d) Any area of a column that extends more than 100 mm beyond the confined core due to the architectural requirements should be detailed in the following manner.

If the contribution of this area to the strength has been considered, then it should have the minimum longitudinal and transverse reinforcement as per IS 13920. However, if this area has been treated as non-structural (Fig. 8.15), the minimum reinforcement requirements should be governed by provisions of minimum longitudinal and transverse reinforcement as per IS 456: 2000. This is so because even when column extensions are considered as non-structural, they contribute to the stiffness of the column. If the extensions are not properly tied to the column core, a severe shaking may cause spalling of this portion, leading to a sudden change in the stiffness of the columns. Therefore, the code requires that such extensions be detailed at least as per IS 456 requirements for the column.

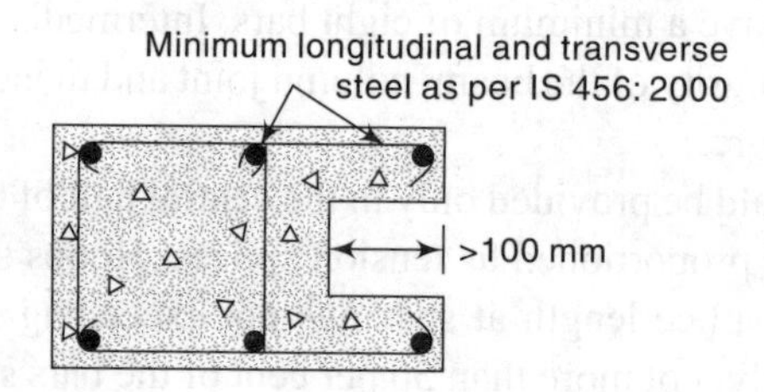

Fig. 8.15 Reinforcement requirement for column with more than 100-mm projection beyond core

8.6.3 Transverse Reinforcement

Transverse reinforcement serves the following four purposes.

(a) It provides shear resistance to the member.
(b) It confines the concrete core and thereby increases ductility.
(c) It provides lateral resistance against buckling to the compression reinforcement.
(d) It prevents loss of bond strength within column vertical bar splices.

Full confinement of the concrete in the columns at beam–column connections may be necessary to ensure the required ductility. Failures to provide transverse reinforcement at the ends where plastic hinges are anticipated results in reduced flexural strength and ductility, as well as degradation of shear resistance. Closely spaced transverse reinforcement is particularly recommended for the unrestrained length of captive columns where inelastic flexure is combined with high shear force. Boundary elements of the wall where significant inelastic action is anticipated should be well confined to provide ductility under axial compression. Columns supporting discontinuous walls should be confined over the entire height. The following are the recommendations for transverse reinforcement using fully confined concrete.

(a) Transverse reinforcement for circular columns should consist of spiral or circular hoops. In rectangular columns, rectangular hoops may be used. A rectangular hoop is a closed stirrup having a 135° hook with a 6 diameter extension (but not less than 65 mm) at each end that is embedded in the confined core [Fig. 8.16(a)].

(b) The parallel legs of the rectangular hoop should be spaced not more than 300 mm centre to centre. If the length of any side of the hoop exceeds 300 mm, a crosstie should be provided [Fig. 8.16(b)]. Alternatively, a pair of overlapping hoops may be provided within the column [Fig. 8.16(c)]. The hooks should engage peripheral longitudinal bars. Consecutive crossties engaging the same longitudinal bars should have their 90° hooks at opposite sides of the flexural member.

(c) Closer spacing of hoops is required to ensure better seismic performance. This should not exceed half the least lateral dimension of the column, except where special confining reinforcement is provided, as per Section 8.7.

(d) The design shear force for columns should be the maximum of the following:
- Calculated factored shear force as per analysis, and
- A factored shear force given by

$$V_u = 1.4\left[\frac{M_u^{bL} + M_u^{bR}}{h_{st}}\right] \tag{8.5}$$

where M_u^{bL} and M_u^{bR} are the moments of resistance of opposite sides of the beams framing into the column from opposite faces (Fig. 8.17) and h_{st} is the storey height. The beam element capacity is to be calculated as per IS 456: 2000.

This provision is based on the strong-column–weak-beam theory. Here, column shear is evaluated based on beam flexural yielding with the expectation that yielding will occur in beams rather than in columns. The factor of 1.4 is based on the consideration that plastic moment capacity of a section is usually calculated by assuming that the stress in flexural reinforcement is $1.25f_y$ as against $0.87f_y$ in the moment capacity calculation.

8.7 Special Confining Reinforcement

Confinement of concrete is essential to provide adequate rotational ductility in potential plastic hinge regions of columns. Lengths of potential plastic hinge

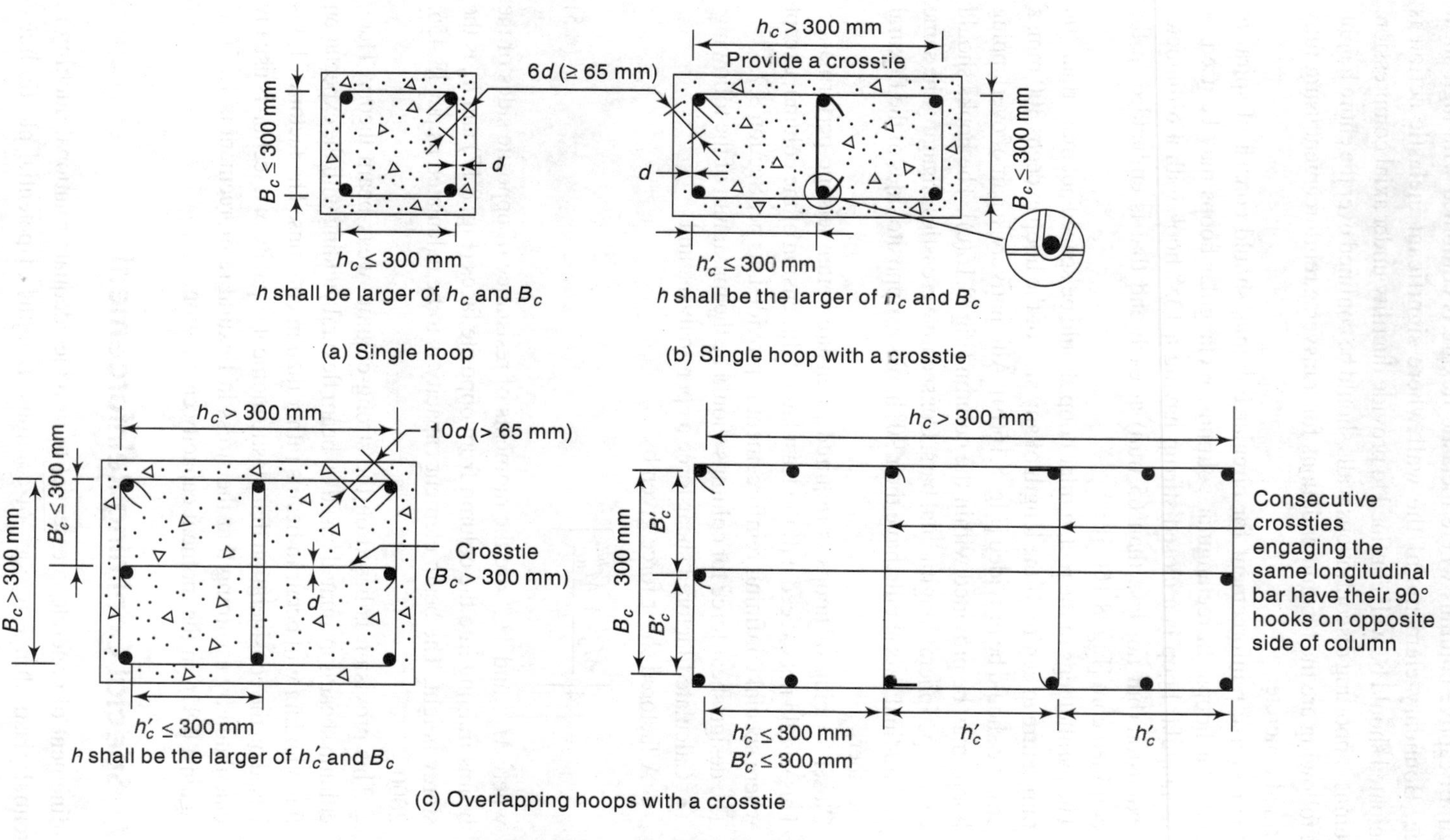

Fig. 8.16 Transverse reinforcement in column

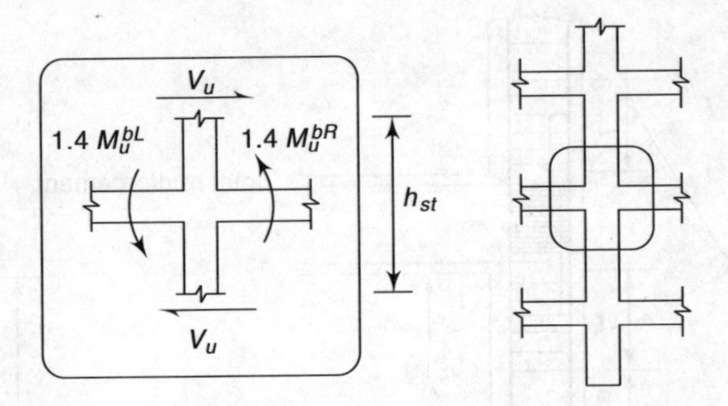

Fig. 8.17 Calculation of design shear force for column

regions in columns are generally smaller than beams partly because column moments vary along the storey height with a relatively large gradient. Therefore, the region of a frame column subjected to tension yielding of reincorcement is limited. The following requirements should be met with, unless a larger amount of transverse reinforcements is required from shear strength consideration.

(a) Special confining reinforcement should be provided over a length L_0 from each joint face towards mid span, and on either side of any section where flexure may occur under the effect of earthquake forces (Fig. 8.18). Hence, special confining reinforcement is provided to ensure adequate ductility and to provide restraint against buckling to the compression reinforcement. The length L_0 should not be less than any of the following:
   - The larger lateral dimension of the member at the section where yielding occurs
   - 1/6 of the clear span of the member
   - 450 mm

(b) During severe shaking, a plastic hinge may form at the bottom of a column that terminates in a footing or a mat. Hence, special confining reinforcement of the column must be extended at least 300 mm into the foundation (Fig. 8.19).

(c) The point of contraflexure is usually in the middle half of the column, except for the column in the top and bottom storeys of a multi-storey frame. When the point of contraflexure, under the effect of gravity and ELs, is not within the middle half of the column, then the zone of inelastic deformation may extend beyond the region that is provided with closely spaced hoop reinforcement. This requires the provision of special confining reinforcement over the full height of the column.

(d) Observations of past earthquakes indicate very poor performance of buildings where a wall in the upper storey terminates on columns in the lower storey. Hence, special confining reinforcement must be provided over the full height of such columns. This implies that columns supporting reactions from

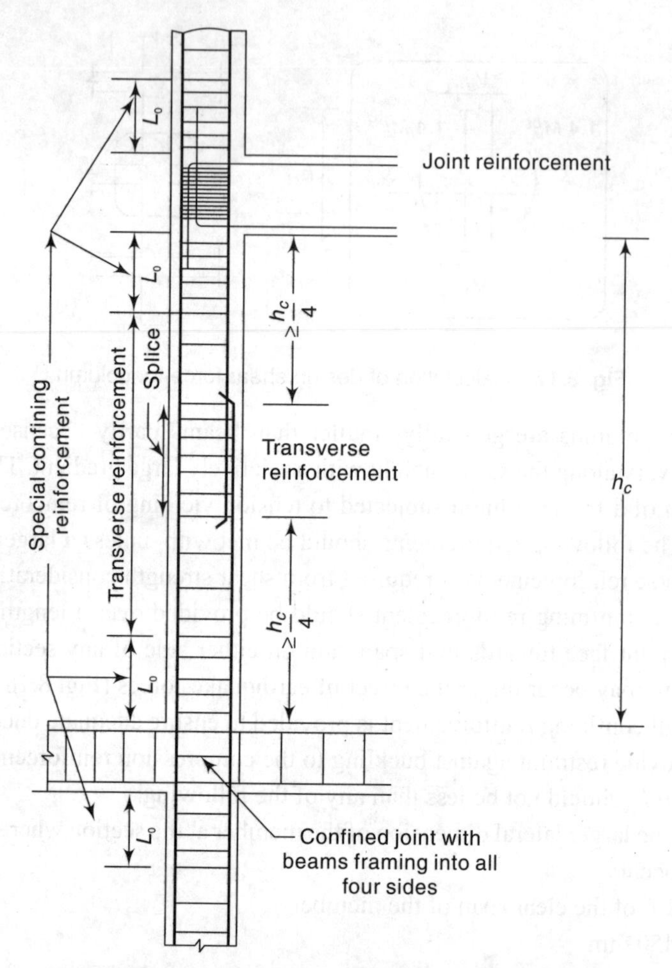

Fig. 8.18 Column and joint detailing as per IS 13920

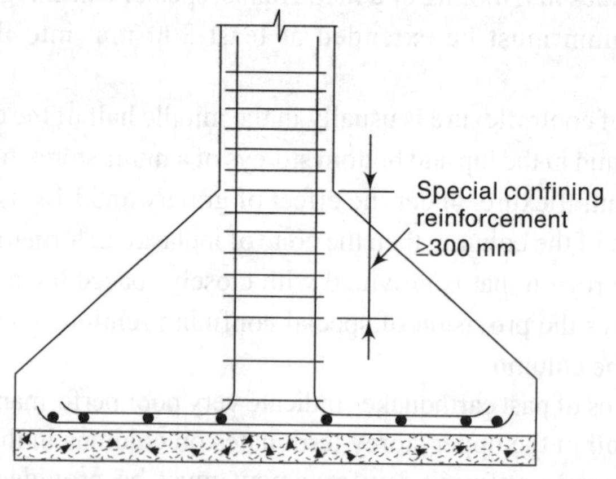

Fig. 8.19 Provision of special confining reinforcement in footing

discontinued stiff members, such as walls, should be provided with special confining reinforcement over their full height (Fig. 8.20). This reinforcement should also be placed over the discontinuity for at least the development length of the largest longitudinal bar in the column. Where the column is supported on the wall, this reinforcement should be provided over the full height of the column; it should also be provided below the discontinuity for the same development length.

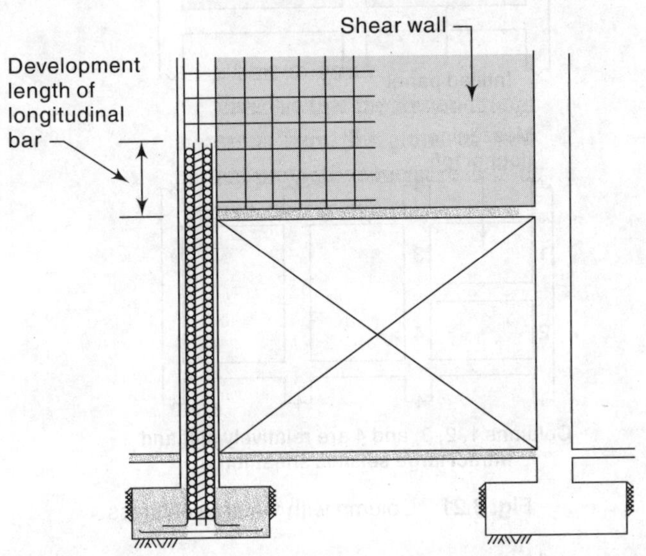

Fig. 8.20 Special confining reinforcements for columns under discontinued walls

(e) Column stiffness is inversely proportional to the cube of the column height. Hence, columns significantly shorter than other columns in the same storey have much higher lateral stiffness, and consequently attract much greater seismic shear force. There is a possibility of brittle shear failure occurring in the unsupported zones of such short columns. This has been observed in several earthquakes in the past. A mezzanine floor or a loft also results in the stiffening of some of the columns, while leaving other columns of the same storey unbraced over their full height. Another example is of semi-basements where ventilators are provided between the soffit of the beam and the top of the wall; here, the outer columns become the 'short columns' as compared to the interior columns; hence, special confining reinforcement is needed over the full height in such columns to give them adequate confinement and shear strength.

(f) Special confining reinforcement should also be provided over the full height of a column which has significant variation in stiffness along its height. This variation in stiffness may be due to the presence of bracing, a mezzanine floor, or an RC wall on either side of the column that extends only over a part of the column height (Fig. 8.21).

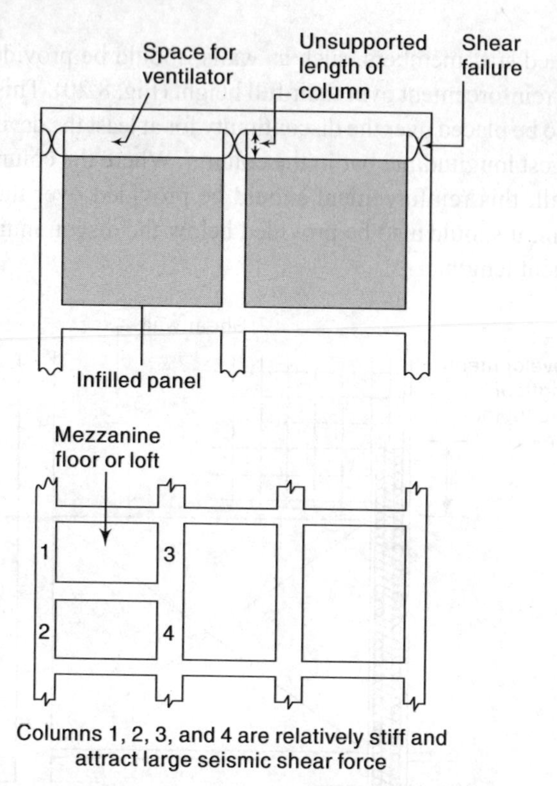

Fig. 8.21 Column with variable stiffness

(g) The spacing of hoops used as special confining reinforcement should not exceed 1/4 of the minimum member dimension, but need not be less than 75 mm nor more than 100 mm. This requirement is to ensure adequate concrete confinement. Restriction of spacing to 75 mm is to ensure proper compaction of concrete. In case of large bridge piers, larger spacing than 100 mm may be allowed.

(h) The area of cross-section A_{sh} of the bar forming circular hoops or spirals, which is to be used as special confining reinforcement should not be less than

$$0.09SD_k \frac{f_{ck}}{f_y}\left[\frac{A_g}{A_k} - 1.0\right]$$

$$\text{or} \quad 0.024SD_k \frac{f_{ck}}{f_y}$$

where A_{sh} is the area of the bar cross-section, S is the pitch of the spiral or the spacing of the hoops, D_k is the diameter of the core measured to the outside of the spiral or hoop, f_{ck} is the characteristic compressive strength of the concrete cube, f_y is the yield stress of steel (of the circular hoop or spiral), A_g is the gross area of the column cross-section, and A_k is the area of the concrete core $\left(\frac{\pi}{4}D_k^2\right)$.

This provision is intended for adequate confining reinforcement to the column. The first equation is obtained by equating the maximum load-carrying capacity of the column, prior to spalling of the concrete, to its axial load-carrying capacity at large compressive strength, with the spiral reinforcement stressed to its useful limit. For very large sections the ratio A_g/A_k tends to be close to 1.0 and, hence, the first equation in this specification gives a very low value of the confining reinforcement. The second equation governs the large section, for instance, the bridge piers.

(h) The area of cross-section A_{sh} of the bar forming a rectangular hoop, which is to be used as special confining reinforcement, should not be less than

$$0.18\, Sh \frac{f_{ck}}{f_y}\left[\frac{A_g}{A_k} - 1.0\right]$$

or $$0.05\, Sh \frac{f_{ck}}{f_y}$$

where h is the longer dimension of the rectangular confining hoop measured to its outer face. It should not exceed 300 mm (Fig. 8.16); A_k is the area of the confining concrete core in the rectangular hoop, measured to its outer dimensions.

The first equation in this provision is intended to provide the same confinement to a rectangular core confined by rectangular hoops as would exist in an equivalent circular column, assuming that rectangular hoops are 50 per cent as efficient as spirals in improving confinement of concrete. The second equation governs the large column section.

Note: The dimension h of the hoop can be reduced by introducing crossties, as shown in Fig. 8.16(b). In this case A_k should be measured as the overall core area, regardless of the hoop arrangement. The hooks or crossties should engage peripheral longitudinal bars.

8.8 Joints of Frames

Quite often, joints are not provided with stirrups because of constriction difficulties. Similarly, in traditional constructions, the bottom beam bars are often not continuous through the joints. Both these practices are not acceptable when the building has to carry lateral loads. The following are the main concerns with the joints:

(a) *Serviceability* Cracks should not occur due to diagonal compression and joint shear.
(b) *Strength* This should be more than that in the adjacent members.
(c) *Ductility* It is not needed for gravity loads, but is required for seismic loads.
(d) *Anchorage* Joints should be able to provide proper anchorage to the longitudinal bars of the beams.
(e) *Ease of construction* Joints should not have congestion of reinforcement.

The lateral restraint of the concrete in the joint is essential to ensure adequate behaviour, especially under repeated, alternating loads. Unless this restraint is provided, the concrete splits under bending moments that are considerably smaller than one would compute if one ignored this phenomenon. A few cycles of loading is sufficient to reduce the capacity of the joint practically to zero. Moreover, large diagonal cracks develop at relatively low stresses.

Beam–column connections of the type shown in Fig. 8.22 exhibit especially poor behaviour. Specifications for joints are based on the fundamental concept that failure should not occur within the joint. It should be strong enough to withstand the yielding of connecting beams (usually) or columns. Special confining reinforcement as required at the end of the column should be provided through the joint as well, unless the joint is confined as follows.

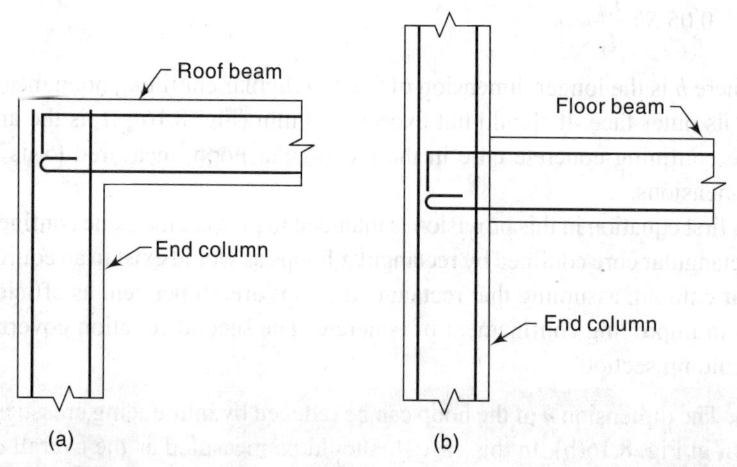

Fig. 8.22 Some beam–column connections in RC structures (not recommended)

Transverse Reinforcement

(a) A joint can be confined by beams/slabs around it, longitudinal bars (from beams and columns passing through the joints), and transverse reinforcement. A member that frames into a face is considered to provide confinement to the joint if at least three-quarters of the face of the joint is covered by the framing member, and if such confining members frame into all faces of the joint.

(b) For a joint that is confined by structural members from all four sides of the joint, transverse reinforcement equal to at least half the special confining reinforcement required at the end of the column should be provided within the depth of the shallowest framing member. The spacing of the hook should not exceed 150 mm.

This provision refers to the wide beams, i.e., when the width of the beam exceeds the corresponding column dimension, as shown in Fig. 8.23. In that case, the beam reinforcement not confined by the column reinforcement

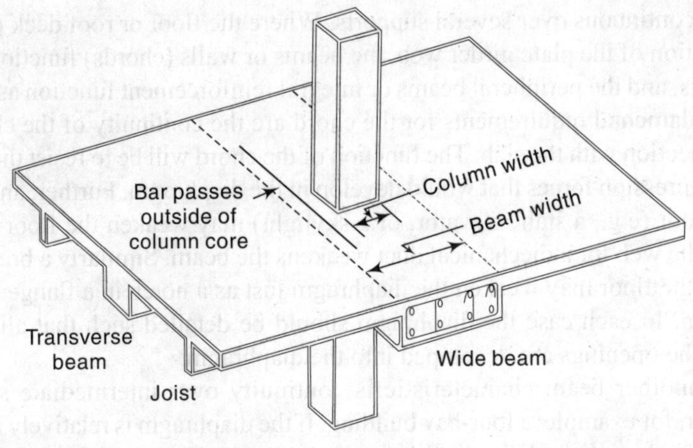

Fig. 8.23 Wide beam

should be laterally supported, either by the girder framing into the same joint or by transverse reinforcement.

(c) In the exterior and corner joints all the 135° hooks of the crossties should be along the outer face of the column. A 135° hook in a crosstie is more effective than a 90° hook to confine core concrete. As the interior face of the beam–column joint is confined by beams, it is preferable to place the crossties such that all the 90° hooks are on the inner side and the 135° hooks at the exterior side of the joint.

8.9 Slabs

Floor slabs forming part of a normal beam-and-slab system are generally designed for gravity loads as flexural members. In addition to their primary function as vertical load-resisting elements, floor slabs also act as elements which distribute earthquake forces into the vertical structural elements; the action is often referred to as *diaphragm action*. The in-plane shear, produced due to earthquake forces acting on the floor, is transferred to stiff earthquake-resisting elements such as shear walls, and is not significant in most building structures. However, reinforcement designed for gravity loads in the slabs forming part of a normal beam-and-slab system will generally be adequate to ensure that the slabs behave satisfactorily both as flexural members and as horizontal diaphragms transmitting earthquake forces. Certain elements, such as flat slabs, waffle slabs, etc., which may form part of the earthquake-resisting framework must, of course, be designed and detailed accordingly.

8.9.1 Diaphragm Action

Horizontal forces at any floor or roof level are distributed to the vertical-resisting elements by using the strength and rigidity of the floor or roof deck to act as a diaphragm. It is customary to consider a diaphragm analogous to a plate girder laid in a horizontal plane, spanning between the vertical resisting element and may be continuous over several supports. Where the floor or roof deck performs the function of the plate girder web, the beams or walls (chords) function as web stiffeners, and the peripheral beams or integral reinforcement function as flanges. The fundamental requirements for the chord are the continuity of the chord and the connection with the slab. The function of the chord will be to resist the tension and compression forces that would develop in the diaphragm. Further, an opening in the floor (e.g., a stair, elevator, or a skylight) may weaken the floor just as a hole in the web for a mechanical duct weakens the beam. Similarly a break in the edge of the floor may weaken the diaphragm just as a notch in a flange weakens the beam. In each case the diaphragm should be detailed such that all stresses around the openings are developed into the diaphragm.

Yet another beam characteristic is continuity over intermediate supports. Consider, for example, a four-bay building. If the diaphragm is relatively rigid, the chords may be designed like the flanges of a beam continuous over the intermediate supports. On the contrary if the diaphragm is flexible, it may be designed as a simple beam spanning between walls with no consideration of continuity; the continuity is simply being neglected. The consequence of the neglect may be some damage where adjacent spans meet.

Another beam analogy applicable to diaphragms is the rigidity of the diaphragm compared to the walls or frame that provide lateral support, and transmit the lateral forces to the ground. A metal-deck roof is relativcly flexible compared to concrete walls, while a concrete floor is relatively rigid compared to steel moment frames.

8.9.2 Ductile Detailing

(a) The minimum bar diameter should be 10 mm.
(b) The minimum content of tension reinforcement in each direction should be 0.15 per cent for high tensile steel, and 0.25 per cent for mild steel. The minimum content of secondary reinforcement should be 0.15 per cent.
(c) For cantilever slabs, bottom steel should be provided to counteract bending tensions, which may occur during earthquakes.
(d) Holes through slabs should be framed with extra steel because of diaphragm action of slabs during earthquakes.
(e) For ground floor or basement slabs, which are designed as ground bearing, special seismic considerations may not exist and it is usual merely to place one layer of nominal steel in each direction to prevent cracking and shrinkage. This is usually placed on top of the slab. Tie steel between column bases may also be placed in the ground slab in some instances, instead of in foundation tie beams.

8.10 Staircases

The staircases in structures may be vulnerable if not detailed properly. When attached rigidly to the floors the flights of the staircase act like braces and cause damage. The following are the three types of stair construction that may be adopted.

Separated staircase One end of the staircase rests on a wall and the other end is carried by columns and beams which have no connection with the floors. The opening at the vertical joints between the floor and the staircase may be covered either with a tread plate attached to one side of the joint and sliding on the other side, or covered with some appropriate material which could crumple or fracture during an earthquake without causing structural damage. The supporting members, columns, or walls are isolated from the surrounding floors by means of separation or crumple sections. A typical example is shown in Fig. 8.24.

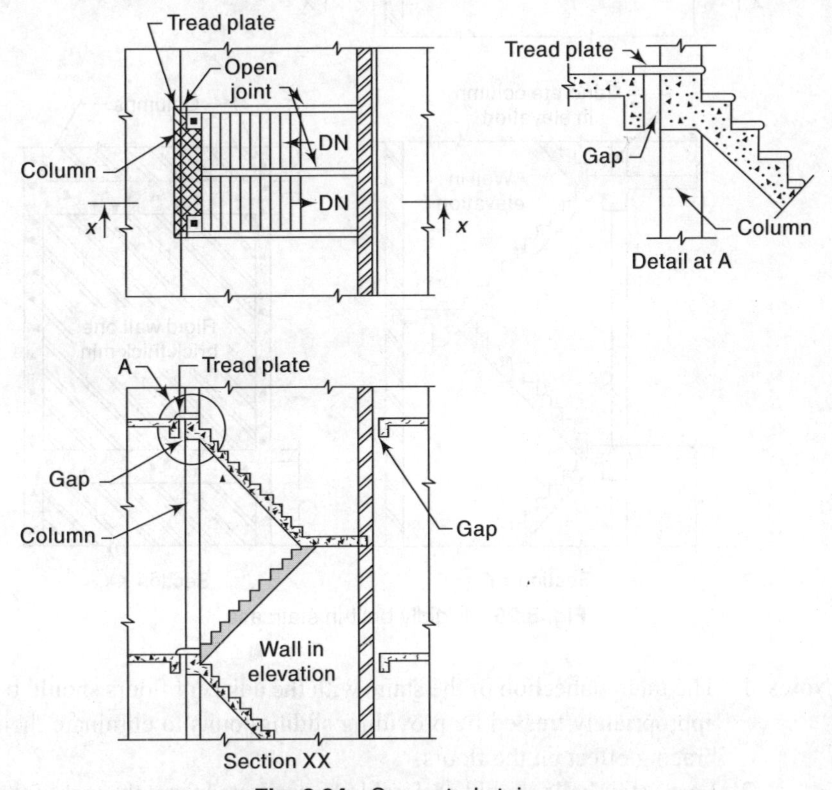

Fig. 8.24 Separated staircase

Built-in staircase When stairs are built monolithically with floors, they can be protected against damage by providing rigid walls at the stair opening. An arrangement, in which the staircase is enclosed by two walls, is given in Fig. 8.25. In such cases, the joints, as mentioned in respect of separated staircase, will not be necessary. The two walls mentioned above, enclosing the staircase, should extend through the entire height of the stairs and to the building foundations.

Staircase with sliding joints In case it is not possible to provide rigid walls around stair openings for built-in staircase or to construct separated staircase, the staircase should have sliding joints so that they will not act as diagonal bracing.

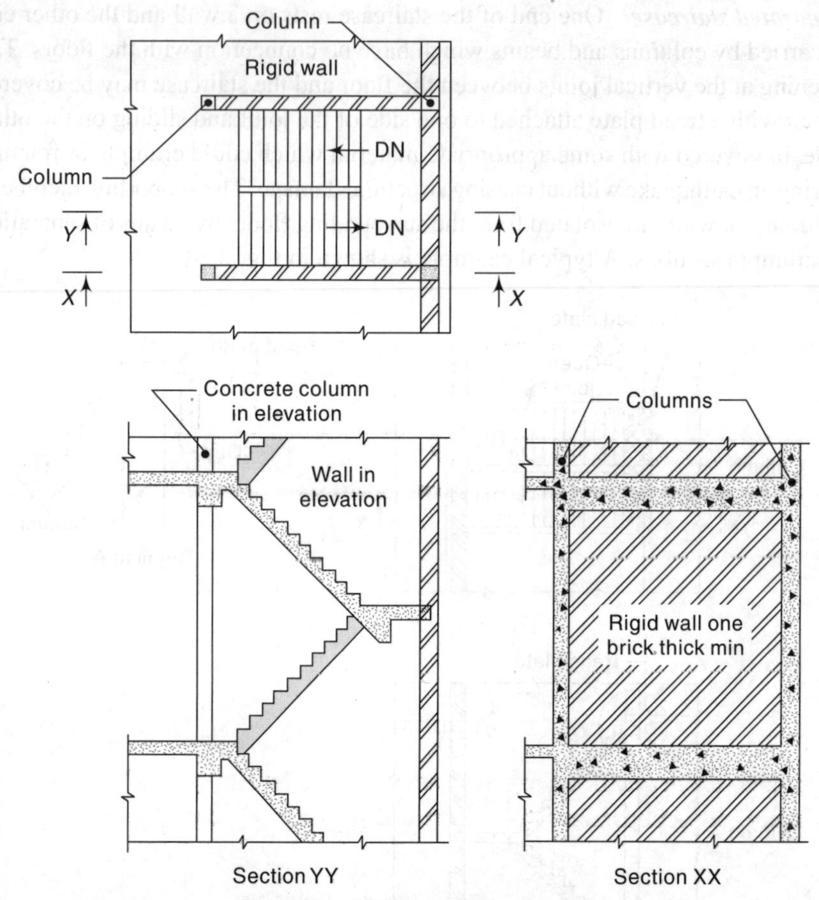

Fig. 8.25 Rigidly built-in staircase

Notes: 1. The interconnection of the stairs with the adjacent floors should be appropriately treated by providing sliding joints to eliminate their bracing effect on the floors.

2. Large stair halls should preferably be separated from the rest of the building by means of separation or crumple sections.

Ductile Detailing

1. Generally, the rules for slabs apply here also. Top steel should be provided at each landing to provide for bending tensions, which may not be apparent from simple analysis.
2. If stairs are part of the horizontal diaphragm or the moment-resisting framework, they should be reinforced accordingly. Due care must then be taken at the change in slope to confine the longitudinal bars.

8.11 Upstands and Parapets

Upstands and parapets should be carefully designed against seismic accelerations, which may considerably exceed those occurring elsewhere in the structure due to resonance. The arrangement of reinforcement at corners and junctions should be as for walls and is shown in Fig. 8.26.

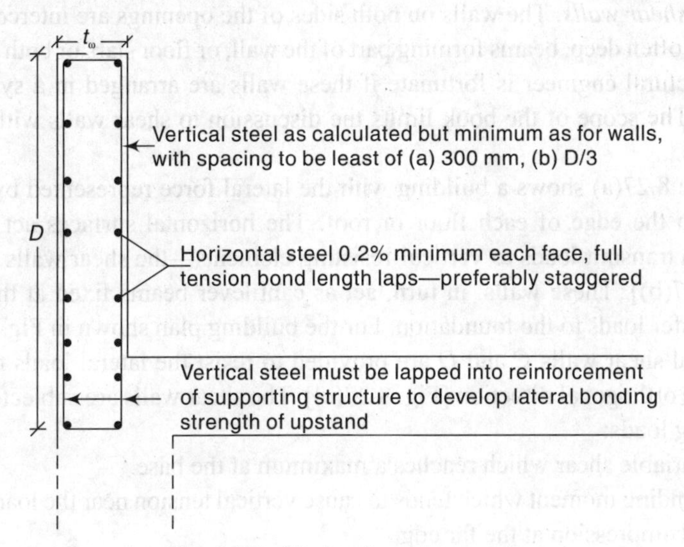

Fig. 8.26 Upstands and parapets

8.12 Shear Walls

The walls, in a building, which resist lateral loads originating from wind or earthquakes are known as *shear walls*. A large portion of the lateral load on a building, if not the whole amount, as well as the horizontal shear force resulting from the load, are often assigned to such structural elements made of RCC. These shear walls, may be added solely to resist horizontal force, or concrete walls enclosing stairways, elevated shafts, and utility cores may serve as shear walls. Shear walls not only have a very large in-plane stiffness and therefore resist lateral load and control deflection very efficiently, but may also help to ensure development of all available plastic hinge locations throughout the structure prior to failure. The other way to resist such loads may be to have the rigid frame augmented by the combination of masonry walls.

The use of shear walls or their equivalent becomes imperative in certain high-rise buildings, if inter-storey deflections caused by lateral loadings are to be controlled. Well-designed shear walls not only provide adequate safety but also give a great measure of protection against costly non-structural damage during moderate seismic disturbances.

The term shear wall is actually a misnomer as far as high-rise buildings are concerned, since a slender shear wall when subjected to lateral force has predominantly moment deflections and only very insignificant shear distortions. High-rise structures have become taller and more slender, and with this trend the analysis of shear walls may emerge as a critical design element. More often than not, shear walls are pierced by numerous openings. Such shear walls are called *coupled shear walls*. The walls on both sides of the openings are interconnected by short, often deep, beams forming part of the wall, or floor slab, or both of these. The structural engineer is fortunate if these walls are arranged in a systematic pattern. The scope of the book limits the discussion to shear walls without any openings.

Figure 8.27(a) shows a building with the lateral force represented by arrows acting on the edge of each floor or roof. The horizontal surfaces act as deep beams to transmit loads to vertical-resisting elements—the shear walls *A* and *B* [Fig. 8.27(b)]. These walls, in turn, act as cantilever beams fixed at their base and transfer loads to the foundation. For the building plan shown in Fig. 8.27(a), additional shear walls *C* and *D* are provided to resist the lateral loads that may act in the orthogonal direction [Fig. 8.27(c)]. The shear walls are subjected to the following loads:

(a) A variable shear which reaches a maximum at the base.
(b) A bending moment which tends to cause vertical tension near the loaded edge and compression at the far edge.
(c) A vertical compression due to ordinary gravity loading from the structure.

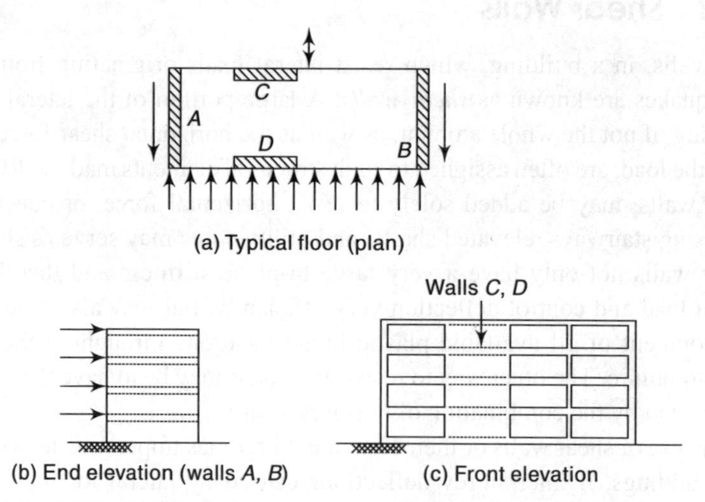

Fig. 8.27 Building with shear walls subject to horizontal loads

8.13 Behaviour of Shear Walls

The behaviour of shear walls, with particular reference to their typical mode of failure is, as in the case of beams, influenced by their proportions as well as their

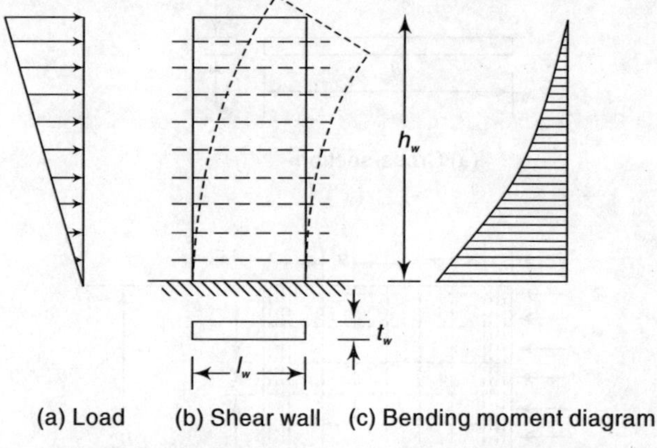

(a) Load (b) Shear wall (c) Bending moment diagram

Fig. 8.28 Behaviour of a cantilever shear wall

support conditions. Low shear walls also known as *squat walls*, characterized by relatively small height-to-length ratios, may be expected to fail in shear just like deep beams. Shear walls occurring in high-rise buildings, on the other hand, generally behave as vertical cantilever beams (Fig. 8.28) with their strength controlled by flexure rather than by shear. Such walls are subjected to bending moments and shears originating from lateral loads, and to axial compression caused by gravity. These may, therefore, be designed in the same manner as regular flexural elements. When acting as a vertical cantilever beam, the behaviour of a shear wall which is properly reinforced for shear (i.e., diagonal tension) will be governed by the yielding of the tension reinforcement located near the vertical edge of the wall and, to some degree, by the vertical reinforcement distributed along the central portion of wall.

It is thus evident that shear is critical for walls with relatively low height-to-length ratios and tall shear walls are controlled mainly by flexural requirements particularly if only uniformly distributed reinforcement is used. Figure 8.29 shows a typical shear wall of height h_w, length l_w, and thickness t_w. It is assumed to be fixed at its base and loaded horizontally along its left edge. Vertical flexural reinforcement of area A_s is provided at the left edge, with its centroid at a distance d_w from the extreme compression face. To allow for reversal of load, identical reinforcement is provided along the right edge. Horizontal reinforcement of area A_h at spacing S_2, as well as vertical reinforcement of area A_v at spacing S_1 is provided as shear reinforcement. Distribution of a minimum reinforcement vertically and horizontally helps to control the width of inclined cracks. Such distributed steel normally is placed in two layers, parallel to both faces of the wall.

Since the ductility of a flexural member such as a tall shear wall can be significantly affected by the maximum usable strain in the compression zone concrete, confinement of concrete at the ends of the shear wall section would improve the performance of such shear walls. Such confinement can take the

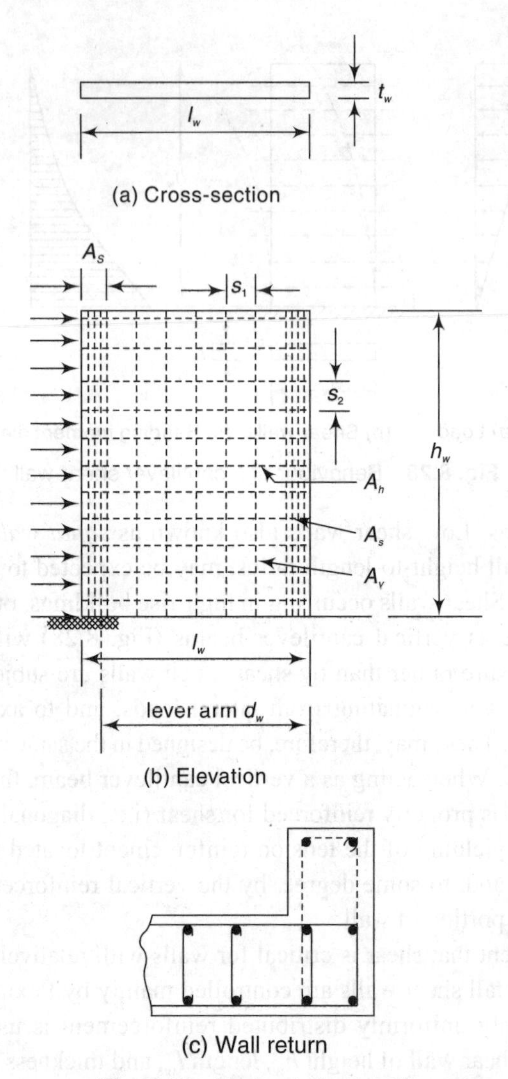

Fig. 8.29 Flanged shear wall

form of enlarged boundary elements with adequate confining reinforcement. In the flanged wall sections, the adjacent parts of the wall will provide lateral support to each other. Confinement may also be obtained from the presence of other walls running at right angles to the shear wall at its ends. In both cases, the additional compression flanges contribute to the increase in ductility. Shear wall sections are often thin and therefore, under reversed cyclic yielding there is a danger of section instability. The wall-return, as shown in Fig. 8.29(c), may usually be necessary between the ground and first floors of a building to increase stability.

It is also required that the vertical (longitudinal) forces resulting from seismic loads are resisted entirely by the boundary elements. It is similar to the design approach used in steel I-beams where flanges resist the flexural stresses and the web (wall panel in the shear walls) carries the entire shear. In shear walls of high-rise structures, enough shear capacity is provided so that a shear failure does not precede a flexural failure.

However, a portion of a shear wall, which interacts with the frames, may behave as a low shear wall, depending upon the proportions of the walls and the location of the point of contraflexure along the height of the wall. The latter is dependent primarily on the relative stiffness of the frame and the shear wall elements in a structure.

8.14 Tall Shear Walls

In multistorey buildings, the shear walls are slender enough and are idealized as cantilevers fixed at base. Their seismic response is dominated by flexure. Because of load reversals, shear wall sections necessarily contain substantial quantities of compression reinforcement. IS 1893: 2002 has laid down the procedure to assess the flexural and shear strengths of tall shear walls and is described in the following subsections.

8.14.1 Flexural Strength

In shear walls, particularly in areas not affected by earthquakes, the strength requirement for flexural steel is not great. Traditionally, the practice is to provide about 0.25 per cent reinforcement (reinforcement ratio of 0.0025 of the gross area) uniformly in both directions over the entire depth as shown in Fig. 8.30(a). Naturally, such an arrangement does not efficiently utilize the steel at the ultimate moment because many bars operate on a relatively small lever arm. Moreover, the ultimate curvature, hence the curvature ductility, is considerably reduced and this arrangement is also uneconomical.

In an efficient shear wall section subjected to considerable moments, the bulk of the flexural reinforcement is placed close to the tensile edge. Because of moment reversals originating from lateral loads, equal amounts of reinforcement are normally required at both extremities [Fig. 8.30(b)]. Thus, a considerable part of bending moment can be resisted by the internal steel couple, and this will result in improved ductility properties. The practice is to provide minimum reinforcement (0.25 per cent) over the inner 80 per cent depth and allocate the remainder of the steel to outer (10 per cent) zones of the section [Fig. 8.30(b)]. As shown by the theoretical moment–curvature relationship in Fig. 8.30(c), this distribution of steel results in an increase in the available strength and ductility.

To increase the ductility of cantilever shear walls at the base, where the overturning moments and axial compression are the largest, the concrete in the compression zone must be confined. The confining steel is provided in the same way as in tied columns and can be extended over that part of the depth l_w, where

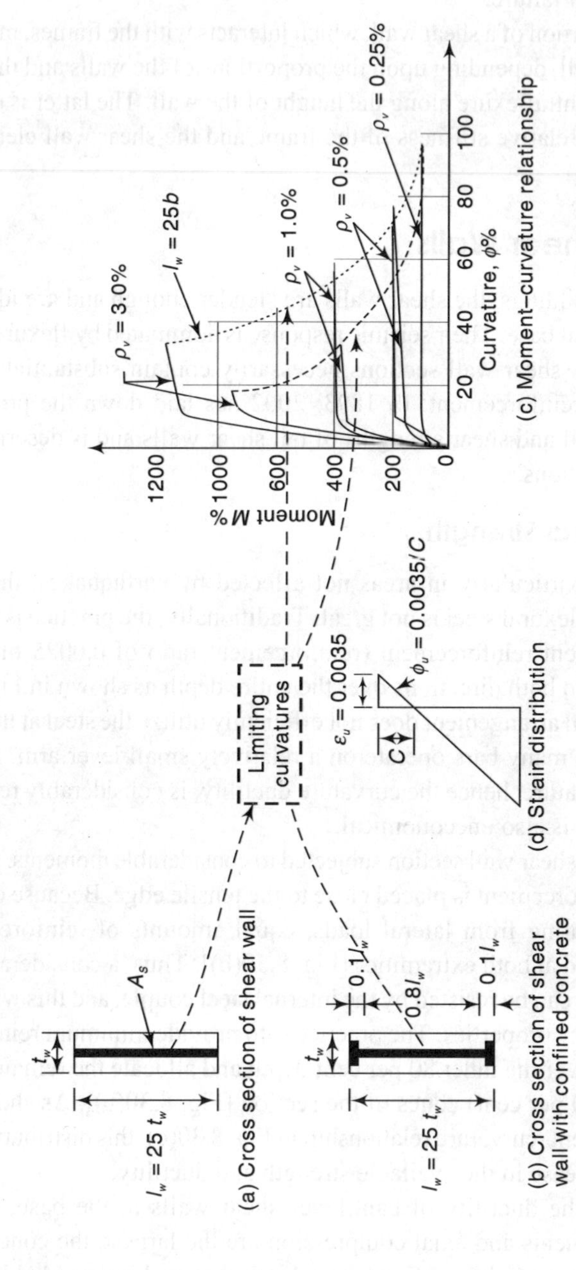

Fig. 8.30 Effect of amount and distribution of vertical reinforcement on ultimate curvature (Cardenas et al. 1973)

concrete strains in excess of 0.0035 are required. This is particularly important over the region of a possible plastic hinge, which may extend over a full storey height or more. The flexural strength of a slender rectangular shear wall section with uniformly distributed vertical reinforcement and subjected to uniaxial bending and axial load may be estimated as follows:

Case I

For $\dfrac{x_u}{l_w} = \dfrac{x_u^*}{l_w}$

$$\frac{M_{ux}}{f_{ck} t_w l_w^2} = \phi\left[\left(1+\frac{\lambda}{\phi}\right)\left(\frac{1}{2} - 0.416\frac{x_u}{l_w}\right) - \left(\frac{x_u}{l_w}\right)^2 \left(0.168 + \frac{\beta^2}{3}\right)\right] \tag{8.6}$$

where

$$\frac{x_u}{l_w} = \frac{\phi + \lambda}{2\phi + 0.36} \tag{8.7}$$

$$\frac{x_u^*}{l_w} = \frac{0.0035}{0.0035 + \dfrac{0.87 f_y}{E_s}} \tag{8.8}$$

$$\phi = \frac{0.87 f_y \rho}{f_{ck}}, \quad \lambda = \frac{P_u}{f_{ck} t_w l_w}$$

$$\rho = \frac{A_{st}}{t_w l_w}, \quad \beta = \frac{0.87 f_y}{0.0035 E_s}$$

where x_u is the depth of the neutral axis from extreme compression flange, x_u^* is the balanced depth of neutral axis, α is the inclination of the diagonal reinforcement in the coupling beam, β is the soil-foundation factor (IS 1893: 2002), ρ is the vertical reinforcement ratio, A_{st} is the area of uniformly distributed vertical reinforcement, E_s is the elastic modulus of steel, and P_u is the axial compression on the wall.

Case II

For $\dfrac{x_u^*}{l_w} < \dfrac{x_u}{l_w} < 1.0$

$$\frac{M_{uv}}{f_{ck} t_w l_w^2} = \alpha_1\left(\frac{x_u}{l_w}\right) - \alpha_2\left(\frac{x_u}{l_w}\right)^2 - \alpha_3 - \frac{\lambda}{2} \tag{8.9}$$

where $\alpha_1 = \left[0.36 + \phi\left(1 - \dfrac{\beta}{2} - \dfrac{1}{2\beta}\right)\right]$

$$\alpha_2 = \left[0.15 + \frac{\phi}{2}\left(1 - \beta - \frac{\beta^2}{2} - \frac{1}{3\beta}\right)\right] \text{ and}$$

$$\alpha_3 = \frac{\phi}{6\beta}\left[\left(\frac{1}{x_u/l_w}\right) - 3\right]$$

The value of (x_u/l_w) to be used in this equation, can be calculated from the following quadratic equation

$$\alpha_1\left(\frac{x_u}{l_w}\right)^2 + \alpha_4\left(\frac{x_u}{l_w}\right) + \alpha_5 = 0$$

$$\alpha_4 = \left(\frac{\phi}{\beta} - \lambda\right) \quad \text{and} \quad \alpha_5 = \left(\frac{\phi}{2\beta}\right)$$

Equations (8.6) and (8.9) have been derived assuming a rectangular wall section of depth l_w and thickness t_w that is subjected to combined uniaxial bending and axial compression. The vertical reinforcement is represented by an equivalent steel plate along the length of the section. The stress–strain curve assumed for concrete is as per IS 456: 2000, whereas that for steel is assumed to be bilinear. Equations (8.6) and (8.9) are given for calculating the flexural strength of the section. Their use depends on whether the section fails in flexural tension or in flexural compression.

8.14.2 Shear Strength

The shear strength of tall shear walls can be assessed in the same way as for beams, with due allowance made for the contribution of axial compression in boosting the share of the concrete shear-resisting mechanism. In doing so, the adverse effect of vertical acceleration induced by earthquakes should also be considered. At the base of the wall, where yielding of the flexural steel is possible in both faces, the contribution of the concrete towards the shear strength should be neglected and shear reinforcement in the form of horizontal stirrups should be provided at least over the possible length of the plastic hinge, to carry all the shear force. The minimum reinforcement of 0.25 per cent in the horizontal direction, when appropriately anchored, is found to be sufficient. The effective depth of the rectangular shear wall can be taken as greater than $0.8l_w$. It must be noted that flanges of the shear wall are not taken into account while calculating the shear strength.

In the potential plastic hinge zone, wide flexural cracks combine with diagonal tension cracks, due to shear. The effect of diagonal cracking on the distribution of flexural shear stresses should be considered in the same way as in beams.

8.14.3 Construction Joints

There are two potential locations in cantilever shear walls where failure by sliding shear can occur. One is a horizontal construction joint and the other is the plastic hinge zone, usually immediately above the foundation level. The inelastic response of mechanisms associated with sliding shear indicates drastic loss of stiffness and strength with reversed cyclic loading. Therefore, sliding shear should be considered as being an unsuitable energy dissipating mechanism in earthquake resistant structures.

Earthquake damage in shear walls is more common at construction joints along which sliding movement may occur (more common in low shear walls, which carry small gravity loads) necessitating efficient vertical reinforcement to check sliding. The shear force that can be safely transferred across a well prepared rough horizontal joint is given by

$$V_j = \mu(P_u + 0.87 f_y A_v) \tag{8.10}$$

where P_u is the factored axial force on the section (positive when producing compression), A_v is the vertical steel to be utilized, and μ is the coefficient of the friction at the joint ($\mu = 1.0$).

For shear walls, gravity loads with 20 per cent reduction to account for negative vertical acceleration are considered

$$V_j = 0.8P_u + 0.87 f_y A_v \tag{8.11}$$

The strength of construction joint is

$$\tau_{vf} = \frac{V_j}{A_g} \tag{8.12}$$

and this must be equal to but preferably greater than the diagonal tension shear strength of the wall.

The steel content across the construction joint is given by

$$\rho_{vf} = \frac{A_v}{A_g}$$

The vertical reinforcement ratio, ρ_v across a horizontal construction joint should not be less than

$$\left(\tau_v - \frac{P_u}{A_g}\right)\frac{0.92}{f_y} \geq 0.0025$$

where τ_v is the factored shear stress at the joint, P_u is the factored axial force (positive for compression), and A_g is the gross cross-sectional area of the joint.

8.15 Squat Shear Walls

In most low-rise buildings, the height of cantilever shear walls is less than their length (i.e., their structural depth). So the technique used to assess the flexural and shear strength for tall cantilever shear walls does not apply here. However, squat shear walls resemble deep beams to some extent. Although the behaviour of squat shear walls is assumed to be analogous to deep beams, there is a difference. In deep beams, arch action prevails because of the type of loading system. The stirrups crossing the main diagonal cracks, which form between the load points and the supports, are not engaged in efficient shear resistance because no compression struts can form between stirrup anchorages. For shear walls, the load is introduced along the joint between floor slabs and walls as a live load. Clearly, no arch action prevails with this type of loading.

Low shear walls normally carry only very small gravity loads. So the beneficial effect of gravity loads in shear walls (shear strength) is absent. However, large internal lever arm provided results for a small flexural steel demand. It is more practical, therefore, to distribute the vertical (i.e., flexural) reinforcement uniformly over the full length of the wall, allowing only a nominal increase at the vertical edges. Also, loss of ductility for seismic loading is not likely to be of great importance.

The crack pattern of a shear wall [Fig. 8.31(a)] reveals the formation of diagonal struts, [Fig. 8.31(b)] hence the engagement of stirrups is necessary. After diagonal cracking, the horizontal shear introduced at the top of a squat shear wall will need to be resolved into diagonal compression and vertical tensile forces. The distributed vertical flexural reinforcement will enable the shear to be transmitted to the foundation. The equilibrium condition of free body marked 2 in Fig. 8.31(b) shows this. The free body marked 1 in Fig. 8.31(c) does not find a support at the foundation level and, therefore, requires an equal amount of horizontal shear reinforcement. In the absence of external vertical compression, the horizontal and vertical steel must be equal to enable 45° compression diagonal to be generated.

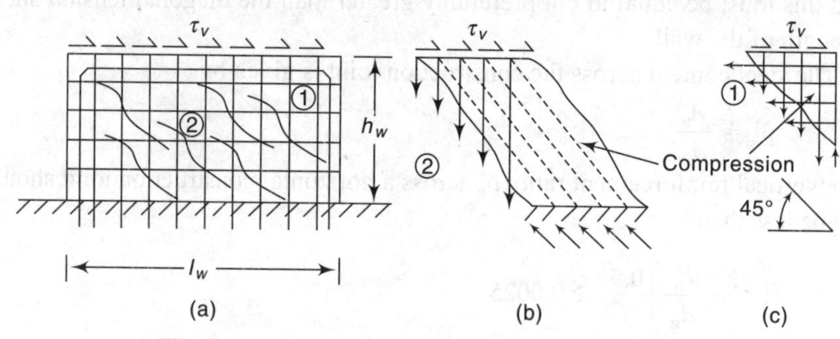

Fig. 8.31 Shear resistance of low-rise shear walls

In the free body diagram [Fig. 8.31(b)], only vertical forces equal to the shear intensity need to be generated to develop the necessary diagonal compression. This steel is called shear reinforcement, even though its principal role is to resist the moment that tends to overturn the free body shown in Fig. 8.31(b). The shear reinforcement for squat shear walls for height-to-length ratios between 0.5 to 2.5 is given by

$$\rho_v = 0.0025 + 0.5\left(2.5 - \frac{h_w}{l_w}\right)(\rho_h - 0.0025) \qquad (8.13)$$

where ρ_v and ρ_h are the vertical and horizontal steel content per unit wall area, V_u is the nominal shear stress transferred across the joint, and V_c is the nominal shear stress taking into account the presence of axial load. The shear reinforcement given by Eqn (8.13) should not be less than 0.025.

8.16 Design of Shear Walls

Shear wall construction is an economical method of bracing buildings to limit damage. For good performance of well designed shear walls, the shear wall structures should be designed for greater strength against lateral loads than ductile reinforced concrete frames with similar characteristics; shear walls are inherently less ductile and perhaps the dominant mode of failure is shear. With low design stress limits in shear walls, deflection due to shear forces is small. However, exceptions to the excellent performance of shear walls occur when the height-to-length ratio becomes great enough to make overturning a problem and when there are excessive openings in shear walls. Also, if the soil beneath its footing is relatively soft, the entire shear wall may rotate, causing localized damage around the wall. Following are the design steps of cantilever shear walls.

General Requirements

(a) The thickness of the shear wall should not be less than 150 mm to avoid unusually thin sections. Very thin sections are susceptible to lateral instability in zones where inelastic cyclic loading may have to be sustained.

(b) The effective flange width for the flanged wall section from the face of web (wall) should be taken as least of
- half the distance to an adjacent shear wall web, and
- one-tenth of total wall height.

(c) The minimum reinforcement in the longitudinal and transverse directions in the plan of the wall should be taken as 0.0025 times the gross area in each direction and distributed uniformly across the cross-section of wall. This helps in controlling the width of inclined cracks that are caused due to shear.

(d) If the factored shear stress in the wall exceeds $0.25\sqrt{f_{ck}}$ or if the wall thickness exceeds 200 mm, the reinforcement should be provided in two curtains, each having bars running in both the longitudinal and transverse directions in the plane of the wall. The use of reinforcement in two curtains reduces fragmentation and premature deterioration of the concrete under cyclic loading.

(e) The maximum spacing of reinforcement in either direction should be lesser than $l_w/5$, $3t_w$, and 450 mm, where l_w is the horizontal length and t_w is the thickness of the wall web.

(f) The diameter of the bars should not exceed one-tenth of the thickness of that part. This puts a check on the use of very large diameter bars in thin wall sections.

Shear Strength

The provisions for shear strength are almost the same as those of RC beams. The increase in shear strength may also be considered. However, for this, only 80 per cent of the factored axial force is considered as effective. This reduction of 20 per cent is made to account for possible effect of vertical acceleration.

(a) The nominal shear stress is

$$\tau_v = \frac{V_u}{t_w d_w} \tag{8.14}$$

where V_u is the factored shear force, t_w is the thickness of web, and d_w is the effective depth of the wall section (may be taken as $0.8l_w$).

(b) The design shear strength of concrete (τ_c) should be as per IS 456: 2000.

(c) The nominal shear stress, τ_v, should not be greater than $\tau_{c,\max}$. The value of $\tau_{c,\max}$ can be found from IS 456: 2000. If $\tau_v < \tau_c$, minimum shear reinforcement of 0.25 per cent should be provided in the horizontal direction. If $\tau_v > \tau_c$, the area of horizontal shear reinforcement A_h at a vertical spacing S_v can be determined from the expression

$$V_{us} = \frac{0.87 f_y A_h d_w}{S_v} \tag{8.15}$$

where V_{us} is the shear force to be resisted by the horizontal reinforcement and is given by

$$V_{us} = V_u - \tau_c t_w d_w \tag{8.16}$$

(e) Uniformly distributed vertical reinforcement not less than the horizontal reinforcement should be provided. This is particularly important for squat walls. When the height-to-width ratio is about 1.0, both the vertical and horizontal reinforcement are equally effective in resisting the shear force.

Flexural Strength

The moment of resistance of short shear walls is calculated as for columns subjected to combined bending and axial load. The procedure for the calculation of moment of resistance M_{uv} of tall rectangular shear walls is as described in Section 8.14.

For walls without boundary elements, the vertical reinforcement is concentrated at the ends of the walls. A minimum of four bars, 12 mm ϕ, arranged in two layers, are provided at each end.

Boundary Elements

These are the portions along the wall edges and may have the same or greater thickness than the wall web. These are provided throughout the height with special confining reinforcement. Wall sections having stiff and well confined boundary elements develop substantial flexural strength, are less susceptible to lateral buckling and have better shear strength and ductility in comparison to plane rectangular walls not having stiff and well-confined boundary elements.

(a) During a severe earthquake, the ends of a wall are subjected to high compressive and tensile stresses. Hence, the concrete needs to be well confined so as to sustain the load reversals without a large deterioration in strength. Thus, the boundary elements are provided along the vertical boundaries of walls, when the extreme fibre compressive stress in the wall due to factored gravity load plus factor earthquake force exceeds $0.2f_{ck}$. The boundary element may be discontinued where the calculated compressive stress becomes less than $0.15f_{ck}$.

(b) The boundary element is assumed to be effective in resisting the design moment due to earthquake-induced forces, along with the web of the wall. The boundary element should have an adequate axial load carrying capacity (assuming short-column action) so as to carry an axial compression equal to the sum of the factored gravity load plus compressive load due to seismic load. The latter may be calculated as

$$P_c = \frac{M_u - M_{uv}}{C_w} \tag{8.17}$$

where M_u is the factored design moment on the entire wall section, M_{uv} is the moment of resistance provided by the distributed reinforcement across the wall section, and C_w is the c/c distance between the boundary elements along the two vertical edges of the wall.

(c) Moderate axial compression results in higher moment capacity of the wall. Hence, the beneficial effect of axial compression by gravity loads should not be fully relied upon in a design, due to the possible reduction in its magnitude by vertical acceleration. When gravity loads add to the strength of the wall, a load factor of 0.8 may be taken.

(d) The percentage of vertical reinforcement in boundary elements should range between 0.8 and 6 per cent (the practical upper limit is four per cent).

(e) During a severe earthquake, boundary elements may be subjected to stress reversals. Hence, they have to be confined adequately to sustain the cyclic loading without a large degradation in strength. Therefore, these should be provided throughout their height.

(f) Boundary elements need not be provided if the entire wall section is provided with special confining reinforcement.

8.17 Restoration and Strengthening

The greatest challenge to the engineer fraternity is to retrofit or rehabilitate the damaged buildings by understanding seismic deficiencies the structures had. At the same time, engineers are equally concerned with the techniques to improve the performance of the buildings having inadequate lateral load-resisting systems.

8.17.1 Restoration

Restoration is the restitution of strength that the building had before the damage occurred. Restoration must be undertaken when there is evidence that structural damage can be attributed to exceptional phenomena that are not likely to happen again and that the original strength provides an adequate level of safety. The main purpose of restoration is to carry out structural repair to load-bearing elements. It may involve cutting portions of elements and rebuilding them or simply adding more structural material so that the original strength is more or less restored. The process may involve inserting temporary supports, underpinning, etc. Some of the approaches are:

(a) Addition of reinforcing mesh on both faces of the cracked wall, holding it to the wall through spikes or bolts, and then covering it suitably with gunite, etc.
(b) Injecting epoxy like material, which is strong in tension, into the cracks in the walls, columns, beams, etc.

Where structural repairs are considered necessary, these should be carried out prior to or simultaneously with the architectural repairs, so that total planning of work could be done in a coordinated manner and wastage is avoided.

8.17.2 Strengthening

The process of *strengthening* involves improving the original strength of the structure. It is carried out when the evaluation of the building indicates that the strength available before the damage was insufficient and restoration alone will not be adequate for resistance of future earthquakes. The extent of the modifications must be determined by the general principles and design methods and should not be limited to increasing the strength of members that have been damaged, but should consider the overall behaviour of the structure. Commonly, strengthening procedures should aim at one or more of the following objectives:

(a) Increasing the lateral strength in one or both directions, by reinforcement or by increasing wall areas or the number of walls and columns.
(b) Giving unity to the structure by providing proper connection between its resisting elements in such a way that inertia forces generated by the vibration of the building can be transmitted to the members that have the ability to resist them. Typical important aspects are the connections between roofs, floors, and walls, between intercepting walls, and between walls and the foundation.

(c) Eliminating features that are sources of weakness or that produce concentration of stresses in some members. Asymmetrical plan distribution of resisting members, abrupt changes of stiffness from one floor to the other, concentration of large masses, and large openings in walls without a proper peripheral reinforcement are examples of defects of this kind.

(d) Avoiding the possibility of brittle modes of failure by proper reinforcement and connection of resisting members. Since its cost may go to as high as 50 to 60 per cent of the cost of rebuilding, the implementation of such strengthening must be well justified.

The strengthening of RC members is a specialized job and should be carried out by a structural engineer according to calculations. The following are a few ways to strengthen a RC member:

(a) Reinforced concrete columns can best be strengthened by jacketing and by providing an additional cage of longitudinal and lateral tie reinforcement around the columns and casting a concrete ring (Fig. 8.32). The desired strength and ductility can thus be built up.

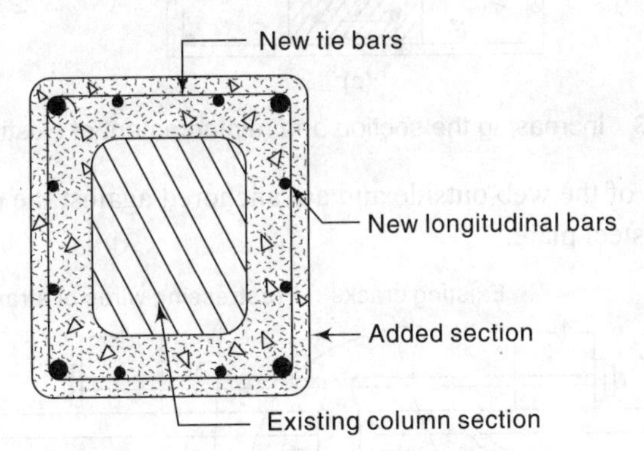

Fig. 8.32 Jacketing a concrete column section

(b) Jacketing of an RC beam can also be carried out in the same way as described for reinforced concrete columns. The structural capacity is enhanced by increasing the section with RCC [Figs 8.33(a) and 8.33(b)] when there are no limitations on beam depths; Fig. 8.33(c) demonstrates the procedure used when there is such a limitation. For holding the stirrups in this case, holes will have to be drilled through the slab. A similar technique could be used for strengthening RC shear walls.

(c) Reinforced concrete beams can also be strengthened by applying prestress to it so that opposite moments are caused to those applied. The longitudinal prestress increases the capacity to resist shear and flexure and forces new diagonal tension cracks to be inclined reducing the ductility of the beam. The loss of ductility can be compensated for by providing the beam with prestressed exterior transverse reinforcement (Fig. 8.34). The wires run on

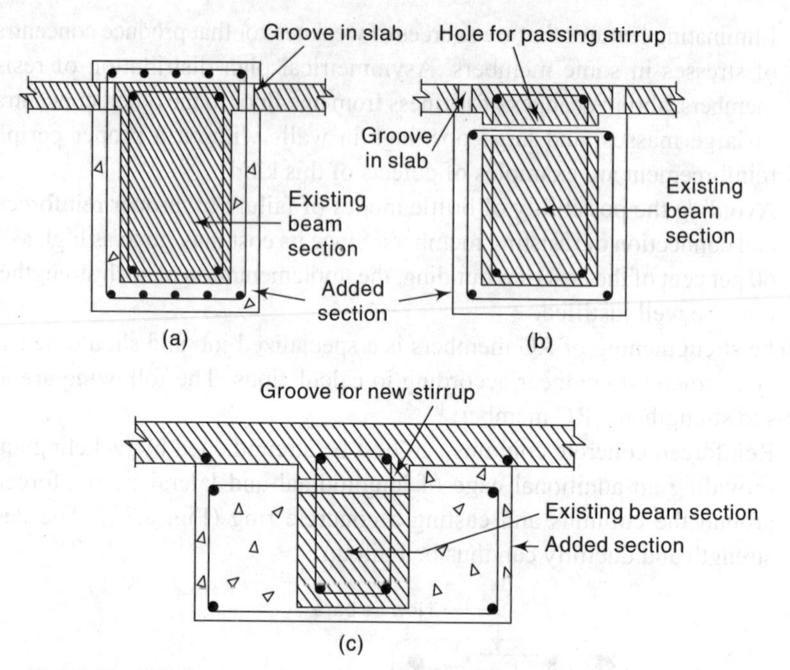

Fig. 8.33 Increasing the section and reinforcement of existing beams

both sides of the web outside and are anchored against the end of the beam through a steel plate.

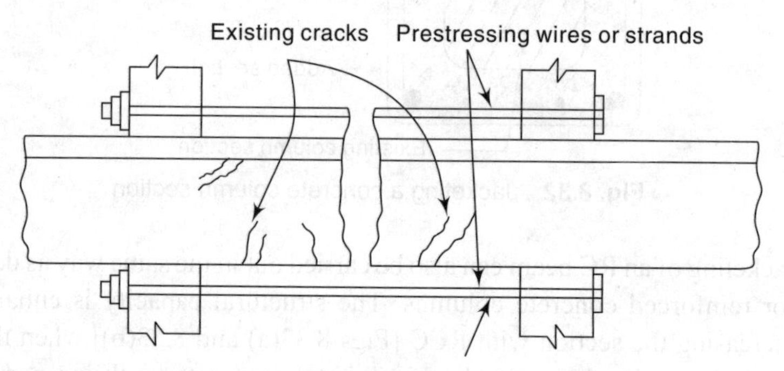

Fig. 8.34 External longitudinal prestressing

(d) In case of severely damaged RC members, it is possible that the reinforcement would have buckled or elongated, or excessive yielding may have occurred. Such elements can be repaired by replacing the old portion of the steel with new steel using butt wielding or lap welding. Splicing by overlapping is risky. If repairs have to be done without removal of the existing steel, the best approach depends upon the space available in the original member. Additional stirrup ties are to be added in the damaged portion before concreting, so as to confine the concrete and include the longitudinal bars to prevent their buckling in the future.

In some cases it may be necessary to anchor additional steel into existing concrete. A common technique for providing the anchorage consists of drilling a hole larger than the bar diameter. The hole is then filled with epoxy expanding cement, or other high strength grouting material. The bar is punched into place and held there until the grout has set.

(e) Strengthening of beams, columns and slabs using fibre-reinforced plastic (FRP) sheets is gaining popularity these days. The FRP sheets are glued to the concrete surfaces. FRP increases the strength of the member in bending, shear, and compression, but its effect on the stiffness is not positive.

The affected columns are completely covered with FRP sheets either by banding the RC column with continuous FRP straps glued on the concrete surface using epoxy resin, or by encasement using FRP sheets glued onto the concrete surface.

For beams, the FRP sheets are glued either on the lower faces of the beam under repair (for strengthening of the tension zone), or on the vertical sides of the beam near the supports (for shear strengthening). This process should be preceded by crack repair with epoxy resin. The glued sheets may then be protected by welded wire mesh and cement plaster or shotcrete.

8.18 Prestressed Concrete Construction

Although prestressed concrete is well established in bridge constructions, it is less widely used in building structures; in practice the use of prestressed concrete frames for seismic resistance is uncommon. This is true in non-seismic areas as well. Its main use in buildings is for floor and cladding components, which are not required to resist seismic forces in a ductile manner. The comparative neglect of prestressed concrete for building structures has occurred partly for constructional and economic reasons, and in earthquake areas it has also occurred because of divergent opinions on the effectiveness of prestressed concrete in resisting earthquakes. Very little data is available for proper assessment of seismic response characteristics of prestressed concrete, further preventing its use. Earthquake response of prestressed concrete structures is significantly higher than RC, and so design earthquake forces are stipulated to be approximately 20 per cent higher than those normally specified for RC structures. This is mainly because of the lack of research data. Codes of practice give little detailed guidance on the seismic design of prestressed concrete and official attitudes are, therefore, cautious towards its use in building structures. Other concerns are the lower energy dissipation characteristics, reduced ductility, and difficulty in predicting the ultimate moment capacity under reversed loading. The use of ungrouted (unbonded) tendons is not recommended. Such tendons are used mainly to balance gravity loads and, if used, the non-prestressed reinforcement should also be used to provide seismic resistance.

The principles of RCC design are equally applicable to prestressed concrete frames. The design checks are required at two levels of earthquake:

(i) Serviceability limit state related to a moderate earthquake.
(ii) Ultimate limit state related to a severe earthquake.

The requirements for the above two states are:

(i) For a moderate earthquake there must be no loss of prestress. For this to occur, using elastic theory, the strain in prestressing steel should exceed neither the limit of proportionality, nor the strain at transfer.
(ii) For a severe earthquake, analysis should take into account elasto-plastic deformation and the ultimate limit state and verify that the structure is safe from collapse.

8.18.1 Specifications

Since the Bureau of Indian Standards is silent on prestressed concrete, some of the specifications for frame members as given by other codes are presented below.

Beams To guarantee sufficient ductility, the steel ratio should not be greater than 0.2 and the flexural cracking load should not be larger than the ultimate flexural strength. Shear reinforcement should be provided so that flexure failure precedes shear failure. The region of a potential plastic hinge is taken to be $2h$, where h is the depth. Stirrups should be included in this region to ensure concrete confinement, prevent buckling of reinforcing bars, and act as shear reinforcement. Stirrups spacing should not be greater than 150 mm, $d/4$, or six longitudinal bar diameter, whichever is the smallest.

Columns Ultimate flexural strength should not be smaller than the flexural cracking moment. Shear reinforcement should be used to ensure that flexural failure precedes shear failure. Transverse reinforcing bars should be provided in any potential plastic hinge region, with spacing not greater than one-fifth of the column width, six longitudinal bar diameters, or 200 mm, whichever is the smallest.

Connections Since a joint core is subjected to high diagonal tension under earthquake loading, prestressing steel should not be anchored at the joint core. For a prestressed beam framed into an exterior column, prestressing steel may be anchored in a concrete stub attached on the outer surface of the column. Shear design of a joint core should follow the design specifications for RC joints. Prestressing steel placed at mid-depth of a beam is effective in resisting the diagonal tension of the joint core.

8.18.2 Characteristics

Idealized forms of hysteresis diagrams for prestressed concrete are shown in Fig. 8.35. Some of the important characteristics of prestressed concrete are described below.

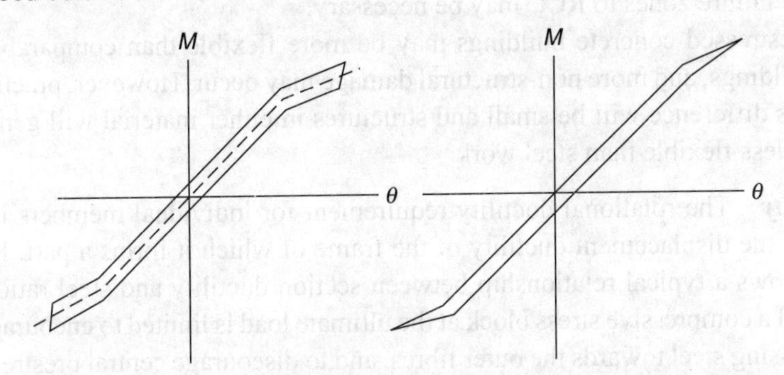

Fig. 8.35 Idealized moment-rotation diagrams for prestressed concrete

Damping The values of damping have been found to increase with amplitude and when the member has been subjected to forces sufficient to cause cracking, so that damping increases in the post-earthquake situation. The range of damping values from the literature is:

(a) For elastic conditions (uncracked): 0.01
(b) For elastic conditions (cracked): 0.02–0.03
(c) For inelastic conditions: 0.03–0.07

Prestressed concrete versus reinforced concrete Prestressed concrete can be used in the primary earthquake-resistant structural systems and steel buildings subjected to strong ground motion, but should be designed carefully so that the structure possesses sufficient strength and ductility.

(a) The narrowness of the hysteresis loops (Fig. 8.35) reflects that the amount of hysteresis energy dissipation of prestressed concrete is relatively small compared to steel or RCC. Further, the capacity of prestressed concrete to store elastic energy is higher than that of a comparable RC member. Prestressed concrete buildings have approximately 40 per cent greater displacement and lower damping. However, prestressed concrete exhibits a greater elastic recovery so that the damage can be expected to be less for moderate earthquakes.
(b) Prestressed concrete in comparison to RCC lacks compression steel, and therefore its performance is impaired once concrete crushing begins. To impart ductility to the prestressed concrete members, ordinary reinforcing bars should be used together with prestressing steel.
(c) Prestressed concrete undergoes relatively more uncracked deformation and relatively less deformation in the cracked state, as compared to RCC. This

implies that prestressed concrete structures exhibit less structural damage in moderate earthquakes.

(d) With regards to structural repairs and restoration, there are obvious difficulties in restoring the prestress to sections of replaced concrete, and conversion of the failure zones to RCC may be necessary.

(e) Prestressed concrete buildings may be more flexible than comparable RC buildings, and more non-structural damage may occur. However, practically, this difference will be small and structures in either material will generally be less flexible than steel work.

Ductility The rotational ductility requirement for individual members is 3 to 5 times the displacement ductility of the frame of which it forms a part. Figure 8.36 shows a typical relationship between section ductility and steel ratio. The depth of a compressive stress block at the ultimate load is limited to encourage the prestressing steel towards the outer fibres, and to discourage central prestressing. The rotational capacity or ductility of prestressed concrete is affected by the following factors:

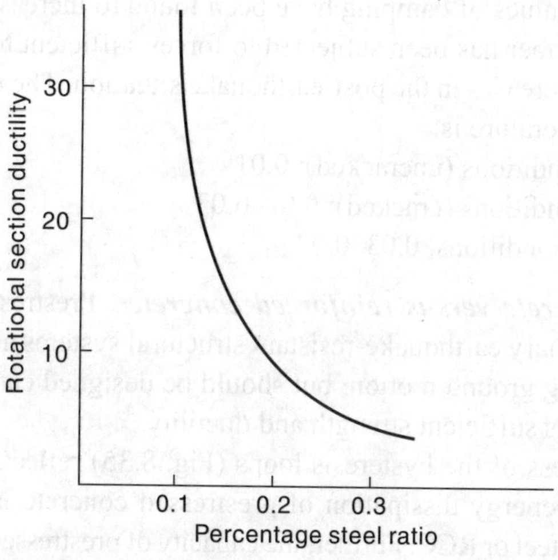

Fig. 8.36 Rotational ductility and steel ratio for prestressed concrete members

(a) *Prestressing steel content* The ductility decreases markedly with increasing prestressing steel content.

(b) *Transverse steel content* An increase in the transverse steel content has little effect on the ductility of beams with moderate prestress.

(c) *Distribution of prestressing steel* At positions of moment reversal where the greatest ductility requirements exist, the required distribution of prestress will usually be nearly axial. It has been shown that a single axial tendon produces a less ductile member than that achieved by multiple tendons placed nearer the extreme fibres. At points in structures where stress reversals do not occur, eccentric prestress may be used. Where no unstressed reinforcement exists, an eccentrically prestressed beam is notably less ductile than a concentrically stressed beam with equal prestressing steel content (Fig. 8.37). The tendons' distribution, as shown in Fig. 8.37(c), is not only as ductile as that shown in Fig. 8.37(b), but also has the advantage that the axial tendon will be practically unharmed by large rotations and would hold the structure together after the tendons near the extreme fibres have failed.

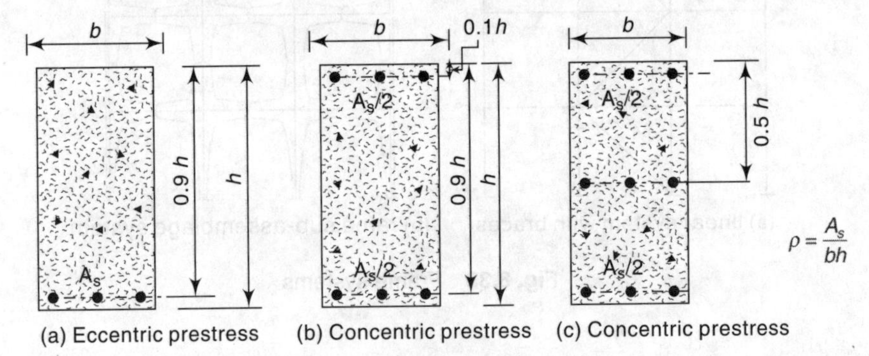

Fig. 8.37 Beams-section ductility

(d) *Axial load* The concrete column section ductility decreases rapidly with increase in column axial load.

8.19 Precast Concrete Construction

Precast concrete is emerging as one of the most popular structural systems because of its economy and quality. However, the overall integrity of the precast system is poor, because making the connections sufficiently strong and ductile poses challenges. In order to overcome the connection problem, partial precasting is often done. For example, precast beams may be used with in situ columns or precast walls may be used with in situ floors, or vice versa. To build earthquake-resistant precast structures, the design must follow the rules used for RC structures. In addition, connections should be carefully designed to be strong as well as ductile, while site connections should be located in low stress regions.

Classification

The precast system is classified into the following two types.

Frame system The frame system is further divided into two groups—the linear system and the frame sub-assemblage system (Fig. 8.38). In the linear system, precast column and beam elements are assembled at the construction site. In the frame sub-assemblage system, components such as T, cruciform, H, Π, and hollow panel frames are assembled at the site. It is very difficult in a linear system to ensure sufficient strength and ductility at beam–column connections and such systems are not suitable as moment-resisting structures. The linear system, therefore, is often combined with cast-in-place shear walls or steel braces. In the frame sub-assemblage system, selection of joint locations is more flexible and connections are usually located in low stress regions such as the inflection points in columns.

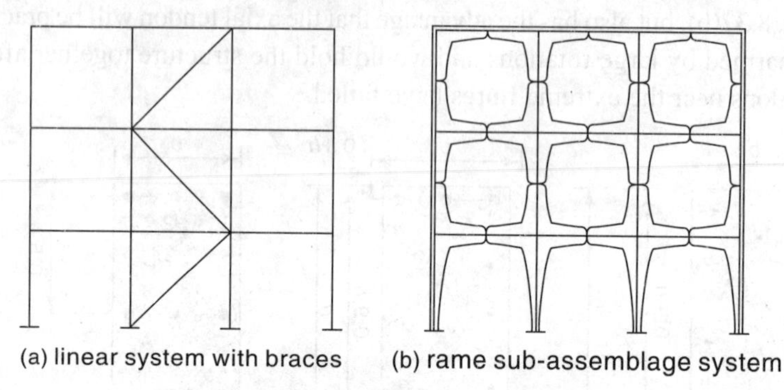

Fig. 8.38 Frame systems

Panel system The panel system has many variations including the small panel system, large panel system, and the box-room system (Fig. 8.39). For all of these, the strength and ductility of the connections are the critical design considerations.

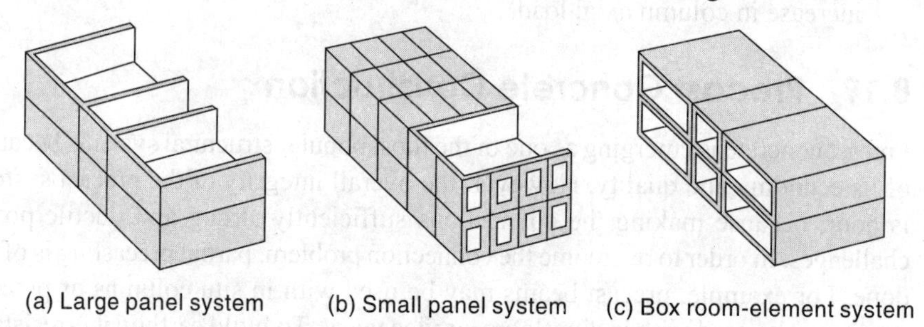

Fig. 8.39 Panel systems

Precast columns, beams, and panels are sometimes designed so that their ends or edges extend into cast-in-place joints. By this means, better integrity can be achieved between precast elements.

Connections

Some typical details of precast concrete member connections are shown in Figs 8.40, 8.41, 8.42, and 8.43 (the member reinforcement is not shown). Good connection details in linear systems are needed to ensure sufficient strength and ductility of joints where the stress level is high. Joints in the sub-assemblage system are relatively easy to design because they are usually located in the vicinity of inflection points of precast beams and columns.

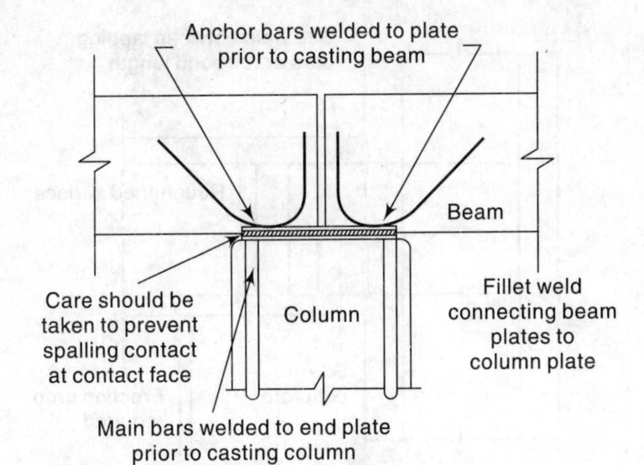

(a) Site welded (low moment capacity)

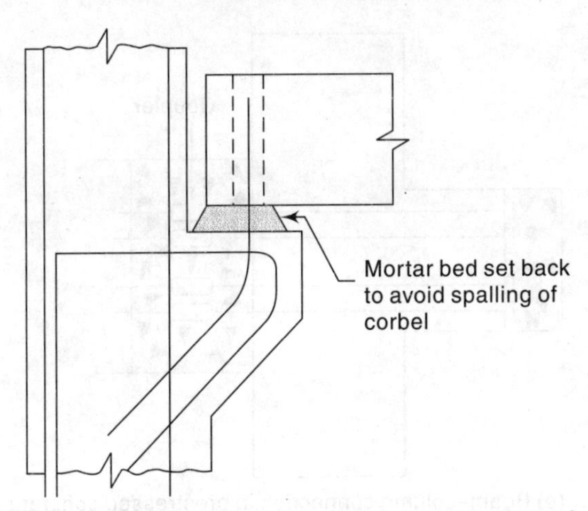

(b) Site grouted (low moment capacity) poor in horizontal shear

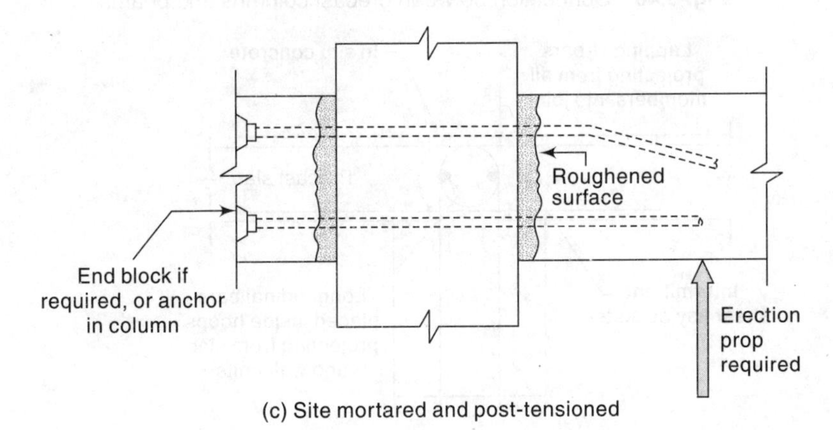

(c) Site mortared and post-tensioned

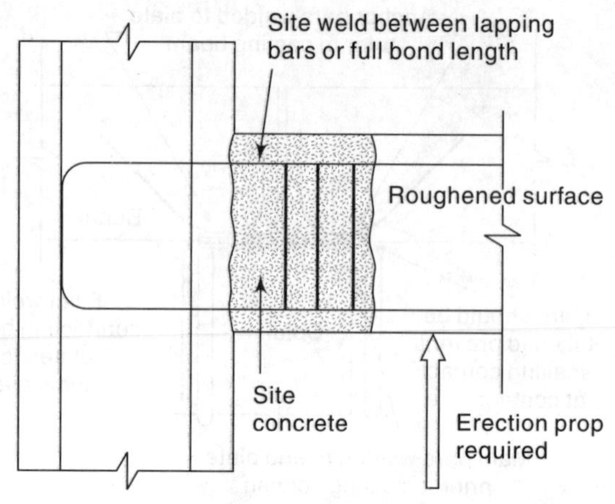

(d) Site concrete and welded and links fixed

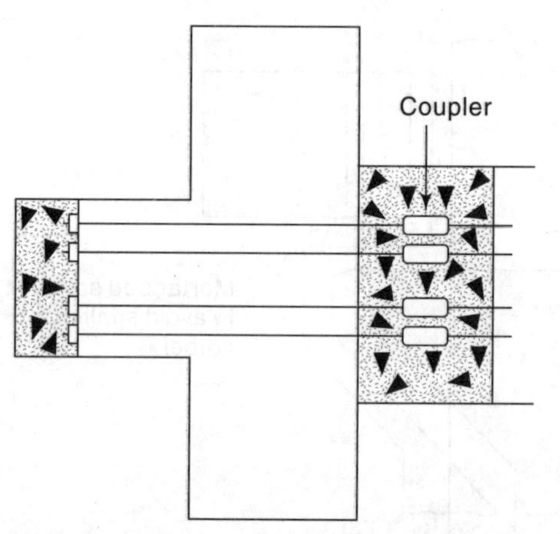

(e) Beam–column connection in prestressed concrete

Fig. 8.40 Connection between precast columns and beams

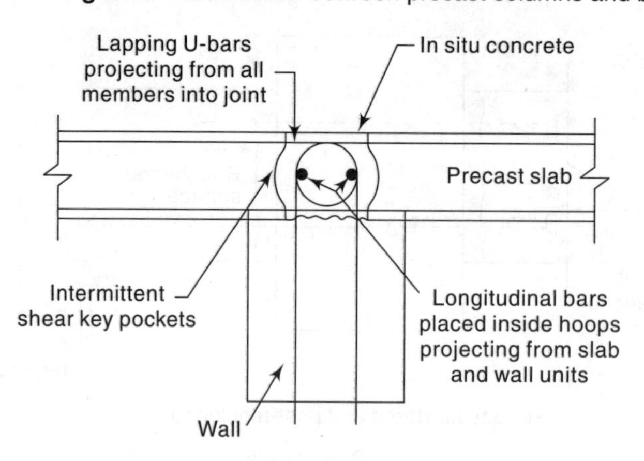

(a) Site concrete and reinforcement

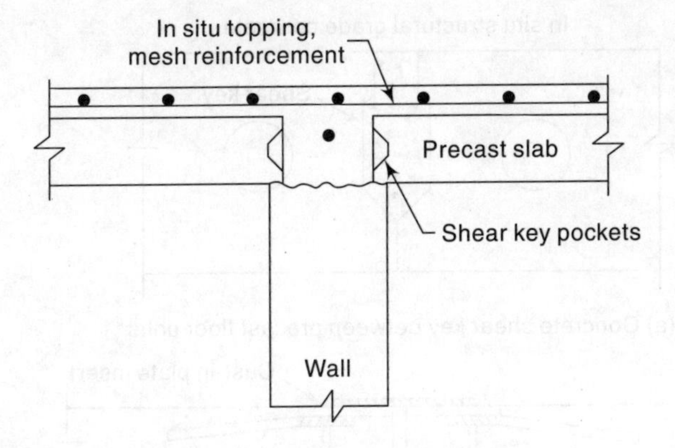

(b) Site concrete and reinforcement

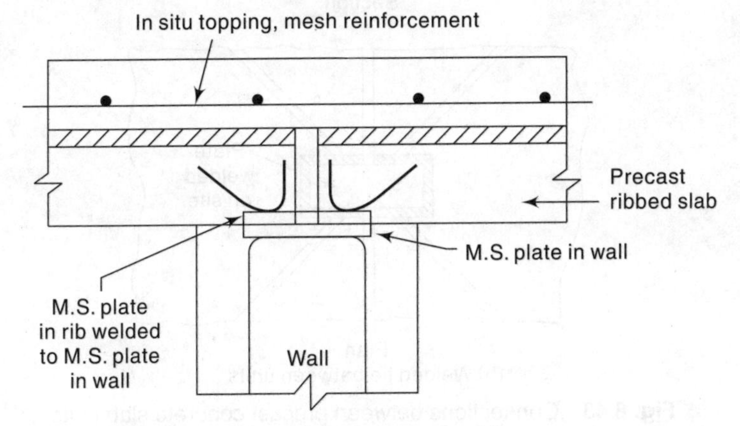

(c) Site concrete, reinforcement, and welding

Fig. 8.41 Connections between precast floors and walls

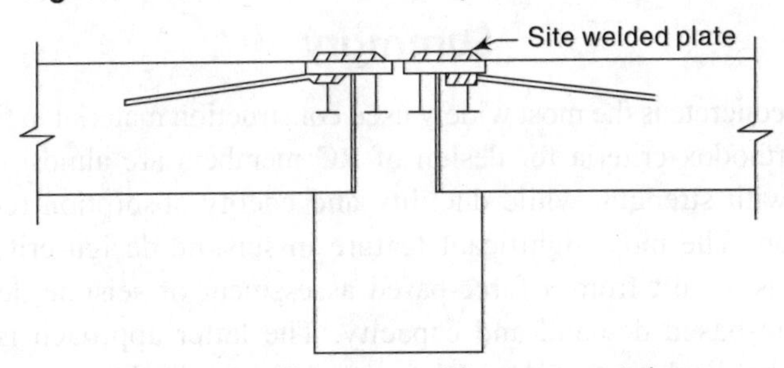

(a) Connection of a precast slab to a concrete beam

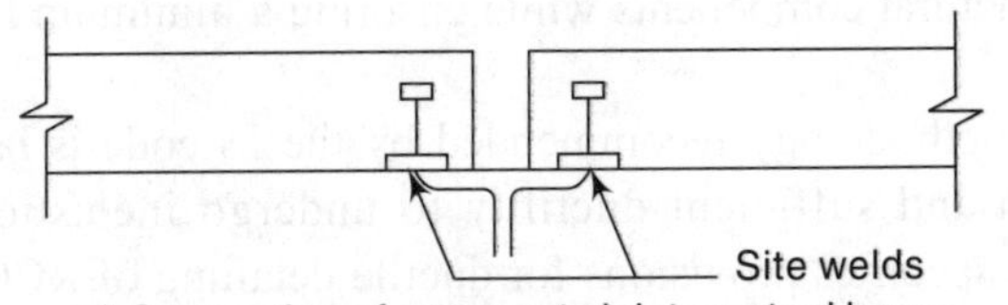

(b) Connection of a precast slab to a steel beam

Fig. 8.42 Connections between precast slab

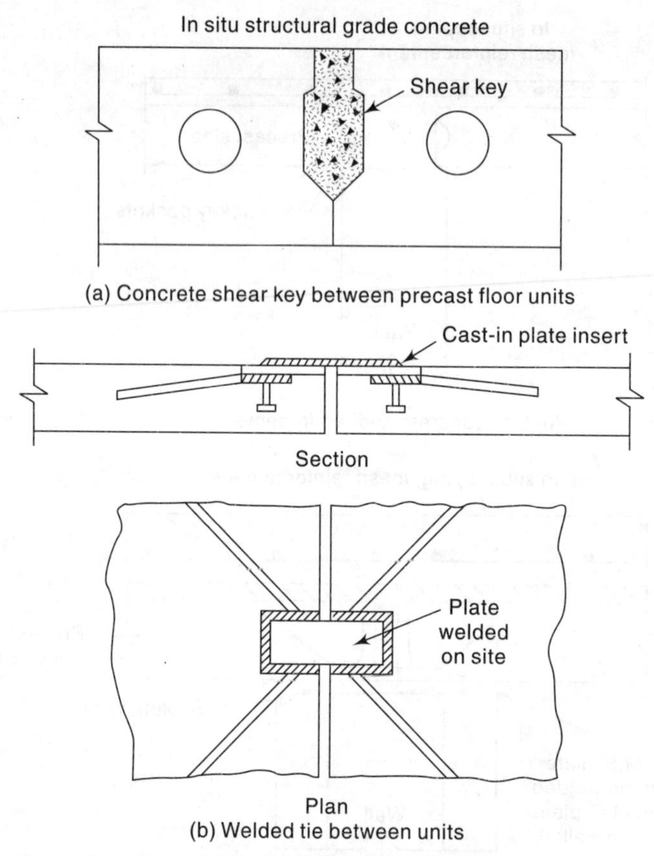

Fig. 8.43 Connections between precast concrete slab units

Summary

Reinforced concrete is the most widely used construction material in the building industry. Orthodox criteria for design of RC members are almost exclusively concerned with strength, while ductility and energy absorption receive little consideration. The most significant feature in seismic design criteria design philosophy is a shift from a force-based assessment of seismic demand to a displacement-based demand and capacity. The latter approach is based on comparing the displacement demand due to inelastic displacement capacity of the primary structural components while ensuring a minimum level of inelastic capacity.

The design methodology recommended by the IS code is based on enough lateral strength and sufficient ductility to undergo inelastic deformations. IS 13920: 1993 specifies provisions for ductile detailing of RCC structures and the limits placed thereon. The specifications and guidelines laid down in IS 1893: 2002 and IS 13920: 2002 and their explanations to achieve ductility and improved detailing have been discussed in this chapter alongwith reasons. The draft code IS 1893 (Parts I and II) have also been referred to wherever relevant.

The fundamental principles of earthquake-resistant design applicable to RC members are outlined. Shear walls provide an important lateral load transfer mechanism. Shear wall framed buildings are an ideal choice for resisting lateral loads and since, as of today, these are a favourite with engineers, the design of shear walls has been discussed in detail. The techniques used to restore the original strength of the seismically affected members and strengthen the RC structural elements are described.

The use of prestressed concrete for earthquake-resistant design is discussed. Although the reduced weight of the building attracts low seismic force, it has certain associated problems. The limitations of prestressed concrete construction in building structures subjected to seismic loads are highlighted. Since the Bureau of Indian Standards is silent on the use of prestressed concrete in building construction, some specifications from other codes have been presented for use. One of the major difficulties with precast and prestressed constructions in earthquake-prone areas is the provision of adequate ductility in prefabricated structural systems; joint connections exhibit limited ductility. The quantification of ductility demands and availability of ductility in precast concrete systems is still under active research. The ductility demand placed on the connectors plays an important role in the evaluation of such systems. The actual ductility requirements of prestressed elements under seismic loading have yet not been conclusively determined, and the research on the joint details has been relatively neglected.

Precast concrete construction may be of the RC or prestressed concrete type. Accordingly, its response can be inferred from the response of RC or prestressed members. Special attention is required for the connections of precast members. Very little information is available on joints, in precast assemblies, where yielding is anticipated.

At the end of the chapter ductile detailing of the connections of pre-cast concrete frame and panels is introduced. A number of examples have been solved to illustrate the design principles outlined in the chapter.

Solved Problems

8.1 A fixed-ended RC beam of rectangular section has to carry a distributed live load of 20 kN/m in addition to its own weight and a dead load of 25 kN/m. The maximum bending moment and shear force due to the earthquake are 60 kNm and 40 kN respectively. Centre-to-centre distance between supports is 6 m. Design the beam using M-20 grade concrete and Fe-415 steel.

Solution

Design of section

Assume a trial section of the beam with the dimensions:

Width $b = 300$ mm

Overall depth $D = 600$ mm

Effective depth $d = D -$ effective cover

$= 600 - 40 = 560$ mm

Distributed load due to self-weight $= 0.3 \times 0.6 \times 25 = 4.5$ kN/m

Total dead load $W_D = 4.5 + 25 = 29.5$ kN/m

Live load $W_L = 20$ kN/m

Maximum bending moment due to dead load $M_D = \dfrac{29.5 \times 6 \times 6}{12} = 88.5$ kNm

Maximum bending moment due to live load $M_L = \dfrac{20 \times 6 \times 6}{12} = 60$ kNm

Seismic moment, $M_E = 60$ kNm

Maximum shear force due to dead load $V_D = \dfrac{29.5 \times 6}{2} = 88.5$ kN

Maximum shear force due to live load $V_L = \dfrac{20 \times 6}{2} = 60$ kN

Seismic design shear $V_E = 40$ kN

Factored moment, $M_U = 1.5(M_D + M_L)$

or

$1.2(M_D + M_L + M_E)$, whichever is more

Therefore, the factored moment is given as

$$M_u = \begin{bmatrix} 1.5 \times (88.5 + 60) = 222.75 \text{ kNm} \\ \text{or} \\ 1.2 \times (88.5 + 60 + 60) = 250.2 \text{ kNm} \end{bmatrix}$$

Factored shear force V_u is given as

$$V_u = \begin{bmatrix} 1.5(V_D + V_L) \\ \text{or} \\ 1.2(V_D + V_L + V_E), \text{ whichever is greater} \end{bmatrix}$$

Therefore, V_u can be calculated as

$$V_u = \begin{bmatrix} 1.5(88.5 + 60) = 222.75 \text{ kN} \\ \text{or} \\ 1.2(88.5 + 60 + 40) = 226.2 \text{ kN} \end{bmatrix}$$

For Fe-415 steel and M-20 concrete

$f_{ck} = 20$ N/mm^2

$f_y = 415$ N/mm^2

$x_{\text{ulim}} = 0.48$d

$R_{\text{lim}} = 2.76$

$$d_{req} = \sqrt{\frac{M_u}{R_{lim} b}}$$

$$= \sqrt{\frac{250.2 \times 10^6}{2.76 \times 300}}$$

$$= 549.7 \text{ mm}$$

This value of d_{req} is less than d and, thus, it is all right.

For 20 ϕ bars and clear cover 30 mm

$$d = 600 - 30 - \frac{20}{2}$$

$$= 560 \text{ mm.}$$

Area of steel required

$$M_U = 0.87 f_y A_{st} d\,[1 - (A_{st} f_y / bdf_{ck})]$$

Substituting the values of M_U, f_y, f_{ck}, b, and d, we get

$$250.2 \times 10^6 = 0.87 \times 415 \times A_{st} \times 560[1 - A_{st} \times 415/300 \times 560 \times 20]$$

$$\therefore \quad A_{st} = 1525 \text{ mm}^2$$

$$A_{st} = 314.16 \times 5 = 1570.8 > 1525 \text{ mm}^2\text{, which is all right.}$$

As per IS 13920, for ductility requirement

$$\text{Minimum percentage of steel } \rho_{min} = 0.24\sqrt{\frac{f_{ck}}{f_y}}$$

$$= 0.24\sqrt{\frac{20}{415}} = 0.0526\ \%$$

Maximum percentage of steel $\rho_{max} = 2.5\ \%$

$$\text{As per IS 456: 2000, minimum steel required} = \frac{0.85bd}{f_y} = \frac{0.85 \times 300 \times 600}{415}$$

$$= 368.67 \text{ mm}^2$$

$$\text{Percentage of steel provided} = \frac{1570.8 \times 100}{(300 \times 560)} = 0.94\ \%\text{, which is all right.}$$

Check for shear

Design shear force $V_u = 226.2$ kN

$$\text{Nominal shear stress } \tau_v = \frac{V_u}{bd} = \frac{226.2 \times 1000}{(300 \times 560)}$$

$$= 1.35 \text{ N/mm}^2$$

8.2. Design the reinforcement for a column of size 450 mm × 450 mm, subjected to the following forces. The column has an unsupported length of 3.0 m and is braced against side sway in both directions. Use M-25 grade concrete and Fe-415 steel.

	Dead load	Live load	Seismic load
Axial load (kN)	1000	800	550
Moment (kNm)	50	40	100

Solution

Given parameters:

Width of column $b = 450$ mm

Depth of column $D = 450$ mm

$f_{ck} = 25$ MPa

$f_y = 415$ MPa

Factored load $P_u = 1.5 \times (1000 + 800)$
$= 2700$ kN

or $P_u = 1.2 \times (1000 + 800 + 550)$
$= 2820$ kN (whichever is more)

Factored moment $M_u = 1.5 \times (50 + 40)$
$= 135$ kNm

or $M_u = 1.2\ (50 + 40 + 100)$
$= 228$ kNm (whichever is more)

Therefore, the factored load and factored moment are 2820 kN and 228 kNm, respectively.

Longitudinal reinforcement

Assuming an effective cover d' = clear cover + $\phi/2 = 40 + 25/2 = 52.5$ mm

$$\frac{d'}{D} = \frac{52.5}{450} = 0.11 \approx 0.10$$

$$\frac{P_u}{f_{ck}bD} = \frac{2820 \times 1000}{25 \times 450 \times 450} = 0.56$$

and $$\frac{M_u}{f_{ck}bD^2} = \frac{228 \times 1000 \times 10^3}{25 \times 450 \times 450 \times 450} = 0.10$$

Let us assume equal reinforcement on all the four sides.

Percentage of steel required $\rho = 3\%$ (Chart 44 of SP:16)

$$\text{Area of steel } A_{sc} = \frac{3 \times 450 \times 450}{100} = 6075 \text{ mm}^2$$

Provide 8 bars of 25 mm ϕ and 8 bars of 20 mm ϕ.

Thus $A_{st} = 491 \times 8 + 314.16 \times 8 = 6441.28 \text{ mm}^2 > 6075 \text{ mm}^2$, which is all right.

Confining reinforcement

Special confining reinforcement should be provided over a length l_0 from each joint face, towards mid height. The length l_0 should not be less than

(a) Larger lateral dimension of member = 450 mm

(b) One-sixth of clear span = 3000/6 = 500 mm

(c) 450 mm

Therefore, $l_0 = 500$ mm

Area of confining reinforcement should not be less than

$$A_{sh} \geq \frac{0.18\, Sh\, f_{ck}\left[\dfrac{A_g}{A_k} - 1.0\right]}{f_y}$$

Using confining reinforcement bars to be of 10 φ, $A_{sh} = 78.5$ mm^2
Diameter of core $D_k = 450 - 2 \times 40 + 2 \times 10 = 390$ mm
Area of core $A_k = 390 \times 390 = 152100$ mm^2
Gross area $A_g = 450 \times 450 = 202500$ mm^2

$$h = \frac{390}{2} = 195 \text{ mm}$$

$$78.5 \geq \frac{0.18 \times S \times 195 \times 25\left(\dfrac{202500}{152100} - 1.0\right)}{415}$$

$\Rightarrow$ $S \leq 112.03$ mm c/c

Also, $A_{sh} \geq 0.05\, Sh \dfrac{f_{ck}}{f_y}$

$$78.5 \geq 0.05\, Sh \frac{25}{415}$$

$\Rightarrow$ $Sh = 133.65$ mm

Also, the maximum permissible spacing S should not exceed one-fourth of the minimum dimension of the member or 100 mm, whichever is less.

$S = (1/4) \times 450$ or 100 mm

Therefore, provide confining reinforcement of 10 φ bars at 100 mm c/c in a length of 500 mm.

Design of transverse reinforcement

Diameter of bar should not be less than one-fourth of the diameter of the main bar or 6 mm, whichever is more.

Therefore, the diameter of the transverse reinforcement bars $= \frac{1}{4} \times 25$ or 6 mm.
Let us provide 10 φ lateral ties.
Spacing of lateral ties should not be more than half the least lateral dimension

$$= \frac{450}{2} = 225 \text{ mm}$$

Provide 10 φ lateral ties at 225 mm c/c in 2 m length.

8.3 A 10-storey building has plan dimensions as shown in Fig. 8.44. Two shear walls are to be provided in each direction to resist the seismic forces. The axial load on the each shear wall is 5700 kN due to both dead and live loads. The height between floors is 3.0 m. The dead load per unit area of the floor, which consist of floor slab, finishes, etc., is 4 kN/m^2 and the weight of partitions on floor is 2 kN/m^2. The intensity of live load on each floor is 3 kN/m^2 and on roof is 1.5 kN/m^2. The soil below the foundation is hard and the building is located in Delhi.

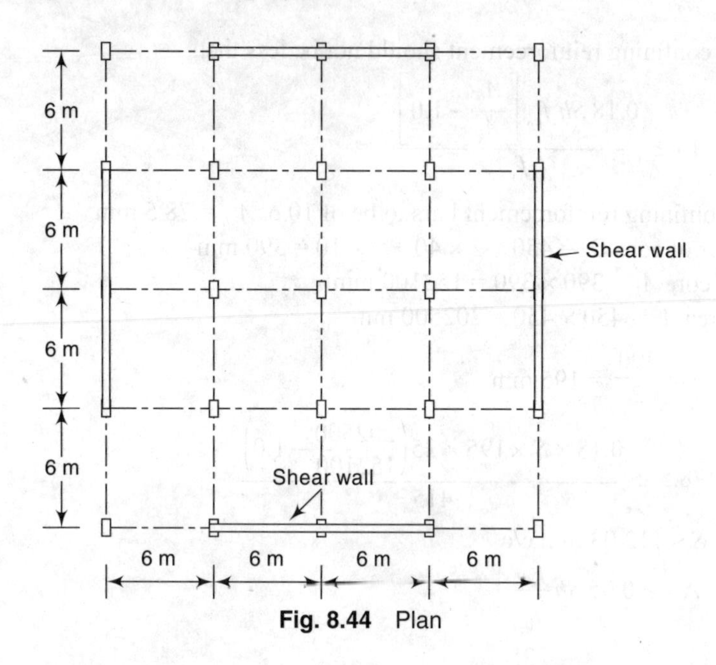

Fig. 8.44 Plan

Determine the seismic forces and shears at different floor levels. Also, design the ductile shear wall to resist the seismic forces using M-25 grade concrete and TOR steel (Fe-415). Assume unit weight of concrete as 25 kN/m^3, and the beams and the columns with cross-sections 600 mm × 300 mm.

Solution

Seismic weight of the building

As per the code provisions, the percentage of design live load to be considered for the calculation of earthquake forces is 25% for the floors, and live load for the roof is not to be accounted for.

Hence, the effective weight at each floor will be $= 4.0 + 2.0 + 0.25 \times 3 = 6.75$ kN/m^2 and that at the roof $= 4.0$ kN/m^2.

Weight of 40 beams, each of 6 m span, at each floor and roof

$$= 0.3 \times 0.6 \times (6 \times 40) \times 25$$
$$= 1080 \text{ kN}$$

Weight of 25 columns at each floor $= 0.3 \times 0.6 \times 2.4 \times 25 \times 25 = 270$ kN

Weight of columns at roof $= \dfrac{1}{2} \times 270 = 135$ kN

Plan area of building is 24 m × 24 m = 576 m^2

Equivalent load at roof level $= 4 \times 576 + 1080 + 135 = 3519$ kN

Equivalent load at each floor $= 6.75 \times 576 + 1080 + 270 = 5238$ kN

Seismic weight of the building, $W = 3519 + 5238 \times 9 = 50661$ kN

Base shear

The fundamental natural period of vibration T for the buildings having shear walls is given by

$$T = \frac{0.09h}{\sqrt{d}} = \frac{0.09 \times 30}{\sqrt{24}} = 0.551 \quad (d\text{, the plan dimension} = 24 \text{ m})$$

Building is situated in Delhi, i.e., in Zone IV.
Zone factor $Z = 0.24$, importance factor $I = 1.0$, response reduction factor $R = 4.0$
For 5% damping and type I soil, average response acceleration coefficient $S_a/g = 1.81$

Design horizontal seismic coefficient $A_h = \frac{ZIS_a}{2Rg}$

$$= \frac{0.24 \times 1.0 \times 1.81}{2 \times 4} = 0.0543$$

Base shear $V_B = A_h W = 0.0543 \times 50661 = 2750.9$ kN

Lateral loads and shear forces at various floor levels

Design lateral force at floor i

$$Q_i = V_B \frac{W_i h_i^2}{\sum_{J_i=1}^{n} W_j h_j^2}$$

Lateral loads and shear forces at different floor level are given in Table 8.1.

Table 8.1 Calculation of lateral loads and shear

Mass No.	W_i (kN)	h_i (m)	$W_i h_i^2$	$\frac{W_i h_i^2}{\sum W_i^2 h_i^2}$	Q_i (kN)	V_i (kN)
1	3519	30.0	3167100	0.1907	524.7	524.7
2	5238	27.0	3818502	0.2299	632.6	1157.3
3	5238	24.0	3017088	0.1817	499.8	1657.1
4	5238	21.0	2309958	0.1391	382.8	2039.9
5	5238	18.0	1697112	0.1022	281.3	2321.2
6	5238	15.0	1178550	0.0709	195.2	2516.4
7	5238	12.0	754272	0.0454	124.9	2641.3
8	5238	9.0	424278	0.0255	70.3	2711.6
9	5238	6.0	188568	0.0114	31.5	2743.1
10	5238	3.0	47142	0.0028	7.8	2750.9
			$\sum W_i^2 h_i^2 = 16602570$			

Bending moment and shear force

Two shear walls are provided in each direction to resist the seismic forces. Therefore, the lateral forces acting on one shear wall will be half the calculated shears and is as shown in Fig. 8.45. The shear wall will be designed as a cantilever fixed at the base and free at the top.

Maximum shear force at base $V = 1375.45$ kN

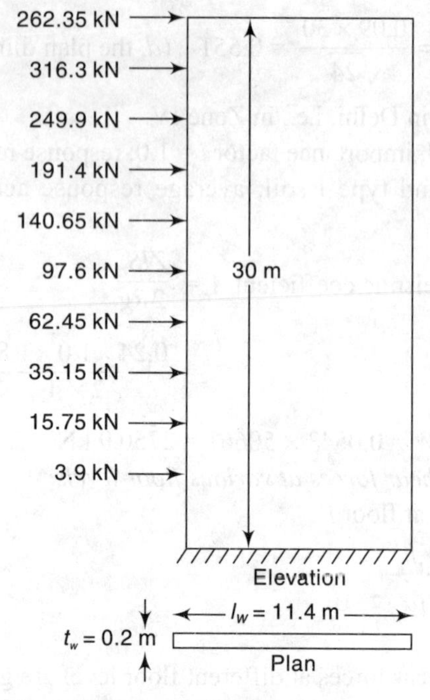

Fig. 8.45 Lateral forces on shear wall

Maximum bending moment at base, $M = (3.9 \times 3) + (15.75 \times 6) + (35.15 \times 9) + (62.45 \times 12) + (97.6 \times 15) + (140.65 \times 18) + (191.4 \times 21) + (249.9 \times 24) + (316.3 \times 27) + (262.35 \times 30)$

$= 31595.25$ kNm

Taking partial safety factor = 1.5

Factored shear force $V_u = 1.5 \times 1375.45 = 2063.2$ kN

Factored bending moment $M_u = 1.5 \times 31595.25 = 47392.9$ kNm

Factored axial load $P_u = 1.5 \times 5700 = 8550$ kN

Flexural strength

$f_{ck} = 25$ N/mm²; $f_y = 415$ N/mm²; $E_s = 2.0 \times 10^5$ N/mm²

Length of wall $l_w = 11.4$ m, and thickness of wall $t_w = 0.2$ m

Providing uniformly distributed vertical reinforcement ratio $\rho = 0.25\%$

$$\phi = \frac{0.87 f_y \rho}{f_{ck}} = \frac{0.87 \times 415 \times 0.0025}{25} = 0.03611$$

$$\lambda = \frac{P_u}{f_{ck} l_w t_w} = \frac{8550 \times 1000}{25 \times 11400 \times 200} = 0.15$$

$$\beta = \frac{0.87 f_y}{0.0035 E_s} = \frac{0.87 \times 415}{0.0035 \times 2 \times 10^5} = 0.5158$$

$$\frac{x_u}{l_w} = \frac{\phi + \lambda}{2\phi + 0.36} = \frac{0.03611 + 0.15}{2 \times 0.03611 + 0.36} = 0.4306$$

$$\frac{x_u^*}{l_w} = \frac{0.0035}{0.0035 + \dfrac{0.87 f_y}{E_s}} = \frac{0.0035}{0.0035 + \dfrac{0.87 \times 415}{2 \times 10^5}} = 0.6597$$

$$\frac{x_u}{l_w} < \frac{x_u^*}{l_w}$$

Hence, moment of the resistance

$$M_u = f_{ck} t_w\, l_w^2 \phi \left[\left(1 + \frac{\lambda}{\phi}\right)\left(\frac{1}{2} - 0.416 \frac{x_u}{l_w}\right) - \left(\frac{x_u}{l_w}\right)^2 \left(0.168 + \frac{\beta^2}{3}\right)\right]$$

$$= 25 \times 200 \times (11400)^2 \times 0.03611 \left[\left(1 + \frac{0.1500}{0.03611}\right)(0.5\right.$$

$$\left. - 0.416 \times 0.4306) - (0.4306)^2 \left(0.168 + \frac{0.5158^2}{3}\right)\right]$$

$$M_u = 37777.48 \text{ kNm} < 47392.9 \text{ kNm}$$

Balance moment to be resisted by the edge reinforcement in each shear wall

$= (47392.9 - 37777.48)$

$= 9615.42$ kNm

Effective depth of wall $d_w = 0.9\, l_w = 10260$ mm

$$\text{Area of steel } A_{st} = \frac{M_u}{0.87 f_y d_w} = \frac{9615.42 \times 10^6}{0.87 \times 415 \times 10260} = 2595.69 \text{ mm}^2$$

Equal amount of reinforcement is provided on the vertical edges of the wall which will act like the flanges of a steel beam.

Provide 10 Nos. 20 ϕ bars in two layers in the wall at each end.

Thus A_{st} provided at the ends = 314.15 × 10 = 3141.5 mm^2

(Minimum area of steel required in the shear wall = 0.0025 × 11400 × 200 = 5700 mm^2)

This minimum reinforcement is provided in the vertical direction for a length of wall $0.8\, l_w = 9120$ mm.

Area of minimum reinforcement per metre length of wall = 0.0025 × 1000 × 200 = 500 mm^2

Maximum permissible spacing = $\dfrac{l_w}{5}$ or $3t_w$ or 450 whichever is less

= 450 mm c/c

Provide 10-mm ϕ bars at 300 mm c/c in the vertical direction in two layers.

Check for shear

Factored shear force $V_u = 2063.2$ kN

Nominal shear stress $\tau_v = \dfrac{V_u}{t_w d_w} = \dfrac{2063.2 \times 1000}{200 \times 10260} = 1.0\ \text{N/mm}^2$

$< 0.25\sqrt{f_{ck}}$

i.e., $0.25\sqrt{25} = 1.25\ \text{N/mm}^2$

which is as expected

Permissible shear stress for M-25 grade concrete and steel ratio $\rho = 0.25\%$

$\tau_c = 0.36\ \text{N/mm}^2$

Since $\tau_v > \tau_c$, the area of horizontal shear reinforcement A_h at a vertical spacing S_v is given by

$$S_v = \frac{0.87 f_y A_h d_w}{V_{us}}$$

Using 10-ϕ two legged horizontal stirrups

$$S_v = \frac{0.87 \times 415 \times 78.5 \times 2 \times 10260}{2063200 - 0.36 \times 10260 \times 200}$$

$$S_v = 439 \text{ mm c/c}$$

Thus the reinforcement provided in horizontal direction

$= \dfrac{78.5 \times 2 \times 1000}{439} = 358\ \text{mm}^2$, which is less than the minimum reinforcement.

Hence provide minimum specified reinforcement of 0.25 per cent of the gross area of the wall, in horizontal direction.

Provide 10-ϕ bars at 300 mm c/c as horizontal reinforcement on both the faces and in the full height of the wall.

The detailing of the shear wall is shown in Fig. 8.46.

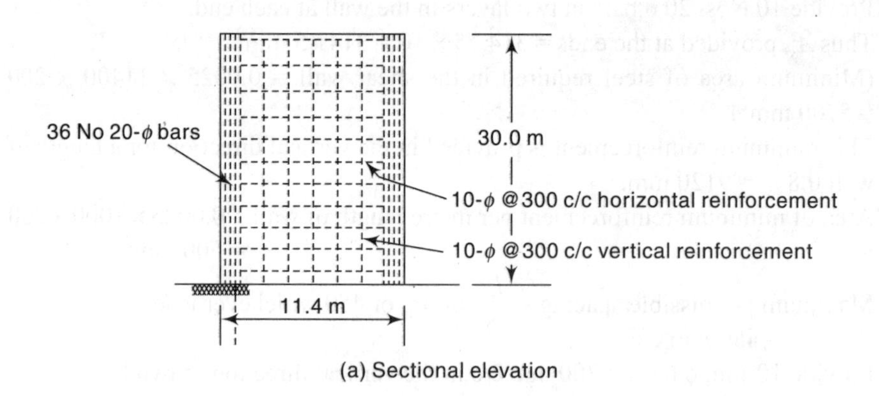

(a) Sectional elevation

Fig. 8.46 (*Contd*)

(Contd)

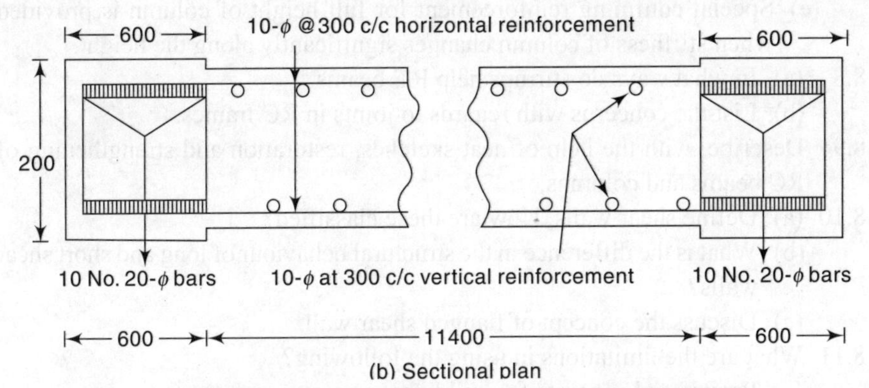

(b) Sectional plan

Fig. 8.46 Reinforcement detail in shear wall

Exercises

8.1 State the advantages of using concrete over brick masonry for buildings in seismic areas. What are the limitations of using concrete in buildings and how are these overcome?

8.2 What are the possible damages to RC buildings in earthquake-prone regions?

8.3 Write short notes on the following:
 (a) Bond between reinforcing bars and concrete
 (b) Effect of transverse reinforcement
 (c) Buckling of reinforcing bars

8.4 Discuss briefly the following types of failures of RC buildings:
 (a) Ductile failure
 (b) Flexural failure
 (c) Failure of joints

8.5 What are the principles of earthquake-resistant design of RC buildings?

8.6 Write notes on the following for in situ concrete detailing.
 (a) Concrete quality
 (b) Reinforcement quality
 (c) Splices
 (d) Anchorage
 (e) Confinement

8.7 Give reasons for the following with regard to RC members subjected to seismic forces:
 (a) Depth of beam should not be more than one-fourth of the clear span.
 (b) Tension steel ratio on any face of beam should not be less than $0.24\sqrt{f_{ck}}/fy$
 (c) Positive steel at a joint face must be at least equal to half the negative steel at that face.

(d) Width of column should not be too small.
(e) Special confining reinforcement for full height of column is provided when stiffness of column changes significantly along the height.

8.8 (a) In what ways do stirrups help RC beams.
(b) List the concerns with regards to joints in RC frames.

8.9 Describe, with the help of neat sketches, restoration and strengthening of RC beams and columns.

8.10 (a) Define shear walls. How are these classified?
(b) What is the difference in the structural behaviour of long and short shear walls?
(c) Discuss the concept of flanged shear wall.

8.11 What are the limitations in using the following?
(a) Prestressed concrete for buildings
(b) Precast concrete members in building construction

8.12 Write short notes on the following properties of prestressed concrete:
(a) Damping
(b) Ductility

8.13 Precast concrete elements have proved their quality. For what reasons has their use in building construction still not gained popularity?.

8.14 Draw neat sketches for the connections of the following precast concrete members.
(a) Column-to-beam
(b) Slab-to-slab panels
(c) Slab-to-beam

8.15 Design a rectangular RC beam of 6 m span supported on a RC column to carry a point load of 100 kN in addition to its own weight. The moment due to seismic force is 5.01 kNm and shear force is 32 kN. Use M-20 grade concrete and Fe-415 steel.

8.16 Design a circular RC column for the following loads, using M-20 grade concrete and Fe-415 steel, and height of column 3.2 m:

	DL	LL	EL
Axial load (kN)	1200	600	360
Moment (kNm)	80	50	150

8.17 Design a rectangular beam for 8 m span to support a dead load of 10 kN/m and a live load of 12 kN/m inclusive of its own weight. Moment due to earthquake load is 100 kNm and shear force is 80 kN. Use M-20 grade concrete and Fe-415 steel.

8.18 Design a rectangular column for the following load combinations, using M-30 grade concrete and Fe-415 steel, and height of column 3.1 m.

	DL	LL	EL
Axial load (kN)	1800	800	800
Moment M_x (kNm)	180	60	200
Moment M_y (kNm)	120	75	175

8.19 Design a shear wall for a 12 storey building for the following data.

- Storey shear at different levels are as follows:

Storey No.	1	2	3	4	5	6	7	8	9	10	11	12
Storey shear (kN)	5	10	30	80	140	200	360	500	700	850	950	900

- Storey height = 3.2 m
- Length of shear wall = 7.5 m
- Seismic weight of building = 60×10^3 kN
- Axial load on shear wall = 3×10^3 kN
- Building is situated in Mumbai
- Use M-20 grade concrete and Fe-415 steel

CHAPTER 9

Steel Buildings

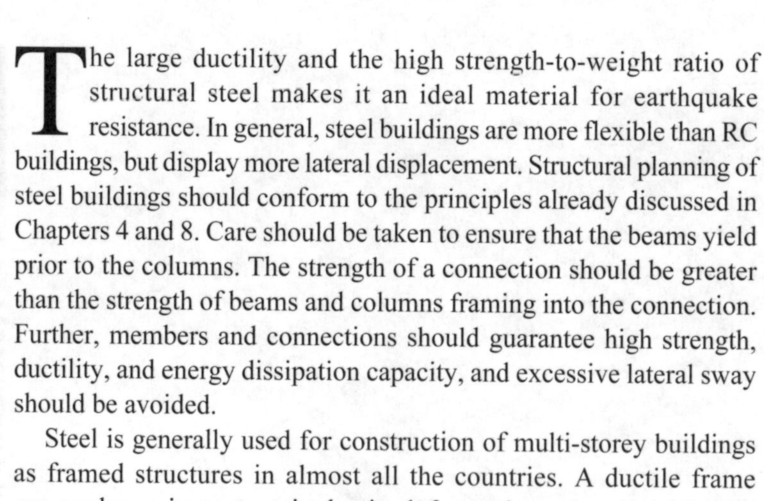

The large ductility and the high strength-to-weight ratio of structural steel makes it an ideal material for earthquake resistance. In general, steel buildings are more flexible than RC buildings, but display more lateral displacement. Structural planning of steel buildings should conform to the principles already discussed in Chapters 4 and 8. Care should be taken to ensure that the beams yield prior to the columns. The strength of a connection should be greater than the strength of beams and columns framing into the connection. Further, members and connections should guarantee high strength, ductility, and energy dissipation capacity, and excessive lateral sway should be avoided.

Steel is generally used for construction of multi-storey buildings as framed structures in almost all the countries. A ductile frame can undergo important inelastic deformations, localized in the neighbourhood of sections with maximum bending moment. These eventually lead to formation and rotation of plastic hinges and redistribution of plastic moments, allowing the structure to resist higher loads than those predicted by the elastic analysis. The steel frames may either be unbraced or braced. Unbraced steel buildings are ductile and possess large energy dissipation capacity, but tend to deform greatly, causing serious damage to non-structural elements during small-to-medium size earthquakes. On the other hand, braced frames can resist large amounts of lateral forces and have reduced lateral deflection and thus reduced $P-\Delta$ effect. However, a uniform distribution of bracing throughout the structure is desirable.

Although steel is highly ductile, inelastic ductility is not necessarily retained in the finished structure. Care must, hence, be taken during design and construction to avoid losing this property. Considerable care is also needed to check failures due to instability and brittle fracture to ensure the development of full ductility and energy dissipation capacity under earthquake loading.

The causes of instability are as follows:

(a) *Local buckling of plate elements (e.g., web, flange)* A steel member containing plate elements with a large width-to-thickness ratio is unable to reach its yield strength, because of prior local buckling. Even if the yield strength is attained, ductility will be inadequate. Under cyclic loading, the strength and ductility decrease with increasing width-to-thickness ratio, and local buckling of web causes further degradation.

(b) *Flexural buckling of long columns and braces* Long columns may fail by buckling. This mode of instability is sudden and can occur when the axial load in a column reaches a certain critical value. In most cases, the stress in the column may never reach the yield. Even a small lateral force under such conditions will produce a substantial deflection leading to instability and the phenomenon is called *flexural buckling*. The capacity of slender columns is, therefore, limited by the stiffness of the member, rather than by strength of the material. The lateral stiffness of frames, therefore, is increased by bracing the frames. However, buckling of braces is a potential source of instability of steel frames. Steel bracing dissipates considerable energy by yielding under tension, but buckle without much energy dissipation in compression. Therefore, the energy dissipation capacity of concentrically braced frames is markedly less, due to buckling of braces, than that of the moment frames.

(c) *Lateral-torsional buckling of beams* During moderate-to-strong shaking of the ground, additional forces are developed in various members of a structure. For a beam loaded in flexure, the load bearing side (generally the top) carries the load in compression, whereas the non-load bearing side (generally the bottom) will be in tension. If the beam is not supported in the opposite direction of bending, and the flexural load increases to a critical limit, the beam will fail due to local buckling on the compression side. In wide-flange sections, designed for flexure only, if the top flange buckles laterally, the rest of the section will twist, resulting in a failure mode known as *lateral-torsional buckling*.

(d) *P-Δ effects in frames subjected to large vertical loads* If the lateral stiffness is not high enough, the building as a whole, or one or more storeys can fail due to the P-Δ effect. This is due to the secondary effect on shears and moments of the frame members, caused by the action of vertical loads, which interact with the lateral displacement of the building resulting from seismic forces (Appendix X).

(e) *Uplift of braced frames* Earthquakes have a vertical component of movement in addition to the traditionally considered horizontal effects. The stresses produced due to vertical motion are generally considered not to be significant to cause instability. However, due to the horizontal component of movement, the overturning moments produce additional longitudinal stresses in walls and columns and additional upward (uplifting) and downward (thrust) forces in foundations causing instability.

(f) *Connection failure* The failure of bolted and welded connections are discussed in Section 9.6.

The causes of brittle failure in steel buildings are as follows:

(a) Brittle failure is more frequent in welded steel structures, particularly those that are fillet welded, than it is in structures connected by mechanical fasteners. This is due to a combination of possible weld defects, high residual stresses, and stress concentration, which reduces the possibility of crack arrest.

(b) Tension failure at net sections of bolted or riveted connections.

(c) Lamellar tearing of plates in which, the 'through-thickness' (Fig. 9.1) strain due to weld metal shrinkage is large and highly restrained.

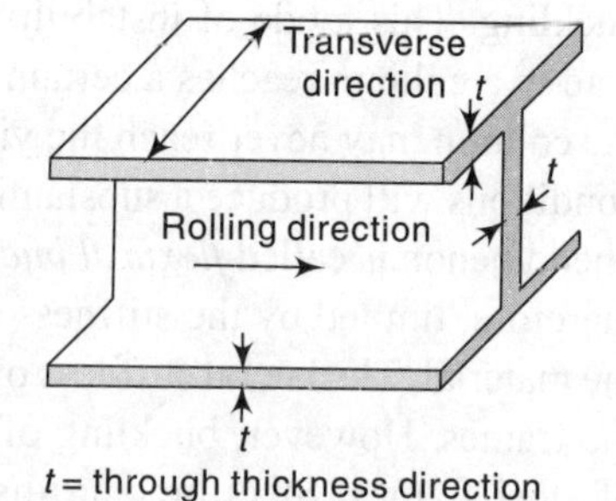

Fig. 9.1 Through thickness in rolled I-section

From the above discussion, it is evident that the main considerations to achieve adequate performance of steel buildings are:

(a) The use of sufficiently ductile steel.

(b) The ductile design and fabrication of framed members and connections.

(c) All forms of instability, especially the excessive sway leading to higher levels of damage to non-structural components, and to higher secondary stresses due to $P-\Delta$ effect, should be avoided.

(d) Avoidance of all forms of brittle failures.

Further, it may be noted that failure mechanism should provide maximum redundancy, i.e., the possibility of failure by local collapse should be avoided, and all portions of the building should be tied well together.

The relevant Indian code of practice, IS 800: 2007, applicable to the structural use of hot-rolled steel, is largely based on the limit state method. However, working stress design and the plastic design methods are also included in the text of its body. In Section 9.3, the provisions for design of steel buildings for earthquake loads have been outlined.

9.1 Seismic Behaviour of Structural Steel

Behaviour of steel buildings under earthquakes has generally been satisfactory from the point of view of strength. The properties of steel that contribute to the elastic resistance of steel structures during moderate earthquakes are the yield strength and elastic stiffness. However, in major earthquakes, a structure may undergo inelastic deformations and rely on its ductility and hysteretic energy dissipation capacity to avoid collapse.

The stress–strain relationship for steel, shown in Fig. 9.2(a), is usually idealized to the bilinear form, shown by the solid lines in Fig. 9.2(b), although strain hardening (broken lines) is taken into account in some cases. The yield stress f_y and the ultimate stress f_u are used for steel sections or plates, and f_y is used for reinforcing bars. The value of Young's modulus E_s is about 2×10^5 MPa.

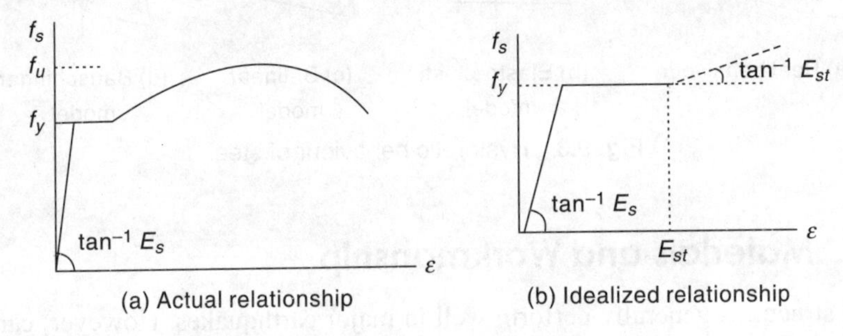

Fig. 9.2 Stress–strain relationship of steel

Steel being a ductile material is equally strong in tension and compression, and is an ideal material for earthquake resistant structures. Ductility allows the structure to undergo large plastic deformations without significant loss of strength. The common grades of mild steel have adequate ductility and perform well under cyclic reversal of stresses. High-strength steels provide higher elastic limits but have low ductility and are, therefore, not recommended. Moreover, high-strength steels result in less areas of member cross-sections and thereby tend to be more prone to local buckling.

Hysteretic energy defined as the energy dissipated by inelastic cyclic deformations is given by the area within the load deformation curve, the *hysteretic curve*. Larger area implies more dissipation of hysteretic energy. Structures having low hysteretic energy dissipation capacities are likely to collapse due to *low cyclic fatigue*, even if the deformations are well below the ultimate deformation. The degradation of strength and stiffness under repeated inelastic cycling is called *low cycle fatigue*. In steel structures, good ductility and energy dissipation capacity can be achieved by using thicker sections to avoid local buckling. This implies that plastic and compact sections should be preferred over semi-compact and slender sections. Since earthquakes produce large deformations and low cycle fatigue, both the ductility and energy dissipation capacity are the prime requirements to resist severe earthquakes.

The hysteretic stress–strain relationship for steel, subjected to alternately repeated loading, is shown in Fig. 9.3(a). The unloading branch shows an incipient slope equal to the elastic slope and is gradually softened owing to the *Bauschinger effect* (Appendix XI). Due to the Bauschinger effect, the plastic deformation of steel increases the tensile yield strength and decreases the compressive yield strength. Some of simple models of hysteretic stress–strain curve are shown in Fig. 9.3(b), (c), and (d).

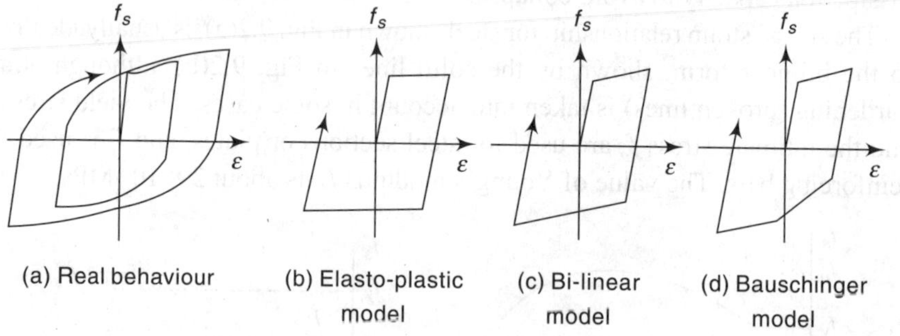

Fig. 9.3 Hysteretic behaviour of steel

9.2 Materials and Workmanship

Steel structures generally perform well in major earthquakes. However, careful detailing and control of material properties are necessary to ensure the development of its full ductility under earthquake loading. The basic steel material must be homogeneous with moderate value of yield stress and of good quality. Steel with minimum yield strength of 340 N/mm^2 is specified for members expecting inelastic action under the effects of the design earthquake. In order to obtain adequate ductility from a steel structure, proper use of the inelastic properties of the steel is to be ensured. To achieve this, the component material must be such that total elongation up to breaking failure is sufficiently large and the ratio of yield stress to ultimate stress is not close to unity. The latter requirement prevents the situation, in which a tension member with bolt holes breaks on a net section, before yielding takes place in a gross section. In general, the steel that is explicitly permitted for use in seismic design is supposed to meet the following requirements:

(a) A ratio of yield stress to tensile stress not greater than 0.85
(b) A pronounced stress–strain plateau at the yield stress
(c) A large inelastic strain capability (e.g., tensile elongation of 20 per cent or greater measured on a gauge length of 5.65 A_0, where A_0 is the cross-sectional area)
(d) Good weldability

In addition to the above characteristics, the steel used should have adequate ductility and consistency of mechanical properties to satisfy seismic requirements, besides satisfying the general requirements, such as adequate notch ductility, freedom from lamination, resistance to lamellar tearing, and good workability.

Ductility Ductility of steel, described as the post-elastic behaviour, allows the structure to undergo large plastic deformations without significant loss of strength. It is expressed as the ratio of the ultimate deformations at an assumed collapse point to the yield deformation. For steel, it may be expressed simply from the results of elongation tests on small samples, or more significantly in terms of moment-curvature or hysteresis relationships.

Notch ductility It is a measure of the resistance of steel to brittle fracture. It is generally expressed as the energy required for fracturing a test piece of a particular geometry. For ductile elements, steel should be of low carbon, and weldable steel should have good notch ductility.

Consistency of mechanical properties Practical realization of the fundamental principle that beams fail before columns is desirable. This requires that the maximum and minimum strengths of members are as nearly equal in magnitude as possible; the standard deviation of strengths should be as small as possible.

Laminations During manufacturing of mild steel plates, hot ingots are pressed together repeatedly to form a uniform layer plate. Sometimes air gets entrapped between two layers, which is not fully removed during rolling operation of the plates. This is known as *lamination*. Thus layering of the steel takes place with little structural connection between the layers. The laminated areas that originate in the casting and cropping procedures for the steel ingots, result in large areas of unbonded steel in the body of a steel plate or section, and may be as much as several square metres in extent. Although the strength is reduced marginally, a significant reduction in heat transfer coefficient at the location of lamination takes place. This is attributed to the air film, which is a bad conductor, sandwiched between two layers of metal. Steel may be screened ultrasonically for lamination before fabrication.

Lamellar tearing It is a tear or stepped crack which occurs under a weld. For lamellar tearing to take place the shrinkage strains on welding must act in the short direction of the plate through the plate thickness. It may be noted that the chances of lamellar tearing increase if the stresses generated on welding act in the through-thickness direction and with the high level of weld metal hydrogen. Typically the cracks appear in solders of several passes, T-butt welds, and in corner welds. These are always associated with points of high stress concentration. Lamellar tearing also occurs at the interface between inclusions and the surrounding material, due to the incapacity of the parent metal to accommodate strains imposed by weld shrinkage in the through-thickness direction. This defect is very common in thick rolled steel plates having poor through-thickness ductility. For example, this failure can occur where butt or fillet welds of 20 mm or over are made on plates of at least 30 mm in thickness, where there is a high degree of restraint. Tearing can occur in planes parallel to the direction of rolling. The fracture surface is fibrous with long parallel sections which are indicative of low parent metal ductility in the through-thickness direction.

The solutions to this problem lie to a limited degree in the selection of the steel with low sulfphur content, or adding alloying elements that control the shape of sulphide inclusions (e.g., rare earth elements, zirconium, or calcium), in inspection procedures as tearing usually occurs during fabrication following cooling of the adjacent weld, and in good detailing of welds. However, since sulfphur levels in steels are nowadays kept typically below 0.005 per cent, lamellar tearing is no longer a problem. The choice of material, joint design, welding process, preheating, etc., helps to reduce the defect. Some of the recommended construction practices to avoid lamellar tearing are shown in Fig. 9.4.

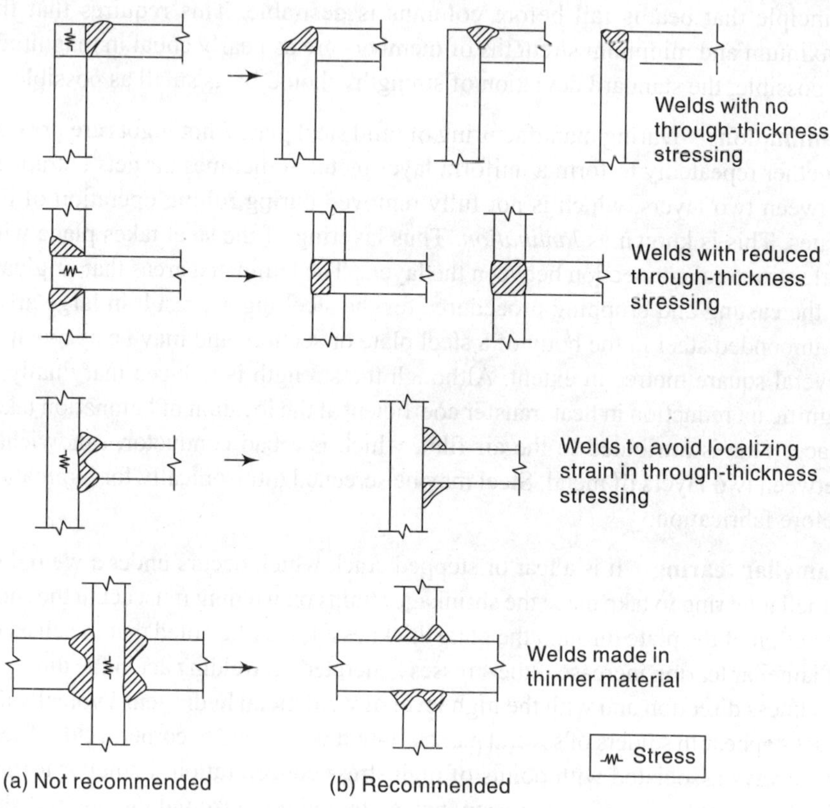

Fig. 9.4 Weld details to avoid lamellar tearing

Workmanship The detailing and fabrication of ductile portions of the structure should consider the possibility of low cycle fatigue—structures responding to earthquakes rarely go through more than 20 cycles of response. Fatigue failure can initiate at notches and cracks which run at right angles to the direction of stress. Inadequate fabrication procedures can drastically reduce ductility. Welding should follow the best standards of quality and inspection. The weld metal should be able to closely match the properties of the parent plates and few material defects should arise. Bolt holes should be drilled and not punched or reamed. Cutting of the sections should be effected by sawing only. The component parts should be assembled and aligned in such a manner that they are neither twisted nor otherwise damaged.

9.3 Steel Frames

On the basis of sensitivity of steel frames to second-order effects in the elastic range, steel frames are classified as sway frames or non-sway frames. A frame in which lateral sway is prevented is called a *non-sway frame*. Further, frames that sway to a small amount and display a negligible P-Δ effect are also called non-sway frames. Both types of the frames can be either concentrically braced or eccentrically braced.

On the basis of inelastic rotation capability, steel moment-resistant frames may be classified as special moment frames (SMFs), intermediate moment frames (IMFs), and ordinary moment frames (OMFs). The three frame types offer three different levels of expected inelastic rotation capability. SMFs are intended to provide for significant inelastic deformations. The majority of the inelastic deformation is intended to take place as rotation in beam 'hinges', with some inelastic deformation permitted in the panel-zone of the column. IMFs are intended to provide inelastic rotation capability that is intermediate between those provided by SMFs and OMFs. It is intended that IMFs will not require the larger plastic rotations expected of SMFs, because of the use of more or larger framing members than for comparably designed SMFs, or because of use in lower seismic zones. OMFs are intended to provide for limited levels of inelastic capability and are thus used in lowest seismic zones and are expected to resist only little inelastic deformation. SMF, IMF, and OMF are designed to accommodate 0.03, 0.02, and 0.01 radians, respectively. The elastic drift of typical moment frames is usually in the range of 0.01 radians and the inelastic rotations of the beams are approximately equal to their inelastic drift. Therefore, these frames can accommodate total drifts in the range of 0.04, 0.03, and 0.02 radians, respectively.

IS 800: 2007, however, classifies the frames as moment frames—special moment frame (SMF) and ordinary moment frame (OMF), and braced frames (BF)—special concentric braced frame (SCBF), ordinary concentric braced frame (OCBF), and eccentrically braced frame (EBF). The inelastic deformation corresponding to a joint rotation, without degradation of strength and stiffness, offered by SMF, OMF, SCBF, and OCBF is 0.04, 0.02, 0.04, and 0.02 radians, respectively. For EBF, the code recommends to refer to the specialist literature. The other requirements and provisions of IS 800: 2007 are presented in Tables 9.1 and 9.2.

Table 9.1 Provisions of IS 800: 2007 for braced frames

S. No	Properties	OCBF	SCBF
1	Limit to withstand inelastic deformation without degradation in strength and stiffness below full yield value corresponding to joint rotation of	0.02 radians	0.04 radians
2	Recommended zones	Can not be used in Zones IV and V	Can be used for any seismic Zone
3	Recommended importance factor	Should not be used for buildings with importance factor > 1, in Zone III	Can be used for any importance factor
4	Bracing provisions:		
	(a) Type of bracing	Only for X-bracing. For other specialist literature.	Only for X-bracing. For other specialist literature.
	(b) Slenderness ratio	< 120	< 120
	(c) Compressive strength	$< 0.8\,p_d$	$< p_d$ p_d = design strength in axial compression
	(d) Tensile strength	Tension braces should be able to resist 30 to 70% of total lateral load	Tension braces should be able to resist 30 to 70% of total lateral load
	(e) Cross-section	Can be plastic, compact or semi-compact only	Plastic section only
	(f) Tack fasteners for built-up braces should furnish slenderness ratio not greater than	0.4 times of most unfavourable slenderness ratio of brace itself	0.4 times of most unfavourable slenderness ratio of brace itself
	(g) Bolted connections	To avoid in middle one-fourth of the clear brace length	To avoid in middle one-fourth of the clear brace length
	(h) The design tensile strength of the brace should be governed by	Gross area yielding and not the net area rupture	Gross area yielding and not the net area rupture
	(i) Steel quality	—	E250B steel (IS 2062)
	(j) End connections should be designed to withstand minimum of following:		
	(i) Tensile force in bracing	$1.2 f_y A_g$	$1.1 f_y A_g$ A_g = gross cross-sectional area of the member

(Contd)

Table 9.1 (*Contd*)

	(ii) Force in the brace due to load combinations	Section 9.9 a) and b)	—
	(iii) Maximum force	Transferred to the brace by the system	Transferred to the brace by the system
	(k) Check	Block shear and rupture for the load determined as above	Check for block shear and rupture for the load determined as above
	(l) Connection should also be able to withstand a moment of	1.2 times the full plastic moment of the braced section about the buckling axis	1.2 times the full plastic moment of the braced section about the buckling axis
5.	Gusset plate for making connection should be checked for	Buckling out of their plane	Buckling out of their plane

Table 9.2 Provisions of IS 800: 2007 for moment frames

S. No.	Properties	OMF	SMF
1.	Limit to withstand inelastic deformation without degradation in strength and stiffness below full yield value corresponding to joint rotation of	0.02 radians	0.04 radians
2.	Recommended zones	Can not be used in Zones IV and V	Can be used for any seismic Zone
3.	Importance factor	Should not be used for building with importance factor > 1, in Zone III	Can be used for any importance factor
4.	Connections	Rigid or semi-rigid	Rigid (Annexure F, IS 800)
	(i) Rigid moment connection is designed for	1.2 times the fully plastic moment of the connected beam or the maximum moment that can be delivered by the beam to the joint due to induced weakness at the end of beam, whichever is less	1.2 times when a reduced beam section is used, its minimum flexural strength should be 0.8 times the full plastic moment of unreduced section.

(*Contd*)

Table 9.2 *(Contd)*

	(ii) Semi-rigid connection to be designed for	Either 0.5 times the full plastic moment of the connected beam or the maximum moment that can be delivered by the system, whichever is less. The design moment should be achieved within a rotation of 0.01 radian (Annexure F of IS 800-2007 may be referred.)	
	(iii) The rigid and semi- rigid connections should be designed to resist a shear resulting from	1.2 DL + 0.5 LL plus shear from moments defined in (i) and (ii) earlier as appropriate	1.2 DL + 0.5 LL plus shear resulting from application of 1.2 M_p at each end of beam causing double curvature bending
5.	Continuity plates	In rigid (fully welded connections) continuity plates (tension stiffener) of thickness ≥ thickness of beam flange should be provided	Continuity plates should be provided in all strong axis welded connections except in end plate connection.
6.	Panel zone	—	In column strong axis connections (beam and column web in the same plane), the panel zone should be checked for shear buckling as per clause 8.4.2 of IS 800: 2007 at the design shear defined at 4(iii). Doubler plates or diagonal stiffness may be used to strengthen the webs
7.	Doubler plates	—	The thickness t of the column web and doubler plates should satisfy $t \geq (d_p + d_f)/90$ d_p = panel zone depth between continuity plate d_f = panel zone width between column flange
8.	Splices	—	To be located in the middle-third height only

(Contd)

Table 9.2 (*Contd*)

9.	Design force for splice	—	Splices should be designed for forces transferred to it. In addition, these should be able to develop at least the nominal shear strength of the smaller connected member and 50% of the nominal flexural strength of the smaller connected section.
10.	Beams and Columns	—	
	(a) Cross-section		Plastic or compact only. At possible hinge locations, only plastic.
	(b) Check		$\sum \frac{M_{pc}}{M_{pb}} \geq 1.2$ where, $\sum M_{pc}$ = sum of moment capacity of the column above and below beam centre line. M_{pb} = sum of moment capacity in the beam at the intersection of beam and column centre line.
	(c) lateral support		To the column at both top and bottom beam flange levels to resist at least 2 per cent of the beam flange strength.

Another type of steel frame, the truss-girder moment frame, has also been used often. However, fracture of web members is prior to or early in the dissipation of energy through inelastic deformations. To overcome this problem, special truss girders have been developed that limit inelastic deformations to a special segment of the truss. As illustrated in Fig. 9.5, the chords and web members (arranged in an X-pattern) of the special segment are designed to withstand large inelastic deformations, while the rest of the structure remains elastic.

9.3.1 Behaviour of Unbraced Frames

The moment-resisting frames (MRF) resist lateral loads because of the partial or fully rigid joints of the frame. Due to their flexibility, they experience large drift

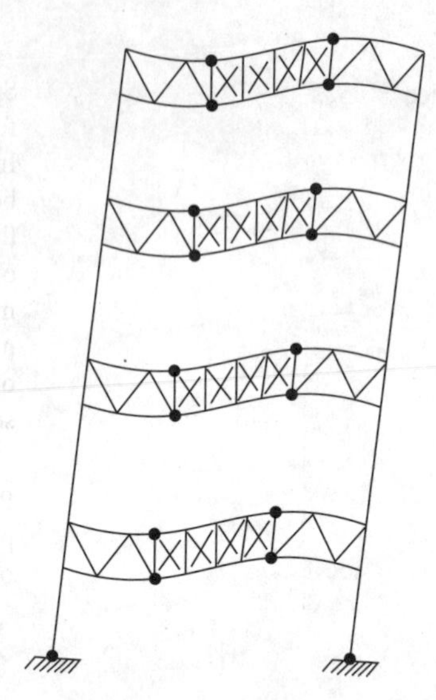

Fig. 9.5 Special truss girder moment frame

especially in multi-storey steel buildings. Moment resisting frames can be designed either to dissipate energy by formation of plastic hinges at the beam ends, or in the connections; the latter is not preferred due to the complexities associated with the analysis and design. In any case, the connections must be strong and display ductile behaviour.

Horizontal load–deflection relationships for a portal frame and for multi-storey frame subjected to constant vertical load and monotonic horizontal load are shown in Fig. 9.6. The frame shown in Fig. 9.6(a) remains in the elastic range until the first plastic hinge forms. Additional load increments are resisted by a deteriorated structure, and lateral displacements increase as the first plastic hinge rotates under constant plastic moment M_p. Lateral stiffness decreases again on formation of a second plastic hinge. The process continues, as the number of plastic hinges increase, until the structure becomes a mechanism which sways under decreasing lateral load. The plastic collapse load corresponding to an overall mechanism is the maximum load, which can theoretically be resisted by a steel frame, and the one which allows energy absorption before failure.

A relatively simple method of analyzing the load–deflection relationship of a frame assumes that the inelastic deformation is concentrated at the plastic hinge and that all other regions remain elastic. The secondary effect of axial force is considered in member stiffness, and incremental relations between horizontal load and deflection are repeatedly determined for the frame in which the plastic hinges are replaced by real hinges, as shown in Fig. 9.6(a). As the strain hardening effect is neglected in this method of analysis, the calculated load–deflection curve lies below the real curve in the large-deflection range. As an alternative to the above

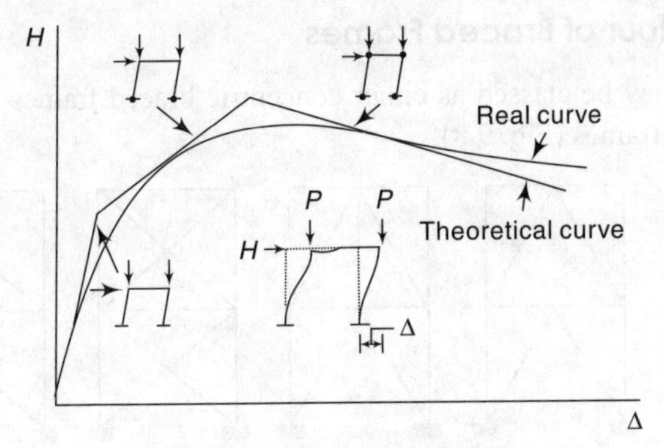

(a) Portal frame under constant vertical load and monotonic horizontal load

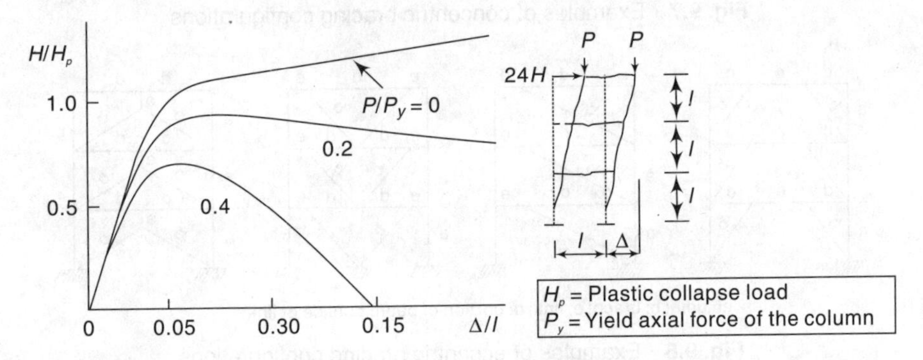

(b) Multi-storey frames under constant vertical load and monotonic horizontal load (Wakabayashi et al. 1969)

Fig. 9.6 Horizontal load–deflection relationship (Wakabayashi et al. 1969)

method, a particular level is separated from the frame at the inflection points in the columns above and below the floor, and a further subdivision is made into sub-assemblages consisting of a column and adjacent beams. The horizontal force-sway curve for each sub-assemblage is determined by using load–deformation curves. The load–deflection curve for the whole frame can be obtained by adding the individual sub-assemblage curves.

For the frame shown in Fig. 9.6(b), its strength under zero vertical loads continues to rise beyond the simple plastic collapse load M_p, because of the strain-hardening effect. The maximum strength of the frame decreases with increasing vertical load as a result of the P-Δ effect, i.e., the effect of the overturning moment produced by the vertical load P and the horizontal deflection Δ. The instability of unbraced frames is influenced by the slenderness ratio of the columns, the ratio of the working axial force to the yield axial force of the columns, and the beam-stiffness ratio.

9.3.2 Behaviour of Braced Frames

Braced frames may be classed as either concentric braced frames (Fig. 9.7) or eccentric braced frames (Fig. 9.8).

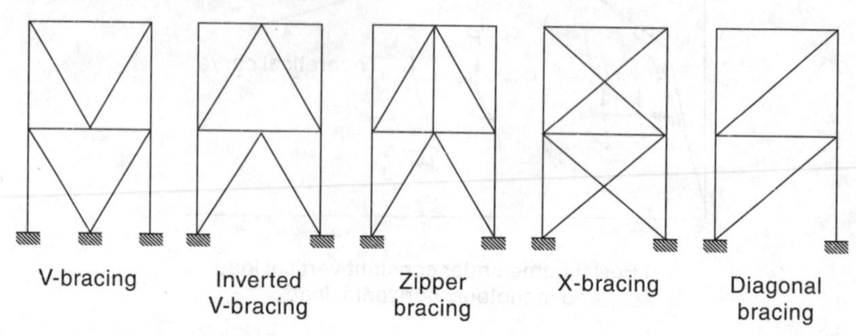

Fig. 9.7 Examples of concentric bracing configurations

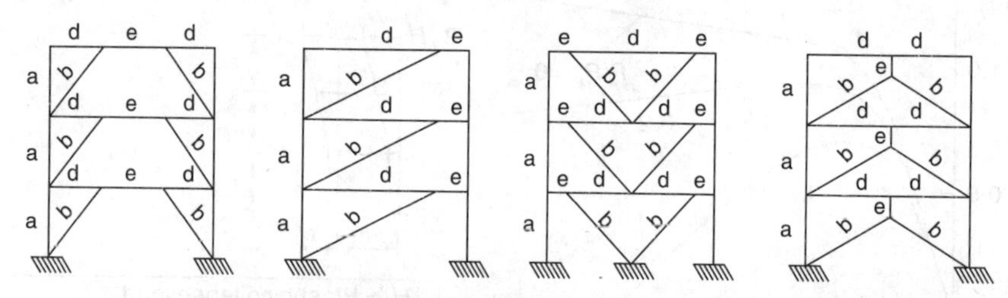

Fig. 9.8 Examples of eccentric bracing configurations

Concentric braced frames

Concentric bracing may be designed to resist the entire seismic load, in which case the braces are used in combination with simple beam-to-column connections—the shear connections. In concentric braced frames (CBFs) having simple connections, the centre lines of the members that meet at a joint intersect at a point to form a vertical truss system that resists lateral forces. Because of their geometry, these frames provide complete truss action with members subjected primarily to axial forces in the elastic range. The concentric bracing may also be designed as a supplementary system in a moment resisting frame.

Concentric braced frames are commonly used to resist wind forces, but suffer from low system ductility for cyclic loads. The poor performance of CBFs is attributed to premature buckling of the braces, which limits their energy dissipation capacity. Moreover, due to their higher stiffness, they attract larger seismic forces. Also, bracing arranged concentrically in the structure poses difficulties in preventing foundation uplift. Since, one diagonal of an opposing pair is always in tension, the possibility of brittle failure is present. Both V- and inverted V-braced frames, often referred to as *chevron braces*, perform poor because of bucking of braces and excessive flexure of beam at mid span where the braces intersect the beam. Buildings with X- or V-braces with zipper perform better and are generally recommended in high seismic zones. However, there is a drawback in the use of cross-braced panels—there being no effective way in which access can be gained through the panel—which places a major restriction on the areas where they can be used. Single diagonal bracing is recommended only when multiple single-diagonal braces are provided along a given brace frame line. *K*-bracing is not permitted at all in high seismic zones.

Earlier, CBFs were supposed to display reliable behaviour by limiting only the global buckling. Cyclic testing of concentric bracing systems, however, shows that larger energy can be dissipated after the onset of global buckling, if brittle failure due to local buckling, stability problems, and connection fractures are prevented. When properly detailed for ductility, diagonal braces sustain large inelastic cyclic deformations without experiencing premature failures. However, during a moderate to severe earthquake, the bracing members and their connections are expected to undergo significant inelastic deformations in the post-buckling range. During a severe earthquake, reversed cyclic rotations occur at plastic hinges in much the same way as they do in beams and columns in moment frames. In fact, braces in a typical CBF can be expected to yield and buckle at rather moderate storey drifts of about 0.3–0.5 per cent. In a severe earthquake, the braces may undergo post-buckling axial deformations 10 to 20 times their yield deformation. Large-storey drifts that can result from early brace fractures can pose excessive demands on columns or their connections.

In order to survive such large cyclic deformations without premature failure, the bracing members and their connections are required to be properly detailed. The improved design parameters, such as limiting the width-to-thickness ratios (to minimize local buckling), closer spacing of stitches, and special design and detailing of end connections, greatly improve the post-buckling behaviour of CBFs. For CBFs, emphasis is placed on increasing brace strength and stiffness, primarily through the use of higher design forces in order to minimize inelastic demand. Accordingly, special concentrically braced frames (SCBFs) have been developed in which all members of the bracing system are subjected primarily to axial forces. SCBFs are intended to exhibit stable behaviour and respond to seismic forces with greater ductility.

Many of the failures reported in CBFs due to strong ground motions have occurred in the connections. Similarly, cyclic testing of specimens designed and detailed in accordance with typical provisions for CBFs has produced connection failures. To achieve adequate performance of CBFs, the following stipulations are necessary:

(a) For double-angle and double-channel braces, closer stitch spacing, in addition to more stringent compactness criteria, is required to achieve improved ductility and energy dissipation. This is especially critical since these braces buckle and large shear forces are imposed on the stitches. The placement of double angles in a toe-to-toe configuration reduces bending strain and local buckling.

(b) For brace buckling out of the plane of single plate gussets, weak axis bending in the gusset is induced by member end rotations. This results in flexible end conditions with plastic hinges at mid-span in addition to the hinges that form in the gusset plate. Satisfactory performance can be ensured by allowing the gusset plate to develop restraint-free plastic rotations. This requires that the free length between the end of the brace and the assumed line of restraint for the gusset be sufficiently long to permit plastic rotations, yet short enough to preclude the occurrence of plate buckling prior to member buckling. A length of two times the plate thickness is generally recommended. Alternatively, connections with stiffness in two directions, such as crossed gusset plate, can be detailed. Forcing the plastic hinge to occur in the brace rather than in the connection plate, results in greater energy dissipation capacity.

(c) For brace buckling in the plane of gusset plates, the end connections should be designed for the forces specified in Table 9.1.

(d) Beams or columns of the frame should not be interrupted at the brace intersections. This provision is necessary to improve the out-of-plane stability of the bracing system at those locations. However, mere continuity of columns or beams at the brace intersections may not be sufficient to provide the required stability. Typical practice is to provide perpendicular framing that engages a diaphragm to provide out-of-plane strength and stiffness, and resistance to lateral-torsional bucking of beams.

Eccentric Braced Frames

Eccentric braced frames (EBFs) represent an economically effective way of designing steel structures for earthquake loading. By selecting a suitable frame stiffness and yield level, it is possible to resist moderate earthquakes elastically, with only moderate displacements, and to resist major earthquakes inelastically. The primary benefit of EBFs is that substantial system ductility can be developed. Moreover, because of the ease with which access can be gained through the plane of the braced panel, they may be located within the building. One potential drawback is the possibility of floor damage near the link beam (Fig. 9.9) during major earthquakes, but in view of the levels of damage normally regarded as acceptable, this is not serious.

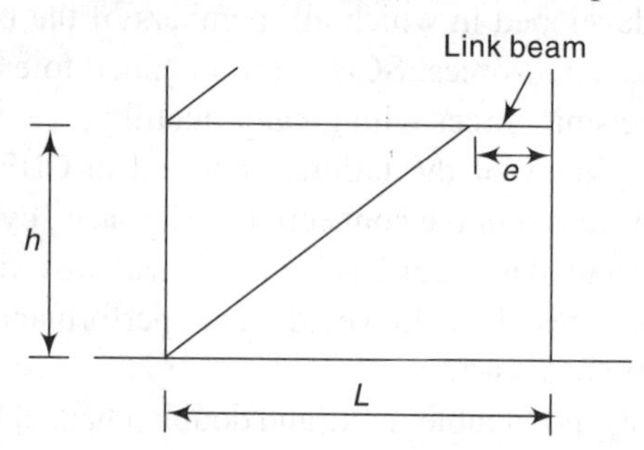

Fig. 9.9 Eccentric braced frame parameters

Eccentrically braced frame is designed by assuming the bracing member to be pin-ended and the beam-to-column connection to be a moment resisting connection. The bracing increases the lateral stiffness of the frame and controls the drift as well. EBF is a framing system in which the forces induced in the braces are transferred either to a column or to another brace through shear and bending in a small segment of beam called the *link*.

Most of the energy in EBFs is dissipated in a link by yielding in shear or flexure. Links longer than twice the depth tend to develop plastic hinges while shorter links tend to yield in shear. The links in EBFs, therefore, act like structural fuses to dissipate earthquake-induced energy in a stable manner. The segment e of the beam (Fig. 9.9) is the ductile link and is designed to carry the earthquake-induced force. The segment of beam outside of e, the brace and the columns are the other links presumed to be brittle and designed to have strength in excess of the strength of the ductile link e to account for the normal uncertainties of material strength and strain-hardening effects at high strains. An eccentrically braced frame dissipates energy by controlled yielding of its link. Therefore, the link needs to be detailed properly so that it has adequate strength and stable energy dissipation characteristics. Holes are not allowed in the web of the link because they affect the inelastic deformation of the link web.

To ensure stable hysteresis, a link must be laterally braced at each end to avoid out-of-plane twisting. Lateral bracing also stabilizes the eccentric bracing and the beam segment outside the link. Both the top and bottom flanges of the link beam must be braced. When detailing a link, full-depth web stiffeners must be placed symmetrically on both sides of the link web at the diagonal brace ends of the link. Further, the link must be stiffened in order to delay the onset of web bucking and to prevent flange local buckling. When cover plates are used to reinforce a link-to-column connection, the link over the reinforced length must be designed such that no yielding takes place in this region. In this context, link is defined as the segment between the end of reinforcement and the brace connection.

Elastic behaviour Whether or not a braced frame acts in conjunction with a moment frame, its stiffness is of great importance. In either case the braced frame will have a major effect on the overall stiffness of the system, thereby determining the level of force that it will be subjected to by moderate earthquakes.

The elastic design parameters of an EBF can be characterized as illustrated in Fig. 9.9. The length assigned to the link, or the 'active link' beam is its clear span. As e/L is varied, the system changes from a moment frame with $e/L = 1$ to a CBF with $e/L = 0$. Parameters such as geometric arrangements and member properties of braces affect considerably the elastic stiffness of braced frames.

Inelastic behaviour The behaviour of EBFs in the inelastic range is dominated by the link members, which may have very high ductility requirements. Thus, the short beam segment, called the link, is intended as the primary zone of inelasticity. Provisions must be made to ensure that cyclic yielding in the links can occur in a stable manner, while the diagonal braces, columns, and the portions of the beam outside of the link remain essentially elastic under the forces that can be generated by fully-yielded and strain-hardened links.

In some bracing arrangements, such as the one illustrated in Fig. 9.10, with links (*a* and *b*) at each end of the brace, the links may not be fully effective. If the upper link has significantly lower design shear strength than the link in the storey below, the upper link will deform inelastically and limit the force that can be delivered to the brace and to the lower link. When this condition occurs, the upper link is termed an *active link* and the lower link is termed an *inactive link*. The presence of potentially inactive links in an EBF increases the difficulty of analysis; an inactive link, in some cases, yields under the combined effect of dead, live, and earthquake loads, thereby reducing the frame strength below that expected. Furthermore, inactive links are required to be detailed and constructed as if they were active. Thus, an EBF configuration that ensures that all the links will be active is recommended. The admissible deformed shapes for two types of EBFs are shown in Fig. 9.11. For some of the EBFs, the relationship between member deformation and structure deformation is shown in Fig. 9.12.

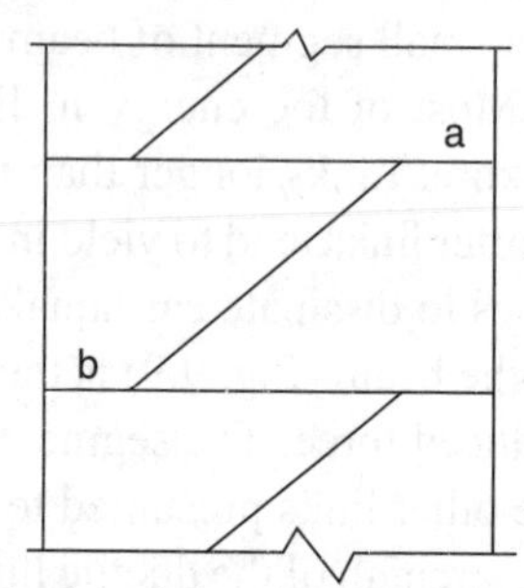

Fig. 9.10 EBF—active and inactive links

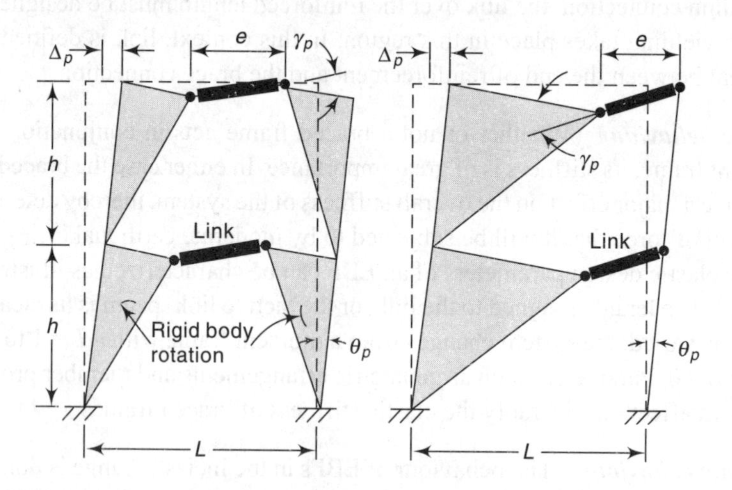

Fig. 9.11 Eccentric braced frame: deformation mechanisms

Note: Link rotation angle γ_p is estimated by assuming that the EBF bay rotates as a rigid body. By geometry, γ_p is related to plastic storey drift angle θ_p, which in turn is related to plastic storey drift = Δ_p/h. Conservatively, Δ_p may be taken to equal design storey drift.

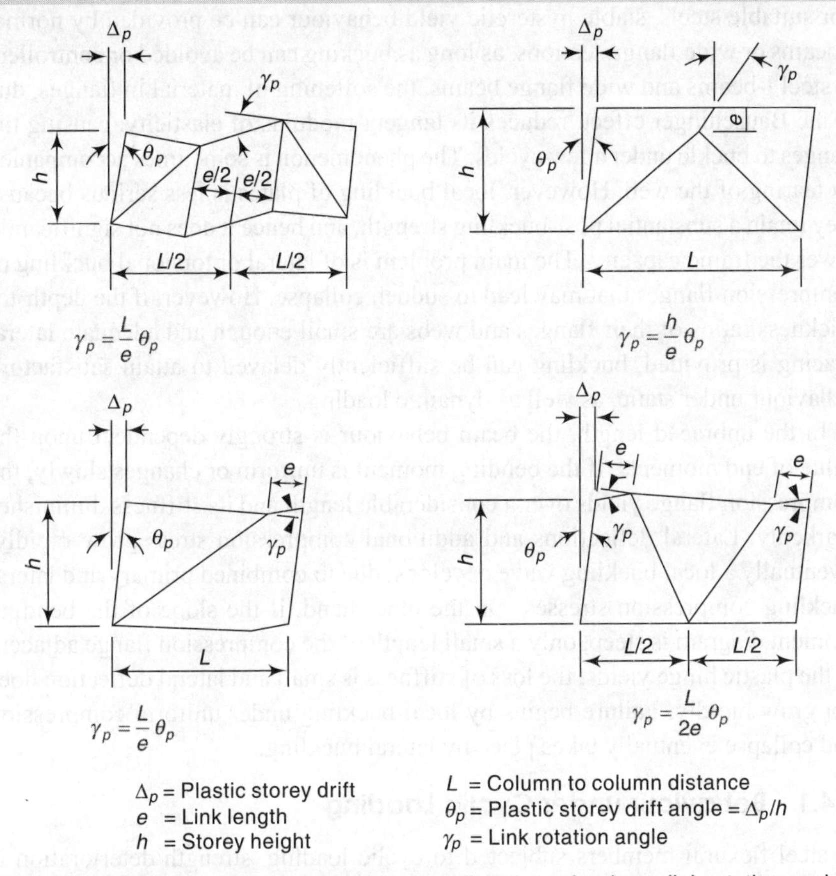

Fig. 9.12 Models for eccentric braced frame collapse mechanisms: link rotation angles

9.4 Flexural Members

For the frames subjected to strong ground shaking, if the columns are weaker in flexure than the framing beams, the plastic hinges form at the base and top of the columns in the storey mechanism. This can lead to very large total drifts, P-Δ instability and total collapse. Moment frames are, therefore, designed on the basis of strong (elastic) columns and weak (inelastic) beams. Thus beams provide for energy absorption and adequate rotation capacity at certain points. For a laterally and torsionally braced beam, the lateral displacements of the compression flange start as soon as the bending moment reaches the M_p value. The beam sections lose their original shape. Lateral buckling and local buckling start after sufficient plastic deformation has taken place under constant plastic moment. Failure of beams is due to instability under continuously increasing lateral deformation and not because of buckling. Very short beams, however, fail by yielding before buckling.

The webs and flanges of beams behave as plates in their buckling performance. For suitable steels, stable hysteretic yield behaviour can be provided by normal I-beams or wide flange sections, as long as bucking can be avoided or controlled; in steel I-beams and wide flange beams, the softening of material in flanges, due to the Bauschinger effect, reduces its tangent modulus of elasticity, causing the flanges to buckle under a few cycles. The phenomenon is sometimes accompanied by tearing of the web. However, local buckling of plates is less serious because they retain a substantial post-buckling strength, and hence it does not significantly lower the frame capacity. The main problem is of lateral or torsional buckling of compression flanges that may lead to sudden collapse. However, if the depth-to-thickness ratios of their flanges and webs are small enough and adequate lateral bracing is provided, buckling can be sufficiently delayed to attain satisfactory behaviour under static, as well as dynamic loading.

In the unbraced length, the beam behaviour is strongly dependent upon the value of end moments. If the bending moment is uniform or changes slowly, the compression flange yields over a considerable length and its stiffness diminishes markedly. Lateral deflections and additional compression stress grow rapidly. Eventually a local buckling wave develops, due to combined primary and lateral buckling compression stresses. On the other hand, if the slope of the bending moment diagram is steep, only a small length of the compression flange adjacent to the plastic hinge yields; the loss of stiffness is small and lateral deflection does not grow rapidly. Failure begins by local buckling under uniform compression and collapse eventually takes place by lateral buckling.

9.4.1 Behaviour under Cyclic Loading

In steel flexural members subjected to cyclic loading, strength deterioration is often caused by cracks in the zone of maximum inelastic deformation because of repeated bending or by local buckling and/or lateral buckling of the web following local buckling of the flange. Hysteresis loops for small rotation amplitude are stable, but strength degradation becomes severe when the rotation amplitude exceeds a value which is less than half of the rotation capacity under monotonic loading. In Fig. 9.13, the typical hysteresis loops of a steel beam are shown, wherein, the decay is mainly due to web buckling.

Flange buckling and lateral-torsional buckling also influence the loss of strength and stiffness of the beams to some extent, and therefore, a shorter, laterally unsupported length must be specified for beams subjected to cyclic loading. Both the flanges of beams should, therefore, be laterally supported, directly or indirectly. In a potential plastic-hinge region, the width-to-thickness ratio of the beam should be kept small, and the lateral braces should be spaced with a small pitch to ensure

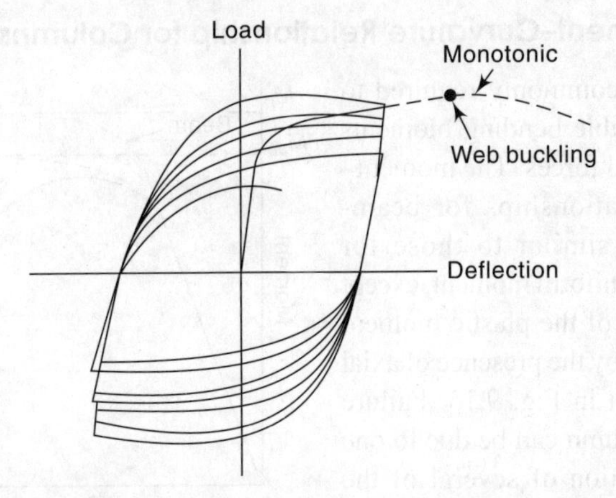

Fig. 9.13 Typical hysteresis loops for a steel beam under cyclic bending

sufficient rotation capacity of the beam. Outside the plastic-hinge regions, beams need only resist external forces (ductility not required), and, therefore, a larger spacing of lateral braces is allowed.

9.5 Frame Members Subjected to Axial Compression and Bending

When a frame is subjected to a horizontal load, the total storey shear is either carried entirely by the bracing, as shown in Fig. 9.14(a), or carried by the columns and bracing in combination, as shown in Fig. 9.14(b). The stress condition for a particular column depends on the load carrying mechanism. Strength reduction under load reversals is very significant if the column is subjected to a large axial force. Further, for a slender column subjected to a large axial force, the column ductility is small, limiting the axial force to less than 60 per cent of the yield axial force. In a potential plastic-hinge region the width-to-thickness ratio and the spacing of lateral braces is limited. In other regions, the width-to-thickness ratio should not be greater than the ratio specified for elastic design, while lateral braces should be so spaced as to provide the required strength.

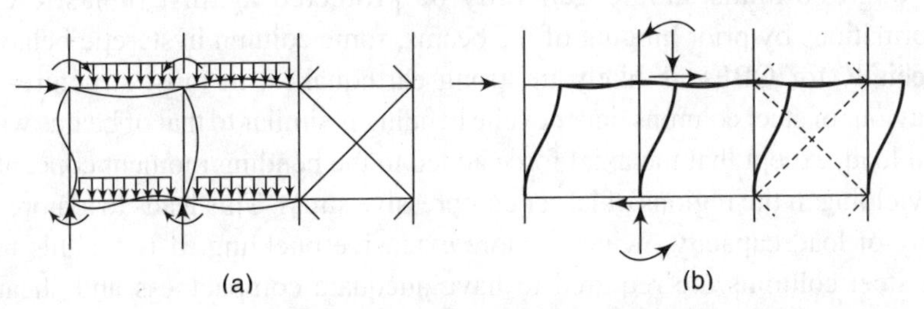

Fig. 9.14 Beam-columns in frame (Wakabayashi 1986)

9.5.1 Moment–Curvature Relationship for Columns

Columns are commonly required to resist appreciable bending moments as well as axial forces. The moment–curvature relationships for beam-columns are similar to those for beams under uniform moment, except that the value of the plastic moment M_p is reduced by the presence of axial load as shown in Fig. 9.15. Failure of a beam-column can be due to one or a combination of several of the following causes:

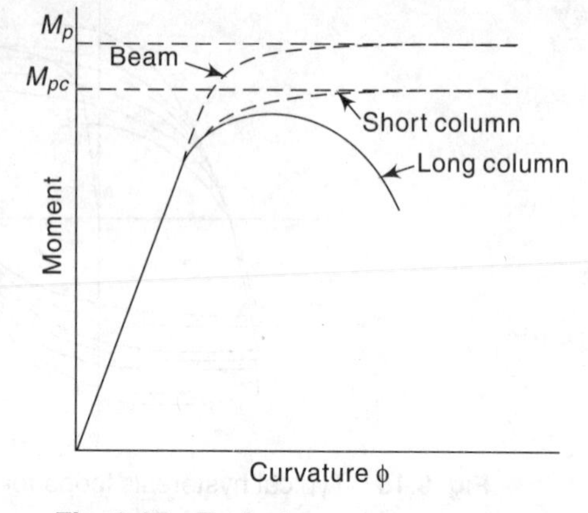

Fig. 9.15 Typical moment–curvature relationships for columns

(a) Plastic hinge formation under axial compression and bending
(b) Instability by axial load and moment interaction
(c) Lateral-torsional buckling
(d) Axial compression buckling by weak axis bending
(e) Local buckling

Building columns usually fail through a combination of the first two factors, by the development of a mechanism under axial load and bending moments magnified by second order effects. Dimensions of these members are such that lateral-torsional buckling is seldom critical and, as bending almost always play an important role, axial compression buckling is not critical either.

Columns of frames built in strongly seismic areas are usually in the low P/P_y range, because they are strongly influenced by bending where, P is the axial force compressive or tensile in the member and P_y is the yield strength of the axially loaded section. Besides, they have small slenderness ratio and bend in double curvature under vertical and earthquake loads.

9.5.2 Behaviour of Columns under Cyclic Loading

Although columns should generally be protected against inelastic cyclic deformations by prior hinging of the beams, some column hysteretic behaviour, especially for CBFs, is likely in strong earthquakes, in most structures. The behaviour of steel columns under cyclic bending is similar to that of beams without axial load, except that the axial force added to the bending moment concentrates the yielding in the regions of larger compressive stress. This leads to a more rapid decay of load capacity owing to more extensive buckling. It is for this reason that steel columns are required to have adequate compactness and shear and flexural strengths in order to maintain their lateral strengths during large cyclic deformations of the frame.

Figure 9.16 shows hysteretic load–deflection relationships for a laterally braced beam-column with a compact section subjected to constant axial force and repeated lateral load. It may be noted that the loops enlarge and the maximum strength increases because total compressive strain extends into the strain-hardening range. However, local buckling causes strength degradation in a non-compact section. The strength of laterally unbraced beam-column initially increases because of the strain-hardening effect, but eventually lateral-torsional buckling causes strength degradation. A wide-flange section of the square type, which is strong in torsion, is usually employed as a column section, but strength degradation still takes place under cyclic loading.

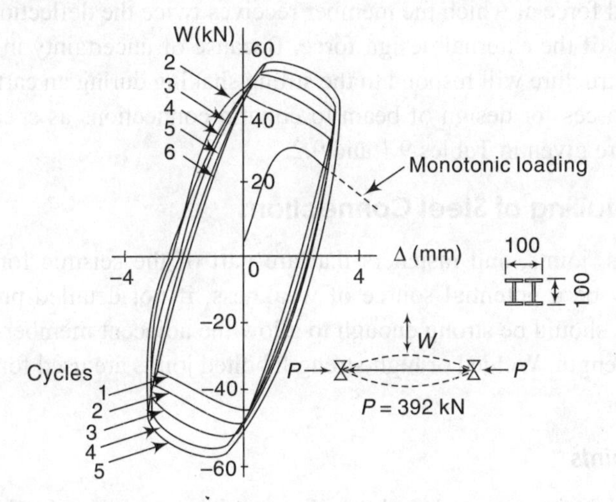

Fig. 9.16 Load–deflection relationships for laterally braced beam-column with a compact section subjected to a constant axial force and repeated lateral load (Takabashi 1971)

9.6 Connection Design and Joint Behaviour

Riveting, bolting, and welding may be employed for making connections either individually or in combination; rivets are rarely used in modern steel construction. A fully bolted connection tends to be very large and expensive. Connections using black bolts tend to slip under cyclic loading causing a reduction in their energy dissipation capacity and hence are avoided; only high strength friction grip bolts are recommended. The most frequently used connection system is a fully welded connection or a combination of welding and bolting. Brittle welding failures are common due to low cycle fatigue and therefore stress concentrations at welds should be avoided. Connections should not be too sensitive to shop or field tolerances, and should minimize the use of highly skilled crafts.

The design of connections in earthquake-resistant structures requires special treatment because of the necessity to accommodate inelastic response in the members. Therefore, to take full advantage of the strength and ductility of the members of a steel frame, the connections should be able to develop at least the full plastic capacity of the members. Since, the behaviour of connections is not as well understood as that of members, some conservativeness in the design of connections relative to members is required. Although studies of inelastic behaviour in connections have shown that some energy absorption is possible, the normal practice is to design connections to remain elastic. A general rule for connection design is that the strength of a connection should not be smaller than the strength of the ends of the member that is framed into the connection. However, if the rotation capacity of a connection is verified to be large by experiment or analysis, the design connection force can be reduced to a value equal to the member-end force at which the member receives twice the deflection computed on the basis of the external design force. Because of uncertainty in the manner in which a structure will respond to the strong shaking during an earthquake, the minimum forces for design of beam to column connections as specified by IS 800: 2007 are given in Tables 9.1 and 9.2.

9.6.1 Detailing of Steel Connections

Connections, joints, and fasteners that are part of the seismic force-resisting system may be a potential source of weakness, if not detailed properly. The connections should be strong enough to allow the adjacent members to develop their full strength. Welded or high-strength bolted joints are used for making the connections.

Welded Joints

To ensure sufficient strength and ductility at a joint, welding should be used so that maximum member strength can be transferred safely to the panel. The highest welding standard should be followed because of the possibility of low cycle fatigue. In addition, details should be designed to avoid stress concentration or lamellar tearing. A typical welded flange plate connection is shown in Fig. 9.17.

But welded joints provide the best earthquake resistance. The best form of load transfer is a complete joint penetration (CJP) butt weld, where the weld material strength is greater but not significantly greater than the parent metal. Partial joint penetration (PJP) butt welds are not recommended. Fillet welded joints are capable of developing the full plastic moment of the members joined, when the loading is monotonic. The following rules are observed in case of fillet welds:

(a) Intermittent welding should be minimized as the ends of runs are stress-raising discontinuities.
(b) The throat thickness should not be less than half the plate thickness.
(c) Tearing stresses in the parent metal should be checked where high-strength electrodes are used and the leg length of the weld is small.

Note: IS 800: 2007 recommends only complete penetration butt welds in frames except for splices.

Failure of welded joints Failure of a welded rigid beam-column connection may occur by yielding or fracture, as a result of high local stress, as shown in Fig. 9.18(a) or, alternatively, by shear yielding of the connection panel, as shown in Fig. 9.18(b). Local stress, developed by the compression and tension forces delivered from the beam flanges, can bring about the following two types of failure.

1. Crippling of the column web due to the compression force delivered from the beam compression flange.
2. Excessive flexural deformation of the column flange followed by fracture of the flange weld in the vicinity of the column web, caused by the tension force delivered from the beam tension flange.

Bolted Joints

The potential for full reversal of design load and the likelihood of inelastic deformations of members and/or connected parts necessitates the use of fully tensioned high-strength bolts in bolted joints in the seismic-force-resisting system. However, earthquake motions are such that slip cannot be prevented in all cases. Accordingly, bolted joints are proportioned as fully-tensioned bearing joints but with faying surfaces prepared for better slip-critical connections. That is, bolted connections can be proportioned with design strength for bearing connections as long as the faying surfaces are still prepared to provide a minimum slip coefficient, μ of 0.33. The resulting nominal amount of slip resistance will minimize damage in more moderate seismic events.

Notes: IS 800: 2007 recommends the following provisions in case of bolted joints:

(a) Use of fully tensioned high-strength friction grip (HSFG) bolts or turned and fitted bolts.

(b) Bolted joints are not supposed to share load in combination with welds on the same faying surface.

Failure of bolted joints Bolted joints are capable of developing the full plastic moment of the connected members, although with local loss of stiffness and energy absorption. To prevent excessive deformations of bolted joints due to slip between the connected plates under earthquake motions, the use of holes in bolted joints in the seismic-force-resisting system is limited to standard holes and short-slotted holes with the direction of the slot perpendicular to the line of force.

9.6.2 Behaviour of Connections under Cyclic Loading

As compared to beam and column elements, relatively few cyclic load tests have been carried out on steel connections. Cyclic bending tests of cantilevers connected to rigid stub columns reveal the following characteristics of connections:

(a) Hysteresis loops of fully welded connections are stable and spindle shaped as shown in Fig. 9.19.

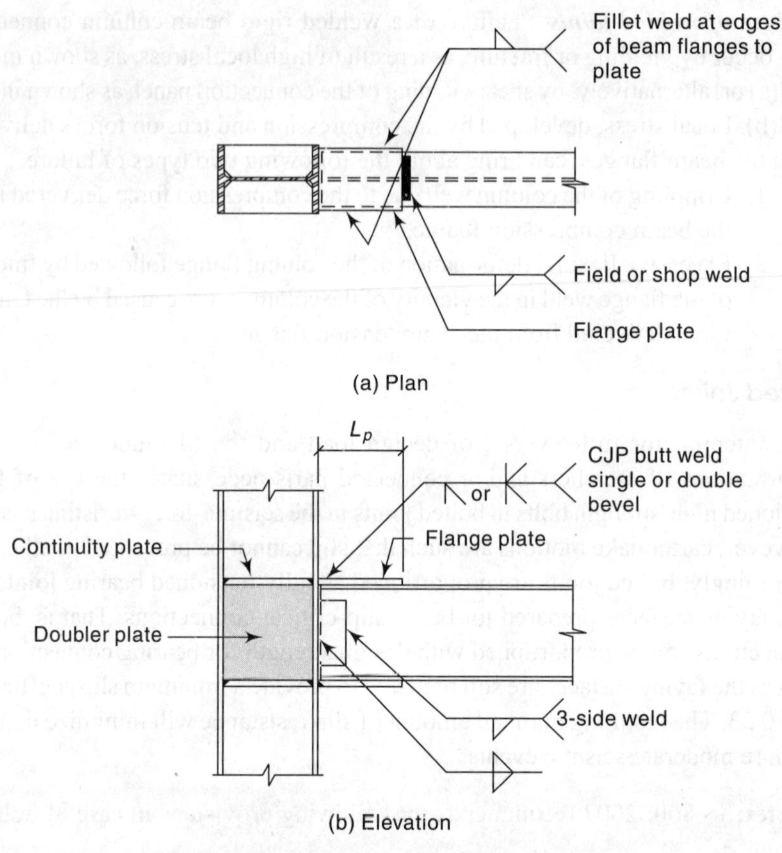

Fig. 9.17 Welded flange plate connection

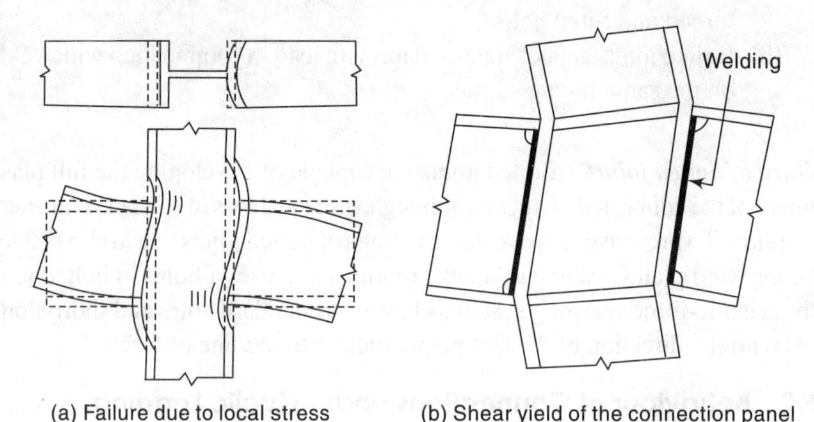

Fig. 9.18 Failure of beam-column connections (Wakabayashi 1986)

(b) Welded flange and bolted web connections, as shown in Fig. 9.20(a), are not fully rigid owing to bolt slippage, but their hysteretic behaviour, in general, is similar to that of fully welded connection.

(c) For connections with welded flange splices [Fig. 9.20 (b)], cracks form early in the splice plate at the end of the fillet weld, and hence ductility is smaller than in cases (a) and (b).

(d) Connections with bolted flange and web splices show hysteresis loops of the slip type because of bolt slippage.

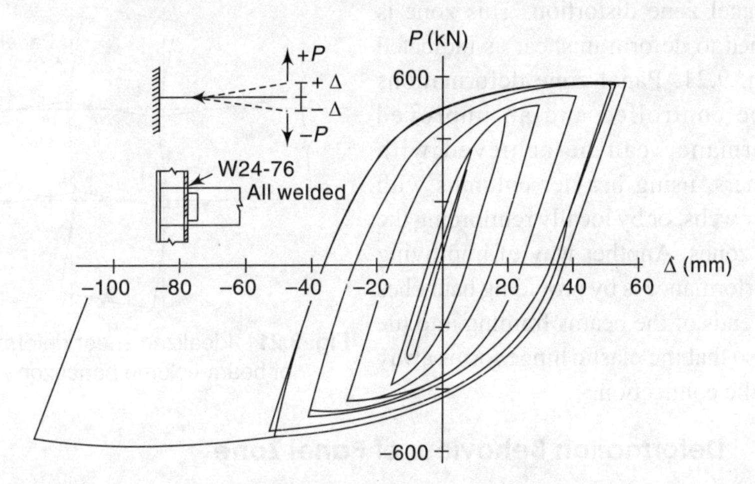

Fig. 9.19 Load–deflection curves for a connection (Krawinkler et al. 1982)

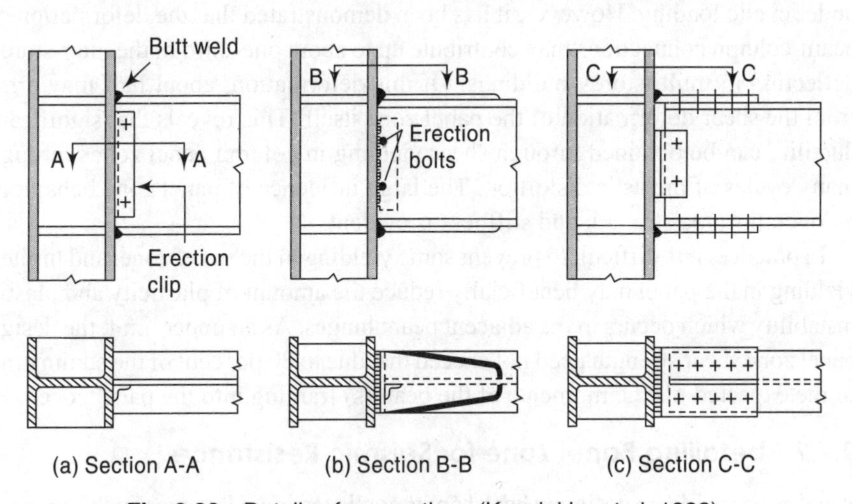

Fig. 9.20 Details of connections (Krawinkler et al. 1982)

9.7 Steel Panel Zones

The panel zone of a connection between two members is the intersection zone common to the two members with their webs lying in a common plane. It is the entire assemblage of the joint at the intersection of beams and columns framing into moment-resistance connection. To design a moment-resisting connection that has a desirable seismic behaviour, there are two choices: (i) to proportion the joint such that shear yielding of the panel zone initiates at the same time as flexural yielding of beam elements, or (ii) to design the joint such that all yielding occurs in the beam. The best performance is likely to be achieved when there is a good balance between beam bending and panel zone distortion. This zone is assumed to deform in shear as indicated in Fig. 9.21. Panel zone deformations can be controlled and an improved performance can be achieved with stiffeners, using heavier columns with thicker webs, or by locally reinforcing the panel zones. Another way of improving the performance is by providing haunches at the ends of the beams framing into the joint, so that the plastic hinge forms away from the connection.

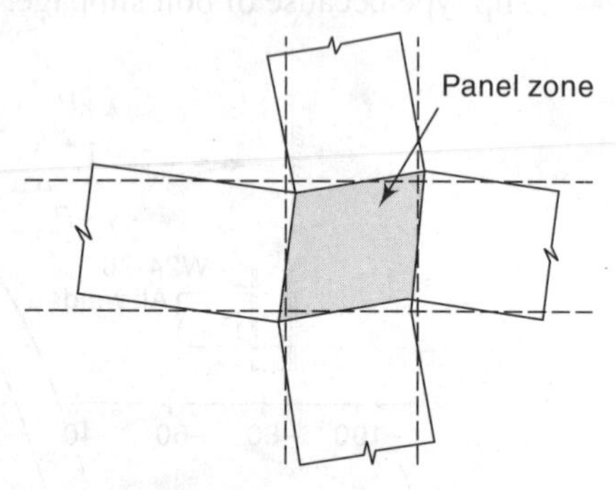

Fig. 9.21 Idealized shear deformation of beam-column panel zone

9.7.1 Deformation Behaviour of Panel Zone

Little is known of the deformation characteristics of panel zones, especially under cyclic loading. However, it has been demonstrated that the deformation of beam-column connections may contribute up to about one-third of the inter-storey deflection in multi-storey buildings. Of this deformation, about half may arise from the shear deformation of the panel zone itself. This reveals that significant ductility can be obtained through shear yielding in column panel zones through many cycles of inelastic distortion. The large influence of panel zone behaviour on overall frame strength and stiffness is evident.

In practice, it is difficult to prevent some yielding in the panel zone, and limited yielding in the panel may beneficially reduce the amount of plasticity and plastic instability which occurs in the adjacent beam hinges. As an upper limit, the design panel zone shear strength need not exceed that due to 80 per cent of the summation of the expected plastic moments of the beam(s) framing into the panel zone.

9.7.2 Detailing Panel Zone for Seismic Resistance

Panel zone should be designed based on its cyclic testing. However, in the absence of design criteria based on cyclic testing, the panel zone should be designed to allow the adjacent members to reach their full strength, while avoiding excessive shear deformation of the panel zone itself. Figure 9.22 shows typical panel zone detailing. Joint panels should have the same sectional shape as that of adjoining columns. They are stiffened by pairs of diaphragms located at the levels of the adjoining beam flanges.

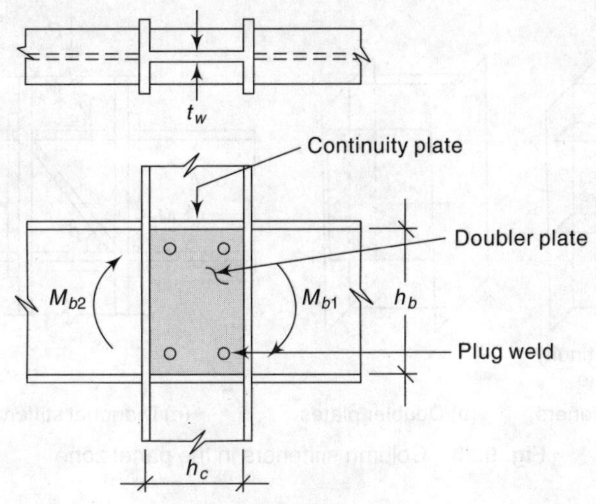

Fig. 9.22 Typical panel zone

9.7.3 Stiffeners in Panel Zone

Stiffener plates are required and provided at the top and bottom of the panel zone, when the column flange thickness is insufficient. These are also known as *transverse stiffeners*. It is normal to place a full-depth transverse stiffener on each side of the column web. These plates are welded to the column flange using complete joint penetration groove weld. An interior column (i.e., one with adjacent moment connections to both flanges) in a moment frame, subjected to seismic forces, receives a tensile flange force on one flange and a compressive flange force on the opposite side. As this stiffener provides a load path for the flanges on both sides of the column, it is commonly called a *continuity plate*. The stiffener also serves as a boundary to the very high stressed panel zone. When the formation of the plastic hinge is anticipated adjacent to the column, the required strength is the flange force that is exerted when the full plastic moment in the beam has been reached, including the effects of overstrength and strain hardening, as well as shear amplification from the hinge location to the column face.

The types of stiffeners used in the panel zone are shown in Fig. 9.23. Stiffening may also be imparted by providing reinforcing plates welded directly to the flange, as shown in Fig. 9.24. The behaviour of the flange will depend on the support, which it derives from stiffeners and is normally analysed on the basis of the yield line theory. The shear forces acting on the panel zone due to lateral loading are shown in Fig. 9.25. It can be seen that the panel zone shear is 2T and the total shear taken by the weld between the stiffener and web is also 2T, Figure 9.26 shows the relation between shear forces and shear distortion for a panel under monotonic loading. The slope of the curve changes in the vicinity of the yield strength and becomes 3 to 8 per cent of the elastic slope. Strength increases well

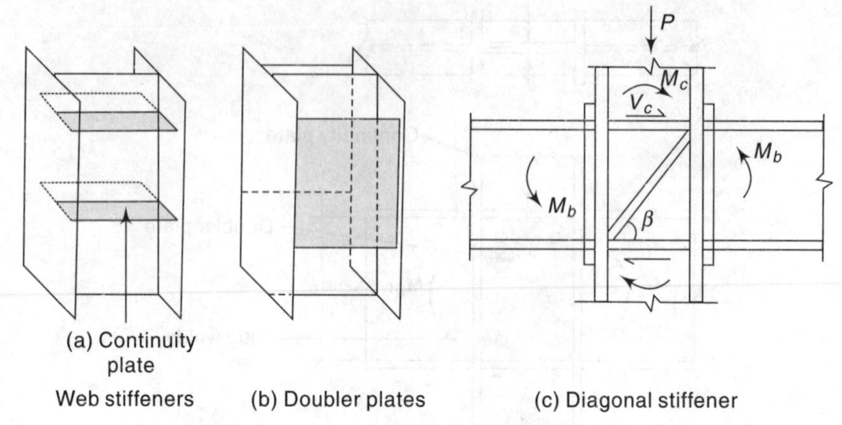

Fig. 9.23 Column stiffeners in the panel zone

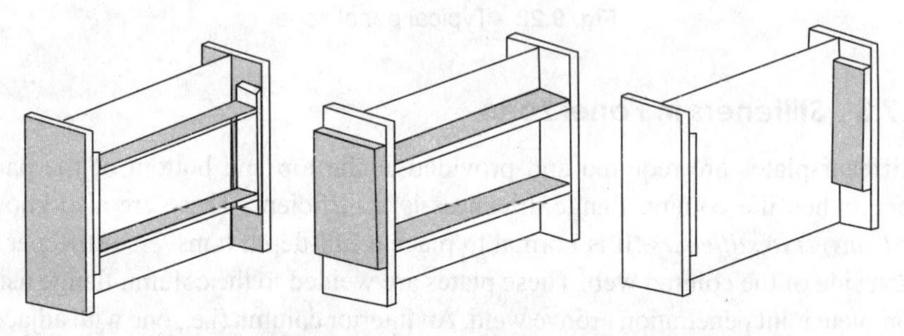

Fig. 9.24 Reinforcing plates to flanges in panel zone

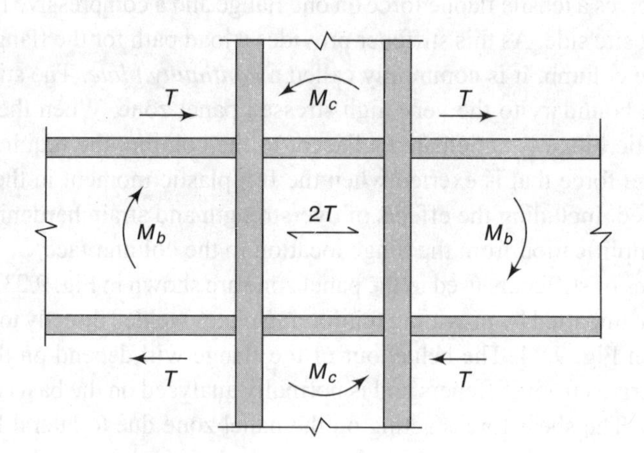

Fig. 9.25 Shear forces in the panel zone

beyond the yield strength and the ductility factor reaches 30 to 40. Extra strength beyond yield strength is provided by the following effects:

(a) Resistance of the boundary elements, i.e., column flanges and diaphragms

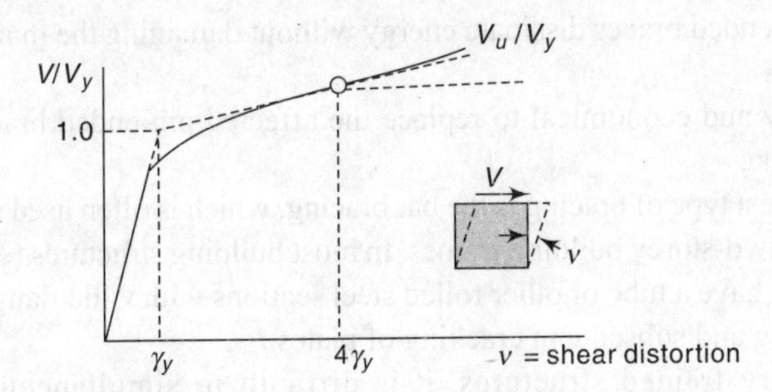

Fig. 9.26 Relationship between shear force and shear deformation for a connection panel under monotonic loading

(b) Resistance of the beam and column webs adjacent to the connection panel
(c) Strain hardening of the panel

Since the extra strength is very large the panel may yield prior to the yielding of the surrounding members. This requires the design of the panel to be based on a higher strength than the yield strength. When panel thickness is not sufficient, the panel is strengthened by doubler plates, which are welded to the column web using plug welds as shown in Fig. 9.22. When placed against the column web, doubler plate should be welded at top and bottom to develop the proportion of the total force that is transmitted to the doubler plate. The doubler plate should also be butt or fillet welded to column flanges to develop its shear strength. Doubler plates may also be placed, in pairs, away from column web [Fig. 9.23(b)] and welded to continuity plates, to develop their share of the total force transmitted to the doubler plate.

It may be noted that in a moment-resisting frame the shear stresses can be critical only when the column is subjected to major axis bending and joint shear is resisted by the column web, necessitating provision of doubler plates. For minor axis bending of columns since the joint shear is carried by column flanges, the resulting shear stresses are seldom critical.

9.8 Bracing Members

Bracing members are used either as part of a lateral load transfer system or to increase the lateral stiffness of a frame. They may be either pin-ended or fixed-ended. Pin-ended braces are better than the fixed-ended braces due to the following reasons.

1. Pin-ended braces are subjected to axial forces only (tension or compression). They are used in pairs in the form of X-braces with one of them in tension and the other in compression. The tension strength of the brace remains relatively unchanged during cycling and presents ductile behaviour. However, the brace in compression fails by global buckling and in subsequent cycles, the buckling strength gets reduced because the member does not remain straight

2. The pin-ended braces dissipate energy without damaging the main structural members
3. It is easy and economical to replace the affected pin-ended braces after an earthquake

The simplest type of bracing is the bar bracing, which is often used in relatively light one- or two-storey building frames. In most building structures (except small ones), braces have a tube or other rolled steel sections with wide flanges to avoid local buckling and subsequent cracking of plates.

In ordinary framed structures, it is difficult to simultaneously satisfy requirements of stiffness under working loads and strength and energy absorption capacity without over or under designing for one of them. If structures are stiff enough to keep lateral displacements under prescribed limits, their strength is generally much higher than required. Lateral stiffness of moderately tall buildings is economically increased using a number of braced frames. These frames restrict deformation of the remaining unbraced frames, with the floor systems acting as horizontal diaphragms. If possible, bracing must be continuous from the bottom of the building to the top of the building. Some of the typical bracing configurations are shown in Figs 9.7 and 9.8.

Braced frames are usually designed as two superimposed systems. The first system is an ordinary rigid frame which supports vertical dead and live loads. The other system is a vertical bracing system, generally regarded as a pin-connected truss, which resists horizontal loads plus P-Δ effects, provides adequate lateral rigidity under working loads, and avoids overall frame buckling under factored vertical loads. Beams and columns of braced bays belong to both systems. In CBFs, the bracing members normally carry most of the seismic storey shear, particularly, if not used as a part of a dual system. The required strength of bracing connections should be adequate so that failures by out-of-plane gusset buckling or brittle fracture of the connections are not critical failure mechanisms.

Eccentric bracing systems, such as those shown in Fig. 9.8, avoid the reduction in energy dissipation capacity due to buckling of the brace. The systems illustrated in the figure ensure large energy-dissipation capacity and good dynamic response. However, a brace, when placed eccentrically to the surrounding column and beams, a torsional moment is induced in those members. Therefore, at the connection between the diagonal brace and beam, the intersection of the brace and beam centre lines should be at the end of the link as shown in Fig. 9.27 or within the length of the link. If the intersection point lies outside the link length, the eccentricity together with the brace axial force produces additional moments in the beam and brace. Further, care must be taken so that no stress concentration is generated in the connections.

The compressive strength of a brace should not be smaller than half of the tensile yield force of the brace. Even if this provision is followed, the deflection in the direction perpendicular to the brace longitudinal axis can be very significantly large once the brace buckles, and damage of some non-structural elements may

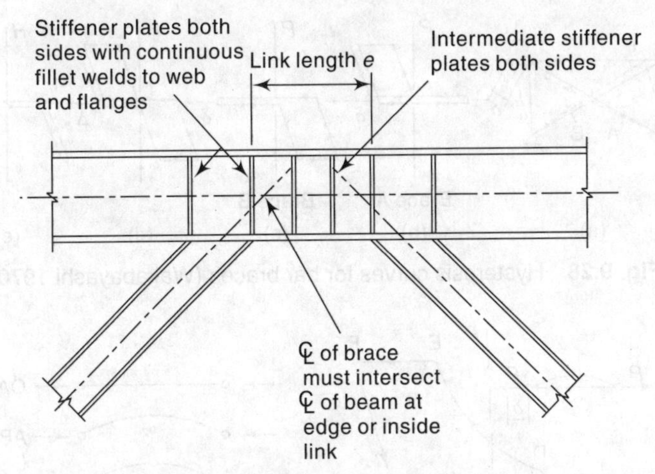

Fig. 9.27 EBF with I-shaped bracing

occur. To safeguard this, IS 800 limits the *kL/r* values for braces as given in Table 9.1. Since braces are subjected predominantly to axial loads, members with solid cross-sections, such as rods and bars, are preferred. Moreover, local buckling of braces is not a problem except when single angle section is used.

The compressive and tensile strengths of braces, as per IS 800: 2007, are given in Table 9.1. A design rule for braces is that the connections should not fracture prior to yielding of the brace. To achieve this goal, axial yield force (i.e., the cross-sectional area of the brace multiplied by yield stress) should be smaller than the strength of the connections. Because of alternating buckling and plastic elongation, the hysteretic behaviour of braces is usually of the degrading type and, therefore, involves little energy dissipation because of the alternating buckling and plastic elongation under load reversals. To allow for this degradation effect, the design earthquake force for braced frames is increased over that specified for unbraced frames.

9.8.1 Behaviour of Bracing under Cyclic Loading

Figure 9.28 shows hysteretic behaviour for a bracing bar under repeated loading, where P and δ are the axial forces induced in a brace and the corresponding elongation, respectively; H_b and Δ are the horizontal loads carried by the braces and the horizontal deflections of the frame, respectively. Since each brace can carry only tension as shown in Fig. 9.28(b) and (c), the total horizontal load carried by the bracing is obtained as shown in Fig. 9.28(d), by summing the horizontal components of the forces inducted in the bars. The hysteresis loop shown in Fig. 9.28(d) indicates that the brace dissipates energy only when it experiences newly developed plastic elongation. The brace does not dissipate energy at all, if it is subjected to repeated loading under constant deflection amplitude as shown in Fig. 9.28(e), and thus it is said in general that the energy-dissipation capacity of the bracing is less than that of the moment frame.

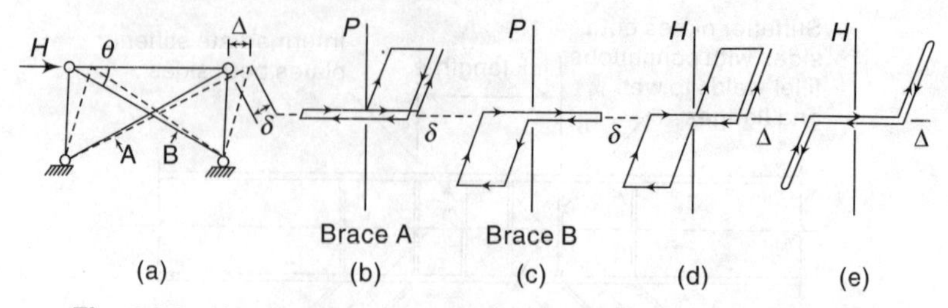

Fig. 9.28 Hysteresis curves for bar braces (Wakabayashi 1970)

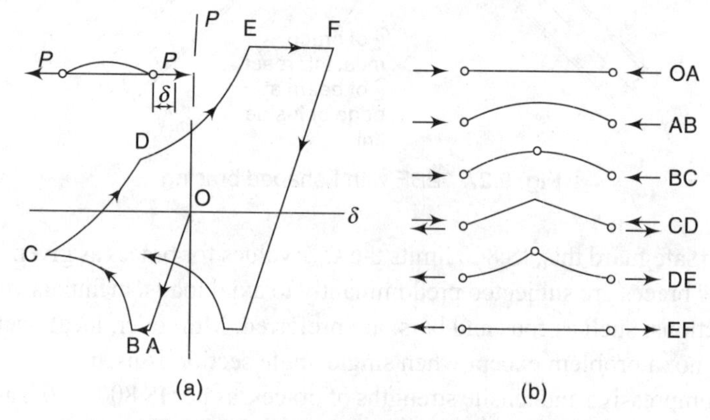

Fig. 9.29 Hysteresis curve for a brace

Hysteresis loops for bracing members which are stockier than the bracing bar become more complex, as shown in Fig. 9.29. Letters indicating various portions of the hysteresis loop in Fig. 9.29(a) correspond to the letters in Fig. 9.29(b), which show various deflected shapes and loading conditions of the brace. The brace is alternately and repeatedly subjected to rotation at the plastic hinge, which forms because of buckling in compression and plastic elongation, following yielding in tension. The relationship between axial force and axial deformation is theoretically determined from the equilibrium condition, the yield condition, and the associated flow rule, all applied at the plastic hinge. The boomerang-shaped hysteresis loop becomes thinner and the energy-dissipation capacity becomes smaller as the slenderness ratio of the brace becomes larger.

The shape of the hysteresis loop varies mainly with the slenderness ratio kL/r and behaviour in the large-deformation range may also be affected by the shape of the cross-section, since local buckling and/or buckling of a single strut of a built-up member may take place. Local buckling often causes cracks in the middle and end portions of the member; the end connection must, therefore, be detailed so as to avoid stress concentration. In addition, the strength of the end connection must be larger than the yield strength of the bracing member; otherwise, a brittle failure of the connection may occur prior to yielding of the member. Braces that have fixed end connections dissipate more energy than those that are pin connected, since buckling requires the formation of three plastic hinges in the brace.

Nonetheless, end connections that can accommodate the rotations associated with brace buckling deformations, while maintaining adequate strength perform well. Where fixed ended connections are used in one axis with pinned connections in the other axis, the effect of the fixity should be considered in determining the critical buckling axis.

9.9 Loads and Load Combinations

IS 800: 2007 recommends the use of IS 1893 (Part I) for assessment of seismic loads, but the response reduction factors to be used are as given in Table 9.3.

Table 9.3 Response reduction factor *R* for building system

S. No.	Lateral load resisting system	R
(i)	*Braced frame systems:*	
	(a) Ordinary concentrically braced frame (OCBF)	4
	(b) Special concentrically braced frame (SCBF)	4.5
	(c) Eccentrically braced frame (EBF)	5
(ii)	*Moment frame systems:*	
	(a) Ordinary moment frame (OMF)	4
	(b) Special moment frame (SMF)	5

The load combinations recommended by IS 800:2007 for limit state design of steel structures are given in Table 9.4.

Table 9.4 Load combinations and partial safety factors (IS 800: 2007, Table 4)

Load combination	DL	Limit state of strength LL		WL/EL	AL
		Leading	Accompanying (CL, SL, etc.)		
DL + LL + CL	1.5	1.5	1.05	–	–
DL + LL + CL+	1.2	1.2	1.05	0.6	
WL/EL	1.2	1.2	0.53	1.2	–
DL + WL/EL	1.5 (0.9)*	–	–	1.5	–
DL + ER	1.2 (0.9)*	1.2	–	–	–
DL + LL + AL	1.0	0.35	0.35	–	1.0

*This value is to be considered when stability against overturning or stress reversal is critical.

Notes:
1. The effects of actions (loads) in terms of stresses or stress resultants may be obtained from an appropriate method of analysis.
2. DL = dead load, LL = imposed load (live load), WL = wind load, SL = snow load, CL = crane load (vertical/horizontal), AL = accidental load, ER = erection load, EL = earthquake load.

As per IS 800: 2007, in addition to the load combinations of Table 9.4, the following load combinations should also be used and examined for seismic design of buildings.

(a) 1.2 DL + 0.5 LL ± 2.5 EL
(b) 0.9 DL ± 2.5 EL

9.10 Ductile Design of Frame Members

During a strong ground motion, a structure is subjected to forces that are much greater than the design forces. It is neither practical nor economically feasible to design a building to remain elastic during major earthquakes. Instead, the structure is designed to remain elastic at a reduced force level. The structure is supposed to sustain post-yield displacement without collapse by prescribing special ductile detailing requirements of the components and their connections. The approaches to design steel frames fall into three principal categories. The first approach is to derive a set of equivalent lateral forces, and to design the structure on the basis of elastic analysis, but to provide appropriate levels of ductility. This is called the working stress design method.

The second approach is to use plastic design methods. This approach is generally suitable for rigidly connected structures of up to four storeys. Care needs to be taken to ensure that local collapse mechanisms, e.g., in the columns, cannot occur and that displacements remain within acceptable levels. In order to obtain adequate ductility from a steel structure, both the members and the connections must be given due consideration. In principle, the use of plastic design is appropriate for earthquake-resistant design of steel structures. However, the following three criteria should be complied with and excessive lateral sway should be avoided.

(a) Excessive alternating stresses in the flanges of beam and column members must be avoided.
(b) Inelastic buckling of columns and bracings must be prevented.
(c) Connections must be designed to allow extensive yielding of the frame members.

The third approach is the capacity design approach. The method based on this approach is called limit state design method and is recommended by IS code. The criterion to be satisfied in this method in the selection of members is that the factored load should be less than or equal to the factored strength. The structural steel members and connections in the steel frames should be designed and detailed to provide adequate strength, stability, and ductility without collapse.

9.11 Retrofitting and Strengthening of Structural Steel Frames

In steel buildings, structural damage refers to degradation of the buildings support system, i.e., the frames, the framing members and connections and braces. Techniques for strengthening and retrofitting of existing/damaged steel buildings will vary according to the nature and extent of deficiencies/damages and the configuration of the structural system. A thorough understanding of existing construction and seismic strengthening/retrofitting objectives acceptable to the owner and the regulatory authority is important before a seismic strengthening or retrofitting scheme is undertaken.

9.11.1 Retrofitting

Any steel member or connection that has been stressed beyond the yield point must be replaced, otherwise the ductility of the structure for resistance to future earthquakes, will be reduced. These critical regions which may develop inelastic deformations are localized and a large part of the structure may be free of such damage. The critical points in a moment-resisting steel frame are likely to be the ends of beams or girders and their connections to columns. The portion of a column extending from just above a beam or girder to just below it, may be critical, although a good design avoids the chance of yielding in the columns because of its possible effect on the stability of the structure. Inelastic deformations in the joint region of a column may significantly add to the joint rotation and the deflection of the frame, i.e., it adds to the ductility of the structure.

Determination of the extent of damage may require the removal of ceilings, plaster, or sprayed-on fireproofing and other finishes, at least in suspected regions, which may be indicated by cracks or other visible damage. A detailed review of the design of the structure should also be made, to help in locating critical areas. Damaged steel members or portions of members can be cut out and replaced by welding. Welding procedures should be carefully planned to prevent the development of residual stresses due to the welding operation. This is particularly important where thick material is involved.

9.11.2 Strengthening

Very little experience or thought has been given to the strengthening of the moment-resisting steel frames of multi-storey buildings for resistance to earthquake forces. This is probably because there are relatively very few multi-storey steel-frame buildings in areas where severe earthquakes have occurred. However, the following measures may be adopted for steel moment frames.

1. The most practicable way to strengthen a moment-resisting steel frame is by adding braces, either X-bracing, if the wall has no windows, or K-bracing or corner bracing, if there are openings. In many cases the columns would probably be adequate for axial stress due to overturning, since they would be entirely or largely relieved of bending stress. However, the braced bent would be stiffer than the original moment-resisting frame and would attract more seismic load. This would increase the overturning moment and axial stress in the columns.

2. The strength and stiffness of existing frames may be increased by welding steel plates or other rolled sections to selected members. Steel columns can be made composite by enclosing them with reinforced concrete, or adding steel plates to them. Stiffening elements may also be added to reduce the expected frame demands. However, if a material must be added to a column to increase its axial load capacity, the resulting section would be very unequally stressed, since the original material must carry the full dead load. It would seldom be practicable to relieve a column of dead load before it is strengthened.
3. To reduce the expected rotation demands, a stiffer lateral force-resisting system may be added. Connections can be modified by adding flange cover plates, vertical ribs, haunches or brackets or by removing beam flange material to initiate yielding away from connection location. Moment-resisting connection capacity can thus be increased by adding cover plates, vertical stiffeners, or haunches.
4. New moment frames, braced frames or shear walls, or infill walls that can reduce the storey drifts, may be added to increase the strength and stiffness of building. The energy dissipation devices may be added to reduce the drift.

Summary

Multi-storey buildings are usually constructed in steel. Steel buildings are flexible but display more lateral displacement as compared to RCC buildings. The need of high strength, ductility, and energy dissipation capacity in members and connections is presented. The framing system of steel structure is of utmost importance. Different types of steel frames that are prevalent are described. In general, steel frames are braced, as in addition to resisting large lateral forces, the lateral deflection and consequently P-Δ effect is reduced. The quality and workmanship has to be of high standards and therefore emphasized. The principles and concepts involved with regards to the steel structure members and their behaviour under seismic loads are discussed. Provisions for design of steel structural members as per code of practice, IS 800: 2007, are presented. The chapter ends with retrofitting and strengthening of steel frames.

Solved Problem

9.1 A four-storey steel office building, shown in Fig. 9.30, is located in seismic zone III on hard soil. The framing system of the building is moment-resisting frames with brick masonry infill panels. Determine the design force in the bracing as per IS 800: 2007 for the building for the following data:

Given data:

Column sections

Ground floor: ISHB 450 @ 872 N/m with 12 mm thick and 250 mm wide cover plate on each flange.

Remaining floors : ISHB 450 @ 872 N/m.

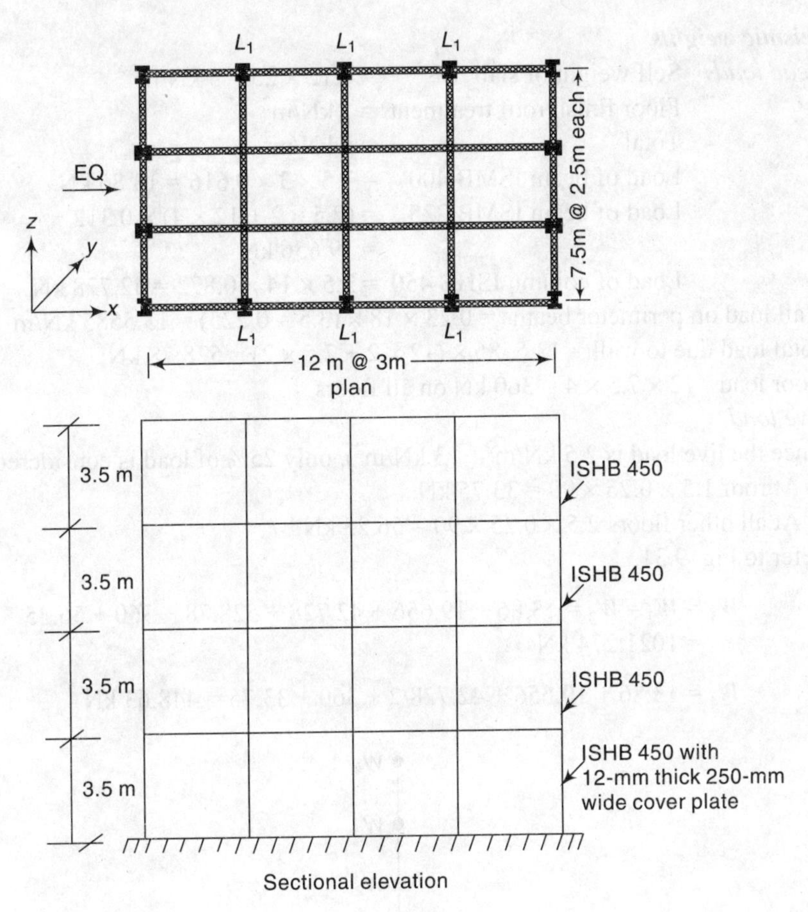

Fig. 9.30

Beam sections

Along 7.5-m intermediate beams (L_1): ISMB 400 @ 616 N/m

All other beams (L_2): ISMB 225 @ 312 N/m

Slab: 120-mm thick RC slab on all floors

Walls: 230-mm thick (unit weight 18 kN/m^3)

Bracing: OCBF

Solution

Zone factor for seismic zone III, $Z = 0.16$ (Table 5.2)

For an office building, importance factor is 1.0 (Table 5.3).

Response reduction factor for a steel frame building with ordinary concentric bracing = 4 (Table 5.4)

Loads

Consider a floor finish of 1 kN/m^2.

Assume the load from the roof treatment to be 1 kN/m^2.

A live load of 2.5 kN/m^2 is considered on all floors except the roof where it is considered to be 1.5 kN/m^2.

Seismic weights

Dead loads Self weight of slab $= 0.12 \times 25 = 3$ kN/m^2

Floor finish/roof treatment $= 1$ kN/m^2

Total $= 4$ kN/m^2

Load of beam ISMB 400 $= 7.5 \times 3 \times 0.616 = 13.86$ kN

Load of beam ISMB 225 $= (7.5 \times 2 + 12 \times 4) \times 0.312$
$= 19.656$ kN

Load of column ISHB 450 $= 3.5 \times 14 \times 0.872 = 42.728$ kN

Wall load on perimeter beams $= 0.23 \times 18 \times (3.5 - 0.225) = 13.5585$ kN/m

Total load due to wall $= 13.5585 \times (12 \times 2 + 7.5 \times 2) = 528.78$ kN

Floor load $=12 \times 7.5 \times 4 = 360$ kN on all floors

Live load

Since the live load is 2.5 kN/m^2 (< 3 kN/m^2), only 25% of load is considered.

At roof $1.5 \times 0.25 \times 90 = 33.75$ kN

At all other floors $2.5 \times 0.25 \times 90 = 56.25$ kN

Refer to Fig. 9.31.

$$W_1 = W_2 = W_3 = 13.86 + 19.656 + 42.728 + 528.78 + 360 + 56.25$$
$$= 1021.274 \text{ kN}$$

$$W_4 = 13.86 + 19.656 + 42.728/2 + 360 + 33.75 = 448.63 \text{ kN}$$

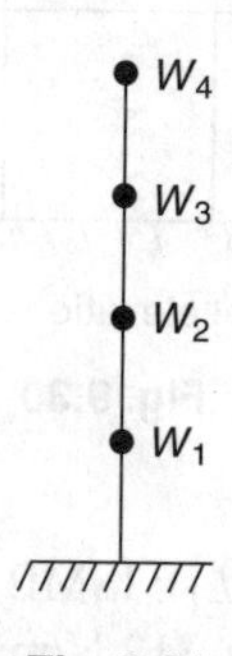

Fig. 9.31

The lateral force at each floor level has been worked out by both the equivalent lateral force procedure and response spectrum procedure and then the bracing has been designed for the critical one.

Equivalent lateral force method

Fundamental natural period of vibration

As the moment-resisting frame system is with brick infill panels, the approximate fundamental natural period of vibration (T_a) in seconds may be estimated by the empirical expression

$T_a = 0.09\ h/\sqrt{d}$ where d in metres, is the base dimension of the building at plinth level along the considered direction of the lateral forces.

$$T_a = 0.09 \times 14/\sqrt{12} = 0.36373 \text{ s}$$

The building is located on hard soil site

$$S_a/g = 2.5$$

The total design seismic base shear V_B along the principal direction is

$$V_B = A_h W$$

where W is the seismic weight of the building (full dead load + appropriate percentage of imposed load).

The design horizontal acceleration seismic coefficient,

$$A_h = \frac{ZIS_a}{2Rg}$$

$$= (0.16 \times 1 \times 2.5)/(2 \times 4) = 0.05$$

$$W = 448.63 + 1021.274 \times 3 = 3512.452 \text{ kN}$$

$$V_B = 0.05 \times 3512.452 = 175.62 \text{ kN}$$

The floorwise calculations of the lateral forces are tabulated below and the forces are shown in Fig. 9.32.

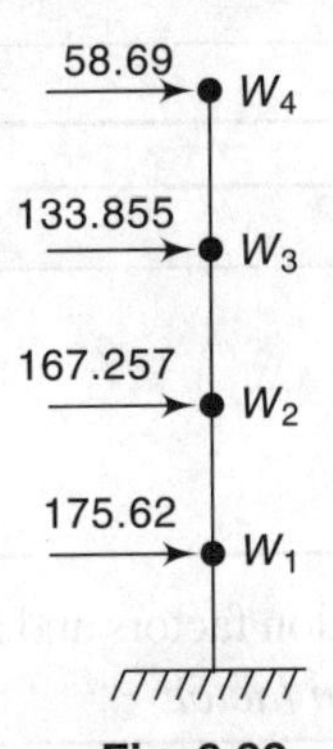

Fig. 9.32

Storey	W_i (kN)	h_i	$W_i h_i^2$	$\frac{W_i h_i^2}{\Sigma W_i h_i^2}$	Q_i	V_B
4	448.63	14	87931.48	0.3342	58.69	58.69
3	1021.274	10.5	112595.458	0.4280	75.165	133.855
2	1021.274	7	50042.426	0.1902	33.402	167.257
1	1021.274	3.5	12510.606	0.0476	8.359	175.62
			$\Sigma W_i h_i^2 =$ 263079.971			

Dynamic analysis

Design lateral force Q_{ik} at each floor in each mode is given by

$$Q_{ik} = A_k \, \phi_{ik} \, P_k \, W_i$$

where A_k = Design horizontal acceleration spectrum value
ϕ_{ik} = Mode shape coefficients at floor i in mode k
W_i = Seismic weight of floor i
P_k = Modal participation factor in the kth mode, given by

$$P_k = \frac{\sum_{i=1}^{n} W_i \phi_{ik}}{\sum_{i=1}^{n} W_i (\phi_{ik})^2}$$

The modal mass of mode k is given by $M_k = \dfrac{\left[\sum_{i=1}^{n} W_i \phi_{ik}\right]^2}{g \sum_{i=1}^{n} W_i [\phi_{ik}]^2}$

where g is the acceleration due to gravity, ϕ_{ik} is the mode shape coefficients at floor i in mode k.

Mode number	1	2	3
Natural period (s)	2.561	0.487	0.192

Storey/Mode	1	2	3
Roof	1	1	1
3rd	0.629	–0.282	–1
2nd	0.296	–0.794	0.425
1st	0.073	–0.385	0.979

The calculation of modal participation factors and lateral loads is tabulated further.

Calculation of modal participation factor

	W_i	ϕ_{ik}	$W\phi_{ik}$	$W\phi_{ik}^2$	T	S_a/g
Storey			Mode 1			
4	448.63	1	448.63	448.63	2.561	039047
3	1021.274	0.629	642.381	404.058		
2	1021.274	0.296	302.297	89.479		
1	1021.274	0.073	74.553	5.442		
		Sum	1467.861	947.609		

M_k = 2273.739, P_k = 1.549

% of total weight 64.733, Participation factor 64.733%

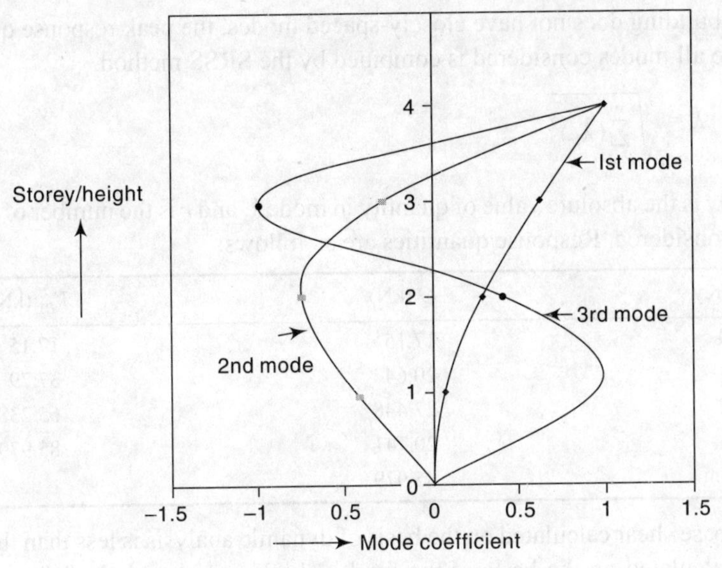

Fig. 9.33 Mode shapes for the building

Mode 2

4	448.63	1	448.63	448.63	0.487	2.0533
3	1021.274	–0.282	–288.80	81.216		
2	1021.274	–0.794	–810.892	643.848		
1	1021.274	–0.385	–393.190	151.378		
		Sum	–1044.252	1325.072		

$M_k = 822.945$, $P_k = -0.788$

% of total weight 23.43, Participation factor 88.163%

Mode 3

4	448.63	1	448.63	448.63	0.192	2.5
3	1021.274	–1	–1021.274	1021.274		
2	1021.274	0.425	434.041	184.467		
1	1021.274	0.979	999.827	978.831		
		Sum	861.224	2633.202		

$M_k = 281.675$, $P_k = 0.327$

% of total weight 8.019, Participation factor 96.182%

Lateral loads

	Mode 1			Mode 2			Mode 3		
Storey	A_h	Q_i	V	A_h	Q_i	V	A_h	Q_i	V
4	0.0078	5.42	5.42	0.0411	–14.529	–14.529	0.05	7.33	7.33
3	0.0078	7.761	13.181	0.0411	9.327	–5.202	0.05	–16.697	–9.367
2	0.0078	3.653	16.834	0.0411	26.262	21.06	0.05	7.097	–2.27
1	0.0078	0.901	17.735	0.0411	12.734	33.794	0.05	16.347	14.077

As the building does not have closely-spaced modes, the peak response quantity λ due to all modes considered is combined by the SRSS method.

$$\lambda = \sqrt{\sum_{k=1}^{r} (\lambda_k)^2}$$

where λ_k is the absolute value of quantity in mode k, and r is the number of modes being considered. Response quantities are as follows:

Storey	Q (kN)	V_{B1} (kN)
4	17.15	17.15
3	20.64	37.79
2	27.448	65.238
1	20.741	85.979
Sum	85.979	

As the base shear calculated on the basis of dynamic analysis is less than the base shear calculated on the basis of the fundamental natural period of vibration, it is necessary to scale the quantities. All the response quantities (member forces, displacements, storey forces, storey shears, and reactions) are multiplied by V_B/V_{B1}, which is given as

$$\frac{V_B}{V_{B1}} = \frac{175.62}{85.979} = 2.0426$$

The final quantities after scaling are tabulated as follows:

Storey	Q (kN)	V_B (kN)
4	35.030	35.030
3	42.159	77.189
2	56.065	133.254
1	42.365	175.62
Sum	175.62	

Design of bracing system (Figs 9.34 and 9.35).

Inclination of main component a with horizontal is 66.8°.

The bracing member b will not carry any force, but is provided for the stability of bracing a (Fig. 9.36).

The lateral force to be resisted by each side bracing is 175.62/2 = 87.81 kN (bracing will be provided on two faces)

Design force for bracing

Partial safety factor for load $\gamma_f = 1.5$

Lateral load = 1.5 × 87.81 = 131.715 kN

Bracing has been provided on both the faces.

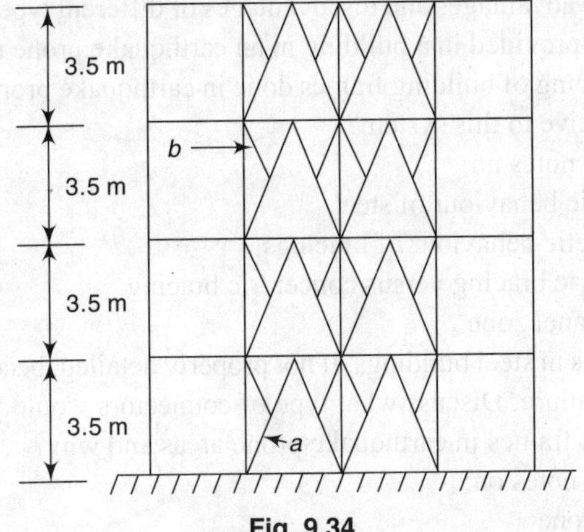

Fig. 9.34

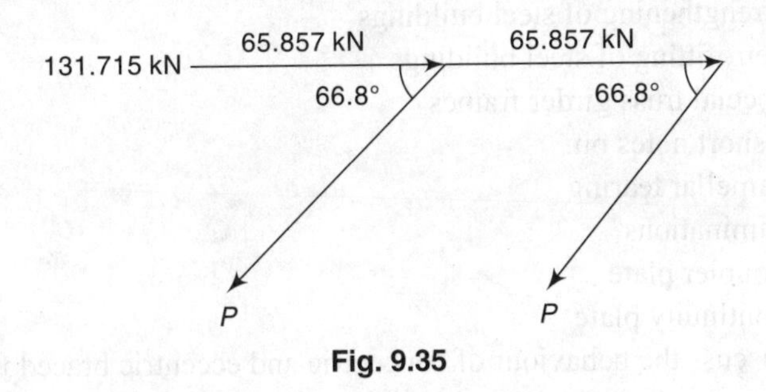

Fig. 9.35

lateral load for each brace = 131.715/2 = 65.857 kN

$$\text{Therefore design force in each brace} = \frac{65.875}{\cos 66.8^\circ} = 167.178 \text{ kN}$$

Exercises

9.1 (a) What are the factors that make steel the most ideal material for earthquake resistance?

(b) State and discuss briefly the considerations for achieving adequate performance of steel buildings.

9.2 What are the causes of instability of steel buildings? Discuss in detail the P-Δ effect.

9.3 Write short notes on:

(a) Secondary effects

(b) Hysteretic behaviour of steel

(c) Lamellar tearing

(d) Concentric and eccentric braced frames

9.4 Discuss the advantages and disadvantages of different types of steel frames that can be provided in a building in an earthquake prone region.

9.5 Why is bracing of building frames done in earthquake prone areas? Is there any alternative to this system?

9.6 Write short notes on:
- (a) Inelastic behaviour of steel
- (b) Hysteretic behaviour of bracings
- (c) Eccentric bracing versus concentric bracing
- (d) Steel panel zone

9.7 Connections in steel buildings, if not properly detailed, become the primary source of failure. Discuss what type of connectors would you recommend for moment frames in earthquake prone areas and why?

9.8 Write short notes on:
- (a) Panel zone
- (b) Types of moment frames
- (c) Strengthening of steel buildings
- (d) Retrofitting of steel buildings
- (e) Special truss girder frames

9.9 Write short notes on:
- (a) Lamellar tearing
- (b) Laminations
- (c) Doubler plate
- (d) Continuity plate

9.10 (a) Discuss the behaviour of concentric and eccentric braced frames.
(b) Discuss about the behaviour of bracing under cyclic loading

CHAPTER

10

Non-structural Elements

Non-structural elements are the architectural, mechanical, and electrical components of a building that directly cater to human needs. Loss or failure of these elements can affect the safety of the occupants of the building and the safety of others who are immediately outside the building. Architectural components include non-bearing walls, partitions, infill walls, parapets, veneers, ceilings, door and window panes, glasses, cladding, etc. Roofing units such as tiles and other individually attached relatively heavy roof elements, stairway and elevator enclosures are also included in architectural components. Mechanical and electrical components include boilers and furnaces, chimneys and smoke stacks, tanks and pressure vessels, machinery, piping systems, communication systems, electrical wire ducts, electrical motors, transformers, lighting fixtures, fire and smoke detection systems, etc. Although, in most buildings, they represent a high percentage of the total cost of the building, the seismic behaviour of non-structures has not received adequate attention and thus effective design specifications are practically non-existent.

It is well recognized that good performance of non-structural elements during earthquakes, especially for buildings such as hospitals, emergency disaster centres, and fire stations, is extremely important. As the use of non-structures in modern buildings is immense, their failure may involve risk of life, financial loss, and the loss of post-earthquake services. The following are some of the consequences of the failure of non-structural elements:

(a) Falling of debris, ceilings, light fixtures, window glass, and exterior wall panels, parapets, and ornamentation. These may not only injure or kill people inside and outside the building but also damage critical mechanical and electrical components.
(b) Collapse of stairways and elevators and damage of exit doors may prevent the escape of people from the building.
(c) Emergency lighting and exit signs may malfunction during and after earthquake.
(d) The fire resistance system may collapse.
(e) Damaged piping systems, boilers, and tanks may explode, or may release steam, inflammable toxic's, or noxious fluids.
(f) Damaged boilers may release materials at a sufficiently high temperatures to cause fires.
(g) Furniture and equipment may overturn.

Damage surveys of earthquakes have shown that in many cases buildings that have suffered only minor structural damages have been rendered un-inhabitable and hazardous to life due to failure of non-structural elements. Therefore, a structural engineer must attempt to combine human safety with economy in the design of non-structural elements. Also, it should be remembered that it is less expensive to make such alterations to a building as are necessary to reduce or prevent non-structural damage, than it is to repair such damage after it has occurred.

10.1 Failure Mechanisms of Non-structures

In general, non-structural elements fail because of either excessive inertia forces applied to them or excessive deflection caused by deformation of the structural system. The inertia force acting on a non-structural element can be predicted if the response of the structure is known at the floor level where the non-structural element is installed. At any level on a multi-storey building, the ground motion will be modified by the motion of the building itself. Generally, the effect is to concentrate the frequency of response around a band close to the natural frequency of the building and to amplify the peak acceleration roughly in proportion to the height, reaching an amplification of perhaps two or three at the roof level. For any contents, which are either very stiff or have a natural frequency of their own close to that of the building, this means that they are subjected to greater forces than they would be if mounted at ground level. Also maximum shear acting on a flexible element may be significantly greater than that acting on a similar rigid element.

Non-structural items that are suspended, such as ceiling systems and light fittings, perform badly. Appendages such as parapets also suffer high levels of damage, especially where they function as SDOF inverted pendulums. Damage also increases on multi-storey structures towards the roof and roof tanks and penthouses are also subjected to high forces.

On the other hand, the non-structural elements subjected to forced deflections because of drift or inter-storey displacement, must be capable of deforming without failure, in accordance with structural deformation. As a design alterative, the element may be detached from the structure so that deformation of the structural system does not affect deformation or force in the element. Windows and cladding elements are frequently connected rigidly to more than one level, and if there is no ductile provision for movement in the connections, they will fail. Some common earthquake failures of non-structural components are given in Table 10.1.

Table 10.1 Common earthquake failures in non-structures

Item	Type of damage
Pumps and boilers	Movement of anchored supports
Tanks	Support failure
Motor generators	Failed isolation supports
Control panels	Overturning of tall units
Piping	Rupture due to excessive movement, failure at bends
Elevators (traction type)	Guide rails break, counter-weights misalign, car misalign
Parapets	Toppling
Concrete/stone cladding	Separation and falling
Windows	Glass breaking, frames detaching
Storage racks	Toppling and/or contents falling
False ceilings	Racking, panels falling
Suspended light fittings	Excessive movements causing damage or falling

10.2 Effect of Non-structural Elements on Structural System

In the normal practice of structural design, non-structural elements are not taken into account. Buildings, however, contain various non-structural elements that influence structural behaviour under earthquakes, which cannot be ignored in some situations. The influence is found to be small if flexible non-structural elements are added to a stiff structural system. For example, when the area of the RC column is extended by more than 100 mm beyond the confined core as shown in Fig. 8.15 due to architectural requirements only (the contribution of this additional area to strength of column has not been considered), this area is treated as a non-structure. In the reversed situation, for example, when exterior or partition walls made of masonry or block concrete are installed in a frame, the influence must be immense. The conceivable effects of non-structural elements on structural behaviour are as follows:

(a) The natural period of the structural system may be shortened resulting in a different input level to the system.

(b) Distribution of storey shear in columns may change, and some columns may sustain more force than that assumed in the original design.
(c) An unsymmetrical arrangement of non-structural walls may cause significant torsion in the system.
(d) Local force may be concentrated if non-structural walls are rearranged non-uniformly in height.

Although the effects of non-structural elements on the behaviour of a structural system are usually considered to be secondary, interactive behaviour between structural and non-structural elements has been reported as the cause of structural failure. In earthquakes, all buildings sway horizontally producing differential movements of each floor relative to the one just below it, called *storey drift* (Fig. 10.1). In addition, this is accompanied by vertical deformations, which involve changes in the clear height, *h*, between floor and beams. Structural engineers have two approaches to deal with these movements.

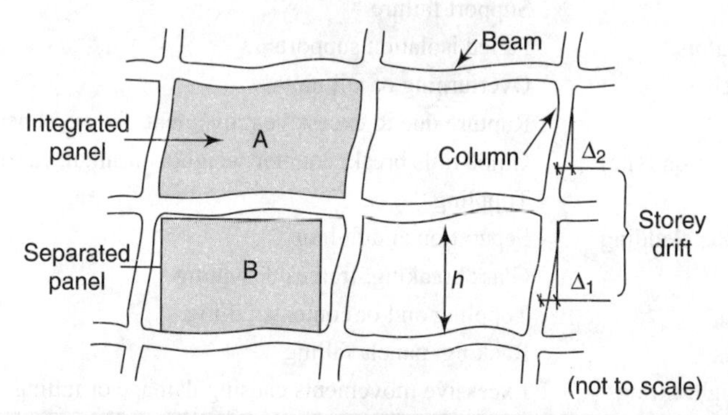

Fig. 10.1 Diagrammatic elevation of structural frame and non-structural infill panels

The first approach is to uncouple the non-structural elements from the structural system. This approach is used almost exclusively in designing tall buildings. In the absence of a reliable computed structural movement, horizontal and vertical movements of 20–40 mm are allowed. This type of construction has two inherent detailing problems. Firstly awkward details may be required to ensure lateral stability of the elements against out-of-plane forces. Secondly sound proofing and fire proofing of the separation gap is difficult. This method is preferred to integral construction when using flexible frames in strong earthquake regions.

In the second approach, non-structural elements can be treated as structural members and their characteristics taken into account in the design. The panels will be in effective contact with the frame such that the frame and panels will have equal drift deformations. Such panels must be strong enough (or flexible enough) to absorb this deformation, and the forces and deformations should be computed properly. This makes insulation of water, noise, or heat more feasible than the first approach. In general, however, stiff and brittle non-structural elements such as wing walls are more vulnerable than a structural member because of their lower deformation capacity; where appreciably rigid materials are used, the panels should be considered as structural elements. Furthermore, hysteretic behaviour of the structural system is very complex when such non-structural elements are included. The complexity often results in a poor understanding of the true response of the structure.

Figure 10.2(a) illustrates examples of the first approach. By providing a slit between the spandrel beam and the column, the column is expected to behave in a ductile manner. The same treatment is possible in wing walls [Fig. 10.2(b)] so that the adjoined beam does not fail by shear mode.

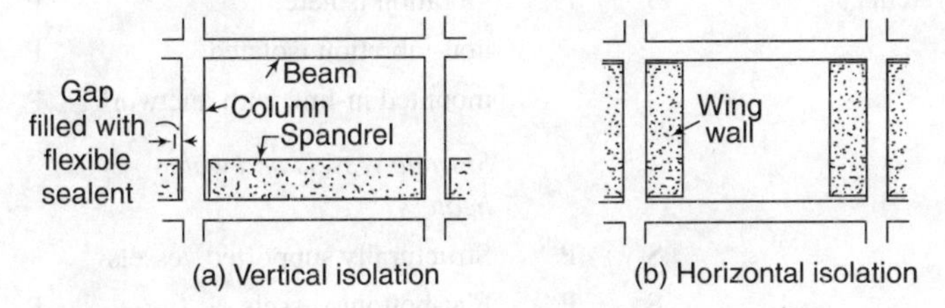

Fig. 10.2 Isolation of non-structural walls from the structural frame

10.3 Analysis of Non-structural Elements

When the non-structural component significantly affects the structural response of the building, the non-structural element should be treated as structural element, and structural provisions should apply to them. Depending upon response sensitivity, non-structural elements can be classified as deformation sensitive, acceleration sensitive, or both deformation and acceleration sensitive. The non-structural elements are classified as acceleration sensitive when they are affected mainly by acceleration of the supporting structure. A mechanical unit anchored to the floor or a roof of a building is a good example. In such a case structural–non-structural interaction due to deformation of the supporting structure is not significant. These components are vulnerable to sliding, overturning, or tilting and, as such, their anchorage or bracing is of prime concern. Non-structural components are regarded as deformation sensitive when they are affected by the supporting structure's deformation, especially the inter-storey drift. Curtain walls and piping systems running floor to floor are some examples of deformation-sensitive components. Many components are both deformation and acceleration sensitive. For example, the exterior skin of a building such as prefabricated panels are both deformation- and acceleration-sensitive. Table 10.2 classifies non-structural elements according to their response sensitivities.

For important non-structural elements, a dynamic analysis should be performed using the floor-response spectra as input to the subsystem. However, for a subsystem that is less important for public safety, an equivalent static analysis can be used to obtain the necessary information for design or strength requirement verification. The two methods of analysis are described in the following subsections.

Table 10.2 Response sensitivity of non-structural components

Architectural component	Sensitivity Acc.	Sensitivity Def.	Mechanical component	Sensitivity Acc.	Sensitivity Def.
Exterior skin			*Mechanical equipment*		
Adhered veneer	S	P	Boilers and furnaces	P	
Anchored veneer	S	P	General manufacturing		
Glass blocks	S	P	and process machinery	P	
Prefabricated panels	S	P	HVAC equipment		
Glazing systems	S	P	vibration isolated	P	
			non-vibration isolated	P	
			mounted in-line with ductwork	P	
Partitions			*Storage vessels and water heaters*		
Heavy	S	P	Structurally supported vessels		P
Light	S	P	Flat-bottom vessels	P	
Interior veneers			*Pressure vessels*	P	S
Stone (including marble)	S	P	*Fire suppression piping*	P	S
Ceramic tile	S				
Ceiling			*Fluid piping (not fire suppression)*		
Directly applied to	P		Hazardous materials	P	S
structure			Non-hazardous materials	P	S
Dropped, furred, Gypsum	P				
Board	S	P	*Ductwork*	P	S
Suspended lath and plaster	S	P			
Suspended integrated ceiling					
Parapets and appendages	P				
Canopies and marquees	P				
Chimneys and stacks	P				
Stairs	P	S			

Acc. = acceleration-sensitive; Def. = deformation-sensitive; HVAC = heat ventilation and airconditioning; P = primary response; S = secondary response

10.3.1 Dynamic Analysis

When a rigid non-structural element is tightly clamped on the floor of a structure, the response of the element is identical with the floor response. The magnification factor, defined as the ratio of the element response to the floor response, is, therefore, unity. When a rigid non-structural element is installed by a flexible connecting device on the floor of a structure, the element response is greater than the floor response. Such behaviour can be represented by a one-mass system with damping. A one-mass system may have as many as six degrees of freedom, but the system can usually be simplified to one with a single degree of freedom. Vibrational characteristics of the part of the structural system where the non-structural element is placed, often represented as the floor acceleration response spectra, can be found by applying time-history response analysis to the structural system. If the connecting device is ductile, the magnification factor derived from elastic response analysis may be relaxed according to ductility.

A long, flexible, non-structural element such as a piping system or a cable tray cannot be simplified to an SDOF system. An analysis of an MDOF system is, therefore, required, with the floor response as input. Equivalent static analysis is usually adequate for the design of such a system, unless its failure is considered to cause crucial damage to the structural system.

10.3.2 Equivalent Static Analysis

Where dynamic analysis is not feasible, it is desirable to establish a suitable equivalent static force. The subsystem is modelled as a separate structure with fixed support conditions. Calculation of the equivalent static forces for acceleration sensitive and displacement sensitive non-structural elements is as follows.

Acceleration Sensitive Non-structural Elements

The design seismic force F_p on the non-structural element is obtained from the following expression

$$F_p = \frac{Z}{2}\left(1+\frac{x}{h}\right)\frac{a_p}{R_p} I_p W_p \tag{10.1}$$

$$\geq 0.10\, W_p$$

where Z is the zone factor, x is the height of attachment of the non-structural element above the foundation, a_p is the component amplification factor (Tables 10.3 and 10.4). [The component of amplification factor represents the dynamic amplification of the component relative to the fundamental period of the structure.] R_p is the component response modification factor (Tables 10.3 and 10.4). [This factor represents ductility, redundancy, and energy-dissipation capacity of the element and its attachment to the structure.] I_p is the importance factor of the non-structural element (Table 10.5), and W_p is the weight of non-structural element.

- F_p is the horizontal force for the vertical non-structural element, and will be the vertical force for the horizontal non-structural element.
- In choosing the values of a_p and R_p, it is expressed that the component will behave as a flexible ($a_p = 2.5$) body. In general, the value of R_p is taken as 1.5, 2.5, and 3.5 for low, limited, and high deformable structures, respectively.

Table 10.3 Amplification factors and response reduction factors for mechanical and electrical components

Mechanical and electrical component or element	a_p	R_p
General mechanical		
Boilers and furnaces	1.0	2.5
Pressure vessels on skirts and free-standing	2.5	2.5
Stacks	2.5	2.5
Cantilever chimneys	2.5	2.5
Others	2.5	2.5
Manufacturing and process machinery		
General	1.0	2.5
Conveyors (non-personnel)	2.5	2.5
Piping system		
High deformability elements and attachments	1.0	2.5
Limited deformability elements and attachments	1.0	2.5
Low deformability elements and attachments	1.0	2.5
HVAC system equipment		
Vibration isolated	2.5	2.5
Non-vibration isolated	1.0	2.5
Mounted in line with ductwork	1.0	2.5
Others	1.0	2.5
Elevator components	1.0	2.5
Escalator components	1.0	2.5
Trussed towers (free-standing or guyed)	2.5	2.5
General electrical		
Distributed systems (bus ducts, conduit, cable tray)	2.5	5.0
Equipment	1.0	1.5
Lighting fixtures	1.0	1.5

Notes: 1. A lower value than that specified in Table 10.3 for a_p is permitted, provided a detailed dynamic analysis is performed, which justifies lower value.
2. The value for a_p shall not be less than 1.0.
3. The value of $a_p = 1.0$ is for equipment generally regarded as rigid and rigidly attached.
4. The value of $a_p = 2.5$ is for flexible components or flexibly attached components.

Table 10.4 Amplification factors and response reduction factors for architectural components

Architectural component or element	a_p	R_p
Interior non-structural walls and partitions		
Plain (unreinforced) masonry walls	1.0	1.5
All other walls and partitions	1.0	2.5
Cantilever elements (unbraced or braced to structural frame, below its centre of mass)		
Parapets and cantilever, interior non-structural walls	2.5	2.5

(*Contd*)

Table 10.4 (*Contd*)

Chimneys and stacks laterally supported by structures	2.5	2.5
Cantilever elements (braced to structural frame above its centre of mass)		
Parapets	1.0	2.5
Chimneys and stacks	1.0	2.5
Exterior non-structural walls	1.0	2.5
Exterior non-structural wall elements and connections		
Wall element	1.0	2.5
Body of wall panel connection	1.0	2.5
Fasteners of the connecting system	1.25	1.0
Veneer		
High deformability elements and attachments	1.0	2.5
Low deformability elements and attachments	1.0	1.5
Penthouses (except when framed by an extension of the building frame)	2.5	3.5
Ceilings (all types)	1.0	2.5
Cabinets		
Storage cabinets and laboratory equipments	1.0	2.5
Access floors		
Special access floors	1.0	2.5
All other	1.0	1.5
Appendages and ornamentations	2.5	2.5
Signboards and hoardings	2.5	2.5
Other rigid components		
High deformability elements and attachments	1.0	3.5
Limited deformability elements and attachments	1.0	2.5
Low deformability elements and attachments	1.0	1.5
Other flexible components		
High deformability elements and attachments	2.5	3.5
Limited deformability elements and attachments	2.5	2.5
Low deformability elements and attachments	2.5	1.5

Notes: 1. A lower value for a_p than that specified in Table 10.4 is permitted provided a detailed dynamic analysis is performed which justifies lower value.

2. The value for a_p shall not less than 1.0.

3. The value of $a_p = 1.0$ is for equipment generally regarded as rigid and rigidly attached.

4. The value of $a_p = 2.5$ is for flexible components or flexibly attached components.

Table 10.5 Importance factor of non-structural elements

Description of non-structural element	I_p
Component containing hazardous contents	1.5
Life safety components required to function after an earthquake (e.g., fire protection sprinkler system)	1.5
Storage racks in structures open to the public	1.5
All other components	1.0

- Mechanical components are often fitted with vibration isolation mounts to prevent transmission of vibrations to the structure. By increasing their flexibility, the vibration isolation mounts can alter the dynamic properties of the components, resulting in a dramatic increase in seismic inertial forces. Therefore, for a component mounted on vibration isolation systems, the design force should be taken as $2F_p$.
- Connections and attachments or anchorage of non-structural elements should be designed for twice the design seismic force required for that non-structural element.

Displacement Sensitive Non-structural Elements

Cladding, staircase, piping systems, sprinkler systems, signboards, etc., are connected to the building at various levels. Equation (10.2) may be used to calculate the seismic relative displacement D_p when the two connection points are on same structure, say A.

$$D_p = \delta_{xA} - \delta_{yA} \tag{10.2}$$

where δ_{xA} and δ_{yA} are the deflections at building levels x and y, respectively, of the structure A due to design seismic load determined by elastic analysis, and are multiplied by the response reduction factor R of the building. Equation (10.2) yields an estimate of the actual structural displacements, as determined by elastic analysis.

The seismic relative displacement D_p should not be taken to be greater than D_{p1} given by

$$D_{p1} = R(h_x - h_y)\frac{\Delta_{aA}}{h_{sx}} \tag{10.3}$$

where h_x is the height of level x to which the upper connection point is attached, h_y is the height of level y to which the lower connection point is attached, Δ_{aA} is the allowable storey drift for structure A, and h_{sx} is the storey height below level x. Equation (10.3) is provided in recognition that elastic displacements are not always defined. The equation allows the use of storey drift limitations.

When the two connection points are on separate structures or structural systems, say A and B, one at height h_x and the other at height h_y, the relative displacement D_p is determined using the following equation:

$$D_p = |\delta_{xA}| + |\delta_{yB}| \tag{10.4}$$

D_p is not required to be taken as greater than D_{p1} given by

$$D_{p1} = R\left[h_x \frac{\Delta_{aA}}{h_{sx}} + h_y \frac{\Delta_{aB}}{h_{sy}}\right] \tag{10.5}$$

where δ_{yB} is the deflection at building level y of structure B due to design seismic load determined by elastic analysis, and multiplied by response reduction factor R of the building, and Δ_{aB} is the allowable storey drift for structure B.

10.4 Prevention of Non-structural Damage

Non-structural earthquake damage is generally caused by excessive lateral movement of the building. The first and most important requirement for all non-structural components is that they are positively anchored to the building structure. For life safety, the objective should be to limit the severity of damage to the components so that they do not slide, topple, rock, or detach themselves from the structure, and fall. Prevention of this type of loss is almost invariably simple and inexpensive. Floor anchorages and angle ties linking the tall items to the structure are usually all that is needed. For higher performance objective, it may be necessary to control damage to the components so that the functionality is not impaired.

10.4.1 Architectural Components

During an earthquake each storey of a building undergoes a shear distortion, which is a horizontal movement of the upper floor of the storey with respect to the lower floor (storey drift). If a partition in the storey is connected to the structure so that it is forced to undergo this same shear distortion, and if this is great enough, the partition will be cross-cracked. This cracking can be prevented if the partition is separated at the top or bottom and at the sides to permit the calculated drift to occur without having the wall involved in the movement. Another solution is to stiffen up the building, mostly by the use of shear walls, so that the shear distortion is not enough to crack the plaster in partitions.

The shear distortion of partitions results in the cracking of door frames so that doors are jammed shut or will not close. When walls surrounding a doorway are subjected to large deformation, the doorway may become jammed. Since doors are vital means of egress, these should be properly designed to remain functional after a strong earthquake. Same is the case with window frames. Proper clearance must be provided as shown in Fig. 10.3(a) and (b).

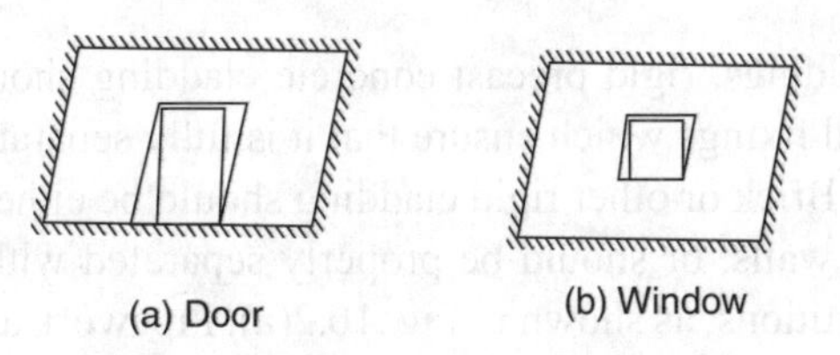

Fig. 10.3 Clearance between windows or doors and walls

Many non-structural components are connected at different levels so that in addition to resisting applied accelerations they must accommodate differential displacements without failure. Windows, for example, frequently need to accommodate inter-storey displacements without failure or fracturing glass. It is shear distortions of a storey that breaks the glass in windows that are rigidly connected to the structure of the building. Breaking of window glass is very dangerous because falling pieces can injure people below. If an expected maximum frame deformation is considered to be small, the glass can be fixed by soft putty. If it is large, a provision for movement of glass within frames to accommodate racking distortions of 0.5 per cent is advisable, and the connection of frames to the structure should provide for yielding of a similar amount.

Protection of window panes from the lateral distortions of the structure has some times been achieved by mounting the window frames on springs that hold them against the structural frame. A detail such as the one shown in Fig. 10.4 has also been used for this purpose. More often, mastics have been used; these retain their plasticity and allow the movement of the panes in the window frames. In every case, there is a need to design against a force perpendicular to the partition or window, whether this force is expected from earthquakes or from wind.

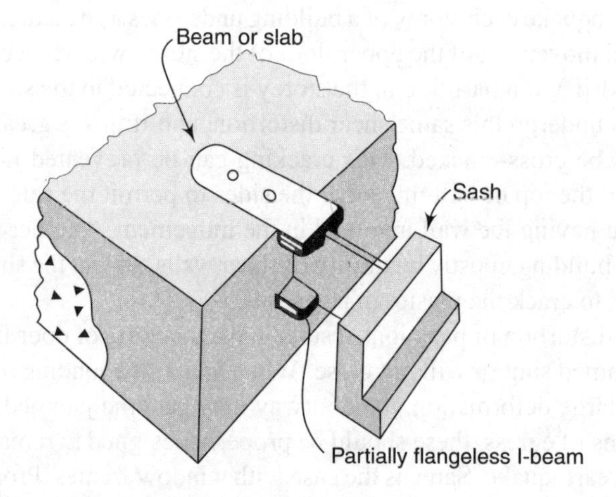

Fig. 10.4 Partial isolation of sash

In flexible buildings, rigid precast concrete cladding should be mounted on specially designed fixings which ensure that it is fully separated from horizontal drift movements. Brick or other rigid cladding should be either fully integral and treated like infill walls, or should be properly separated with details similar to those for rigid partitions, as shown in Fig. 10.2(a). Pipework and ducts traversing movement joints in the building also need to accommodate movement without failure.

Ceilings sometimes fall during an earthquake, but, generally this is not as hazardous as it used to be, because ceilings are now generally made of acoustical tile, not plaster. Tiles are mounted an furring channels or T-members, to which they should be securely fastened. However, if a suspended ceiling is provided, the connections with the suspending members must be properly designed. Detail of such a suspended ceiling system is shown in Fig. 10.5. Care also must be taken so that ceilings do not hit surrounding walls in the course of their horizontal movement—one way of doing this is to provide a gap and sliding cover (Fig. 10.6). Furthermore, design precautions must be taken to prevent ceiling finishes and lighting fixtures from falling to the floor. The lighting fixtures should be secured to the ceiling grid members.

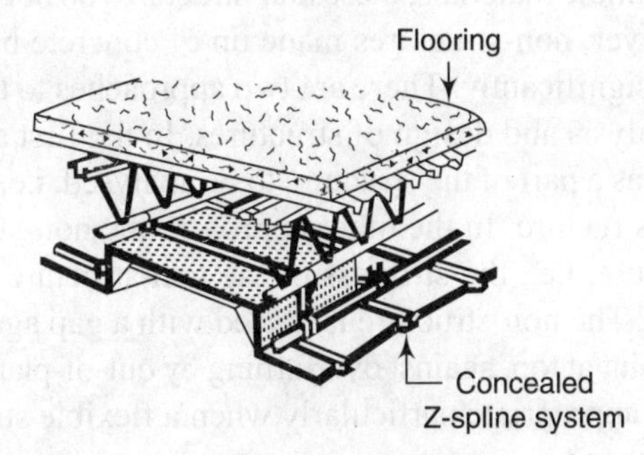

Fig. 10.5 Details of suspended ceiling construction providing movement restraint (Berry 1972)

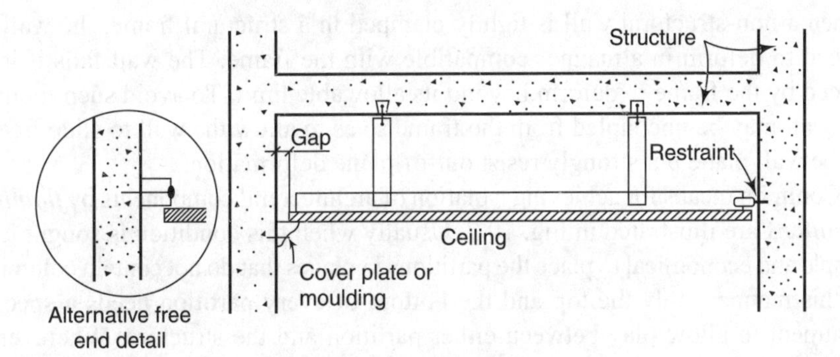

Fig. 10.6 Details at preiphery of suspended ceilings to prevent hammering and excessive movement (Berry 1972)

10.4.2 Mechanical and Electrical Components

All equipments or furnishings that are hung from the ceiling, such as light fixtures, ducts, piping, and heating units, should be braced so that they cannot swing. Long pendant-mounted fluorescent lighting fixtures that are free to swing, usually break loose. Equipment that is not securely fastened to the floor, such as boilers, furnaces, water heaters, storage tanks, and air conditioning units, may move horizontally or fall down due to the rocking motion during earthquake vibration. Generally, utility lines to these units will be broken. Vibration isolation supports are particularly vulnerable to earthquake forces. Storage cabinets and storage racks should be securely anchored to walls or partitions. The basic design requirement is that the services should not fail before the building fails in an earthquake.

10.5 Isolation of Non-structures

Non-structural failures due to earthquakes are of great concern as they affect the loss of human life to a large extent. Non-structures such as cladding, perimeter infill walls, and partitions become structurally very responsive during earthquakes. When made up of flexible materials, these non-structures do not affect the structure significantly. However, non-structures made up of concrete blocks, bricks, etc., affect the structure significantly. There are two approaches to taking care of non-structures in the analysis and design of structures. In the first approach, the non-structures are taken as a part of the structure to be analysed, i.e., the non-structure is made into a real structure. In the other approach, the non-structure is isolated from the real structure, i.e., the stiffness of the non-structure is not included in that of the structure. The non-structure is placed with a gap against the structure, however, with restraint at top, against overturning by out-of-plane forces. Isolation of non-structures is appropriate particularly when a flexible structure is required for low seismic response.

10.5.1 Architectural Components

When a non-structural wall is tightly clamped in a structural frame, the wall is forced to deform in a manner compatible with the frame. The wall fails if it is forced by the frame to deform beyond its allowable limit. To avoid such failure, the wall may be uncoupled from the frame so as to allow the wall to slide freely in the wall plane but strongly resist out-of-plane deformation.

Common means for achieving isolation of architectural components by *floating partitions* are illustrated in Fig. 10.7. Usually when this condition is sought it is simple and economical to place the partitions in planes that do not contain columns. In this manner, only the top and the bottom of every partition needs a special treatment to allow play between either partition and the structure. Wherever a gap between the partition and the structure is to be visible, there is often a need for an element to hide it or to fill it and prevent unsightliness and dust gathering.

Especially in buildings that are repaired and strengthened after suffering earthquake damage, there is sometimes an advantage in using a peripheral metal band of the type shown schematically in Fig. 10.8. If the shearing and normal forces required to make the band yield are chosen properly, it is possible to limit the lateral forces that the structure will transmit to the partition and at the same time take advantage of the capacity of the partition to resist such forces and make use of the energy absorbing capacity of the band.

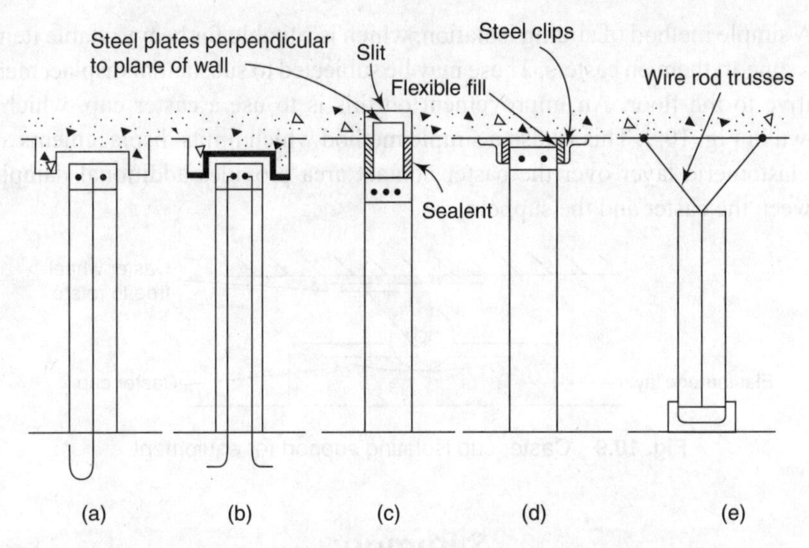

Fig. 10.7 Common solutions for isolation of partitions (Newmark et al. 1971)

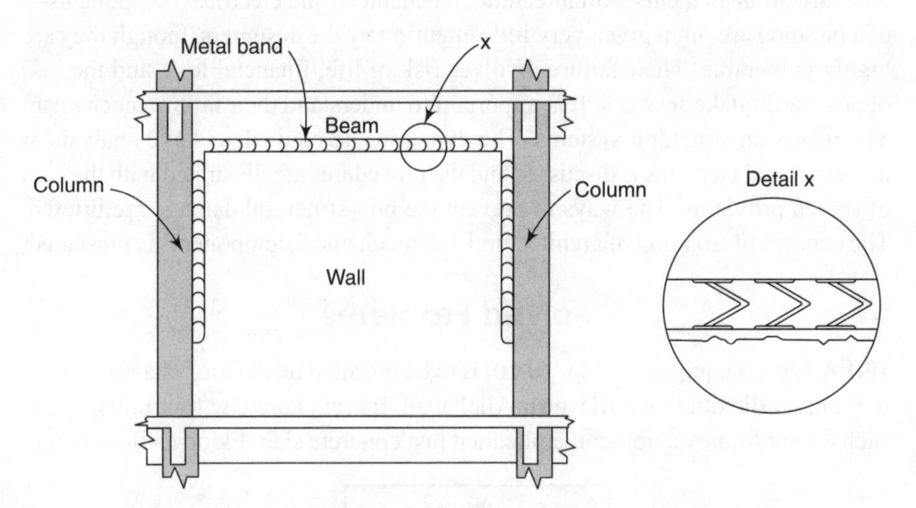

Fig. 10.8 Metal band to protect partitions (Newmark et al. 1971)

10.5.2 Mechanical Components

An efficient way of achieving seismic safety of mechanical components is by isolating them from the structure, using springs, so that the deformation of the structural system does not affect the deformation or force in them. Although a spring supported plan may be vulnerable to large displacements and damage in a major earthquake, properly designed isolation systems will provide a valuable form of protection. The principle used in design is that the base motion is that of the building at the point (or points) of support.

A simple method of sliding isolation, which is suitable for heavy, stable items, is to support them on casters. These may be subjected to substantial displacements relative to the floor. An improvement on this is to use a caster cup, which is shown in Fig. 10.9. This is also a simple method which limits displacements and an elastomeric layer over the caster contact area provides additional damping between the caster and the support.

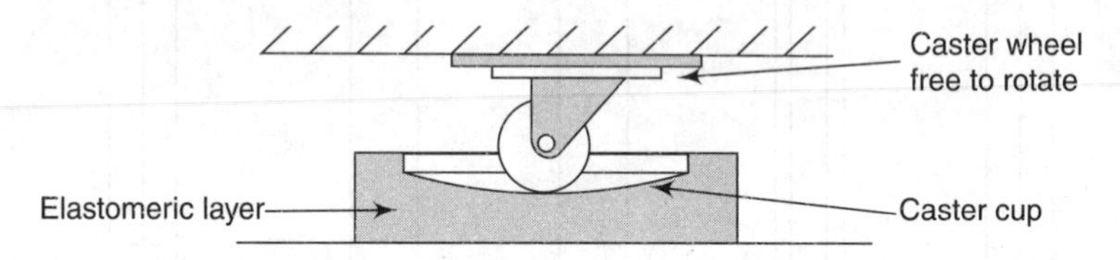

Fig. 10.9 Caster cup isolating support for equipment

Summary

Non-structural elements—architectural, mechanical, and electrical components—of a building are often given very little attention by the designers, though they are highly vulnerable. Their failure involves risk of life, financial loss, and the loss of post earthquake services. It is important to understand their failure mechanism and effects on structural systems. The dynamic and equivalent static analysis of non-structural elements is discussed and the procedures are illustrated with the help of solved problems. The ways to prevent the non-structural damage are briefed. The concept of isolation of architectural and mechanical components is presented.

Solved Problems

10.1 A 120 kN equipment (Fig. 10.10) is to be installed on the roof of a five-storey building in Bhadoi (zone III), near Allahabad. It is anchored by four bolts, one at each corner of the equipment, embedded in a concrete slab. Floor-to-floor height

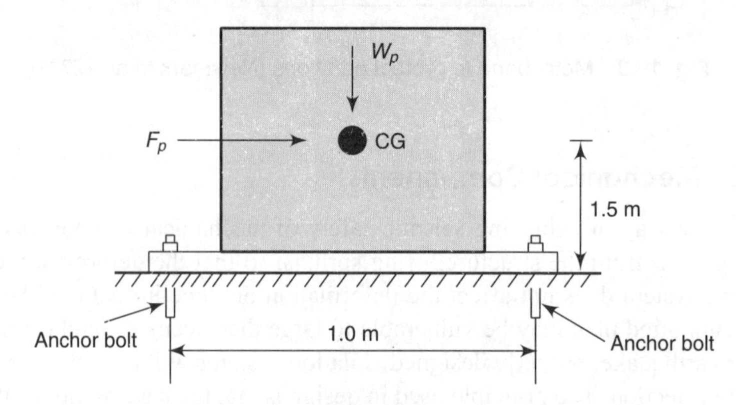

Fig. 10.10 Equipment installed at roof

of the building is 3.1 m for each of the four storeys and for the ground storey it is 4 m. Determine the shear and tension demands on the anchored bolts during earthquake shaking.

Solution

Zone factor $Z = 0.16$ (for Zone III, Table 2 of IS 1893)

Height of point of attachment of the equipment above the foundation of the building

$$x = 4.0 + 3.1 \times 4 = 16.4 \text{ m}$$

Height of the building $h = 16.4$ m

Amplification factor of the equipment $a_p = 1$ (rigid component, Table 10.3)

Response modification factor $R_p = 2.5$ (Table 10.3)

Importance factor $I_P = 1$ (it is not a life-safety component, Table 10.5)

Weight of the equipment $W_p = 120$ kN

The design seismic force,

$$F_p = \frac{Z}{2}\left(1+\frac{x}{h}\right)\frac{a_p}{R_p} I_p W_p$$

$$= \frac{0.16}{2} \times \left(1+\frac{16.4}{16.4}\right) \times \frac{1.0}{2.5} \times 1 \times 120$$

$$= 7.68 \text{ kN} \qquad (0.1\ W_p = 0.1 \times 120 = 12.0)$$

$$< 12.0 \text{ kN}$$

Hence, $F_p = 12.0$ kN

The anchorage of equipment with the building (being flexible) should be designed for two times this force.

Shear per anchor bolt

$$V = 2 \times \frac{F_p}{4} \quad \text{(since there are four bolts)}$$

$$= 2 \times \frac{12}{4}$$

$$= 6 \text{ kN}$$

The overturning moment $M_{ot} = 2 \times 12 \times 1.5$

$$= 36.0 \text{ kNm}$$

This overturning moment is resisted by two anchor bolts provided on either side. Hence, tension per anchor bolt from overturning,

$$F_t = \frac{36.0}{1 \times 2}$$

$$= 18.0 \text{ kN}$$

10.2 A 120-kN electrical generator is to be installed on the third floor of a five-storey hospital building in Bhadoi (Zone III), near Allahabad. It is to be mounted on four flexible vibration isolators (Fig. 10.11), one at each corner of the unit.

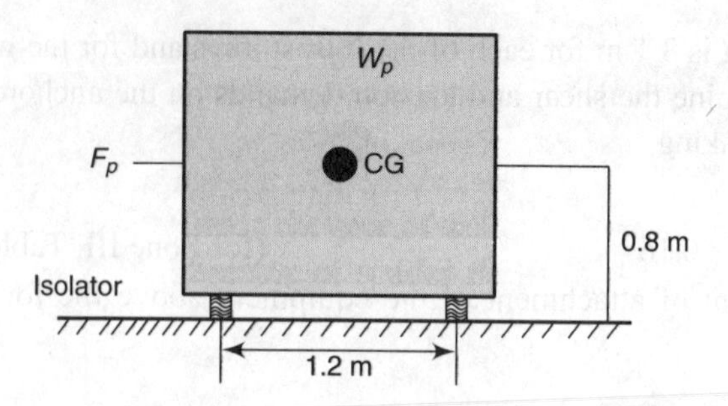

Fig. 10.11 Electrical generator installed on the third floor

Floor-to-floor height of the building is 3.1 m except the ground storey, which is 4.0 m in height. Determine the shear and tension demands on the isolators during earthquake shaking.

Solution

Zone factor, $Z = 0.16$ (for Zone III, Appendix III)

Height of point of attachment of the equipment above the foundation of the building

$$x = 4.0 + 3.1 \times 2$$
$$= 10.2 \text{ m}$$

Height of the building,

$$h = 4.0 + 3.1 \times 4$$
$$= 16.4 \text{ m}$$

Amplification factor for the generator $a_p = 2.5$ (vibration isolated, Table 10.3)
Response modification factor $R_p = 2.5$ (vibration isolator, Table 10.3)
Importance factor $I_P = 1.5$ (it is a life safety component, Table 10.5)
Weight of the generator, $W_p = 120$ kN
The design seismic force on the generator,

$$F_p = \frac{Z}{2}\left(1+\frac{x}{h}\right)\frac{a_p}{R_p} I_p W_p$$

$$= \frac{0.16}{2} \times \left(1+\frac{10.2}{16.4}\right) \times \frac{2.5}{2.5} \times 1.5 \times 120$$

$$= 23.35 \text{ kN} \quad (0.1\, W_p = 0.1 \times 120 = 12.0)$$
$$> 12.0 \text{ kN}$$

Hence, $F_p = 23.35$ kN

Since the generator is mounted on flexible vibration isolator, the design force is doubled.

$$F_p = 2 \times 23.35$$
$$= 46.7 \text{ kN}$$

Shear force resisted by each isolator

$$V = \frac{F_p}{4} \text{ (since there are four isolators)}$$

$$= \frac{46.7}{4}$$

$$= 11.675 \text{ kN}$$

The overturning moment

$$M_{ot} = 46.7 \times 0.8$$

$$= 37.36 \text{ kNm}$$

The overturning moment is resisted by two anchor bolts provided on either side. Hence, tension per anchor bolt from overturning

$$F_t = \frac{37.36}{1.2 \times 2.0}$$

$$= 15.57 \text{ kN}$$

10.3 An electronic signboard is attached to a five-storey building consisting of special moment-resisting frame system in Varanasi (seismic zone III). It is attached by two anchors at heights 12.0 m and 9.0 m. From the elastic analysis under design seismic load, the deflections obtained for the upper and lower attachments of the signboard are 35.0 mm and 28.0 mm, respectively. Find the design relative displacement.

Solution

A signboard is a displacement-sensitive non-structural element, hence it should be designed for seismic relative displacement.
Height of level x to which upper connection point is attached, $h_x = 12.0$ m
Height of level y to which lower connection point is attached, $h_y = 9.0$ m

Deflection at building level x of structure A due to design seismic load = 35.0 mm

Deflection at building level y of structure A due to design seismic load = 28.0 mm

Response reduction factor $R = 5$ (special RCC moment-resisting frame, Table 5.4)

$$\delta_{xA} = 5 \times 35 = 175.0 \text{ mm}$$

$$\delta_{yA} = 5 \times 28 = 140.0 \text{ mm}$$

$$D_p = \delta_{xA} - \delta_{yA}$$

$$= 175.0 - 140.0$$

$$= 35.0 \text{ mm}$$

The connections of the signboard should be designed to accommodate a relative displacement of 35 mm.

Alternatively, assuming that the analysis of building is not possible to assess deflections under seismic loads, one may use the drift limits (this presumes that the building complies with the seismic code).

Maximum inter-storey drift allowance is 0.004 times the storey height (Section 5.13.1),

$$\frac{\Delta_{aA}}{h_{sx}} = 0.004$$

$$D_{p1} = R(h_x - h_y)\frac{\Delta_{aA}}{h_{sx}}$$

$$= 5 \times (12000.0 - 9000.0) \times 0.004$$

$$= 60.0 \text{ mm}$$

The electronic signboard will be designed to accommodate a relative displacement of 60 mm.

Exercises

10.1 What are non-structures? How do these affect the performance of a structural system?

10.2 Discuss briefly the effect of a structural system on the behaviour of a non-structure.

10.3 Draw neat sketches to show the isolation of the following non-structures from the main building:
- (a) Doors and windows
- (b) Partition walls
- (c) Equipment

10.4 What are the common earthquake damages in non-structures? What measures do you suggest to prevent them?

10.5 Write short notes on
- (a) Importance of non-structures in a building
- (b) Failure mechanisms of non-structures
- (c) Consequences of failure of non-structural elements
- (d) Prevention of non-structural damage

10.6 Why is it important to take suitable measures for prevention of non-structural failure rather than to undertake repairs after damage?

10.7 A trussed tower 7 m in height, 1.5 m × 1.5 m in cross-section at base, and 50 kN in weight, for signal transmission, is to be installed on the roof of a six-storey multiplex at Allahabad (seismic Zone III). It is attached by 16 anchored bolts, four at each corner of the tower base, embedded in the concrete blocks. The ground storey is 4.3 m in height, whereas other floor-to-floor heights are 3.0 m each. Calculate the shear and tension demands on the anchored bolts during the earthquake. *Ans:* 1 kN, 3.11 kN

10.8 A glow sign hoarding of 10 m length is to be fixed on the front side of a seven-storey special RC framed building in New Delhi (seismic Zone IV). It is attached by four anchors at 15.0 m and 9.0 m levels, respectively. The deflections at the upper and lower fastening of the glow sign hoarding are 40 mm and 28 mm, from elastic analysis. Determine the design relative displacement of the hoarding. *Ans:* 120 mm

10.9 An airconditioning unit weighing 120 kN is to be installed on the roof of a G + 10 storey building. The dimensions of the unit are shown in Fig. 10.12. The fundamental period of the airconditioning unit is 0.05 s. There are four 24 mm diameter anchor bolts, one at each corner of the unit, embedded in the roof concrete slab up to 180 mm depth. The building is in seismic zone IV. Assume all the storeys of the building to be 3.1 m high except the ground storey, which is 4 m high. Determine the shear on the anchor bolt during earthquake. *Ans:* 9.6 kN

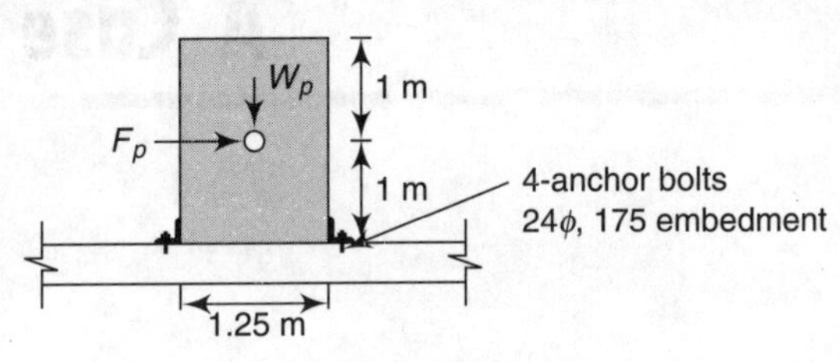

Fig. 10.12

CHAPTER 11

Bhuj Earthquake 2001: A Case Study[1]

The first historical Kutch earthquake to attract international attention was the 1819 Allah Bund earthquake, which created a 6-m high and 6-km wide natural dam across the Puran river, which enters the Rann of Kutch from the north. A lake, 30 km in diameter, lake Sindri, was formed south of the Allah Bund. A lake was also formed north of the Bund in 1819, which drained in 1826 when a torrent broke through several artificial dams on the Puran and cut a gorge through the Bund, flooding regions downstream. Damage to Bhuj and Anjar during the 1819 earthquake was substantial. Damaging earthquakes also occurred in 1845, 1856, 1857, 1864, 1903, 1927, 1940, 1956, and 1970 in the Kutch region, but with less severity ($5 < M < 6$).

The most devastating earthquake, of January 26, 2001, that struck at 8:46 am IST in the Kutch region of Gujarat, India, was an eye opener for structural engineers and designers. The devastation was major in terms of lives lost, injuries suffered, as well as structural collapses and economic losses. The entire Kutch region was extensively damaged and several towns and villages, such as Bhuj, Anjar, Vondh, Gandhinagar, Kandla Port, Morbi, Ahmedabad, Rajkot, and Bhachau, sustained wide-spread destruction. Numerous newly constructed buildings collapsed leading

[1]This case study is mostly adapted from EERI 2002, 'Bhuj India earthquake of Jan 26, 2001', *Reconnaissance Report Earthquake Spectra; supplement to vol 18.*

to extensive causalities. The earthquake is subsequently referred to as the Kutch earthquake or the Bhuj earthquake.

The Bhuj earthquake is considered to be the largest intra-plate earthquake ever recorded. Though the mechanism of the 2001 Bhuj earthquake is currently unresolved, the event apparently occurred on a steeply dipping thrust that did not break the surface. An unusual feature of the event is that aftershocks have occurred at considerable depth, about 20 km, suggesting rupture through much of the lithosphere. The event had reverse motion, with a slight right-lateral component of slip.

The occurrence of the Bhuj earthquake less than 200 years after the severe 1819 event provides further evidence that large intra-plate earthquakes can occur in clusters in regions of the crust where the strain rate is relatively low. The event highlights the potential hazard faced by areas that lie outside more rapidly deforming plate boundary regions. The 2001 Bhuj earthquake has important implications for earthquake hazard, not only in India but also in other parts of the world where the source zones and/or the wave travel paths are similar.

11.1 Earthquake Parameters and Effects

The important parameters and data relating to the 2001 Bhuj earthquake are listed as follows:

Region: Kutch, Gujarat
Date: 26-01-2001
Time: 8:46 am (IST)
Epicentral coordinates: 23.36° N, 70.34° E (near Bhuj)
Hypocentral depth: Between 17 to 22 km, on a fault plane that strikes about 60° N and dips 60° to 70° S with a slip direction of 62°.
Official death toll: 35,000
Persons injured: 1,60,000
Economic losses: US$5 billion
Strong ground shaking: Lasted for about 85 s
Magnitude: M_W 7.7, M_S 7.6, M_b 7.0, and M_L 6.9
Seismic moment: 6.2×10^{28} dyne-cm
Peak ground acceleration: 0.11 (measured at Ahmedabad, 225 km from epicentre)

The failure of buildings in the Bhuj earthquake may be attributed to the geological and geotechnical effects, poor form, inadequate design and detailing, and poor quality of construction. These are described briefly in the following subsections.

11.1.1 Geological Effects

Despite a severe earthquake, no evidence of surface fault rupture or sharp folding has been reported. A zone of ground deformation is reported to occur within alluvial deposits near the northern margin of an anticline along the Mainland fault. The ground deformations include extensional ground cracking (Fig. 11.1) and compressional bulging in a zone over 16 km long and 0.5 km wide near the epicentre. The features are associated with extensive sand boils in extensional cracks (Fig. 11.2). These ground failures have been considered to be related to liquefaction and lateral spreading and not primary fault rupture.

Fig. 11.1 Extensional ground cracking (photo by James Hengesh)

Fig. 11.2 Sand boils in extensional cracks (photo by James Hengesh)

The possibility of secondary tectonic fault rupture in two areas has also been expressed. Near the town of Manfara, north of Bhachau, a northwest-striking rupture about 8 km long has been observed with up to 32 cm of right-lateral displacement. This feature may be a secondary tear fault in the hanging wall of the main thrust fault. At a second location southeast of Chung Dam, a northeast-striking rupture has been found to extend for several kilometres into an area of thin alluvium and locally may thrust bedrock over alluvium by up to 30 cm.

11.1.2 Geotechnical Effects

Most of the buildings that collapsed lie along the old path of the river Sabarmati. The buildings that collapsed in areas west of the river Sabarmati are closely aligned with the old path of the river, just west of the present river path. The south and south-east of the city, especially the Mani Nagar area, where additional collapses were observed, fall between two lakes, indicating the presence of either poor soil conditions or possibly construction on non-engineered fills.

The earthquake produced widespread liquefaction in the Great Rann, Little Rann, Banni Plains, river Kandla, and the Gulf of Kutch. These areas contain low-lying salt flats, estuaries, inter-tidal zones, and young alluvial deposits, which typically have a high susceptibility to liquefaction. Liquefaction was manifest at the surface as sand boils, lateral spreads, and collapse features.

Fig. 11.3 Failure on upstream face of earthen dam

Liquefaction caused damage to several bridges, the Ports of Kandla and Navlakhi, and numerous embankment dams in the epicentral area. Seven medium-size earth dams (Shivlakha, Rudramata, Fategad, Suvi, Kaswati, Tapar, and Chang) and 14 smaller earth dams were damaged during the earthquake. Liquefaction of the foundation soils beneath these dams produced moderate to severe failure of the upstream (Fig. 11.3), and, locally, the downstream faces of the dams.

11.2 Buildings

Indian seismic codes are relatively well developed for buildings, and code provisions are available for different types of construction. However, most buildings in the region have been reported as not conforming to the seismic code provisions. Most government organizations attempt to comply with the code requirements; however, in the private sector it is not so. The earthquake destroyed about 300,000 houses and damaged another 700,000. For the purpose of discussion, the buildings may be classified as non-engineered buildings made with load-bearing masonry walls supporting a tiled roof or RCC slab/roof; and RCC frame buildings with unreinforced masonry infills. A brief introduction to the types of building constructions and the prevailing design and detailing practices (non compliance of the codal specifications) in the major earthquake-affected cities of Gujarat is presented in the following subsections.

11.2.1 Masonry Buildings

Non-engineered construction constitutes over 95 per cent of the building stock in the Kutch region. These houses were either traditional earthen houses constructed with sun-dried clay bricks and wooden sticks (these dwellings were circular in plan and about 4–8 m in diameter with conical roofs and shallow foundations) or made up from different types of masonry as listed below:

(a) Random rubble stones with mud or cement mortar
(b) Small or large cut stones in mud or cement mortar
(c) Burnt-clay bricks in mud or cement mortar
(d) Solid or hollow cement blocks in cement mortar

Among the non-engineered constructions in the Kutch area, very large stone blocks (0.25 m × 0.40 m × 0.60 m) in masonry walls (Fig. 11.4), with mud mortar or low-strength cement mortar were used. These exhibited very poor performance. The quake-affected areas of Kutch and Saurashtra have numerous historical buildings, tombs, minarets, and pagodas in stone masonry. Many of these structures collapsed or sustained heavy damage during the earthquake.

Fig. 11.4 Failure of a typical stone masonry building

In the meizoseismal[2] area, masonry buildings collapsed, causing a large number of causalities. The performance of stone masonry with mud mortar was particularly poor. On the other hand, masonry buildings up to about four storeys did well in Ahmedabad (about 225 km from the epicentre).

The loosening of the stone blocks of the building shown in Fig. 11.4 owing to lack of plumb in construction and to the action of the out-of-plane earthquake forces led to the collapse of the wall, leading to the overall instability of the building and of similar such dwellings.

[2]The places of most severe damage.

Critical Review

The provisions of IS 4326 were not followed. Stone masonry houses in mud mortar without earthquake-resistant features were the most common type of construction. The damages occurred due to one or more of the following reasons:

(i) Structural integrity was not ensured as bands were not provided at any level
(ii) Connections between walls and roofs were not made
(iii) Rafters rested directly on the walls
(iv) No through stones were used to achieve connections between walls

Most buildings having random rubble masonry in cement mortar with reinforced concrete slab used in the construction of single or two-storey residential units with plinth and lintel band, performed very well. The heavy damage to masonry construction resulted from the poor performance of mortar and the use of heavy and loosely formed roofs. The wall–roof interface had nominal sliding and separation, and the walls between plinth and lintel bands sustained shear cracks.

11.2.2 Reinforced Concrete Buildings

Some of the features of the RCC buildings in Gujarat are as follows. Most of the RCC buildings are reported to

(a) be G+4 to G+10 storeys with moment-resisting frames having RCC slabs cast monolithically with beams. Generally, these buildings had few or no infill walls in the ground floor to accommodate commercial establishments and/or vehicular parking. It has been reported that most buildings were designed only for gravity loads and only a few for earthquake forces with ductile detailing practices. The materials used in the construction were M-15 grade concrete for G+4 storey buildings and M-20 grade concrete with Fe 415 reinforcement for taller buildings.

> An almost total absence of infill walls at the ground level creates a very distinct stiffness discontinuity or a soft storey. Virtually all the earthquake-induced deformations in such a building occur in the columns of the soft storey, with the rest of the building basically going along for a ride. If these columns are not designed to accommodate the large deformations, they may fail, leading to catastrophic failure of the entire building, as was the case with many buildings in Ahmedabad and elsewhere.

(b) have roofs-usually RCC slabs of 100–120 mm thickness resting on beams, with 500–650 mm depth (including the slab) and 200–250 mm width. In some cases, the slab was directly cast on columns. The main reinforcement in the slab was 8 mm ϕ at a spacing of 100 mm c/c and the distribution steel was 6 mm ϕ at 150 mm to 200 mm c/c.

> The provisions of ductile detailing of the code have not been followed.

(c) have overhanging covered balconies of about 1.5 m span on higher floors. Heavy beams from the exterior columns of the building to the end of the balcony on the first floor onwards was found to be a common practice. To create more parking spaces at the ground floor and to allow more space on the upper floors a peripheral beam was provided at the end of the erected girder. The upper floor balconies or other constructions were constructed on the peripheral beams. The infill walls, which were present in upper floors and absent in the ground floor, created a floating box-type situation.

The local municipal corporation in Ahmedabad imposes a floor surface index (FSI), which restricts the ground floor area of a building to be no more than a certain percentage of the plot area. It is, however, permitted to cover more area at upper floor levels than at the ground floor level. Thus, most buildings had overhanging covered floor areas at upper floors, with the overhangs frequently ranging up to 1.5 m or more. The columns on the periphery of the upper floors did not continue down to the ground level. The columns at the ground floor level, also, were sometimes not aligned with the columns at the upper levels.

Significant vertical discontinuities are therefore generated in the lateral force-resisting system. The dynamic analysis of a G+4 storey RCC building on floating column shows that such buildings vibrate in a torsional mode, which is undesirable.

(d) have non-uniform column spacing, leading to varying beam spans 2–5 m. In general, the beams were deeper than columns to accommodate large spans and overhangs. For taller buildings, the beam size was found to be similar to the column size. The reinforcement was three to four longitudinal bars of 12 or 16 mm ϕ and transverse reinforcement of 6–8 mm ϕ at c/c of 200–250 mm with 90° hooks at the ends.

A beam that is larger than the column, is against the weak-beam–strong-column principle of earthquake-resistant design. The detailing is not in accordance with the ductile detailing provisions of the code.

(e) often have columns of rectangular cross-sections, with typical dimensions, i.e., 230 × 450 mm for G+4 and varying to 300 × 600–800 mm for more storeys. Longitudinal reinforcement consisted of two rows of four to six bars of 12–18 mm diameter. The longitudinal reinforcement ratio was generally between 1 and 2 per cent of the gross cross-sectional area. Transverse reinforcements was of a single hoop of 6–8 mm diameter having 90° hooks, spaced at 200–250 mm and terminated at the joints. The longitudinal reinforcement was often lap-spliced just above the floor slab. The spacing of transverse reinforcement over the lap splice was the same as that elsewhere in the column.

- There is no sign of special confinement reinforcement and ductile detailing in the columns (Fig. 11.5). Such non-ductile detailing of RCC construction is common. This is a faulty design practice from the seismic point of view.

Fig. 11.5 Column failures due to non-ductile detailing

- The large deformations that take place in soft storey columns also impose extreme shear demands on them. The meagre lateral reinforcement described earlier not only provides poor confinement but also makes the shear strength quite low. As a result, many ground floor columns failed in a brittle-shear mode, or in a combined shear-plus-compression mode, bringing down the supported buildings (Fig. 11.6). Many times, in columns that had not failed, diagonal shear cracking was evident.

Fig. 11.6 Failure due to soft storey

- In addition, column reinforcement is typically spliced right above the floor levels; splice length in columns as well as beams is often insufficient; continuity of beam reinforcement over and into the supports is often also insufficient.

(f) have ground floor columns not cast up to the bottom of the beam. A gap of 200–250 mm was left, called *topi* (Fig. 11.7), to accommodate the beam reinforcement. This type of construction is vulnerable.

Due to the congestion of reinforcement in this region, the compaction of concrete cannot be properly performed, which results in poor quality of concrete and honeycombing.

Fig. 11.7 Poorly compacted concrete due to congestion of reinforcement at the top of columns

(g) have foundations as isolated footing with a depth of about 1.5 m for G+4, and 2.7 to 3.5 m for G+10 structures in private buildings. The plan sizes of footings are usually 1.2 m × 1.2 m, 1.8 m × 1.8 m, or 2.4 m × 2.4 m. There are no tie beams interconnecting the footing, and plinth beams connecting to the column at the ground storey level. The majority of the damaged buildings were founded on deep alluvium where the amplification of motion in the soil seems to have caused large forces in the buildings. The official buildings, however, were provided with raft foundations.

Independent footings without the beams offer poor earthquake resistance.

(h) have elevator cores made of RCC structural walls called shear walls. The shear walls are reported to be typically about 100–150 mm thick with very light reinforcement consisting of two layers of mesh formed with three or four bars of 10 ϕ at vertical and horizontal spacings of about 450 mm. Severe shear cracking of shear walls at the ground level was reported. The shear cracking often did not extend above the ground floor level, reflecting reduced shear demand on the walls. The shear wall core was often connected to the rest of the building only through the floor slabs and, with no beams framing into the shear walls, the anchorage of slab reinforcing into the elevator core was found to be insufficient. As a result, the shear walls got pulled out from portions of the building, leaving them devoid of much lateral resistance.

Such detailing is insufficient to resist the lateral loads at the ground floor level.

(i) have some failures of water tanks constructed over the roof of RCC framed buildings (Fig. 11.8). Water tanks experienced large inertia forces due to the amplification of the ground acceleration along the height of the building.

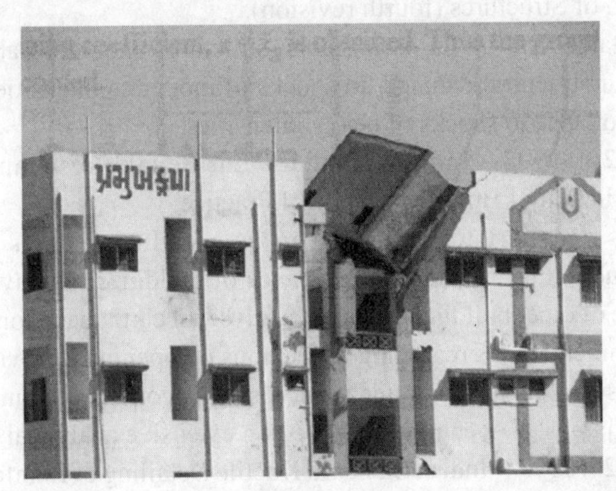

Fig. 11.8 Failure of water tank over the top of a multistorey building (Photo by Jaswant N. Arlekar)

(j) have no sliding joints in masonry staircases. These were built with unreinforced masonry wall enclosures, which failed in most cases.

Further, the observations of failed RCC sections revealed the following:

(a) The concrete disintegrated within the reinforcement cage, and when touched, the concrete felt sandy with little cement.
(b) The 90° hooks opened up, leading to little or no confinement of the concrete.
(c) The concrete cover for the reinforcement was found to be less than 12 mm. Most of the cover was provided by the plaster used to smooth the column surface.
(d) Most of the water supply in the outer part of the city is through ground water, which is salty. Therefore, the presence of salts may have also affected the quality of concrete.

Critical Review

Reinforced cement concrete structures, if not properly designed, detailed, and constructed, prove to be more vulnerable than even non-engineered masonry structures. The satisfactory performance of RCC structures during strong ground motion depends on the ductility built into the structure and the structural overstrength. The required overstrength can be achieved by constructing well the structure and ductility can be achieved by following the provisions of IS 13920. The lack of awareness of the detailing practices amongst professionals and builders, poor forms of structures, and the attitude of building structures at low cost with inferior materials/construction practices has probably contributed to the large-scale devastation at Gujarat.

The following three codes relevant to earthquake-resistant design of RCC structures are in practice in India:

(i) IS 1893: (revised in 2002)—Indian Standard Criteria for Earthquake-resistant Design of Structures (fourth revision).

It states that, as far as possible, structures should be able to respond, without structural damage, to shocks of moderate intensities, and without total collapse to shocks of heavy intensities.

(ii) IS 4326: 1993—Indian Standard Earthquake-resistant Design and Construction of Buildings: Code of Practice.

This code is intended to cover the specified features of design and construction for earthquake resistance of buildings of conventional types. In case of other buildings, detailed analysis of earthquake forces is required. Recommendations regarding restrictions on openings, provision of steel in various horizontal bands, and vertical steel in corners and junctions, in walls and at jambs of openings, are based on extensive analytical work.

(iii) IS 13920: 1993—Indian Standard Ductile Detailing of Reinforced Concrete Structures subjected to Seismic forces: Code of Practice.

This document incorporates the following important provisions that are not covered in IS 4326:

(a) The deficiencies in the design and detailing of RCC structures, as per IS 4326: 1976 were identified based on experiences gained from past earthquakes and were corrected in IS 13920.
(b) Provisions on detailing of beams and columns were revised with an aim of providing them with adequate toughness and ductility so as to make them capable of undergoing extensive inelastic deformations and dissipating seismic energy in a stable manner.
(c) Specifications on seismic design and detailing of RCC shear walls were included.

Beside these, the other significant items incorporated in IS 13920 are as follows:

(a) Material specifications are included for lateral force-resisting elements of frames.
(b) Geometric constraints are imposed on the cross-section for flexural members. Provisions on minimum and maximum reinforcement have been revised. The requirements for detailing of longitudinal reinforcement in beams at joint faces, splices, and anchorage requirements are made more explicit. Provisions are also included for calculation of design shear force and for detailing of transverse reinforcement in beams.
(c) For members subjected to axial load and flexure, dimensional constraints have been imposed on the cross-section. Provisions are included for the detailing of lap splices and for the calculation of design shear force. A comprehensive set of requirements is included on the provision of special confining reinforcement

in those regions of a column that are expected to undergo cyclic inelastic deformations during a severe earthquake.

(d) Provisions have been included for estimating the shear strength and flexural strength of shear wall sections. Provisions are also given for detailing of reinforcement in the wall web, boundary elements, coupling beams, around openings, at construction joints, and for the development, splicing, and anchorage of reinforcement.

Limitations of the codal provisions While the common methods of design and construction have been covered in IS 13920, special systems of design and construction of any plain or RC structure not covered by this code is permitted on production of satisfactory evidence, regarding their adequacy for seismic performance by analysis or tests or both. It is interesting to note that the provisions of IS 13920 apply to RCC structures that satisfy one of the following conditions:

(a) The structure is located in seismic Zone IV or V.
(b) The structure is located in seismic Zone III and has an importance factor of greater than 1.0.
(c) The structure is located in seismic Zone III and is an industrial structure.
(d) The structure is located in seismic Zone III and is more than five storeys high.

The residential buildings in Ahmedabad with G+4 storeys do not fulfil any of the above four conditions and therefore are exempt from the requirements of IS 13920. This requires a serious consideration.

All the above three codes are quite sophisticated. However, code enforcement practically does not exist in India. Central and state governments at times require code compliance for buildings owned by them. For other buildings, code requirements are seldom, if ever, enforced. Local jurisdictions typically do not have a mechanism in place to enforce code requirements. This not only explains much of the damage observed to engineered buildings but also indicates that future earthquakes may cause a lot more loss and devastation due to design flaws of structures.

11.2.3 Precast Buildings

Some single-storey school buildings in the Kutch region were made of large panel precast RCC components for the slab and walls, and precast RCC columns. Approximately one-third such schools in the Kutch region had roof collapses (Fig. 11.9).

> Inadequate connection between the roof panels led to lack of floor-diaphragm action, and insufficient seating and anchorage of the roof panels over the walls and beams led to dislodgement of the precast roof panels from atop the walls.

Fig. 11.9 Collapse of precast components (Photo by Sudhir K. Jain)

Appendices

I. Seismic Zones in India

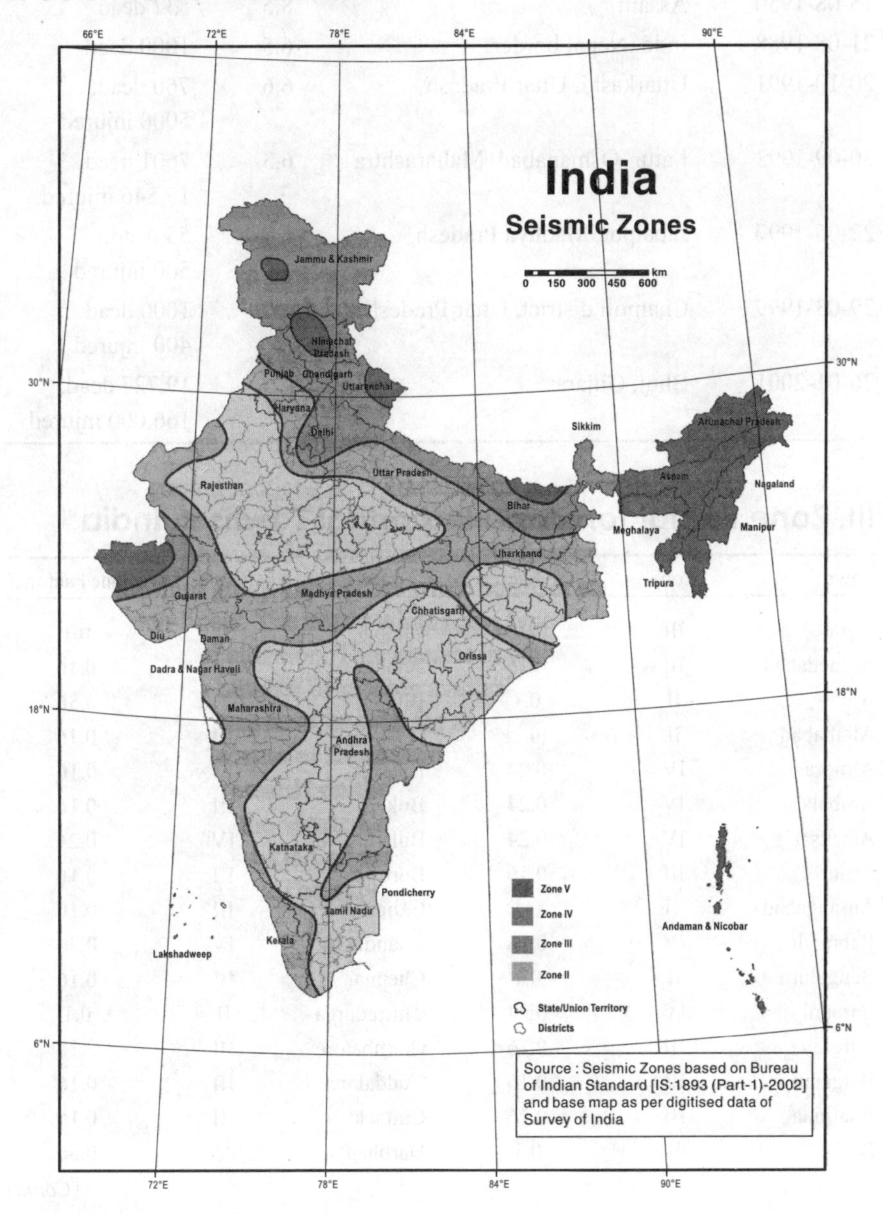

II. Some Significant Earthquakes in India

Date	Location	Magnitude	Causalities
16-01-1819	Kutch, Gujarat	8.0	2000 dead
12-01-1897	Shillong plateau	8.7	1542 dead
04-04-1905	Kangra, Himachal Pradesh	8.0	20,000 dead
15-01-1934	Bihar–Nepal border	8.3	1000 dead, 9000 injured
15-08-1950	Assam	8.5	532 dead
21-08-1988	Indo–Nepal border	6.5	1000 dead
20-10-1991	Uttarkashi, Uttar Pradesh	6.6	760 dead, 5000 injured
30-09-1993	Latur–Osmanabad, Maharashtra	6.3	7601 dead, 15,846 injured
22-05-1997	Jabalpur, Madhya Pradesh	6.0	55 dead, 500 injured
29-03-1999	Chamoli district, Uttar Pradesh	6.8	1000 dead, 400 injured
26-01-2001	Bhuj, Gujarat	7.9	19,727 dead, 166,000 injured

III. Zone Factor for Some Important Towns in India

Town	Zone	Zone Factor, Z	Town	Zone	Zone Factor, Z
Agra	III	0.16	Bhopal	II	0.1
Ahmedabad	III	0.16	Bhubaneswar	III	0.16
Ajmer	II	0.1	Bhuj	V	0.36
Allahabad	II	0.1	Bijapur	III	0.16
Almora	IV	0.24	Bikaner	III	0.16
Ambala	IV	0.24	Bokaro	III	0.16
Amritsar	IV	0.24	Bulandshahr	IV	0.24
Asansol	III	0.16	Burdwan	III	0.16
Aurangabad	II	0.1	Calicut	III	0.16
Bahraich	IV	0.24	Chandigarh	IV	0.24
Bengaluru	II	0.1	Chennai	III	0.16
Barauni	IV	0.24	Chitradurga	II	0.1
Bareilly	III	0.16	Coimbatore	III	0.16
Belgaum	III	0.16	Cuddalore	III	0.16
Bhatinda	III	0.16	Cuttack	III	0.16
Bhilai	II	0.1	Darbhanga	V	0.36

(*Contd*)

(*Contd*)

Darjeeling	IV	0.24	Nagpur	II	0.1
Dehradun	IV	0.24	Nainital	IV	0.24
Delhi	IV	0.24	Nasik	III	0.16
Dharampuri	III	0.16			
Dharwad	III	0.16	Nellore	III	0.16
Durgapur	III	0.16	Osmanabad	III	0.16
Gangtok	IV	0.24	Panjim	III	0.16
Gaya	III	0.16	Patiala	III	0.16
Goa	III	0.16	Patna	IV	0.24
Gorakhpur	IV	0.24	Pilibhit	IV	0.24
Gulbarga	II	0.1	Pondicherry	II	0.1
Guwahati	V	0.36	Pune	III	0.16
Hyderabad	II	0.1	Raipur	II	0.1
Imphal	V	0.36	Rajkot	III	0.16
Jabalpur	III	0.16	Ranchi	II	0.1
Jaipur	II	0.1	Roorkee	IV	0.24
Jamshedpur	II	0.1	Rourkela	II	0.1
Jhansi	II	0.1	Sadiya	V	0.36
Jodhpur	II	0.1	Salem	III	0.16
Jorhat	V	0.36	Shimla	IV	0.24
Kakrapara	III	0.16	Sironj	II	0.1
Kalpakkam	III	0.16	Solapur	III	0.16
Kanchipuram	III	0.16	Srinagar	V	0.36
Kanpur	III	0.16	Surat	III	0.16
Karwar	III	0.16	Tarapur	III	0.16
Kohima	V	0.36	Tezpur	V	0.36
Kolkata	III	0.16	Thane	III	0.16
Kota	II	0.1	Thanjavur	II	0.1
Kurnool	II	0.1	Thiruvananthapuram	III	0.16
Lucknow	III	0.16	Tiruchirappalli	II	0.1
Ludhiana	IV	0.24	Tiruvannamalai	III	0.16
Madurai	II	0.1	Udaipur	II	0.1
Mandi	V	0.36	Vadodara	III	0.16
Mangalore	III	0.16	Varanasi	III	0.16
Monghyr	IV	0.24	Vellore	III	0.16
Moradabad	IV	0.24	Vijayawada	III	0.16
Mumbai	III	0.16	Visakhapatnam	II	0.1
Mysore	II	0.1			
Nagarjunasagar	II	0.1			

IV. Definitions of Irregular Buildings—Plan Irregularities

Torsion Irregularity

It is to be considered when floor diaphragms are rigid in their own plane in relation to the vertical structural elements that resist the lateral forces. Torsional irregularity is considered to exist when the maximum storey drift, computed with design eccentricity, at one end of the structure transverse to an axis is more than 1.2 times the average of the storey drifts at the two ends of the structure.

Re-entrant Corners

Plan configurations of a structure and its lateral force-resisting system contains re-entrant corners, where both projections of the structure beyond the re-entrant corner are greater than 15 per cent of its plan dimension in the given direction.

Diaphragm Discontinuity

Diaphragms with abrupt discontinuities or variations in stiffness, including those having cut-out or open areas greater than 50 per cent of the gross enclosed diaphragm area, or changes in effective diaphragm stiffness of more than 50 per cent from one storey to the next.

Out-of-plane Offsets

Discontinuities in a lateral force-resistance path, such as out-of-plane offsets of vertical elements.

Non-parallel Systems

The vertical elements resisting the lateral force are not parallel to or symmetric about the major orthogonal axes or the lateral force-resisting elements.

Plan irregularities are shown in Fig. IV.1.

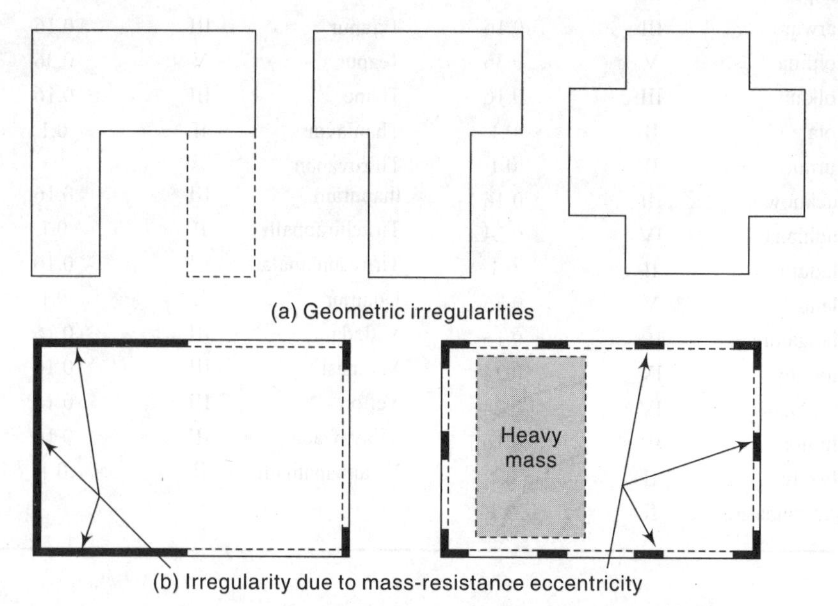

(a) Geometric irregularities

(b) Irregularity due to mass-resistance eccentricity

Fig. IV. 1 (*Contd*)

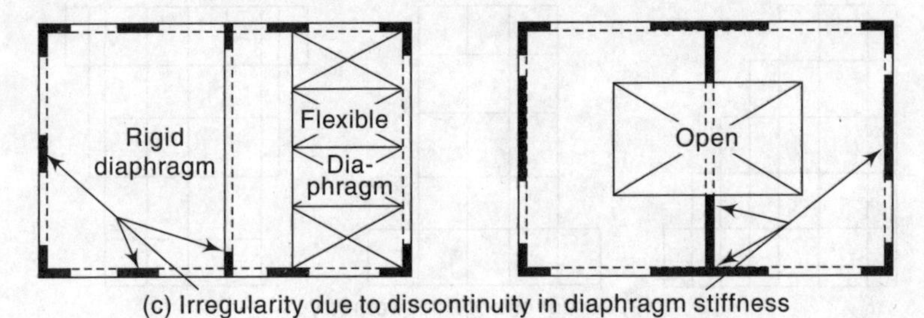

(c) Irregularity due to discontinuity in diaphragm stiffness

Fig. IV. 1 Plan irregularities—vertical components of seismic resisting system

V. Definitions of Irregular Buildings—Vertical Irregularities

Stiffness Irregularity

Soft Storey

A soft storey is one in which the lateral stiffness is less than 70 per cent of that in the storey above or less than 80 percent of the average lateral stiffness of the three storeys above.

Extreme Soft Storey

An extreme soft storey is one in which the lateral stiffness is less than 60 per cent of that in the storey above or less than 70 per cent of the average stiffness of the three storeys above. For example, buildings on stilts will fall under this category.

Mass Irregularity

Mass irregularity should be considered to exist where the seismic weight of any storey is more than 200 per cent of that of its adjacent storeys. The irregularity need not be considered in the case of roofs.

Vertical Geometric Irregularity

Vertical geometric irregularity should be considered to exist where the horizontal dimension of the lateral force-resisting system in any storey is more than 150 per cent of that in its adjacent storey.

In-plane Discontinuity in Vertical Elements Resisting Lateral Force

An in-plane offset of the lateral force-resisting elements greater than the length of those elements.

Discontinuity in Capacity—Weak Storey

A weak storey is one in which the storey lateral strength is less than 80 per cent of that in the storey above. The storey lateral strength is the total strength of all seismic force-resisting elements sharing the storey shear in the considered direction.

(Vertical irregularities are shown in Fig. V.1.)

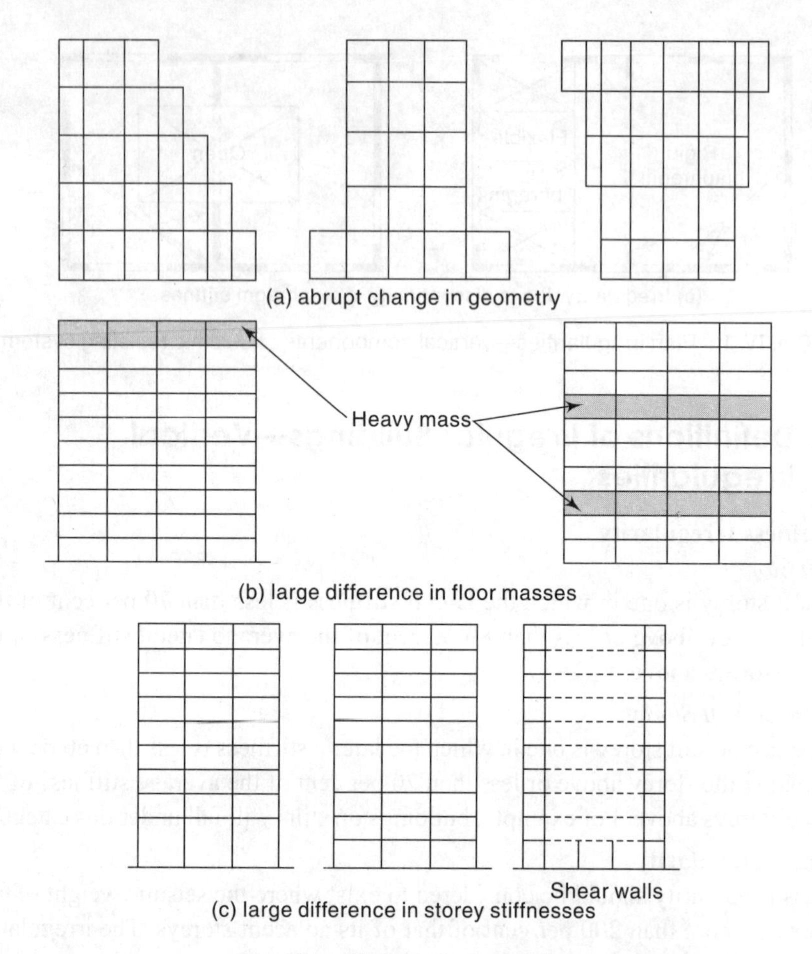

Fig. V.1 Elevation irregularities

VI. Determination of Natural Frequencies and Mode Shapes

The differential equation governing the free vibration of a linear n-DOF system is

$$\mathbf{m}\ddot{\mathbf{x}} + \mathbf{k}\mathbf{x} = \mathbf{0} \tag{1}$$

where, the mass matrix $\mathbf{m}$ and the stiffness matrix $\mathbf{k}$ are both symmetric ($n \times n$) and $\mathbf{x}$ is the n-dimensional column vector of generalized coordinates.

Since there is no damping, the free vibrations described by the solution of the above system of equations are periodic. Damping in real structures is very small and assumption of no damping is justified. Thus, when free vibrations at a single frequency are initiated for a particular system, the ratio of any two dependent variables is independent of time, leading to normal-mode solution of the form

$$x(t) = \phi e^{i\omega t} \tag{2}$$

where, ω is frequency of vibration and φ is n-dimensional vector of constants, called mode shapes.

Values of ω are natural frequencies; each natural frequency has at least one corresponding mode shape. Since the differential equations, represented by Eqn (1), are linear and homogeneous, their general solution is a linear superposition of all possible modes.

From Eqns (1) and (2)

$$(-\omega^2 \mathbf{m}\phi + \mathbf{k}\phi)e^{i\omega t} = 0 \tag{3}$$

For any real value of i, $e^{i\omega t} \neq 0$

Hence, $-\omega^2 \mathbf{m}\phi + \mathbf{m}\phi = 0$ (4)

$$[\mathbf{k} - \omega^2 \mathbf{m}]\{\phi\} = 0 \tag{5}$$

$$\left|\mathbf{k} - \omega^2 \mathbf{m}\right| = 0 \tag{6}$$

The expression for determinant yields algebraic equation of nth order. This equation is known as *characteristic equation*. The positive roots of this equation are known as *natural frequencies* ω_i or eigen values. Each eigen value has a distinct shape and is known as *natural mode shape* φ or eigen vector.

Note: Since the algebraic complexity of the solution grows exponentially with the number of DOFs, the only choice left for MDOF system is the use of numerical techniques.

VII. Horizontal Seismic Coefficient (α_o)

Zone	Horizontal seismic coefficient (α_o)
II	0.02
III	0.04
IV	0.05
V	0.08

VIII. Importance Factor (*I*)

Structure	Value of importance factor, I^1
Dams (all types)	3.0
Containers of inflammable or poisonous gases or liquids	2.0

Important service and community structures, such as hospitals; water towers and tanks; schools; important bridges; important power houses; monuments; emergency buildings such as telephone exchange and fire bridges; large assembly structures such as cinemas, assembly halls, and subway stations	1.5
All others	1.0

[1]The values of importance factor, I, given in this table are for guidance. A designer may choose suitable values depending on the importance of the structure based on economy, strategy, and other considerations.

IX. Soil-foundation Factor (β)

Type of soil mainly constituting the foundation	Value of soil-foundation factor (β) for			
	Bearing piles resting on soil type I or raft foundation	Friction piles, combined footing, RCC footing with the tie-beams	Isolated RCC footings without tie-beams or strip foundation	Well foundation
Type I rock or hard soils with $N > 30$	1.0	1.0	1.0	1.0
Type II medium soils with N between 10 and 30	1.0	1.0	1.2	1.2
Type II soft soils with $N < 10$	1.0	1.2	1.5	1.5

N is the standard penetration value according to IS 2131

X. Second-order Effects (*P*-Δ Effects)

Most structural systems under the action of seismic forces, because of their inelastic response, sustain large horizontal displacements resulting in the creation of large secondary effects. In evaluating overall structural frame stability, in general, it is necessary to consider the P-Δ effects. The moment induced by the P-Δ effects is a secondary effect and may be ignored when it is less than 10 per cent of the primary action of lateral loads. However, the effect need not be considered where the storey drift ratio does not exceed $0.02/R$, where R is the response reduction factor. Consider the frame shown in Fig. X.1. When this frame, for some external reason (an earthquake in this case), is displaced by Δ, each of the two $W/2$ column loads can be analysed into an axial force on the column with a value of $W/2$ and a horizontal one,

$$\Delta H_1 = \frac{\Delta}{h}\frac{W}{2}$$

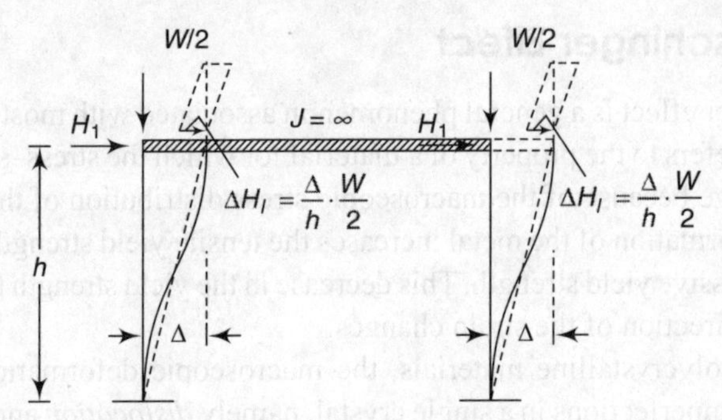

Fig. X.1 Deflected shape of a typical frame dipicting secondary effects

Thus the floor is loaded with an additional (second-order) horizontal force equal to

$$\Delta H = \Delta H_1 + \Delta H_1 = \frac{\Delta}{h} W \tag{X.1}$$

In the case of seismic action, the displacement Δ is equal to Δ_{el} which results from the seismic loading of the code, multiplied by a behaviour factor q of the structure

$$\Delta = \Delta_{el} q$$

Therefore, the additional shear force of the storey, because of the second-order effect, is equal to

$$\Delta V = \frac{\Delta_{el} q}{h} W \tag{X.2}$$

(a) For $\theta = \frac{\Delta V}{V} = \left(\frac{\Delta_{el} q}{h}\right)\left(\frac{W}{V}\right) \leq 0.10$ (X.3)

where θ is the ratio of ΔV and V.
A second-order analysis is not required.

(b) For $0.10 \leq \theta \leq 0.20$ (X.4)

the *P*-Δ effect must be taken into account. In this case an acceptable approximation could be to increase the relevant seismic action effects by a factor equal to $1/(1-\theta)$.

(c) For $0.20 \leq \theta$ (X.5)

the lateral stiffness of the system must be increased.

In the above equations, V is the shear force of the storey due to seismic action, Δ_{el} the relative lateral displacements of the top in relation to the bottom of the storey, also known as *inter-storey drift*, q the behaviour factor of the structure, h the storey height, and W is the total gravity load above the storey under consideration. A high degree of lateral stiffness must be provided for the structural system, so that at least second-order effects are prevented.

XI. Bauschinger Effect

Bauschinger effect is a general phenomenon associated with most polycrystalline metals. It refers to the property of a material for which the stress–strain characteristics change because of the macroscopic stress distribution of the material. The plastic deformation of the metal increases the tensile yield strength and decreases the compressive yield strength. This decrease in the yield strength takes place only when the direction of the strain changes.

In the polycrystalline materials, the macroscopic deformation is related to structural imperfections in a single crystal, namely *dislocation* and its movement. When one part of a crystal slides upon its adjacent part, it is known as *deformation.* As a result of partial slip, the crystal distortion, called dislocation, occurs. With continued external loading, propagation of slip occurs and the dislocation starts moving through the crystal in the direction of slip.

The mechanism for the Bauschinger effect is related to the dislocation of structure in the cold worked metals. When a crystal is subjected to an applied force, the dislocation will try to move in the direction of maximum shear stress to lower down the potential energy of the system. Hence, under the action of external loading, newly produced and existing dislocations move along the slip planes until they reach grain boundary. The defects (grain boundaries, solute atoms, etc.) in the crystals act as barriers and may fully stop or slow down the moving dislocations. As the deformation occurs, the dislocations accumulate at the barriers and produce dislocation pile-ups and tangles. The dislocations normally cannot cross the boundary and therefore pile-up occurs by the boundary. Two theories arc used to describe the Bauschinger effect, based on the cold work structure, and are as follows:

1. The local back stresses that may be present in the metal assist the movement of dislocations in the reverse direction, resulting in the lower yield strength.
2. When the strain direction is reversed, dislocations of opposite sign may be produced from the same source that produced the slip, causing dislocations in the initial direction. The opposite-signed dislocations move on the slip planes in opposite direction and may attract each other and annihilate. Since strain hardening is related to an increased *dislocation density*, a reduction in the number of dislocations results in reduced strength.

Bibliography

Abe, K. 1973, 'Tsunami and mechanism of great earthquakes', *Physics of the Earth Planet Interiors*, vol. 7, pp. 143-53.

Aboutaha, R.S., M.D. Engelhardt, J.O. Jirsa, and M.E. Kreger 1996, 'Retrofit of concrete column with inadequate lap splices by the use of rectangular steel jackets', *Earthquake Spectra*, vol. 12, no. 4.

Acharaya, H.K. 1979, 'Regional variations in the rupture-length magnitude relationships and their dynamical significance', *Bulletin of the Seismological Society of America*, vol. 69, pp. 2063-84.

ACI 318M-99/318RM-99, 'Building code requirements for structural concrete (318M-99) and commentary (318RM-99)', American Concrete Institute, Farmington Hills, Michigan.

Agarwal, P. 1999, 'Experimental study of seismic strengthening and retrofitting measures in masonry buildings', *PhD Thesis*, Department of Earthquake Engineering, University of Roorkee, Roorkee.

Agarwal, P. and M. Shrikhande 2006, *Earthquake Resistant Design of Structures*, Prentice-Hall, New Delhi.

Aki, K. 1988, 'Local site effects on strong ground motion', *Proceedings, Earthquake Engineering and Soil Dynamics II-Recent Advances in Ground Motion Evaluation*, ASCE, Geotechnical Special Publication No. 20, pp. 103-55.

Aki, K., B.H. Chin, and K. Kato 1992, 'Seismological and geotechnical studies of local V site effects on strong and weak motions', *Proceedings of the International Symposium on the Effects of Surface Geology on Seismic Motion*, ESG 1992, Odawara, Japan, Pages I: 97-110, IASPEI/IAEE Joint Working Group on ESG, Association for Earthquake Disaster Prevention, Tokyo, Japan.

Alcocer, S.M. 1992, 'Rehabilitation of RC frame connections using jacketing', *Tenth World Conference on Earthquake Engineering*, Madrid, Spain.

Alcocer, S.M. et al. 1996, 'Retrofitting of confined masonry walls with welded wire mesh', *Eleventh World Conference on Earthquake Engineering*, Acopulco, Mexico.

Ambraseys, N.N. 1988, 'Engineering seismology', *Earthquake Engineering and Structural Dynamics*, vol. 17, pp. 1-105.

American Concrete Institute 1971, Building code requirements for reinforced concrete (ACI 318-71), *ACI Standard*, pp. 318-71.

Amrhein, J.E. 1972, *Reinforced Masonry Engineering Handbook*, Masonry Institute of America, Los Angeles.

Anderson, J.C. 2001, 'Dynamic response of structures', In Farzad Naeim (Ed.), *The Seismic Design Handbook*, 2nd ed., Kluwer Academic Publisher, The Netherlands.

Arias, A. 1970, 'A measure of earthquake intensity', In R.J. Hansen (Ed.), *Seismic Design for Nuclear Power Plants*, MIT Press, Cambridge, Massachusetts, pp. 438-69.

Armstrong, I.E. 1972, 'Capacity design of reinforced concrete frames for ductile earthquake performance', *Bulletin of the New Zealand Society for Earthquake Engineering*, vol. 5, no. 4.

Arnold, C. 1991, 'The seismic response of nonstructural elements in building', *V Bulletin of the New Zealand National Society for Earthquake Engineering*, vol. 24, no. 4.

Arnold, C. 1998, 'Architectural aspects of seismic resistant design', *Eleventh* V *World Conference on Earthquake Engineering.*

Arnold, C. and E. Elsesser 1980, 'Building configuration: problems and solutions', *Seventh World Conference on Earthquake Engineering.*

Base, G.D. and J.B. Read 1965, 'Effectiveness of helical binding in the compression zone of concrete beams', *ACI Jill*, vol. 62, pp. 763-81.

Bath, M. 1966, 'Earthquake energy and magnitude', *Physics and Chemistry of the Earth*, Ahren, L.H. Press, pp. 115-65.

Benero, V.V., J.E. Anderson, H. Krawinkler, E. Miranda et al. 1991, 'Design guidelines or ductility and drift limits', *Report No. UCB/EERC-91/15*, Earthquake Engineering Research Centre, University of California.

Berg, G.V. 1989, *Elements of Structural Dynamics*, Prentice Hall, Englewood Cliffs, New Jersey, 268 pp.

Berry, O.R. 1972, 'Architectural seismic detailing', *State of the Art Report No. 3*. Technical Committee No. 12, Architectural-Structural Interaction IABSE-ASCE International Conference on Planning and Design of Tall Buildings, Lehigh University, 1972 (Conference Preprints, Reports vol. 2).

Bertero, V.V. 1986, 'Implication of recent earthquakes and research on earthquake resistant design and construction of buildings', *Report No. UCB/EERC-86/03*, Earthquake Engineering Research Centre, University of California.

Beskos, D.E. and S.A. Anagnostoulos 1997, 'Advances in earthquake engineering' *Computer Analysis and Design of Earthquake Resistant Structures: A Handbook*, Computational Mechanics Publications, Southampton, UK.

Blackwell, F.N. 1970, 'Earthquake protection for mechanical services', *New Zealand Engineering*, vol. 25, no. 10, pp. 271-75.

Blakeley, R.W.G. 1973, 'Prestressed concrete seismic design', *Bull. New Zealand Society for Earthquake Engineering*, vol. 6, no. 1.

Blakeley, R.W.G. and R. Park, 1971, 'Ductility of prestressed concrete members', *Bull. New Zealand Society for Earthquake Engineering*, vol. 4, no. 1, pp. 145-70.

Blakeley, R.W.G., and R. Park 1973, Prestressed concrete sections with cyclic flexure, *J. Struct. Div., Am. Soc. Civ. Eng.*, vol. 99, no. 8, pp. 1717-42.

Blakely, R.W.G., A.W. Charleson, H.C. Hitchcock, L.M. Megget, M.J.N. Priestley, R.D. Sharp, and R.I. Skinner 1979, Recommendation for the design and construction of base isolated structures, *Bull. N. Z. Nat. Soc. Earthquake Eng.*, vol. 12, no. 2, pp. 136-57.

Blume, J.A. 1970, 'The motion and damping of buildings relative to seismic response spectra', *Bull. Seismological Society of America*, vol. 60, no. 1, pp. 231-59.

Bolt, B.A. 1970, 'Causes of earthquakes', In R.L. Wiegel (Ed.), *Earthquake Engineering*, Chap. 2, Prentice-Hall, Englewood Cliffs, NJ.

Bolt, B.A. 1988, *Earthquakes*, W.H. Freeman and Company, New York, 282 pp.

Bolt, B.A. 1989, 'The nature of earthquake ground motion', In F. Naeim (Ed.), *The Seismic Design Handbook*, Van Nostrand Reinhold, New York.

Booth, E. 1994, 'Concrete Structures in Earthquake Regions', *Longman Scientific and Technical*, Longman Group UK Limited.

Bouwkamp, J.G. 1961, 'Behaviour of window panels under in-plane forces', *Bull. Seismological Society of America*, vol. 51, no. 1, pp. 85-103.

Bruce, A. Bolt 2004, *Earthquakes*, Fifth Edition, W.H. Freeman and Company, New York.

BSSC 1994A, *NEHRP Recommended Provisions for the Development of Seismic Regulations for New Buildings, Part I: Provisions*, Building Seismic Safety Council, Federal Emergency Management Agency, Washington, DC.

BSSC 1994B, *NEHRP Recommended Provisions for the Development of Seismic Regulations for Wind Buildings, Part II: Complimentary*, Building Seismic Safety Council, Federal Emergency Management Agency, Washington, DC.

Buckle, I.G. 2000, *Passive Control of Structures for Seismic Loads*, Twelfth World Conference on Earthquake Engineering, New Zealand.

Bullen, K.E. 1953, *An Introduction to the Theory of Seismology*, Cambridge University Press, London.

Bullen, K.E. and B.A. Bolt 1985, *An Introduction to the Theory of Seismology*, Cambridge University Press, Cambridge.

Burns, R.J. 1965, 'An approximate method of analysing coupled shear walls subject to triangular loading', *Proc. 3rd World Conference on Earthquake Engineering*, New Zealand, m, IV-123 to IV-14O.

Campbell, K.W. 1981, 'Near source attenuation of peak horizontal acceleration', *Bull. Seismological Society of America*, vol. 71, pp. 2039-70.

Cardenas, A.E., J.M. Hanson, W.G. Corley, and E. Hognestad 1973, 'Design provisions for shear walls', *ACI Journal*, vol. 70, no. 3, pp. 221-30.

Cardenas, A.E. and D.D. Magura 1973, 'Strength of high-rise shear walls-rectangular cross-sections, *Response of Multistory Concrete Structures to Lateral Forces*, ACI special Publication 36, pp. 119-150,.

Carter, D.P. and H.B. Seed 1988, 'Liquefaction potential of sand deposits under low levels of excitation', *Report UCB/EERC-88/11*, Earthquake Engineering Research Centre, University of California, Berkeley, 309 pp.

Casagrande, A. 1976, 'Liquefaction and cyclic mobility of sands: a critical review', *Harvard Soil Mechanics Series*, vol. 88, Harvard University, Cambridge, MA.

Castro, G. 1991, 'On the behaviour of soils during earthquakes-liquefaction', *Proceedings, NSF/EPRI Workshop on Dynamic Soil Properties and Site Characterization*, EPRI NP-7337, vol. 2, Electric Power Research Institute, Palo Alto, CA, pp. 1-36.

Castro, G. and S.J. Poijlos 1977, 'Factors affecting liquefaction and cyclic mobility', *Journal of the Geotechnical Engineering Division*, ASCE, vol. 106, no. GT6, pp. 501-06.

CEB 1996, *RC Frame Under Earthquake Loading-State of the Art Report*, Thomas Telfoford.

Chandra, B., S.K. Thakkar, S. Basu, A. Kumar, M. Shrikhande, J. Das, P. Agarwal, and M.K. Bansal 1993, 'Strong motion records', *Earthquake Spectra*, In R.W. Clough and J. Penzien (Eds.), *Dynamics of Structures*, 2nd ed., Supplement A to vol. 18, pp. 53-66, 2002.58, McGraw-Hill, New York.

Chopra, A.K. 1995, *Dynamics of Structures*, Prentice Hall, Englewood Cliffs, New Jersey, 729 pp.

Chopra, A.K. 2005, *Earthquake Dynamics of Structures: A Primer*, 2nd ed. ..Earthquake Engineering Research Institute, nicee, at IIT Kanpur, pp. 128.

Chopra, A.K. 2007, *Dynamics of Structures: Theory and Applications to Earthquake Engineering*. 3rd ed. A.K. Chopra, Pearson Education publishing, pp. 912

Chopra, A.K., D.P. Clough, and R.W. Clough 1973, 'Earthquake resistance of buildings with a 'soft' first storey', *Earthquake Engineering and Structural Dynamics*, vol. 1, no. 4, pp. 347-55.

Chowdhury, R.N. 1978. *Slope Analysis*, Elsevier, New York, 423 pp.

Christian, J.T., J.M. Roes set, and C.S. Desai 1977, 'Two- and three-dimensional dynamic analyses', In C.S. Desai and J.T. Christian (Eds.), *Numerical Methods in Geotechnical Engineering*, McGraw Hill Book Company, New York, pp. 683-718.

Clough, R.W. 1970, 'Earthquake response of structures', In R.L. Wiegel (Ed.), *Earthquake Engineering*, Chap. 12, Prentice Hall, Englewood Cliffs, New Jersey, pp. 307-34.

Clough, R.W. and J. Penzien, 1975, *Dynamics of Structures*, McGraw-Hill, New York, 634 pp.

Colaco, J.P. 1971, 'Preliminary design of shear walls for tall buildings', *ACI Journal*.

Corley, W.G. and J.M. Hanson 1973, 'Design of earthquake-resistant structural walls', *Proc. 5th World Conference on Earthquake Engineering, Rome*, vol. 1, pp. 933-6.

Craig, R.R. Jr. 1990, *Structural Dynamics*, John Wiley, New York, 1981.

Dowrick, D.J. 1970, In V.P. Drnevich and F.E. Richart Jr. (Eds.), *Earthquake Resistance Design*, A Wiley Interscience Publication, John Wiley.

DeMets et al. 2002-03, 'Ductile detailing of reinforced concrete structures subjected to seismic forces', *Current Plate Motions*, vol. 101, Edition 1.2, Bureau of Indian Standards, New Delhi, pp. 425-78, 1990.13920.

DEQ 1988, *Damage Survey Report on Bihar-Nepal Earthquake of August 21, 1988*, Department of Earthquake Engineering, University of Roorkee, Roorkee.

DEQ 2000, *A Report on Chamoli Earthquake of March 29, 1999*, Department of Earthquake Engineering, University of Roorkee, Roorkee.

DEQ 2000, *Jabalpur Earthquake of May 22, 1997: Reconnaissance Report*, Department of Earthquake Engineering, University of Roorkee, Roorkee.

Desai, C.S. and J.F. Abel 1972, *Introduction to the Finite Element Method*, Van Nostrand Reinhold, New York.

Dowrick, D.J., 'Modern construction techniques for earthquake areas', *Earthquake Engineering, Proc. 4th European Symposium on Earthquake Engineering, London, 1972*, Bulgarian National Committee on Earthquake Engineering, Sofia, pp. 287-300.

Drydale, R.G., A.A. Hamid, and L.R. Baker, 1994, *Masonry Structures-Behaviour and Design*, Prentice Hall, Englewood Cliffs, New Jersey.

Dubey, R.N., S.K. Thakkar, and P. Agarwal, 2002, 'Performance of masonry building during Bhuj earthquake', *12th Symposium on Earthquake Engineering*, IIT Roorkee.

EERI 2002, 'Bhuj, India Earthquake of January 26, 2001: Reconnaissance Report', *Earthquake Spectra*, Supplement to vol. 18.

Emilio, R., *Design of Earthquake Resistant Structures*, Pentech Press, London.

Endo, T. et al. 1984, 'Practices of Seismic Retrofit of Existing Concrete Structures in Japan', *Eighth World Conference on Earthquake Engineering*, San Francisco.

FEMA-306, 'Evaluation of earthquake damaged concrete and masonry wall buildings', *ATC-43 Project*, Applied Technology Council, California.

FEMA 172 1992, *NEHRP Handbook for Seismic Rehabilitation of Existing Buildings*, Building Seismic Safety Council, Washington.

Finn, W.D.L., P.L. Bransby, and D.J. Pickering 1970, 'Effect of strain history on liquefaction of sands', *Journal of the Soil Mechanics and Foundations Division*, ASCE, vol. 96, no. SM6, pp. 1917-34.

Finn, W.D.L., R.H. Ledbetier, and G. Wu 1994, 'Liquefaction in silty soils: Design and analysis', *Ground Failures under Seismic Conditions*, Geotechnical Special Publication 44, ASCE, New York, pp. 51-76.

Fintel, M., *Handbook of Concrete Engineering*, *Multi-storey Structures* (ch. 10) by M. Fintel, *Earthquake Resistant Structures* (ch. 12), by Aranaldo T. Derecho and Mark Fintel, Van Nostrand Reinhold Company.

Florin, V.A. and P.L. Ivanov 1961, 'Liquefaction of saturated sand soil', *Proceedings*, *5th International Conference on Soil Mechanics and Foundation Engineering*, Paris.

Geli, L., P.Y. Bard, and B. Jullien 1988, 'The effect of topography on earthquake ground motion: A review and new results', *Bulletin of the Seismological Society of America*, vol. 78, pp. 42-63.

Gioncu, V. and P.M. Mazzolani 2002, *Ductility of Seismic Resistant Steel Structures*, Spon Press, New York.

Goel, R.K., *Performance of Buildings during the January 26, 2001 Bhuj Earthquake*, Earthquake Engineering Research Institute, California.

Gould, P.L 1965, 'Interaction of shear wall-frame system in multistorey buildings', *Journal of ACI*, vol. 62, no. 1, pp. 45-70.

Goyal, A., R. Sinha, M. Chaudhari, and K. Jaiswal 2004, 'Performance of Reinforced Concrete Buildings in Ahmedabad during Bhuj Earthquake January 26, 2001, *Workshop on Recent Earthquakes of Chamoli and Bhuj,* vol. I, Roorkee, India.

Grimn, C.T., 'Masonry cracks: a review of the literature', *Masonry: Materials, Design, Construction, and Maintenance*, pp. 257-80.

GSI 1992, *Uttarkashi Earthquake, October 20, 1991*, Geological Survey of India, Special Publication No. 30.

GSI 1995, *Uttarkashi Earthquake*, Geological Survey of India.

Gutenberg, B. 1945, 'Magnitude determination for deep-focus earthquakes', *Bulletin of the Seismological Society of America*, vol. 35, pp. 117-130.

Gutenberg, B. 1956, 'Magnitude determination for deep focus earthquakes', *Bulletin of the Seismological Society of America*, vol. 35, pp. 117-30.

Gutenberg, B. and C.F. Richter 1936, 'On seismic waves (third paper)', *Gerlands Bietraege zur Geophysik*, vol. 47, pp. 73-131.

Gutenberg, B. and C.F. Richter 1942, 'Earthquake magnitude, intensity, energy, and acceleration', *Bulletin of the Seismological Society of America*, vol. 32, pp. 163-91.

Gutenberg, B. and C.F. Richter 1945, *Seismicity of Earth and Related Phenomenon*, Princeton University Press, Princeton, New Jersey.

Gutenberg, B. and C.F. Richter 1954, *Seismicity of the Earth and Related Phenomena*, Princeton University Press, Princeton, New Jersey, p. 310.

Gutenberg, B. and C.F. Richter 1956, 'Earthquake magnitude: intensity, energy, and acceleration', *Bulletin of the Seismological Society of America*, vol. 46, pp. 104-45.

Gutenberg, B., and C.F. Richter 1965, *Seismicity of the Earth*, Hafner, New York.

Guevara, L.T. and L.E. Garcia 2005, 'The captive and short column effect earthquake', *Spectra*, vol. 21, no. 1, pp. 141-60.

Hank, T.C. and H. Kanamori, 1979, 'A moment magnitude scale', *JGR*, vol. 84, pp. 2348-50.

Hart, G.C. and K. Wong 2001, *Structural Dynamics for Structural Engineers*, John Wiley, New York.

Hatanaka, M. 1952, 'Three-dimensional consideration on the vibration of earth dams', *Journal of the Japanese Society of Civil Engineers*, vol. 37, no. 10.

Hou, S. 1968, 'Earthquake simulation models and their applications', *Report R68-17*, Department of Civil Engineering, Massachusetts Institute of Technology, Cambridge, Massachusetts.

Housner, G.W. 1947, 'Characteristics of strong motion earthquakes', *Bulletin of the Seismological Society of America*, vol. 37, no. 1, pp. 19-31.

Housner, G.W. 1952, 'Spectrum intensities of strong motion earthquakes', *Proceedings of the Symposium of Earthquake and Blast Effects on Structures*, Earthquake Engineering Research Institute, Los Angeles, California, pp. 21-36.

Housner, G.W. 1959, 'Behaviour of structures during earthquakes', *Journal of the Engineering Mechanics Division*, ASCE, vol. 85, no. EMI4, pp. 109-29.

Housner, G.W. 1970, 'Design spectrum', In R.L. Wiegel (Ed.), *Earthquake Engineering*, Chap. 4, Prentice-Hall, Englewood Cliffs, New Jersey pp. 73-106.

Housner, G.W. 1973, 'Important features of earthquake ground motion', *Proc. 5th World Conference on Earthquake Engineering, Rome*, 1, CLIX-CLXVIII.

Housner, G.W. 1975, 'Measures of severity of earthquake ground shaking', *Procedings of the US National Conference on Earthquake Engineering*, Earthquake Engineering Research Institute, Ann Arbor, Michigan, pp. 25-33.

Hudson, D.E. 1956, 'Response spectrum techniques in engineering seismology', *Proceedings of the First World Conference on Earthquake Engineering*, vol. 4 Earthquake Engineering Research Institute, Los Angeles, California, pp. 1-12.

Humar, J.L. 1990, *Dynamics of Structures*, Prentice Hall.
Hurty, W.C. and M.P. Rubinstein 1967, *Dynamics of Structures*, Prentice-Hall, New Delhi.
IAEE 2001, *Guidelines for Earthquake Resistant Non-engineered Construction*, ACC Limited, Thane.
Idriss, I.M. 1991, 'Earthquake ground motions at soft soil sites', *Proceedings, 2nd International Conference on Recent Advances in Geotechnical Earthquake Engineering and Soil Dynamics*, vol. III, pp. 2265-71.
IS-456 2000, *Plain and Reinforced Concrete-Code of Practice*, Bureau of Indian Standards, New Delhi.
IS-1893 2002, *Indian Standard Criteria for Earthquake Resistant Design of Structures*, Part 1, BIS, New Delhi.
IS-1905 1985, *Code of Practice for Structural Use of Unreinforced Masonry*, Bureau of Indian Standards, New Delhi.
IS-3935 1993, *Repair and Seismic Strengthening of Buildings-Guidelines*, Bureau of Indian Standards, New Delhi.
IS- 4326 1993, *Earthquake Resistant Design and Construction of Buildings Code of Practice*, Bureau of Indian Standards, New Delhi.
IS-13827 1993, *Indian Standard Guidelines for Improving Earthquake Resistance of Earthen Buildings*, BIS, New Delhi.
IS-13828 1993, *Indian Standard Guidelines for Improving Earthquake Resistance of Low Strength Masonry Buildings*, New Delhi.
IS-13920 1993, *Ductile Detailing of Reinforced Concrete Structures Subjected to Seismic Forces-Code of Practice*, Bureau of Indian Standards, New Delhi.
IS-13935 1993, *Ductile Detailing of Reinforced Concrete Structures Subjected to Seismic Forces*, Bureau of Indian Standards, New Delhi.
ISET 1994, 'Damage Report of the Latur-Osmanabad Earthquake of September 30, 1993, *Bulletin of Indian Society of Earthquake Technology*, vol. 31, no. 1.
Ishihara, K. 1993, Liquefaction and flow failure during earthquakes, *Geotechnique*, vol. 43, no. 3, pp. 351-415.
Ishihara, K. and Y. Oshimine 1992, 'Evaluation of settlements in sand deposits following liquefaction during earthquakes', *Soils and Foundations*, vol. 32, no. I, pp. 173-88.
Joyner, W.B. and D.M. Boore 1982, 'Prediction of earthquake response spectra', *Proceedings, 51st Annual Convention of the Structural Engineers of California*, also *USGS Open-File Report 82*977, 16 pp.
Joyner, W.B. and D.M. Boore 1993, 'Method for regression analysis of strong motion data', *Bulletin of the Seismological Society of America*, vol. 83, pp. 469-87 (Errata in 1994).
Kadir, M.R.A. 1974, 'The structural behaviour of masonry infill panel in framed structures', *Ph.D. Thesis*, University of Edinburgh.
Kahn, L.W. 1984, 'Shotcrete retrofit for unreinforced brick masonry', *Eighth World Conference on Earthquake Engineering*, vol. 1, San Francisco.
Kanamori, H. 1972, 'Mechanism of Tsunami earthquake', *Physics of the Earth Planet, Interiors*, vol. 6, pp. 246-59.
Kanamori, H. 1977, 'The energy release in great earthquakes', *Tectonophysics*, vol. 93, pp. 185-199.
Kanamori, H. 1983, 'Magnitude scale and quantification of earthquakes', *Tectonophysics*, vol. 93, pp. 185-99.
Karbhari, V.M. 2001, 'Use of FRP composite materials in the renewal of civil infrastructure in seismic region', *Second MCEER Workshop on Mitigation of Earthquake Disaster by Advanced Technologies (MEDAT-2), Technical Report MCEER-Ol-0002.*
Kato, B. 1974, 'A design criteria of beam-to-column joint panels', *Bull. New Zealand National Society for Earthquake Engineering*, vol. 7, no. 1, pp. 14-26.

Katsumata, H. and Y. Kobatake, 1996, 'Seismic retrofit with carbon fibres for reinforced concrete columns', *Eleventh World Conference on Earthquake Engineering*, Paper No. 293.

Kawamura, S., R. Sugisaki, K. Ogura, S. Maezawa, S. Tanaka, and A. Yajima 2000, 'Seismic isolation retrofit in Japan', *Twelfth World Conference on Earthquake Engineering*, Paper No. 2523.

Kausel, E., R.V. Whitman, J.P. Morray, and E. Elsabee 1978, 'The spring method for embedded foundations', *Nuclear Engineering and Design*, vol. 48, pp. 377-392.

Kearey, P. and F.J. Vine 1990, *Global Tectonics*, Blackwell, Oxford, 302 pp.

Key David, *Earthquake Design Practice for Buildings*, Thomas Telford, Lone.

Khan, F.R. and J.A. Sbarounis 1964, 'Interaction of shear wall and frames',*Proceedings of ASCE,* vol. 90 (St3), pp. 285-335.

Khouri, N.Q. 1984, 'Dynamic properties of soils', *Masters Thesis*,Department of Civil Engineering, Syracuse University, Syracuse, New York.

Kramar, S.L. 1996, *Geotechnical Earthquake Engineering*, Prentice Hall, New Jersey.

Krawinkler, H. and B. Alavi 1998, 'Development of improved design procedures for near 'Fault Ground Motions', *SMIP98, Seminar on Utilization of Strong Motion Data*, Oakland, California.

Krawinkler, H., and E.P. Popov 1973, 'Hysteretic behaviour of reinforced concrete rectangular and T-beams', *Proc. 5th World Conference on Earthquake Engineering, Rome,* vol. 1, pp. 249-258.

Krishna, J. 1958, 'Earthquake engineering problems in India', *Journal of Institution of Engineers*, India.

Krishna, J. 1959, 'Seismic zoning of India', In H.L. Sally (Ed.), *Earthquake Engineering Seminar*, University of Roorkee, India, pp. 24-31.

Lai, S.P. 1982, 'Statistical characterization of strong motions using power spectral density function', *Bulletin of the Seismological Society of America*, vol. 72, no. 1, pp 259-74.

Lay, T. and T.C. Wallace 1995, *Modern Global Seismology*, Academic Press, San Diego, 521 pp.

Lee, K.L. and A. Albaisa 1974, 'Earthquake induced settlements in saturated sands, *Journal of the Soil Mechanics and Foundations Division*, ASCE, vol. 100, no. GT4.

Liauw, T.C. 1972, 'An approximate method of analysis for infilled frame with or without openings', *Building Science*, vol. 7, Pergamon Press, pp. 223-38.

Lomnitz-Adler and C. Lomnitz 1979, 'A modified form of the Gutenberg-Richter magnitude-frequency law', *Bulletin of the Seismological Society of America*, vol. 63, pp. 1999-2003.

Lynn. A.C., J.P. Moehle, S.A. Mahin, and W.T. Holmes 1996, 'Seismic evaluation existing reinforced concrete building columns', *Earthquake Spectra*, vol. 12, no. 4.

Luco, J.E. and H.L. Wong 1986, 'Response of a rigid foundation to a spatially random ground motion, *Earthquake Engineering and Structural Dynamics*, vol. 14, no. 6, pp. 891-908.

Machida, A., J. Moehle, P. Pinto, and N. Matsumoto 1999, 'Ductility consideration for single element and for frame structures', In T. Tanabe (Ed.), *Proceedings of Comparative Performance of Seismic Design Codes for Concrete Structures*, vol. 1, Elsevier Science.

Macleod, I.A. 1990, *Analytical Modelling of Structural Systems*, Ellis Horwood, England.

Madhekar, M.S. and S.K. Jain 1993, 'Seismic behaviour design and detailing of RC shear walls, Part II: design and detailing'. *Indian Concrete Journal*, vol. 67, no. 9, pp. 451-457.

Mallick, D.V. and R.J. Severn 1968, 'Dynamic characteristics of infilled frames', *Proc. Institution of Civil Engineers*, vol. 39, pp. 261-87.

Martin, G.R., W.D.L. Finn, and H.B. Seed 1975, 'Fundamentals of liquefaction under cyclic loading', *Journal of the Geotechnical Engineering Division*, ASCE, vol. WI, no. GT5, pp. 423-38.

Matthiesen, J. 1982, 'Recommendations concerning seismic design of zonation', *Critical-Aspects of Earthquake Ground Motion and Building Damage Potential*, A TC 10-1, Applied Technology Council, Redwood City, California, pp. 213-46.

Mcguire, R.K. 1977, 'Seismic design spectra and mapping procedures using hazard analysis based directly on oscillator response', *Journal of Earthquake Engineering and Structural Dynamics*, vol. 5, pp. 211-34.

Mcguire, R.K. 1978, 'Seismic ground motion parameter relations', *Journal of the Geotechnical Engineering Division*, ASCE, vol. 104, no. GT4, pp. 481-90.

Medhekar, M.S. and S.K. Jain 1993, 'Seismic behaviour design and detailing of RC shear walls, Part 1: Behaviour and strength', *Indian Concrete Journal*, vol. 67, no. 7, pp. 311-18.

Medvedev, S.V. and V. Sponheur 1969, 'Scale of seismic intensity', *Proceedings, 4th World Conference on Earthquake Engineering*, Santiago, Chile, pp. 143-53.

Megget, L.M. 1974, 'Cyclic behaviour of exterior of reinforced beam-column joints', *Bull, New Zealand National Society for Earthquake Engineering*, vol. 7, no. 1, pp. 27-47.

Meirovitch, L. 1980, *Computation Methods in Structural Dynamics*, Sijthoff and Noordhoff, Alphen aan den Rijn, The Netherlands.

Meli, R. 1973, 'Behaviour of masonry walls under lateral loads', *Proc. 5th World Conference on Earthquake Engineering, Rome*, vol. 1, pp. 853-62.

Michael, A. Cassaro, and Enrique Martinez Romero', The Mexico Earthquakes- 1985', American Society of Civil Engineers, New York.

Mitchell, J.K. and D.-J. Tseng, 1990, 'Assessment of liquefaction potential by cone penetration resistance', In J.M. Duncan (Ed.), *Proceedings H. Bolton Seed Memorial Symposium*, Berkeley, California, vol. 2, pp. 335-50.

Miranda, E. and V.V. Bertero 1994, 'Evaluation of strength reduction factors for earthquake-resistant design', *Earthquake Spectra*, vol. 10, no. 2, pp. 357-79.

Miyamoto, H.K. and R.E. Scholl 1996, 'Seimic rehabilitation of a non-ductile soft storey concrete structure using viscous damper', *Eleventh World Conference on Earthquake Engineering*, Acapulco, Mexico.

Modena, C. 1994, 'Repair and upgrading techniques of unreinforced masonry structures utilized after fruli and compania/basilicata earthquakes', *Earthquake Spectra*, vol. 10, no. 1.

Mogam, I.T. and K. Kubu 1953, The behaviour of soil during vibration, *Proceedings, 3rd International Conference on Soil Mechanics and Foundation Engineering*, Zurich, vol. I, pp. 152-55.

Mohraz, B. and F.E. Elghadamsi 1989, 'Earthquake ground motion and response spectra', In F. Naeim (Ed.), *The Seismic Design Handbook*, Van Nostrand Reinhold, New York, pp. 32-80.

Moehle, J.P. and S.A. Mahin 1991, 'Observation of the behaviour of reinforced concrete buildings during earthquake', SP-I27, In S.K. Ghosh (Ed.), *Earthquake Resistant Concrete Structures Inelastic Response and Design*, American Concrete Institute Publication.

Murphy, J.R and L.J. O'Brien 1977, 'The correlation of peak ground acceleration amplitude with seismic intensity and other physical parameters', *Bull. of the Seismological Society of America*, vol. 67, no. 3, pp. 877-915.

Murty, C.V.R. 2005, 'IITK-BMTPC Earthquake Tips: Learning Earthquake Design and Construction', *National Information Centre of Earthquake Engineering*, IIT Kanpur, India.

Mylonakis, G. and C. Syngros 2002, 'Discussion of response spectrum of incompatible acceleration, velocity and displacement histories', *Earthquake Engineering and: Structural Dynamics*, vol. 31, pp. 1025-31.

Naelm, F. and M. Lew 1995, 'On the use of design-spectrum compatible time histories', *Earthquake Spectra*, vol. II, no. I, pp. 111-27.

Narayan, J.P. and D.C. Rai 2001, 'An observational study of local site effects in the Chamoli earthquake', *Proceedings of Workshop on Recent Earthquakes of Chamoli and Bhuj*, pp. 273-9.

Narayan, J.P., M.L. Sharma, and A. Kumar 2002, 'A seismological report on the January 26, 2001 Bhuj, India Earthquake', *Seismological Research Letters*, vol. 73, pp. 343-55.

Nateghi, F. and B. Shahbazian 1992, 'Seismic evaluation, upgrading and retrofitting structures: recent experiences in Iran', *Tenth World Conference on Earthquake Engineering*, Madrid, Spain.

National Research Council 1985, *Liquefaction of Soils During Earthquakes*, National Academy Press, Washington, DC, 240 pp.

Nau, J.M. and W.J. Hall 1984, 'Scaling methods for earthquake response spectra', *Journal of Structural Engineering Division*, ASCE, vol. 110, no. 7, pp. 1533-48.

NEHRP 1997, 'Recommended provisions for seismic regulation for new buildings and other structures', *Technical Report*, Building Safety Council for Federal Emergency: Management Agency, Washington, DC.

Newman, A. 2001, *Structural Renovation of Buildings-Methods, Details and Design Example,* McGraw-Hill, USA.

Newmark, N.M. 1959, 'A method of computation for structural dynamics', *Journal of Engineering Mechanics Division*, ASCE, vol. 85, pp. 67-94.

Newmark, N.M. 1970, 'Current trends in the seismic analysis and design of high-rise structures', In R.L. Wiegel (Ed.), *Earthquake Engineering*, Chap. 16, Prentice-Hall, pp. 40-24.

Newmark, N.M. 1973, *A Study of Vertical and Horizontal Earthquake Spectra*, N.M. Newmark Consulting Engineering Services, Directorate of Licensing, U.S. Atomic Energy Commission, Washington, DC.

Newmark, N.M. and W.J. Hall 1973, 'Procedures and criteria for earthquake-resistant design', *Building Practices for Disaster Mitigation*, Building Science Series 46, US Department of Commerce, Washington, DC, pp. 209-36.

Newmark, N.M. and W.J. Hall 1978, 'Development of criteria for earthquake resistant design', *Report NUREG/CR-0098*, Nuclear Regulatory Commission, Washington, DC, 49 pp.

Newmark, N.M. and W.J. Hall 1982, 'Earthquake spectra and design', *Technical Report*, Earthquake Engineering Research Institute, Berkeley, California.

Newmark, N.M. and E. Rosenblueth 1971, *Fundamentals of Earthquake Engineering*, Prentice Hall, New Jersey.

New Zealand Timber Research and Development Association 1973, 'Plywood design for seismic areas', *T.R.A.D.A.*, *Timber and Wood Products Manual*, Section-I, 12 pp.

NZS 4230 1990, 'Code of practice for the design of masonry structures (Part 1)', Standard Association of New Zealand, Willington.

Oldham, R.D. 1906, 'The constitution of the interior of the earth, as Revealed by earthquakes', *Quarterly Journal of Geological Society of London*, vol. 62, pp. 456-75.

Osawa, Y., T. Morishita, and M. Murakam, 'On the damage to window glass in reinforced concrete buildings during the earthquake of April 20, 1965', *Bull. Earthquake Research Institute*, University of Tokyo, vol. 43, pp. 819-27.

Otani, S. 2004, 'Earthquake resistant design of reinforced concrete buildings-past and future', *Journal of Advanced Concrete Technology*, vol. 2, no. 1, Japan Concrete Institute, pp. 3-24.

Ovikova, E.I. and M.D. Trifunac 1994, 'Duration of strong ground motion in terms of earthquake magnitude, epicentral distance, site conditions, and site geometry', *Earthquake Engineering and Structural Dynamics*, vol. 23, pp. 1023-43.

Ovikova, E.I. and M.D. Trifunac 1993, 'Modified Mercalli intensity scaling of the frequency dependent duration of strong ground motion', *Soil Dynamics and Earthquake Engineering*, vol. 12, pp. 309-22.

Page, A.W. 1982, 'Concentrated loads on solid masonry walls-a parametric study and design recommendations', *Proceedings of Institution of Civil Engineers*, Part 2, vol. 85, pp. 271-89.

Pais, A. and E. Kausel 1998, 'Approximate formulas for dynamic stiffnesses of rigid foundations', *Soil Dynamics and Earthquake Engineering*, vol. 7, pp. 213-26.

Park, R., and T. Paulay 1973, 'Behaviour of reinforced concrete external beam-column joints under cyclic load', *Proc. 5th World Conference on Earthquake Engineering, Rome*, vol. 1, pp. 772-81.

Paulay, T. 1972, 'Some aspects of shear wall design', *Bull. New Zealand Society for Earthquake Engineering*, vol. 5, no. 3, pp. 89-105.

Penelis, G.G. and A.J. Kappos 1997, 'Earthquake-resistant concrete structures', *E & FN SPON*, An imprint of Chapman & Hall.

Penzien, J. and M. Watabe 1975, 'Characteristics of 3-dimensional earthquake ground motions', *Earthquake Engineering and Structural Dynamics*, vol. 3, pp. 365-73.

Pfrang, E.O., C.P. Siess,. and M.A. Sozen 1964, 'Load-moment-curvature characteristics of R.C. cross-sections', *A.C.I. J.*, vol. 61, pp. 763-78.

Popov, E.P and R.B. Pinkney 1969, 'Cyclic yield reversal in steel building connections', *J. Structural Division*, ASCE, vol. 95, no. ST3, pp. 327-53.

Popov, E.P. and R.M. Stephen 1972, 'Cyclic loading of full size steel connections', *American Iron Alloy Steel Institute, Steel Research for Construction Bulletin No.* 21.

Popov, E.P. 1973, 'Experiments with steel members and their connections under repeated loads', *Preliminary Report of Title Symposium on Resistance and Ultimate Deformability of Structures Acted on by Well Defined Repeated Loads*, IABSE, Lisbon, pp. 125-35.

Poston, W.R. 1997, 'Structural concrete repair: general principles and a case study', In Edward G. Nawy (Ed.), *Concrete Construction Engineering Handbook*, Chap. 19, CRC Press, New York.

Poulos, S.J., G. Castro, and I.W. France 1985, 'Liquefaction evaluation procedure', *Journal of Geotechnical Engineering*, ASCE, vol. 3, no. 6, pp. 772-92.

Raven, E. and O.A. Lopez 1996, 'Regular and irregular plan shape buildings in seismic regions: approaching to an integral evaluation', *Eleventh World Conference on Earthquake Engineering.*

Reid, H.E 1911, 'The elastic rebound theory of earthquakes', *Bulletin of Department of Geology*, University of Berkeley, vol. 6, pp. 413-44.

Richter, C.E. 1935, 'An instrumental earthquake magnitude scale', *Bulletin of the Seismological Society of America*, vol. 25, pp. 1-32.

Richter, C.F. 1958, *Elementary Seismology*, Freeman, San Francisco.

Richart, F.E., J.R. Hall, and R.O. Woods 1970, *Vibrations of Soils and Foundations*, Prentice-Hall, Englewood Cliffs, New Jersey.

Riddell, R. and J.E.D.L. Llera 1996, 'Seismic analysis and design: current practice and future trends', *Eleventh World Conference on Earthquake Engineering*, Mexico.

Robert L. Wiegel, *Earthquake Engineering*, Prentice-Hall, Englewood Cliffs, NJ.

Robinson, W.H. 1996, 'Latest advances in seismic isolation', *Eleventh World Conference on Earthquake Engineering*, Acapulco, Mexico.

Rodriguez, M. and R. Park 1991, 'Repair and strengthening of reinforced concrete building for seismic resistance', *Earthquake Spectra*, vol. 7, no. 3.

Salse, E.A.B. and M. Fintel 1973, 'Strength, stiffness and ductility properties of slender shear walls', *Proc. 51th World Conference on Earthquake Engineering*, Rome, vol. 1, pp. 919-28.

Satake, K. 2002, In Lee et al. (Eds.), *Tsunamis*, *International Handbook of Earthquake and Engineering Seismology-Part B*, pp. 437-51.

Savarensky, Y.F. and D.P. Klrnos 1955, *Elements of Seismology and Seismometry*, State Press of Technical-Theoretical Literature, Moscow, 543 pp.

Schneider, R.R. and W.L. Dickey 1994, *Reinforced Masonry Design*, 3rd ed., Prentice Hall, NJ.

Schwegler, G. and P. Kelterbom 1996, 'Earthquake resistance of masonry structures strengthen with fibre composites', *Eleventh World Conference on Earthquake Engineering*, Acopulco, Mexico.

Seed, H.B. and I.M. Idriss 1982, *Ground Motions and Soil Liquefaction During Earthquakes*, Earthquake Engineering Research Institute, Berkeley, California, 134 pp.

Seed, H.B., I.M. Idriss, and F.W. Kiefer 1969, 'Characteristics of rock motions during earthquakes', *J. Soil Mechanics and Foundation Division, ASCE* 95, no. SM5, pp. 1199-218.

Seed, H.B. and K.L. Lee 1965, 'Studies of liquefaction of sands under cyclic loading conditions', Report TE-65-65, Department of Civil Engineering, University of California, Berkeley.

Seed, H.B. and M.L. Silver 1972, 'Settlement of dry sands during earthquakes', *Journal of the Soil Mechanics and Foundations Division*, vol. 98, no. SM4, pp. 381-97.

Seed, H.B., C. Ugas, and J. Lysmer 1976, 'Site dependent spectra for earthquake-resistant design', *Bulletin of the Seismological Society of America*, vol. 66, pp. 221-43.

Shrikhande, M., J.D. Das, M.K. Bansal, A. Kumar, S. Basu, and B. Chandra 1998, 'Analysis of strong motion records from Dharmsala earthquake of April 26, 1986', *Proceedings of the Eleventh Symposium on Earthquake Engineering*, Department of Earthquake Engineering, University of Roorkee, India, pp. 281-85.

Shrikhande, M., J.D. Das, M.K. Bansal, A. Kumar, S. Basu, and B. Chandra 2001, 'Strong motion characteristics of Uttarkashi earthquake of October 20, 1991 and its engineering significance', In O.P. Varma (Ed.), *Research Highlights in Earth System-Science: Seismicity*', vol. 2, Indian Geological Congress, Roorkee, India, pp. 337-42.

Shepherd, R. 1967, 'Determination of seismic design loads in a framed structure', *New Zealand Engineering*, vol. 22, no. 2, pp. 56-61.

Silva, W.J. 1988, *Soil Response to Earthquake Ground Motion*, EPRl Report NP-5747, Electric Power Research Institute, Palo Alto, California.

Singhal, A. 1971, 'Elastic earthquake resistance of multi-storey buildings', *The Structural Engineer*, vol. 49, no. 9, pp. 397-412.

Singhal, A., P.R. Bose, A. Bose, and V. Prakash 2001, 'Destruction of multistoreyed buildings in Kutch earthquake of January 26, 2001', *Workshop on Recent Earthquakes of Chamoli and Bhuj*, vol. II, Roorkee, India.

Smith, S.W. 1976, 'Determination of maximum earthquake magnitude', *Geophysical Research Letters*, vol. 3, no. 6, pp. 351-54.

SP-16 1980, *Design Aids for Reinforced Concrete to IS: 456-1978*, Bureau of Indian Standards, New Delhi.

SP-34 1987, *Handbook on Concrete Reinforcement and Detailing*, Bureau of Indian Standards, New Delhi.

SP-20 (S&T) 1991, *Handbook on Masonry Design and Construction*, Bureau of Indian Standards, New Delhi.

Spencer, B.F. Jr., and T.T. Soong 1999, 'New applications and development of active, semi-active, and hybrid control techniques for seismic and non-seismic vibration in the USA', Proceedings of International Post-SMiRT Conference Seminar on Seismic Isolation, Passive Energy Dissipation, and Active Control of Vibration of Structures; August 23-25; Cheju, Korea.

Srivastav, S.K. 2001, 'Bhuj earthquake of January 26, 2001-Some pertinent questions', *International Conference on Seismic Hazard with Particular Reference to Bhuj Earthquake of January 26, 2001*, New Delhi.

Stafford-Smith, B. 1996, 'Behaviour of square infilled frames', *Journal of the Structural Division*, Proceedings of ASCE, vol. 91, no. ST, pp. 381-403.

Steven L. Kramer 2003, *Geotechnical Earthquake Engineering*, Prentice-Hall International Series.

STP 992 1988, In Harry A. Harris (Ed.), *Masonry: Materials, Design, Construction and Maintenance*, ASTM, Philadelphia, PA.

Stratta, J.L. and J. Feldman 1971, 'Interaction of infill walls and concrete frames during earthquakes', *Bulletin of the Seismological Society of America*, vol. 61, no. 3, pp. 609-12.

Sugano, S. 1981, 'Seismic strengthening of existing reinforced concrete buildings in Japan', *Bulletin of the New Zealand National Society for Earthquake Engineering*, vol. 14, no. 4.

Surtees, J.O. and A.P. Mann 1970, 'End plate connections in plastically designed structures', *Conference on Joints in Structures, Institution of Structural Engineers and the University of Sheffield.*

Takanashi, K. 1973, 'Inelastic lateral buckling of steel beams subjected to repeated and reversed loadings', *Proc. 5th World Conference on Earthquake Engineering*, Rome, vol. 1, pp. 795-98.

Tally, N. 2001, *Design of Reinforced Masonry Structures*, McGraw-Hill.

Teran, A. and J. Ruiz 1992, 'Reinforced concrete jacketing of existing structures', *Tenth World Conference on Earthquake Engineering*, Madrid, Spain.

Teal, E.J. 1968, 'Structural steel seismic frames-drift ductility requirements', *Proc. 37th Annual Convention Structural Engineers Association of California.*

Thakkar, S.K., R.N. Dubey, and P. Agarwal 1996, 'Damages and lessons learnt from recent Indian earthquakes', *Symposium on Earthquake Effects on Structures, Plant and Machinery*, New Delhi.

Thomson, W.T. 1988, *Theory of Vibration*, 3rd ed., CBS Publishers, New Delhi.

Timber Engineering Company 1956, *Timber Design and Construction Handbook*, F.W. Dodge Corporation, New York.

Tomazevic, M. 1999, *Earthquake Resistant Design of Masonry Buildings*, Imperial College Press, London.

Tomazevic, M. 2000, *Earthquake-Resistant Design of Masonry Buildings*, Imperial Colleges Press, London.

Toomath, S.W. 1968, 'Architectural details for earthquake movement', *Bulletin of New Zealand Society for Earthquake Engineering*, vol. 1, no. 1, 7 pp.

Trifunac, M.D. and A.G. Brady 1975, 'A study on the duration of strong earthquake ground motion', *Bull. of the Seismological Society of America*, vol. 65, pp. 581-626.

Tso, W.K., E. Pollner, and A.C. Heidebrecht, 'Cyclic loading on externally reinforced masonry walls', *Proc. 5th World Conference on Earthquake Engineering*, Rome, vol. 1, pp. 1177-86.

Turner, E. 2004, 'Retrofit Provisions in the International Existing Building Code', *13th World Conference on Earthquake Engineering*, Vancouver, B.C., Canada.

UNDP/UNIDO Project RER/79/015 1983, 'Repair and strengthening of reinforced concrete, stone and brick masonry buildings', *Building Construction Under Seismic. Conditions in the Balkan Regions*, vol. 5, United Nations Industrial Development Programme, Austria.

U.S. Geological Survey 1975, *The Interior of the Earth*, U.S. Government Printing Office, Washington, DC.

Vanmarcke, E.H. and S.P. Lai 1977, *Strong Motion Duration of Earthquakes*, Report R77,16, Massachusetts Institute of Technology, Cambridge, MA.

Vann, W.P., L.E. Thompson, L.E. Whalley, and L.D. Ozier 1974, 'Cyclic behaviour of rolled steel members', *Proc. Fifth World Conf. Earthquake Eng.*, Rome, vol. 1, pp. 1187-93.

Velestos, A.S. and L.W. Meek 1974, 'Dynamic behaviour of building foundation systems', *Earthquake Engineering and Structural Dynamics*, vol. 3, pp. 121-38.

Velestos, A.S. and V.V. Nair 1975, 'Seismic interaction of structures on hysteretic foundations', *Journal of Structural Engineering*, ASCE, vol. 101, pp. 109-29.

Vukazich, S.E. 1998, *The Apartment Owner's Guide to Earthquake Safety*, San Jose State University.

Wells, D.L. and K.J. Coppersmith 1994, 'New empirical relationships among magnitude, rupture length, rupture width, rupture area and surface displacement', *Bulletin of the Seismological Society of America*, vol. 84, no. 4, pp. 974-1002.

Whitman, R.V., J.M. Biggs, J. Brennan, C.A. Cornell, R. de. Neufville, and E.H. Vanmarcke 1974, 'Seismic design analysis', *Structures Publication No.* 381, Massachusetts Institute of Technology, 33 pp.

Williams, A. 2003, *Seismic Design of Buildings and Bridges*, University of Oxford, New York.

Williams, D., and J.E. Scrivener 1973, 'Response of reinforced masonry shear walls to static and dynamic cyclic loading', *Proc. 5th World Conference on Earthquake Engineering*, Rome, vol. 2, pp. 1491-94.

Woods, R.D. 1978, 'Measurement of dynamic soil properties', *Proceedings, Earthquake Engineering and Soil Dynamics Specialty Conference*, ASCE, Pasadena, California, vol. I, pp. 91-178.

Wrburton, G.B. 1976, *The Dynamical Behaviour of Structures*, 2nd ed., Pergamon Press.

Wyllie, L.A. 1996, 'Strengthening strategies for improved seismic performance', *Eleventh World Conference on Earthquake Engineering*, Acopulco, Mexico.

Yang, C.Y. 1986, *Random Vibration of Structures*, John Wiley and Sons, New York, 295 pp.

Zahrah, T.P. and W.J. Hall 1984, 'Earthquake energy absorption in SDOF structures', *Journal of Structural Engineering, ASCE*, vol. 110, no. 8, pp. 1757-72.

Index

RELATED TITLES

DESIGN OF STEEL STRUCTURES (with CD)

[9780195676815]

N. Subramanian, *Consulting Engineer, Maryland, USA*

This book provides an extensive coverage of the design of steel structures in accordance with the latest code of practice for general construction in steel (IS 800: 2007).

Key Features

- Covers topics such as materials, concepts, loading, analysis, design, and fire and corrosion resistance
- Includes important tables and figures from the Indian Standard codes (IS 800: 2007 and IS 875: 1987) for easy reference
- Includes a CD containing computer programs for design and additional chapters on advanced topics

RAILWAY ENGINEERING, 2E

[9780198083535]

Satish Chandra, *Professor, Dept of Civil Engineering, Indian Institute of Technology Roorkee*, and **M.M. Agarwal**, *retired as Chief Engineer, Northern Railway*

This second edition provides an exhaustive coverage of all aspects of railways, from fundamental concepts to modern technological developments.

Key Features

- New chapter on the Dedicated Freight Corridor Project and other recent developments in Indian Railways
- A complete procedure on how to calculate stresses in different components of a railway track
- Updated statistical data on Indian Railways wherever relevant

REMOTE SENSING AND GIS, 2E (with CD)

[9780198072393]

Basudeb Bhatta, *Course Coordinator of Computer Aided Design Centre, Jadavpur University, Kolkata*

This second edition gives an exhaustive coverage of optical, thermal, and microwave remote sensing, global navigation satellite systems, digital photogrammetry, visual image analysis, digital image processing, spatial and attribute data model, and planning, implementation, and management of GIS.

Key Features

- Includes topics such as classification of remote sensing, geometry of aerial photographs, airborne vs space-borne radar, and urban applications of remote sensing
- Describes the latest remote sensing satellites (GeoEye-1, WorldView-1, WorldView-2, Cartosat-2A and 2B, Oceansat-2) and launches such as Chandrayaan-1 and GSAT
- Includes a CD containing latest version of PCI Geomatica and additional sample data

Other Related Titles

9780195671537	Santhakumar: *Concrete Technology*
9780195694833	Sarkar and Saraswati: *Construction Technology*
9780198086352	Gangopadhyay: *Engineering Geology, 2e*
9780195694611	Ojha, Berndtsson, and Bhunya: *Engineering Hydrology*
9780195682724	Muthu Shoba Mohan: *Principles of Architecture*
9780198069188	Thandavamoorthy: *Structural Analysis*
9780198085423	Subramanian: *Surveying and Levelling, 2e*

Visit us at www.oup.co.in and oupinheonline.com